Advances in Dielectrics

Series editor

Friedrich Kremer, Leipzig, Germany

Aims and Scope

Broadband Dielectric Spectroscopy (BDS) has developed tremendously in the last decade. For dielectric measurements it is now state of the art to cover typically 8–10 decades in frequency and to carry out the experiments in a wide temperature and pressure range. In this way a wealth of fundamental studies in molecular physics became possible, e.g. the scaling of relaxation processes, the interplay between rotational and translational diffusion, charge transport in disordered systems, and molecular dynamics in the geometrical confinement of different dimensionality—to name but a few. BDS has also proven to be an indispensable tool in modern material science; it plays e.g. an essential role in the characterization of Liquid Crystals or Ionic Liquids and the design of low-loss dielectric materials.

It is the aim of "Advances in Dielectrics" to reflect this rapid progress with a series of monographs devoted to specialized topics.

Target Group

Solid state physicists, molecular physicists, material scientists, ferroelectric scientists, soft matter scientists, polymer scientists, electronic and electrical engineers.

More information about this series at http://www.springer.com/series/8283

Friedrich Kremer · Alois Loidl
Editors

The Scaling of Relaxation Processes

Springer

Editors
Friedrich Kremer
Molekülphysik, Peter-Debye-Institut für Physik der weichen Materie
Universität Leipzig
Leipzig
Germany

Alois Loidl
Institut für Physik
Universität Augsburg
Augsburg
Germany

ISSN 2190-930X ISSN 2190-9318 (electronic)
Advances in Dielectrics
ISBN 978-3-030-10254-8 ISBN 978-3-319-72706-6 (eBook)
https://doi.org/10.1007/978-3-319-72706-6

Printed on acid-free paper

This Springer imprint is published by the registered company Springer International Publishing AG part of Springer Nature
The registered company address is: Gewerbestrasse 11, 6330 Cham, Switzerland

Preface

During the past two decades, the experimental capabilities to study materials in an extraordinary wide spectral range at largely varying temperatures as well as pressures have enormously developed. This is especially true for Broadband Dielectric Spectroscopy (BDS) which nowadays covers the whole frequency range from 10^{-6} Hz up to the far infrared (FIR) *without any gap*. In addition, other methods like photon correlation spectroscopy (PCS), nuclear magnetic resonance (NMR), viscosimetry and mechanical spectroscopy and even calorimetry have also become broadband. Consequently, knowledge concerning the scaling of relaxation processes has tremendously grown, and it is now common to combine different techniques in order to determine the distinct correlations and their mutual interactions in a material under study. This development has been nicely exemplified for amorphous systems being characterized by glassy dynamics which is extended from very long (10^6 s) to short (picoseconds) timescales, including the structural α-relaxation (dynamic glass transition), secondary (slow and fast) β-relaxations, and the Boson peak in the far infrared.

It is well established that structural relaxations roughly follow a Vogel–Fulcher–Tammann type of thermal activation, while secondary relaxations, at least below the glass transition temperature, can be described by an Arrhenius-like temperature dependence. Fast processes in the GHz regime seem to follow scaling predictions, while the boson peak and intramolecular vibrations are only weakly temperature dependent. The striking similarities of frequency and temperature dependencies for the large class of supercooled liquids lead to the development of scaling approaches and theories that can model these universalities. However, often a closer inspection reveals severe discrepancies and deviations.

It is the objective of this issue of "Advances in Dielectrics" on the "Scaling of Relaxation Processes" to summarize the current knowledge and the enormous amount of high-quality data on glassy dynamics of supercooled liquids and to discuss it with respect to the often competing theoretical concepts.

Leipzig, Germany
Augsburg, Germany
March 2018

Friedrich Kremer
Alois Loidl

Contents

The Scaling of Relaxation Processes—Revisited 1
Friedrich Kremer and Alois Loidl

Glassy Dynamics: From Millihertz to Terahertz 23
P. Lunkenheimer and Alois Loidl

Glassy Dynamics as Reflected in the Inter- and Intra-molecular Interactions 61
Friedrich Kremer, Wilhelm Kossack and Arthur Markus Anton

Universality of Density Scaling 77
Andrzej Grzybowski and Marian Paluch

Scaling of Suprastructure and Dynamics in Pure and Mixed Debye Liquids 121
Sebastian Peter Bierwirth, Jennifer Bolle, Stefan Bauer, Christian Sternemann, Catalin Gainaru, Metin Tolan and Roland Böhmer

Dynamic Heterogeneities in Binary Glass-Forming Systems 173
D. Bock, Th. Körber, F. Mohamed, B. Pötzschner and E. A. Rössler

Depolarized Dynamic Light Scattering and Dielectric Spectroscopy: Two Perspectives on Molecular Reorientation in Supercooled Liquids 203
J. Gabriel, F. Pabst, A. Helbling, T. Böhmer and T. Blochowicz

Relaxation Processes in Liquids and Glass-Forming Systems: What Can We Learn by Comparing Neutron Scattering and Dielectric Spectroscopy Results? 247
Arantxa Arbe and Juan Colmenero

The Scaling of the Molecular Dynamics of Liquid Crystals as Revealed by Broadband Dielectric, Specific Heat, and Neutron Spectroscopy 279
Andreas Schönhals, Bernhard Frick and Reiner Zorn

The Calorimetric Glass Transition in a Wide Range of Cooling Rates and Frequencies 307
T. V. Tropin, J. W. P. Schmelzer, G. Schulz and C. Schick

Dipolar Correlations in 1,4-Polybutadiene Across the Timescales: A Numerical Molecular Dynamics Simulation Investigation 353
Mathieu Solar and Wolfgang Paul

Linear Viscoelasticity of Polymers and Polymer Nanocomposites: Molecular-Dynamics Large Amplitude Oscillatory Shear and Probe Rheology Simulations 375
Theodoros Davris, Alexey V. Lyulin, Arlette R. C. Baljon, Victor M. Nazarychev, Igor V. Volgin, Sergey V. Larin and Sergey V. Lyulin

Index 405

The Scaling of Relaxation Processes—Revisited

Friedrich Kremer and Alois Loidl

Abstract Glassy dynamics covers the extraordinary spectral range from 10^{+13} to 10^{-3} Hz and below. In this broad frequency window, four different dynamic processes take place: (i) the primary or α-relaxation, (ii) (slow) secondary relaxations (β-relaxations), (iii) fast absorption processes in the GHz and (iv) the boson-peak in the THz range. The dynamic glass transition is assigned to fluctuations between structural substates and scales well with the calorimetric glass transition temperature. It shows a similar temperature dependence as the viscosity and fluctuations of the density or heat capacity. The temperature dependence of the mean relaxation rate of the dynamic glass transition follows at first glance the empirical Vogel–Fulcher—Tammann law, albeit a further analysis unravels clear-cut deviations. The (slow) secondary relaxations are assigned to librational relaxations of molecular subgroups hence having a straightforward molecular assignment. They may also show up as a wing on the high-frequency side of the dynamic glass transition. The fast absorption processes at GHz frequencies can formally be described within the framework of the mode-coupling theory (MCT). The boson-peak resembles the Poley absorption and originates from overdamped oscillations. In this chapter, especially the first three contributions will be discussed in detail and compared with existing theoretical models.

F. Kremer (✉)
Molekülphysik, Peter-Debye-Institut für Physik der weichen Materie, Universität Leipzig, Linnéstr. 5, 04103 Leipzig, Germany
e-mail: kremer@physik.uni-leipzig.de

A. Loidl
University of Augsburg, Experimental Physics V, Universitätsstrasse 2, 86135 Augsburg, Germany
e-mail: alois.loidl@physik.uni-augsburg.de

F. Kremer and A. Loidl (eds.), *The Scaling of Relaxation Processes*,
Advances in Dielectrics, https://doi.org/10.1007/978-3-319-72706-6_1

1 Introduction

The glassy state is ubiquitous in inorganic and organic matter. It is characterized by the lack of long range order and shows a refined dynamics including processes spanning a spectral range from 10^{+13} to 10^{-3} Hz and below. Despite concentrated efforts [1–9], a common theoretical understanding of the glassy state does not exist and a variety of different and often controversial views compete. The glassy state is furthermore reflected in many different physical quantities, e.g. the heat capacity, the viscosity, the mechanical moduli, the density, ultrasonic absorption, magnetization, the complex index of refraction and the complex dielectric function. Hence, a multitude of experimental techniques have been employed to study glassy materials, such as frequency-dependent and differential scanning calorimetry [10], dynamic mechanical spectroscopy [11], ultrasound attenuation [12], light [13] and neutron scattering [14], NMR spectroscopy [15] and especially broadband dielectric spectroscopy [16–38].

The mean relaxation rate $\nu(T)$ of the α-relaxation is characterized by the empirical Vogel–Fulcher–Tammann (VFT)-equation [39–41]:

$$\nu(T) = \frac{1}{2\pi\tau(T)} = \upsilon_\infty \exp\left[\frac{-DT_0}{T - T_0}\right] \tag{1}$$

where $\upsilon_\infty = (2\pi\tau_\infty)^{-1}$ is the high temperature limit of the relaxation rate, D is a constant, and T_0 denotes the Vogel–Fulcher temperature. The "fragility" parameter D [42] describes hereby the deviation from an Arrhenius-type temperature dependence

$$\nu(T) = \nu_\infty \exp\left(\frac{-E_A}{kT}\right) \tag{2}$$

where E_A is the activation energy and k the Boltzmann constant. At the calorimetric glass transition T_g, the mean relaxation rate $\nu(T_g)$ and the viscosity $\eta(T_g)$ have reached typical values of ~10^{-3} Hz and ~10^{13} Poise, respectively. In general, T_0 is found to be approximately 40 K below T_g. Thus, the change in the dynamics of the glass-forming processes spans more than 15 decades.

The divergence of Eq. (1) at $T = T_0$ is also supported by the so-called Kauzmann paradox occurring in the entropy determined by measurements of the specific heat [43, 44]: if the entropy of the supercooled liquid is extrapolated to low temperatures, it seems to become identical to that of a crystal or even smaller. In some theories (like the Gibbs–Di Marzio model [45] for polymers), the Kauzmann paradox is resolved by a phase transition. But the physical meaning of the divergence of $\nu(T)$ at $T = T_0$ remains unclear. Because of the universality of Eq. (1), T_0 is considered as a characteristic temperature, where the mean relaxation rate extrapolates to zero, albeit little evidence could be found for a dynamic divergence [46].

Qualitatively, glassy dynamics is often discussed as fluctuation of a molecule in the cage of its neighbours. The librational motions of the latter give rise to fast secondary β-relaxations which take place on a time scale of 10^{-10}–10^{-12} s, while the reorientations of the molecules forming the cage are assigned to the dynamic glass transition or α-relaxation obeying a VFT-temperature dependence. This relaxation process must have cooperative character; i.e. the fluctuations of the molecules forming the "cage" cannot be independent from each other. The extension of the size of such "cooperatively rearranging domains" [3, 6, 7, 45] is one of the central (and controversial) topics of glass research.

The relaxation function of the α-relaxation is usually broadened. Its high-frequency side exhibits often two power laws. In the case of glycerol, this was observed already by Davidson and Cole [47] and interpreted as caused by high-frequency vibrations. It is nowadays established for a variety of glass-forming (low molecular weight and polymeric) materials [s. also the chapter of P. Lunkenheimer and A. Loidl and F. Kremer et al. in this book] and considered to be the high-frequency contribution of a secondary relaxation.

Many systems show additionally a slow secondary β-relaxation (with an Arrhenius-type temperature dependence). This process being observed for relaxation rates ~<10^8 Hz can often be assigned to intramolecular fluctuations. But there are several examples like the low molecular weight liquid ortho-terphenyl (OTP) [19, 20] or the main chain polymer poly(ethylene terephthalate) (PET) [38] where such an interpretation is not immediately obvious. Therefore, it was suggested by Goldstein and Johari [11, 12] that the slow β-relaxation "is intrinsic to the nature of the glassy state" [12]. In the THz regime a further molecular process is observed, the "boson peak" [4] which has similarities with the Poley absorption [48], which was interpreted as being caused by strong local fields exerted on a molecule by its immediate neighbours in the glassy state (Fig. 1).

In detail in this chapter, the following questions will be addressed: (i) is there a scaling function which describes the temperature dependence of the mean relaxation rate in the entire spectral range from 10^{+11} to 10^{-3} Hz and below? (ii) How does the relaxation time distribution function change with temperature, or in other words, is time–temperature superposition in general valid for (dielectric) relaxation processes? (iii) How does the strength $\Delta\varepsilon$ of a relaxation process change with temperature in the course of the dynamic glass transition? (iv) What is the molecular origin of the "high-frequency" wing, sometimes termed excess wing, which is observed in the dynamic glass transition of many (low molecular weight and polymeric) systems? (v) Is there a model-free characteristic temperature, where glassy dynamics undergoes a change?

2 Theories Describing the Scaling of Relaxation Processes in Glassy Systems

Numerous approaches [49–69] have been developed to describe the dynamics of glassy systems. In the following, two most important approaches are briefly described.

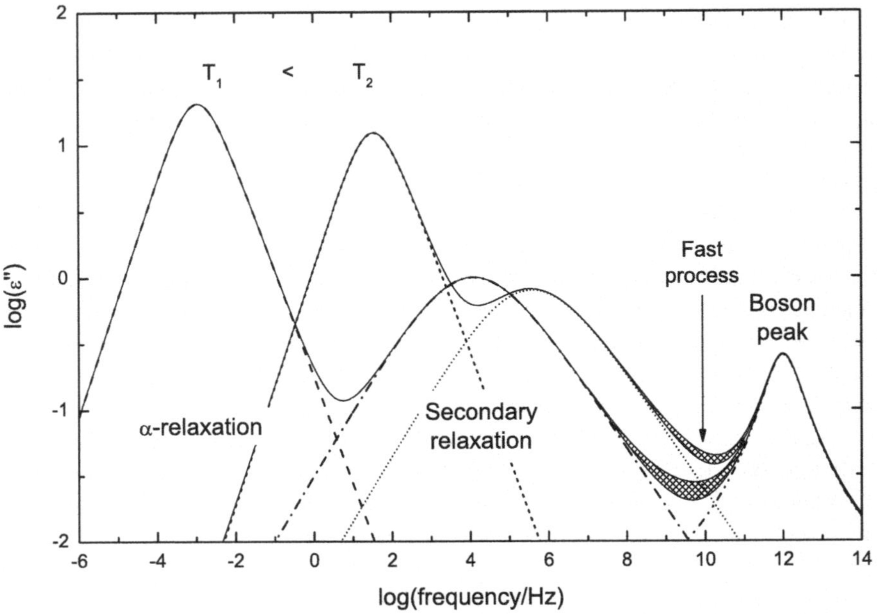

Fig. 1 Scheme of the dynamical processes taking place in the spectral range between 10^{-6} and 10^{14} Hz, (i) the α-relaxation, (ii) (slow) secondary relaxations, (iii) fast absorption processes and (iv) the boson peak. The temperature shift is depicted for two different temperatures $T_1 < T_2$

The experimentally observed VFT-dependence (Eq. (1)) can be founded by two approaches: the Adam–Gibbs model [52] and the free volume theory as developed by Doolittle [53] and Cohen and Turnbull [54, 55]. The latter is based on four assumptions:

(i) A local volume V is attributed to a molecule or polymer segment.
(ii) The difference $V_f = V - V_c$ can be considered as "free", if V is larger than a critical value V_c.
(iii) If for the free volume $V_f \geq V^* \approx V_M$ holds, molecular transport takes place in jumps over a distance corresponding to the size of the molecule V_M. V^* is the minimal free volume required for a jump of a molecule.
(iv) The molecular rearrangement of free volume does not require free energy.

Following Boltzmann statistics, a molecule or a polymer segment carries out positional jumps only if the necessary free volume is provided. Hence for the jump rate $1/\tau$

$$\frac{1}{\tau} \sim \int_{V^*}^{\infty} \exp\left[-\frac{V_f}{V_f}\right] dV_f \sim \exp\left[-\frac{V^*}{V_f}\right]. \tag{3}$$

is obtained where $\overline{V}_f$ is the averaged free volume. Assuming that the relative averaged free volume $\bar{f} = \overline{V}_f/V$ (V: total volume) depends linearly on temperature

$$\bar{f} = f_g + \alpha_f(T - T_g) \tag{4}$$

while $f^* = \frac{V^*}{V}$ is temperature-independent results in a VFT-equation. α_f is the thermal expansion coefficient of the free volume and f_g the relative free volume at T_g. Comparison with Eq. (1) delivers

$$DT_0 = \frac{f^*}{\alpha_f}, \quad T_0 = T_g - \frac{f_g}{\alpha_f} \tag{5}$$

At the temperature T_0, the volume $\overline{V}_f$ vanishes. Within this approach, no inherent length scale is involved and all transport properties should have the same temperature dependence because the jump between holes is the only transport mechanism. Cohen and Grest [55] extended this approach by considering solid- and liquid-like clusters in a percolation approach.

The model of Adam and Gibbs [52] suggests the existence of "Cooperatively Rearranging Regions (CRR)" being defined as the smallest volume which can change its configuration independent from neighbouring regions. It relates the relaxation time to the numbers of particles (molecules for a low molecular liquid, segments for a polymer) $z(T)$ per CRR by

$$\frac{1}{\tau} \sim \exp\left[-\frac{z(T)\,\Delta E}{k\,T}\right] \tag{6}$$

where ΔE is a free energy barrier for one molecule. $z(T)$ can be expressed by the total configurational entropy $S_c(T)$ as $z(T) = S_c(T)/(N\ k_B \ln 2)$ where N is the total number of particles, k_B the Boltzmann constant and ln 2 the minimum entropy of a CRR assuming a two-state model. Using thermodynamic considerations, $S_c(T)$ can be linked to the change of the heat capacitance Δc_p at the glass transition by

$$S_c(T) = \int_{T_2}^{T} \frac{\Delta c_p}{T}\, dT \tag{7}$$

With $T_2 = T_0$ and $\Delta c_p \approx C/T$ from Eqs. (6) and (7), the VFT-dependence follows. At T_0, the configurational entropy vanishes and the size of a CRR diverges as $z(T) \sim \frac{1}{C(T-T_0)}$. The Adam–Gibbs model does not provide information about the absolute size of the CRR at T_g.

Donth [3, 6, 7] suggested a thermodynamic fluctuation model leading to a expression which connects the height of the step in c_p and the temperature fluctuation δT of a CRR at T_g with the correlation length ξ as

$$\xi^3 \sim V_{\text{CRR}} = \frac{k\,T_g^2\,\Delta(1/c_p)}{\rho(\delta T)^2}. \tag{8}$$

where ρ is the density and $\Delta(1/c_p)$ the step of the reciprocal specific heat (if $c_V \approx c_p$ is assumed). Experimentally, δT can be extracted from the width of the glass transition [6, 7] or from thermal heat spectroscopy measurements [56, 57].

Within the fluctuation approach for the temperature dependence of ξ

$$\xi(T) \sim \frac{1}{(T - T_0)^{2/3}} \tag{9}$$

is obtained. A similar equation was derived by Kirkpatrick and Tirumalai [58] using scaling arguments.

Based on the Adam–Gibbs equation (6) and an expression proposed by Waterton [59] as early as 1932, Mauro et al. [60] suggested an approach, which avoids the divergence of the VFT-formula (1) at $T = T_0$

$$\nu(T) = \upsilon_\infty \exp\left[\frac{K}{T}\exp\left(\frac{C}{T}\right)\right] \tag{10}$$

K and C are related to activation energies deduced through a "physical realistic model for configurational entropy based on a constraint approach".

In comparing viscous liquids with spin glasses, Souletie and Bertrand [61] suggested for the mean relaxation rate

$$\tau^{-1} \sim \left[\frac{(T - T_c)}{T}\right]^{\gamma} \tag{11}$$

where $\gamma > 0$ and T_c are constants.

The shoving model developed by Dyre et al. [62] is based essentially on three assumptions.

1. The activation energy is mainly elastic energy.
2. This elastic energy is mainly located in the surroundings of the flow event.
3. The elastic energy is mainly shear elastic energy.

It relates the mean relaxation rate to the mean square vibrational displacement $\langle u^2\rangle(T)$ and a characteristic molecular length a, which is assumed to be constant.

$$\nu(T) = \upsilon_\infty \exp\left(-\frac{a^2}{\langle u^2\rangle(T)}\right) \tag{12}$$

In [63, 64], it is shown that the temperature dependence of the shear modulus dominates the temperature dependence, leading to

$$\langle u^2\rangle(T) \propto \frac{T}{G_\infty(T)} \tag{13}$$

where $G_\infty(T)$ is the elastic shear modulus. The shoving model does not make a specific prediction of the temperature dependence of the mean relaxation rate, except that it cannot diverge at any finite temperature. The model, however, relates two independently measurable quantities in a prediction that has been confirmed for several glass-forming liquids; see for example, the review of the experimental situation given in [65].

The mode-coupling theory (MCT) [9, 66–69] is a hard sphere model based on a generalized nonlinear oscillator equation

$$\frac{d^2\Phi_q(t)}{dt^2} + \Omega^2\Phi_q(t) + \zeta\frac{d\,\Phi_q(t)}{dt} + \Omega^2\int_0^t m_q(t-\tau)\frac{d\,\Phi_q(\tau)}{d\tau}d\tau = 0 \tag{14}$$

where $\Phi(t)_q$ is the normalized density correlation function defined as

$$\Phi_q(t) = \frac{\langle\Delta\rho_q(t)\Delta\rho_q(0)\rangle}{\langle\Delta\rho_q^2\rangle} \tag{15}$$

$\Delta\rho_q(t)$ are density fluctuations at a wavevector q, Ω is a microscopic oscillator frequency, and ς describes a frictional contribution. The first three terms of Eq. (14) describe a damped harmonic oscillator; the fourth term contains a memory function $m_q(t-\tau)$. As a consequence, the total frictional losses in the system become time-dependent.

In order to solve Eq. (14), an ansatz for $m_q(t)$ is required. Already a simple Taylor expansion of m leads to a relaxational response of Φ_q having some similarity with the dynamic glass transition [66, 67]. Assuming $m_q(t) = v_1\Phi_q(t) + v_2\Phi_q^2(t)$ (F_{12}-model, [67]) delivers a two-step decrease of the correlation function $\Phi_q(t)$. The faster contribution is interpreted in terms of a (fast) β-relaxation while the slower component to the dynamic glass transition (α-relaxation). At a critical temperature T_c, the relaxation time diverges; this is interpreted as a phase transition from an ergodic ($T > T_c$) to a non-ergodic ($T < T_c$). Furthermore, MCT (in the idealized version) makes the following predictions:

(i) for $T > T_c$ the relaxation time τ_α of the α-relaxation scales according to

$$\tau_\alpha \sim \eta \sim \left[\frac{T_c}{T-T_c}\right]^\gamma \tag{16}$$

where γ is a constant.

(ii) the relaxation function of the α-relaxation can be described by

$$\Phi_q(t) \sim \exp\left[-\left(\frac{t}{\tau_\alpha}\right)^{\beta_{KWW}}\right] \tag{17}$$

with ($0<\beta_{KWW}<1$), where Φ_0 is the amplitude of the α-relaxation. For $T>T_c$, the relaxation time distribution should be temperature-independent; i.e. time–temperature superposition should hold.

(iii) above and close to the critical temperature T_c, the minimum of the susceptibility $\left(\varepsilon''_{min}, \omega_{min}\right)$ between the α-relaxation and the β-relaxation should follow a power law

$$\varepsilon''_{min} \sim \left|\frac{T - T_c}{T_c}\right|^{1/2} \tag{18}$$

Glassy dynamics spans a time scale of more than 15 decades. In order to unravel the evolution of the temperature dependence in detail, it is most advantageous to calculate the derivatives of the mean relaxation rate with respect to 1/T of the different theoretical approaches. By that, one obtains for the VFT-equation (Eq. (1)) the Arrhenius dependence (Eq. (2)), the Mauro equation (Eq. (10)), the approach by Souletie and the MCT (Eq. (11)) for $T>T_c$ the following expressions:
VFT:

$$\frac{\mathrm{d}\log\nu}{\mathrm{d}(1/T)} = -(DT_0)\cdot\log e\cdot\left(1-\frac{T_0}{T}\right)^{-2} \tag{19}$$

Arrhenius:

$$\frac{\mathrm{d}\log\nu}{\mathrm{d}(1/T)} = \frac{-E_A}{\mathrm{k}}\cdot\log e \tag{20}$$

Mauro:

$$\frac{\mathrm{d}\log\nu}{\mathrm{d}(1/T)} = K\cdot\log e\cdot\exp\left(\frac{C}{T}\right)\cdot\left(\frac{C}{T}+1\right) \tag{21}$$

Souletie:

$$\frac{\mathrm{d}\log\upsilon}{\mathrm{d}(1/T)} = \gamma\cdot\log e\cdot\frac{T_C\cdot T}{T_c - T} \tag{22}$$

MCT:

$$\frac{\mathrm{d}\log\upsilon}{d(1/T)} = \gamma\cdot\log e\cdot\frac{T^2}{T_C - T} \tag{23}$$

Hence in a plot of the *differential* quotient $\mathrm{d}(-\log\nu/\mathrm{d}(1/T))^{-1/2}$ versus $1/T$, the VFT-dependence shows up as a straight line. The derivative plots enable to analyse in detail the scaling with temperature (Fig. 2). This is especially true for the high temperature regime. By that the *difference* quotient, $\Delta(-\log\nu)/\Delta(/1/T)$ of the experimental data can be determined and compared with the analytical derivatives.

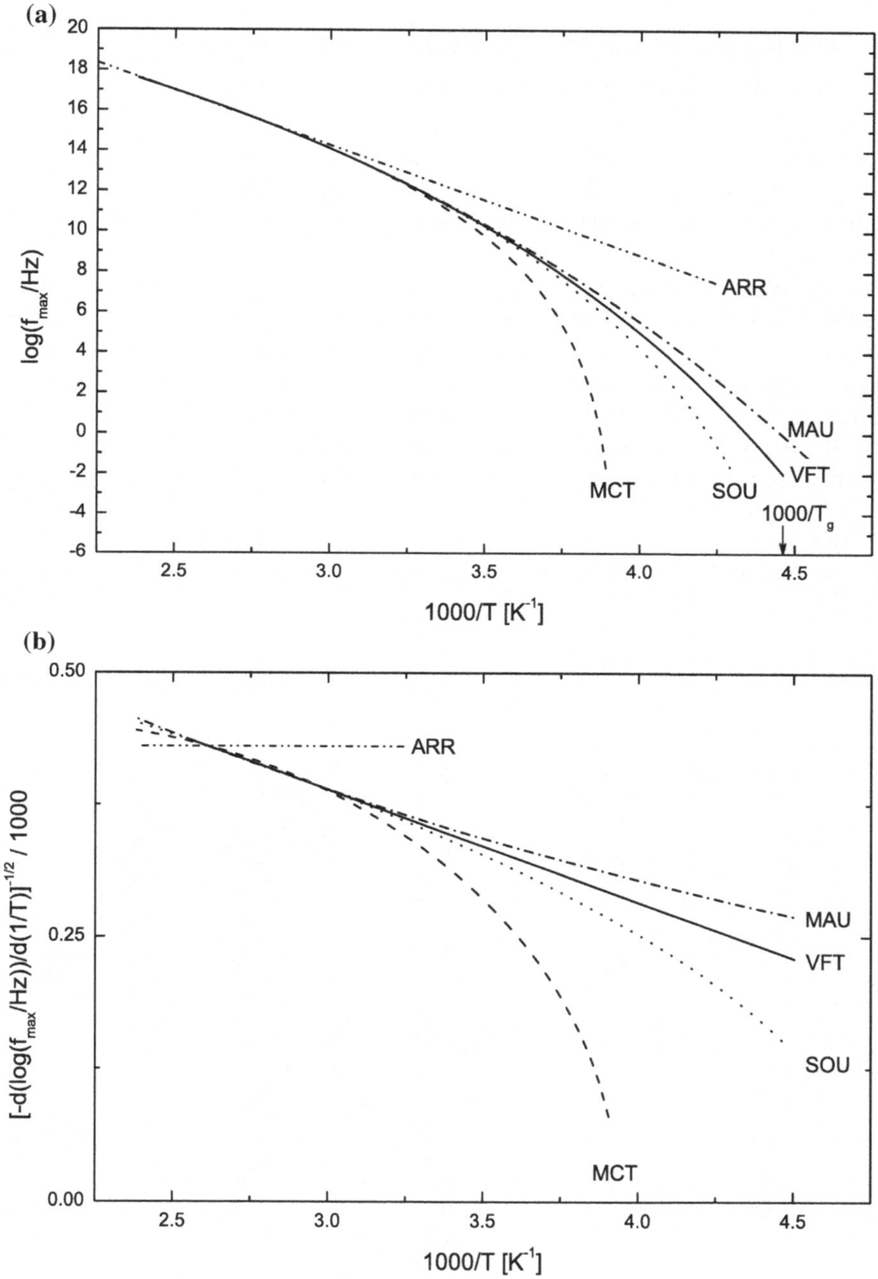

Fig. 2 **a** The scaling behaviour as predicted by the Arrhenius equation (Eq. 2), the Vogel-Fulcher-Tammann equation (VFT) (Eq. 1), the Mauro approach (Eq. 10), that of Souletie (Eq. 11) and of the mode-coupling theory (MCT) (Eq. 16). The glass transition temperature Tg as the temperature, where the mean relaxation rate according to the VFT-function has reached a value of 10^{-2} Hz is indicated. **b** Differential quotient $(-d(\log(\nu)/(d(1/T))^{-1/2} \times 100$ for the functionalities shown in **a**

3 The Scaling of the Dynamic Glass Transition in Low Molecular Weight and Polymeric Organic Glasses

Salol is one of the most explored organic glass-forming liquids. It is considered as a van der Waals glass, despite the fact that it can form H-bonds, presumably mainly within the same molecule. In Fig. 3, dielectric measurements [70] extended over a broad spectral range from about 10^{-2} Hz up to 10^{11} Hz are displayed for temperatures 211 and 361 K. The charts are characterized by a pronounced dynamic glass transition (α-relaxation) having an excess wing, which appears as a second power law on the high-frequency flank of the α-relaxation [71]. The latter is interpreted as a submerged slow secondary relaxation showing up as a shoulder with a significant curvature for a sample aged at 211 K for 6.5 days as discussed in detail in ref. [72]. For frequencies $\nu > 10^{10}$ Hz, a shallow loss minimum is found; it can be interpreted in terms of the fast β-relaxation of the mode-coupling theory (s. below) but also other explanations have been proposed [73].

The spectra can be described by a superposition of a Havriliak–Negami (HN) and Cole–Cole (CC) [16] function for the primary α-process or for the secondary β-process, respectively:

$$\varepsilon^*_{total}(\omega) = \varepsilon_\infty + \frac{\Delta\varepsilon_{HN}}{(1 + (i\omega\,\tau_{HN})^{\beta_{HN}})^{\gamma_{HN}}} + \frac{\Delta\varepsilon_{CC}}{(1 + (i\omega\,\tau_{CC})^{\beta_{CC}})} \tag{24}$$

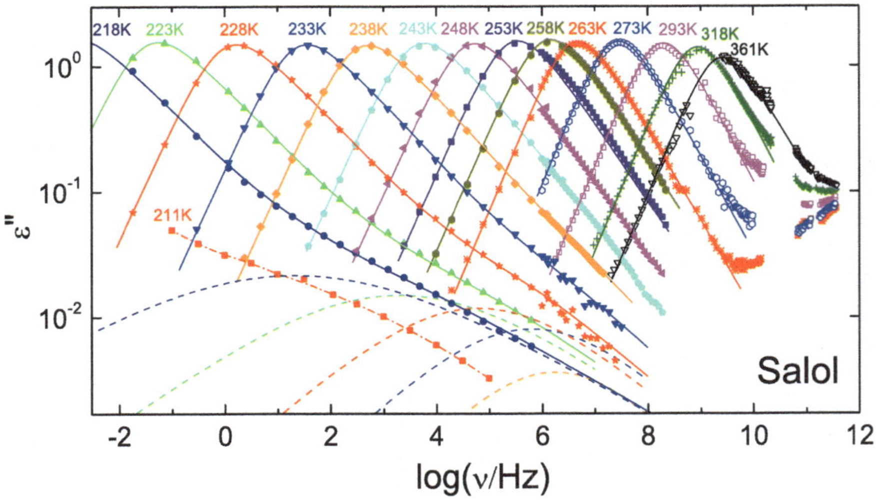

Fig. 3 Dielectric loss as a function of frequency for a series of temperatures from 211 K up to 361 K for salol. The solid lines are fits with a Havriliak–Negami (HN) function for $T \geq 243$ K and with the sum of a HN and Cole–Cole (CC) function for $T \leq 238$ K. The dashed lines show the CC components. The dash-dotted line through the 211 K data is a guide to the eyes. Taken and modified from [70] with kind permission of The European Physical Journal (EPJ)

where $\Delta\varepsilon_{HN}$ and $\Delta\varepsilon_{CC}$ are the relaxation strengths, τ_{HN} and τ_{CC} the relaxation times, and β_{HN}, γ_{HN} and β_{CC} the spectral width parameters of the HN and CC function, respectively, and ω is the circular frequency. For temperatures $T \geq 243$ K, the secondary β-peak has completely merged with the α-peak.

By fitting the dielectric spectra with the empirical relaxation function of Eq. (24), an activation plot is obtained, where the mean relaxation rate versus the inverse temperature is displayed (Fig. 4a). The charts at temperatures >300 K can be equally well described by the Arrhenius equation, the formula suggested by Mauro (Eq. (10)) and Souletie (Eq. (11)) and the MCT-ansatz (Eq. (16)). Comparing the experimentally determined difference quotients with the derivatives of the different scaling functions with respect to 1/T however proofs that *none* of the suggested formulae describes the data within the limits of experimental accuracy in the entire temperature range and that it is furthermore *not* possible to describe the experimental data adequately by use of *one* VFT-function or to replace the VFT-dependence by an Arrhenius function as one might expect from the raw data in Fig. 4a. This is supported as well by an analysis [74] based on the second derivative of the temperature dependence of the structural relaxation time $\tau_\alpha(T)$ with respect to T_g/T.

Glycerol (Fig. 5a/b) is an H-bond forming liquid. Its mean relaxation rate shows a pronounce VFT-dependence; the data for temperatures ≥270 K seem to follow equally well a VFT-function or dependencies as suggested by Mauro, Souletie or the MCT. But from the derivative plot (Fig. 5b) again one must conclude that none of the suggested formulae fits the temperature dependence correctly within the limits of experimental accuracy. Similarly as for salol, two VFT-equations (VFT1 and VFT2) are required to describe the data within experimental accuracy in the entire temperature range. From the derivative plots, it can be deduced that at temperatures above 270 K neither the Arrhenius equation nor the MCT-ansatz is adequate.

The dynamic glass transition for propylene glycol, tripropylene glycol and its polymeric counterpart poly(propylene glycol) having a mean molecular weight of $M_w = 4000$ g/mol are compared in Fig. 6a. Both charts display a VFT-dependence, but due to the connectivity of the chain for the latter the relaxation is slower, especially at lower relaxation rates. In the derivative plots (Fig. 6b), it is shown again that a single VFT-dependence is *not* sufficient to describe the data adequately in the entire temperature range.

A dielectric relaxation process is not only characterized by the relaxation rate but also by its dielectric strength and by the shape of the relaxation time distribution function. According to the Debye formula, the product $T\Delta\varepsilon$ should be independent on temperature besides the weak temperature effect on number density of dipoles. Instead one observes (Fig. 7a) for all materials that $T\Delta\varepsilon$ increases with decreasing temperature; this might be interpreted as caused by a growing length scale, where polar fluctuations become more cooperative and hence its *effective* dipole moment increases. The temperature dependencies suggested by the MCT $T\Delta\varepsilon \sim (T_c - T)^{1/2}$ for $T < T_c$ and $T \times \Delta\varepsilon \approx$ const. for $T > T_c$ are not fulfilled. However, one has to be aware that the reported values of delta epsilon in most cases exhibit large experimental uncertainties and sometimes differ considerably when reported by different

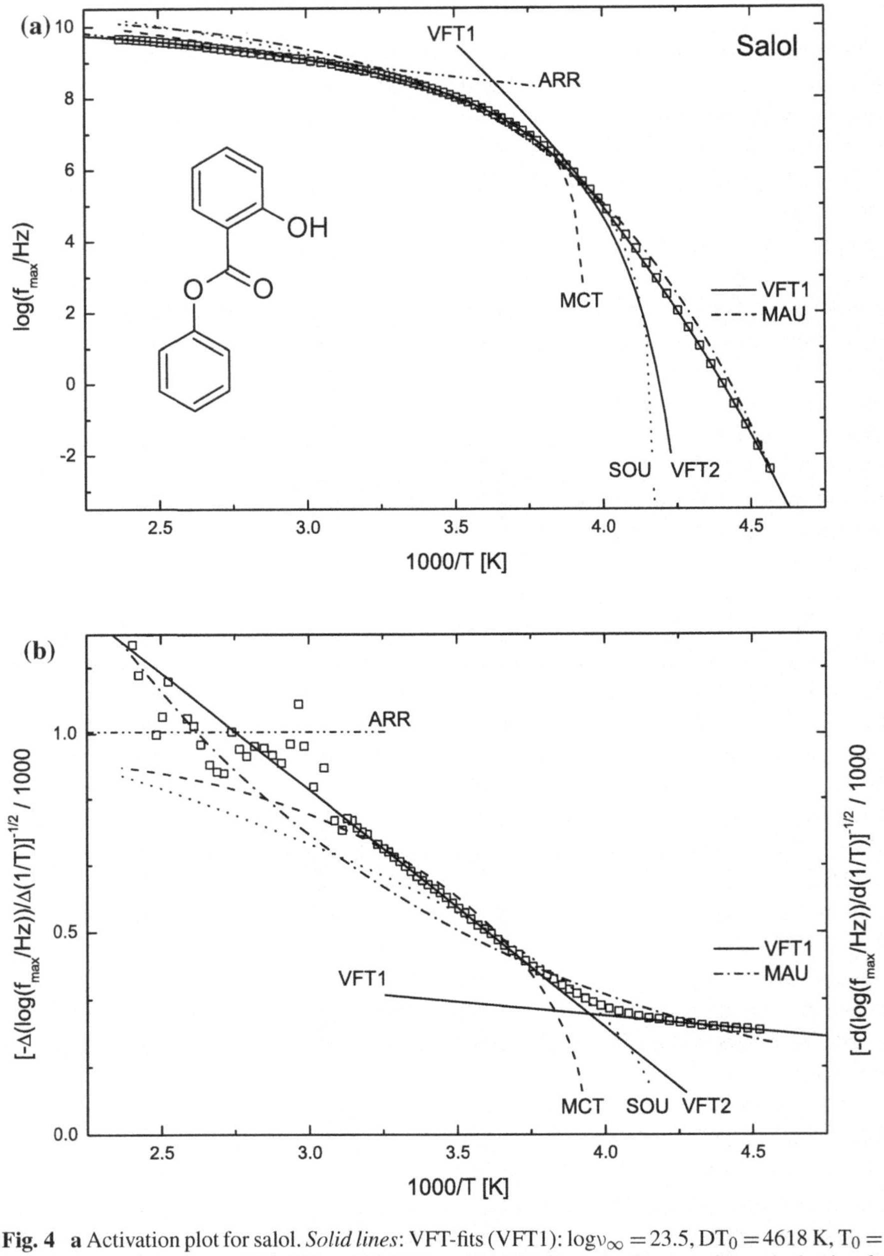

Fig. 4 **a** Activation plot for salol. *Solid lines*: VFT-fits (VFT1): $\log\nu_\infty = 23.5$, $DT_0 = 4618$ K, $T_0 = 141.6$ K; (VFT2) $\log\nu_\infty = 10.4$, $DT_0 = 333$ K, $T_0 = 224.7$ K. *Dash double dotted line*: Arrhenius-fit $\log\nu_\infty = 12.1$, $E_A/k_B = 2283$ K. *Dashed line*: MCT fit $\log\nu_\infty = 10.4$, $\gamma = 2.6$, $T_c = 254$ K; *dotted line*: Souletie fit $\log\nu_\infty = 12.1$, $\gamma = 5.25$, $T_c = 239$ K; *dash-dotted line*: Mauro fit $\log\nu_\infty = 10.5$, $K = 17.1$ K, $C = 1301$ K. The data are taken from [37b, 75]; the error bars are smaller than the size of the symbols if not indicated otherwise. **b** Difference quotient $(-\Delta(\log(\nu_{max}))/\Delta(1000/T))^{-1/2}$ versus 1000/T for the data shown in **a**. For comparison, the differential quotients for the VFT-equation and the temperature dependencies as suggested by the mode-coupling theory (MCT), Souletie (SOU), and Mauro theory (MAU) using the fit parameters shown in **a**

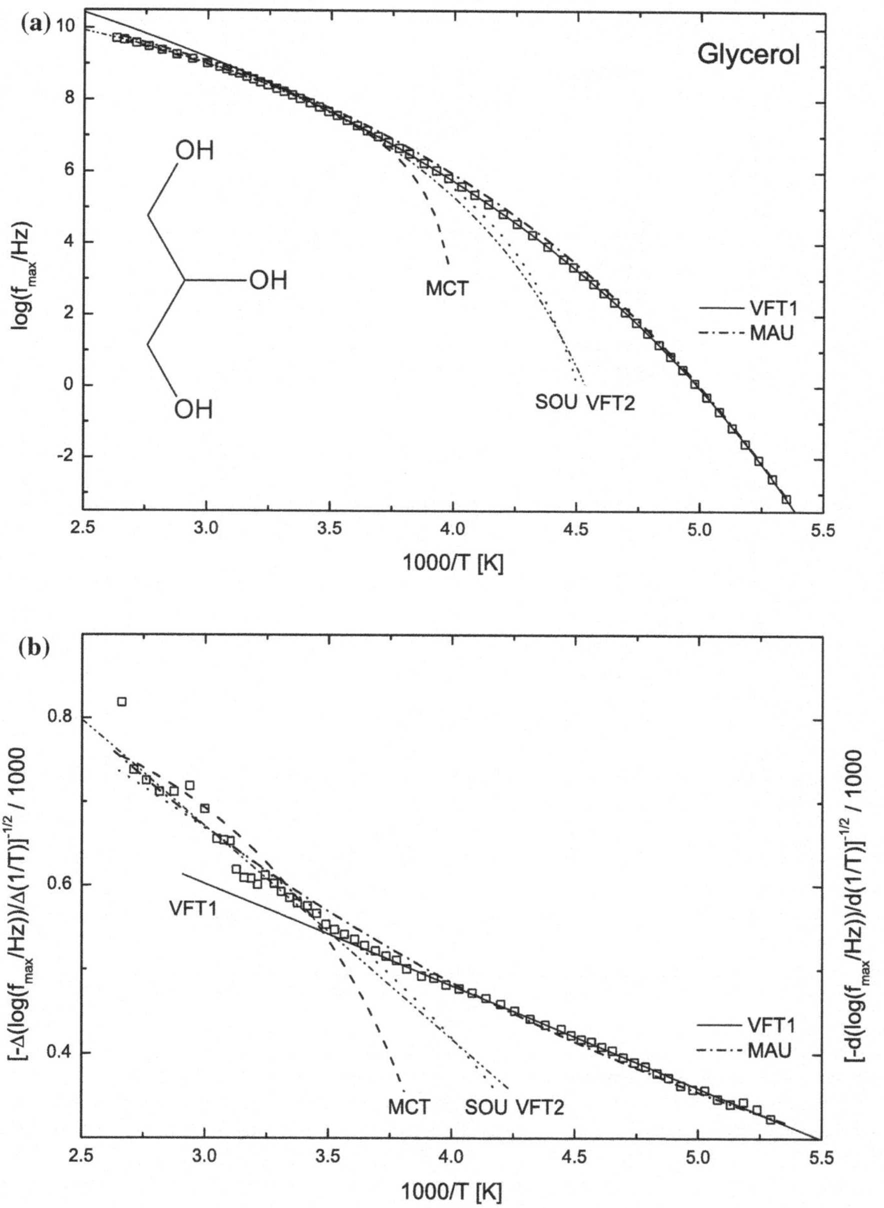

Fig. 5 **a** Activation plot for glycerol. *Solid line*: VFT-fit (VFT1): $\log\nu_\infty = 14.3$, $DT_0 = 2448$ K, $T_0 = 126.0$ K. *Dash double dotted*: VFT-fit (VFT2): $\log\nu_\infty = 12.0$, $DT_0 = 1331$ K, $T_0 = 183.1$ K. *Dashed line*: MCT fit with $\log\nu_\infty = 10.4$, $\gamma = 3.65$, $T_c = 248.8$ K. *Dotted line*: Souletie fit with $\log\nu_\infty = 12.8$, $\gamma = 3.69$, $T_c = 215.1$. *Dash-dotted line*: Mauro fit with $\log\nu_\infty = 12.8$, $K = 517$ K, $C = 471$ K. Data taken from [37b, 75]. The *error bars* are smaller than the size of the symbols if not indicated otherwise. **b** Experimentally determined difference quotient $(-\Delta(\log(\nu_{max}))/\Delta(1000/T))^{-1/2}$ versus $1000/T$. The lines describe the fits shown in **a**. For comparison, the differential quotients for the VFT-fits and the temperature dependencies as suggested by the mode-coupling theory (MCT), Souletie (SOU), and Mauro (MAU) theory using the fit parameters shown in **a**

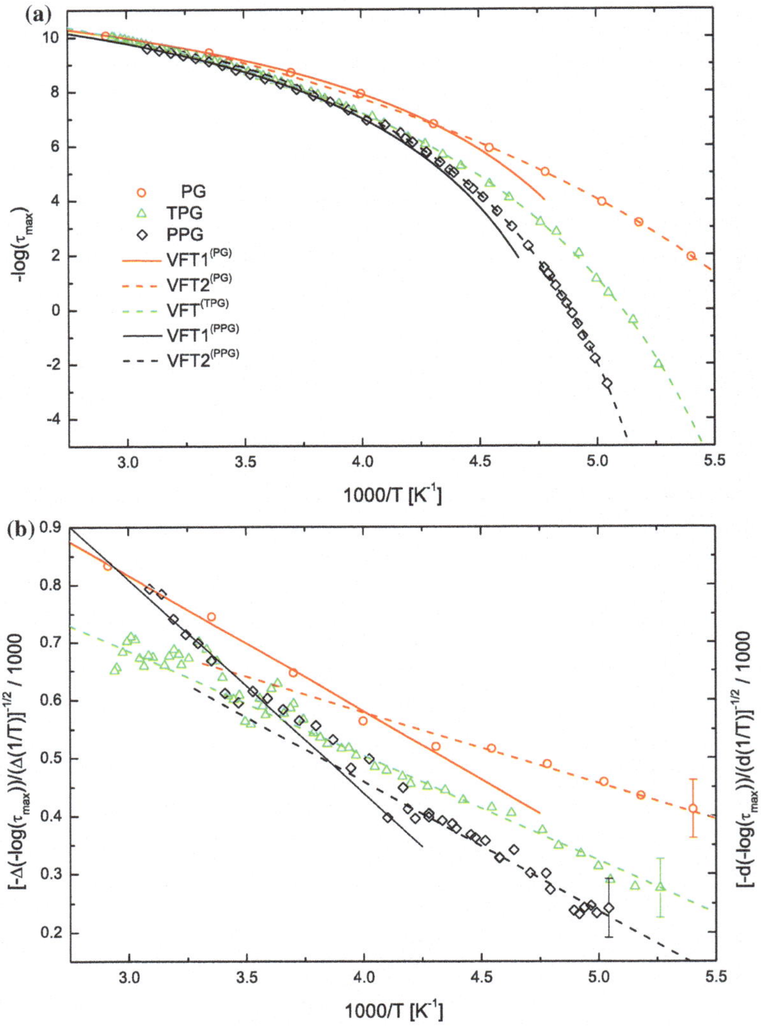

Fig. 6 **a** Activation plot for propylene glycol (open circles), tripropylene glycol (open triangles) and the polymeric pendant ($M_W = 2000$ g/mol) poly(propylene glycol) (open diamonds). The error bars are smaller than the size of the symbols if not indicated otherwise. Solid lines (VFT_1): VFT-fits with $\log\nu_\infty = 12.1$, $DT_0 = 793$ K, $T_0 = 166$ K for propylene glycol and $\log\nu_\infty = 12.1$, $DT_0 = 833$ K, $T_0 = 179$ K for poly(propylene glycol)). Dashed lines (VFT_2): VFT-fits for the lower temperature range with $\log\nu_\infty = 14.1$, $DT_0 = 1956$ K; $T_0 = 115$ K for propylene glycol, $\log\nu_\infty = 13.1$, $DT_0 = 1343$ K; $T_0 = 151$ K for tripropylene glycol, and $\log\nu_\infty = 12.8$, $DT_0 = 1041$ K, $T_0 = 169$ K for polypropylene glycol. **b** Difference quotient $(-\Delta(\log(\nu_{max}))/\Delta(1000/T))^{-1/2}$ versus $1000/T$. For comparison the differential quotients for the VFT-fits using the fit parameters from **a** are depicted. The data for propylene glycol and poly(propylene glycol) are taken from [16] and for tripropylene glycol from [75]

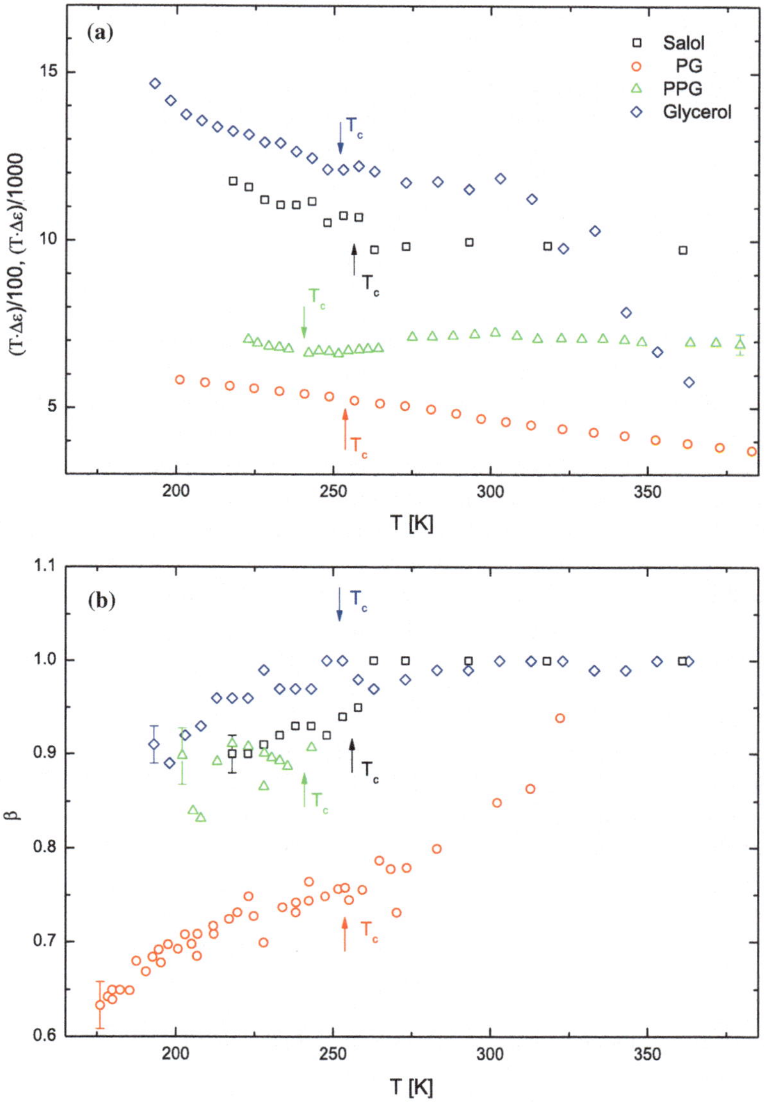

Fig. 7 **a** The product of relaxational strength Δε and temperature T, $(T\Delta\varepsilon)$ versus T; for salol, propylene glycol (PG), poly(propylene glycol) (PPG) and glycerol as indicated. For salol and PPG, TΔε is normalized by 100 and for PG and glycerol by 1000. The error bars are smaller than the size of the symbols if not indicated otherwise. The critical temperatures Tc of the MCT are indicated for the different materials. The data for salol are taken from [76], for PG from [16], for PPG from [77] and for glycerol from [78]. **b** Shape parameter β from the Cole–Davidson function for the materials shown in **a**

groups. There are also some reports which show at least a rough agreement with the predictions of MCT [76].

For all examined materials shown in Fig. 7b, the shape parameter β of the Cole—Davidson function shows a strong temperature dependence. This holds in general for the vast majority of glass-forming (low molecular weight and polymeric) materials and proves that relaxation processes do *not* obey the rule of time–temperature superposition which is often employed in mechanical spectroscopy.

Schönhals [79] analysed the scaling of the dynamic glass transition for a variety of glassy materials and suggested to display the two *measured* quantities, the relaxation strength versus the mean relaxation rate. By that, he found unambiguously that as different materials as salol, glycerol, propyleneglycol, dipropylenglycol, tripropylenglycol and poly(propylene glycole), a pronounced change in the slope of the correlation between the two dependent quantities exists. This crossover takes place at a mean relaxation rate of about 10^8 Hz and marks perhaps the beginning of cooperative dynamics. For all materials, the relaxation strength increases strongly with decreasing temperature. Extrapolated to high temperatures, the mean relaxation rate is in the range between 10^{11} and 10^{13} Hz which is typical for highly activated librational fluctuations. The fact that a crossover temperature T_B exists can be interpreted in several ways; (i) T_B and the critical temperature T_c of the MCT have some resemblance, hence the crossover might reflect a transition from an ergodic to a non-ergodic state. (ii) T_B can be also comprehended as the onset of a cooperative dynamics as suggested by Donth [6, 7]. It is characterized by cooperatively rearranging domains having a size $\xi(T)$ which increases with decreasing diameter. At the calorimetric glass transition temperature T_g, a value between 2 and 3 nm can be estimated based on multiple studies [80] of glassy dynamics in nanometric confinement (Fig. 8).

The mode-coupling theory makes detailed predictions for the minimum region between the "microscopic peak" and the dynamic glass transition following a master function:

$$\varepsilon''(\omega) = \frac{\varepsilon''_{\min}}{a+b}\left[b\left(\frac{\omega}{\omega_{\min}}\right)^a + a\left(\frac{\omega_{\min}}{\omega}\right)^b\right] \tag{25}$$

with temperature-independent exponents a and b being interrelated as

$$\frac{\Gamma^2(1+b)}{\Gamma(1+2b)} = \lambda = \frac{\Gamma^2(1-a)}{\Gamma(1-2a)} \tag{26}$$

where Γ is the Γ-function. The exponents can be as well determined from the temperature dependence of the frequency of the minimum of the susceptibility and of the frequency of the maximum $\omega_{\max}$ of the dynamic glass transition

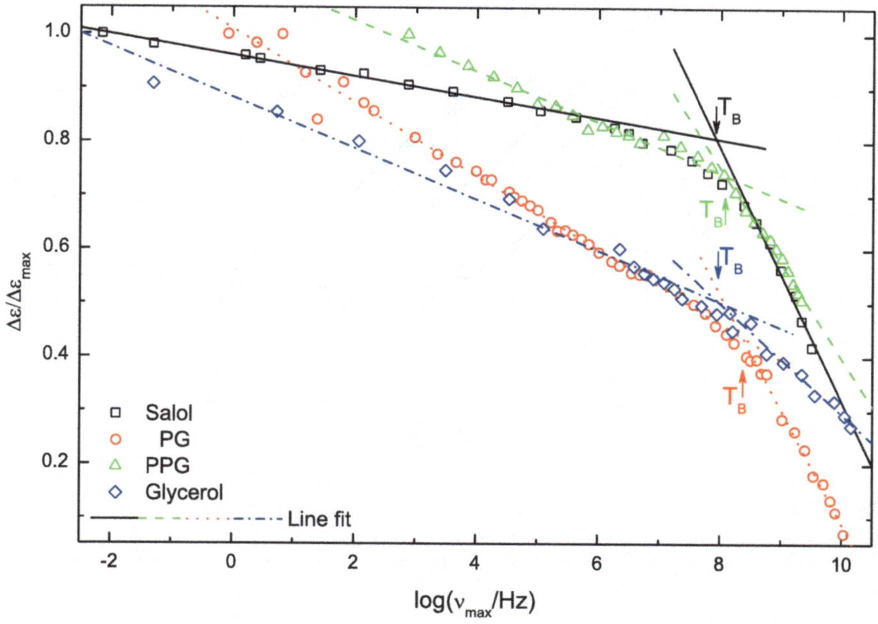

Fig. 8 Relaxational strength Δε, normalized with its maximum value versus the mean relaxation rate log ν_{max} for salol, propylene glycol (PG), poly(propylene glycol) (PPG) and glycerol as indicated. At the temperature T_B, the slope of the correlation between Δε and ν_{max} changes. The data for salol are taken from [22], for PG and PPG from [16] and for glycerol from [81]

$$\omega_{\min} \sim \left|\frac{T - T_c}{T_c}\right|^{\frac{1}{2a}} \tag{27}$$

$$\omega_{\max} \sim \left|\frac{T - T_c}{T_c}\right|^{\left(\frac{1}{2a}+\frac{1}{2b}\right)} \tag{28}$$

Carrying out such an analysis delivers for glycerol a value of $a = 0.325$ and $b = 0.63$. For the lowest temperatures, the increase towards the boson peak approaches a power law $\varepsilon'' \sim \nu^3$ as indicated by the dashed line in Fig. 9. The inset demonstrates for two temperatures that the simple superposition ansatz of, Eq. (25), is not sufficient to describe the shallow minimum.

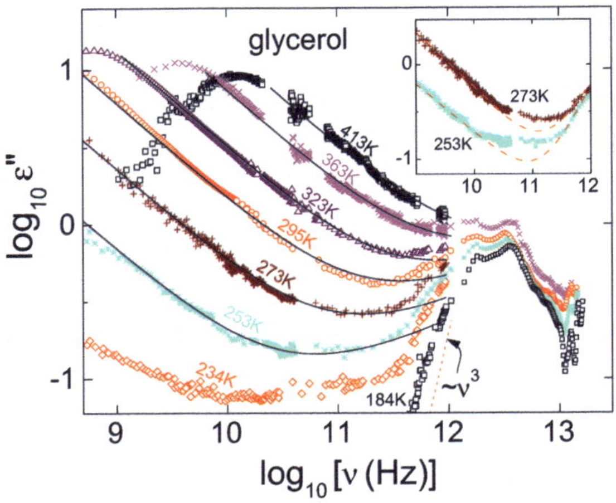

Fig. 9 Dielectric loss of glycerol in the minimum and boson peak region. The solid lines are fits with the MCT prediction, Eq. (10), with $a = 0.325$, $b = 0.63$ for glycerol. For the lowest temperatures, the increase towards the boson peak approaches power laws $\varepsilon \sim \nu^3$ for glycerol as indicated by the dashed line. Note that, in contrast to PC, the boson peak seems to be superimposed to the shallow minimum in glycerol. The inset demonstrates for two temperatures that the simple superposition ansatz, Eq. (9), is not sufficient to explain the shallow minimum. Taken from [82] with permission

4 Conclusions

In the spectral range between 10^{-3} and 10^{13} Hz, four dynamic processes take place in the dynamic glass transition, slow and fast secondary relaxations and the boson-peak. The questions formulated in the introduction can be answered in detail:

(i) Is there a scaling function which describes the temperature dependence of the mean relaxation rate in the entire spectral range from 10^{+13} to 10^{-3} Hz and below?
For all materials under study, *none* of the suggested scaling functions is able to describe the observed temperature dependence of the mean relaxation rate in the entire spectral range. The analysis of the data using derivative plots reveals furthermore, that even in the high-frequency limit an Arrhenius dependence does not describe the measurements within the limits of experimental accuracy. The empirical Vogel–Fulcher–Tammann dependence turns out to be a coarse-grained description only within a limited temperature range. There is no indication pointing towards a divergence at the Vogel temperature T_0.

(ii) How does the relaxation time distribution function change with temperature or in other words, is time–temperature superposition in general valid for (dielectric) relaxation processes?
The relaxation time distribution function with its shape parameters β and γ shows a pronounced temperature dependence. Hence, time–temperature superposition is *not* valid in general.

(iii) How does the strength $\Delta\varepsilon$ of a relaxation process change with temperature in the course of the dynamic glass transition?
The relaxation strength $\Delta\varepsilon$ of the dynamic glass transition decreases with increasing temperature, an effect which can be not explained by the temperature dependence of the density. Instead an increasing length scale of the dynamic glass transition seems to be likely resulting in an increased effective dipole moment.

(iv) What is the molecular origin of the "high-frequency" wing which is observed in the dynamic glass transition of many (low molecular weight and polymeric) systems?
Several glass formers show a high-frequency wing; it is considered as a slow secondary relaxation which might be coupled to the dynamic glass transition.

(v) What is the assignment of the "fast secondary relaxation"?
In the spectral range between 10^9 and 10^{12} Hz, a fast secondary relaxation is observed. It can be quantitatively described by the MCT.

(vi) Is there a characteristic temperature, where glassy dynamics undergoes a change?
As suggested by A. Schönhals, one observes by displaying the correlation between the two dependent variables, relaxation strength and mean relaxation rate—without any assumptions—a transition at about 10^8 Hz. This might be interpreted as the onset of cooperative dynamics with decreasing temperature.

Acknowledgements Support by M. Anton in preparing some of the figures is highly acknowledged.

References

1. 1th international discussion meeting on relaxation in complex systems. J Non-Cryst Solids 131–133:1–1285 (1991); 2th international discussion meeting on relaxation in complex systems. J Non-Cryst Solids 172–174:1–1457 (1994); 3th international discussion meeting on relaxation in complex systems. J Non-Cryst Solids 235–237:1–814 (1998); 4th international discussion meeting on relaxation in complex systems. J Non-Cryst Solids 307–310:1–1080 (2002); 5th international discussion meeting on relaxation *in* complex systems. J Non-Cryst Solids 352:4731–5250 (2006); 6th international discussion meeting *on* *r*elaxation in *c*omplex systems. J Non-Cryst Solids 357:241–782 (2011); 7th international discussion meeting on *r*elaxation in *c*omplex systems (2013); 8th international discussion meeting on relaxation in complex systems (2017)
2. Wong J, Angell CA (1976) Glass: structure by spectroscopy. Marcel Dekker, New York
3. Donth EJ (1981) Glasübergang. Akademie Verlag, Berlin
4. Zallen R (1983) The physics of amorphous solids. Wiley, New York
5. Elliott SR (1990) Physics of amorphous materials. Longman Scientific & Technical, London
6. Donth EJ (1992) Relaxation and thermodynamics in polymers, glass transition. Akademie Verlag, Berlin
7. Donth EJ (2001) The glass transition. Springer, Berlin
8. Ngai K (2011) Relaxation and diffusion in complex systems. Springer, Berlin
9. Götze W (2012) Complex dynamics of glass-forming liquids—a mode-coupling theory. Oxford Scientific Publications, Oxford

10. Cheng SZD (ed) (2002) Handbook of thermal analysis and calorimetry. Elsevier Science B.V
11. Hecksher T, Torchinsky DH, Klieber C, Johnson JA, Dyre JC, Nelson KA (2017) PNAS 114:8715
12. Jeon YH, Nagel SR, Bhattacharya S (1986) Phys Rev A 34:602
13. Pecora R (ed) (1985) Dynamic light scattering, applications of photon correlation spectroscopy. Springer
14. Frick B, Richter D (1995) Science 267:1939–1945
15. Schmidt-Rohr K, Spiess HW (1994) Multidimensional solid-state NMR and polymers. Academic Press, London
16. Kremer F, Schönhals A (eds) (2003) Broadband dielectric spectroscopy. Springer
17. Williams G, Watts DC (1970) Trans Faraday Soc 66:80
18. Williams G, Watts DC, Dev SB, North AM (1971) Trans Faraday Soc 67:1323
19. Johari GP, Goldstein M (1970) J Chem Phys 53:2372
20. Johari GP (1976) In: Goldstein M, Simha R (eds) The glass transition and the nature of the glassy state. Ann New York Acad Sci 279:117
21. Johari GP (1986) J Chem Phys 85:6811
22. Dixon PK, Wu L, Nagel SR, Williams BD, Carini JP (1990) Phys Rev Lett 65:1108
23. Dixon PK (1990) Phys Rev B 42:8179
24. Dixon PK, Menon N, Nagel SR (1994) Phys Rev E 50:1717
25. Lunkenheimer P, Gerhard G, Drexler F, Böhmer R, Loidl A (1995) Z Naturforsch 50A:1151
26. Lunkenheimer P, Pimenov A, Schiener B, Böhmer R, Loidl A (1996) Europhys Lett 33:611
27. Lunkenheimer P, Loidl A (1996) J Chem Phys 104:4324
28. Lunkenheimer P, Pimenov A, Dressel M, Gonscharev Yu G, Böhmer R, Loidl A (1996) Phys Rev Lett 77:318
29. Lunkenheimer P, Pimenov A, Loidl A (1997) Phys Rev Lett 78:2995
30. Lunkenheimer P, Pimenov A, Dressel M, Schiener B, Schneider U, Loidl A (1997) Progr Theor Phys Suppl 126:123
31. Lunkenheimer P, Schneider U, Brand R, Loidl A (1999) In: Tokuyama M, Oppenheim I (eds) Slow dynamics in complex systems: Eighth Tohwa University International Symposium. AIP, New York, AIP Conf Proc 469:433
32. Schönhals A, Kremer F, Schlosser E (1991) Phys Rev Lett 67:999
33. Schönhals A, Kremer F, Hofmann A, Fischer EW, Schlosser E (1993) Phys Rev Lett 70:3459
34. Schönhals A, Kremer F, Stickel F (1993) Phys Rev Lett 71:4096
35. Schönhals A, Kremer F, Schlosser E (1993) Progr Colloid Polym Sci 91:39
36. Schönhals A, Kremer F, Hofmann A, Fischer EW (1993) Phys A 201:263
37. Stickel F, Fischer EW, Schönhals A, Kremer F (1994) Phys Rev Lett 73:293632, b. Stickel F, Fischer EW, Richert R (1995) J Chem Phys 102:6521
38. Hofmann A, Kremer F, Fischer EW, Schönhals A (1994) In: Richert R, Blumen A (eds) Disorder effects on relaxational processes. Springer, Berlin, Chap. 10:309
39. Vogel H (1921) Phys Z 22:645
40. Fulcher GS (1923) J Am Ceram Soc 8:339
41. Tammann G, Hesse W (1926) Z Anorg Allg Chem 156:245
42. Angell CA (1985) In: Ngai KL, Wright GB (eds) Relaxations in complex systems. NRL, Washington, DC: 3
43. Kauzmann W (1942) Rev Mod Phys 14:12
44. Kauzmann W (1948) Chem Rev 43:219
45. Gibbs JH, DiMarzio EA (1958) J Chem Phys 28:373
46. Hecksher T, Nielsen AI, Olsen NB, Dyre JC (2008) Nat Phys 4(9):737–741
47. Davidson DW, Cole RH (1951) J Chem Phys 19:1484
48. Poley JPh (1955) J Appl Sci B4:337
49. Angell CA (1997) Physica D 107:122
50. Stillinger FH (1995) Science 267:1935
51. Debenedetti PG, Stillinger FH (2000) Nature 410:259
52. Adam G, Gibbs JH (1965) J Chem Phys 43:139

53. Doolittle AK (1951) J Appl Phys 22:1471
54. Cohen MH, Turnbull D (1959) J Chem Phys 31:1164
55. Cohen MH, Grest GS (1979) Phys Rev B 20:1077
56. Donth E, Hempel E, Schick Ch (2000) J Phys Cond Mat 12:L281
57. Donth E, Huth H, Beiner M (2001) J Phys: Cond Mat 13:L451
58. Kirkpatrick TR, Tirumalai D (1989) Phys Rev A 40:1045
59. Waterton SCJ (1932) Soc Glass Technol 16:244
60. Mauro JC, Yue Y, Ellison AJ, Gupta PK, Allan DC (2009) PNAS 106(47):19780–19784
61. Souletie H, Bertrand D (1991) J Phys (Paris) 51:1627
62. Dyre JC (2006) Rev Mod Phys 78(3):953–972
63. Dyre JC, Olsen NB (2004) PRE 69:042501
64. Dyre JC, Olsen NB, Christensen T (1996) PRB 53:2171
65. Hecksher T, Dyre JC (2015) J Non-Cryst Solids 407:14
66. Leutheuser E (1984) Phys Rev A 29:2765
67. Bengtzelius U, Götze W, Sjölander A (1984) J Phys C 17:5915
68. Götze W (1985) Z Phys B 60:195
69. Götze W, Sjögren L (1992) Rep Prog Phys 55:241
70. Bartoš J, Iskrová M, Köhler M, Wehn R, Šauša O, Lunkenheimer P, Krištiak J, Loidl A (2011) Eur Phys J E 34:104
71. Schneider U, Brand R, Lunkenheimer P, Loidl A (2000) PRL 84:5560
72. Lunkenheimer P, Wehn R, Riegger Th, Loidl A (2002) J Non-Cryst Solids 307–310:336–344
73. Lunkenheimer P, Schneider U, Brand R, Loidl A (2000) Contemp Phys 41:15
74. Novikov VN, Sokolov AP (2015) PRE 92:062304
75. Lunkenheimer P, Kastner S, Köhler M, Loidl A (2010) PRE 81:051504
76. Lunkenheimer P, Wehn R, Köhler M, Loidl A (2018) J Non-Cryst Solids 492:63
77. Schönhals A (1995) Habilitation thesis. Technical University Berlin
78. Hofmann A (1993), Dissertation, University Mainz
79. Schönhals A (2001) EPL 56:815–821
80. Kremer F (ed) (2014) Dynamics in geometrical confinement. Springer, Berlin
81. Lunkenheimer P, Loidl A (2002) Chem Phys 284:205–219
82. Lunkenheimer P, Loidl A (2002) Glassy dynamics beyond the alpha-relaxation. In: Kremer F, Schönhals A (eds) Broadband Dielectric Spectroscopy. Springer, Berlin Chapter 5

Glassy Dynamics: From Millihertz to Terahertz

P. Lunkenheimer and Alois Loidl

Abstract In this article, we review broadband dielectric spectroscopy in supercooled liquids, in many cases covering more than 15 decades in frequency and a wide range of temperatures from the low-viscosity liquid to the rigid sub-T_g glass. The access to this extremely broad frequency window allows a detailed study of the complexity of glassy freezing and glassy dynamics in a large variety of materials. Dielectric spectroscopy not only documents the enormous slowing down of the structural relaxation when approaching the glass transition, but also reveals a variety of further relaxation processes, which are important to understand the physics of the transition from a supercooled liquid into a rigid glass. After a short introduction, mainly focusing on long-term experiments on glasses and on the classification of glass formers into strong and fragile, we shortly discuss some basics of relaxation and conductivity contributions when viewed via dielectric spectroscopy. We provide some prototypical examples of dielectric loss spectra covering a large frequency and temperature regime. The glass formers shown can be categorized into two classes, type A and type B. The latter reveal well-defined Johari–Goldstein secondary relaxations, which lead to peaks in the dielectric loss at least at low temperatures. The former exhibit an excess wing, showing only a change of slope of the high-frequency flank of the structural-relaxation loss peaks. Then, we exemplify the phenomenology of glassy dynamics as revealed by these broadband spectra: The structural relaxation, the Johari–Goldstein relaxation, the appearance of a fast process as proposed by the mode-coupling theory, and the boson peak, a well-defined feature in the dielectric loss at THz frequencies, are discussed in detail. In a further chapter, we focus on the importance of sub-T_g experiments: Aging experiments and a possible experimental evidence of the Gardner transition are discussed. Finally, we summarize the experimental dielectric results documenting the universality of glassy freezing, which can be directly derived from these measurements.

P. Lunkenheimer · A. Loidl (✉)
Experimental Physics V, Center for Electronic Correlations and Magnetism,
University of Augsburg, 86135 Augsburg, Germany
e-mail: alois.loidl@physik.uni-augsburg.de

P. Lunkenheimer
e-mail: peter.lunkenheimer@physik.uni-augsburg.de

F. Kremer and A. Loidl (eds.), *The Scaling of Relaxation Processes*,
Advances in Dielectrics, https://doi.org/10.1007/978-3-319-72706-6_2

Keywords Dielectric spectroscopy · Glassy freezing
Deborah number · Structural relaxation · Relaxation time · Non-Arrhenius behavior · Johari–Goldstein relaxation · Excess wing · Fast process
Boson peak · Aging · Gardner transition

1 Introduction

Glasses are of outstanding importance in our daily life. The origins of glassmaking probably reach back to around 3000 BC where first glass beads were manufactured. Since then, glass formation plays a dominant role in many different areas: Not only for bottles, vessels, or windows in all types of architecture—from churches to skyscrapers—but also for polymers, biopolymers, metallic glasses, pharmaceuticals, optical cables, or fiber composites. Grain boundaries are a characteristic fingerprint of crystalline materials. The lack of grain boundaries in glasses, e.g., results in optical transparency in the case of silicate glasses or in superior mechanical stability in metallic glasses and is an outstanding property of the glassy state. Due to supercooling, glasses have a broad temperature window for processing. On the contrary, crystalline materials liquefy or solidify nearly instantaneously. It is a general belief that any substance can be supercooled into a glassy low-temperature state, either being small enough or being cooled fast enough. Sometimes liquids even do not crystallize at equilibrium. Binary mixtures of hard spheres with different radii are prototypical examples of the latter scenario [1].

It is a long-standing mystery that a glass is a solid without any long-range order and at the same time is a liquid that cannot flow. The question still has to be answered if the glassy state is a genuine state of matter [2]. When a liquid transforms into glass, its viscosity increases by almost 20 orders of magnitude and so does the average molecular relaxation time. However, the structure, when viewed via the pair correlation function, remains almost unchanged. How can an ensemble of atoms or molecules attain rigidity without breaking symmetry? When a liquid crystallizes in a first-order phase transition, symmetry is broken and ergodicity is broken: Disordered configurations are no longer available. If a liquid transforms into glass, ergodicity is broken, but the symmetry obviously remains the same, and disordered configurations are frozen on almost infinite times. Of course, if supercooling can be viewed as a purely kinetic phenomenon, critical temperatures would be shifted to 0 K and whether a material behaves as a liquid or solid is a function of time, only. The Deborah number D, which measures the relation of relaxation time to observation time, is the only important quantity. If $D \ll 1$, the material behaves as a liquid, for $D \gg 1$ as a solid [3]. Indeed, as expressed in the Old Testament in a song by the prophet Deborah (Judges 5:5), one might argue that, given enough time, all matter flows. In the Vulgate (the Latin version of the Bible), one can read: "*Montes fluxerunt a facie Domini*". Of course, this hypothesis can hardly be proven experimentally. Taking materials with relaxation times characterized by much shorter timescales can illustrate this behavior. Silly putty is an extreme example of such a material: It bounces like an elastic solid when dropped to a hard surface, but flows when put on a table for a couple of minutes under the action of gravity.

It has been speculated that old window glasses also will flow under the action of gravity and, hence, will be thicker at the lower edge when compared to the upper edge. This fact in some cases was verified in old medieval glasses, like the window glasses of the cathedral in Augsburg from the end of the eleventh century, showing the prophets Hosea, Moses, Daniel, Jonas, and King David. However, even if so, it cannot be a proof of glass flow on long times. Zanotto [4, 5] (see also the comment by Pasachoff [6]) calculated the flow of silica glasses and concluded that at elevated temperatures far above room temperature, the glass would move a visible amount in 800 years. However, at room temperature, any glass flow would appear on incredibly long times only, far exceeding human timescales. These calculations simply tell us that medieval artisans were clever enough to mount large window glasses with the thicker end at the bottom.

The pitch drop experiment probably is the world's longest-term laboratory experiment to probe this liquid–solid duality of matter. This experiment was started in 1927 by Thomas Parnell at the University of Queensland to document that solid materials, in fact, on very long timescales behave as liquids: At elevated temperatures, a sample of pitch was filled into a funnel, which then was sealed and was allowed to settle for three years. Then, the neck was cut allowing the pitch to flow (Fig. 1) [7]. The first drop fell in December 1938 while the latest drop—number 9—fell in April 2014. In the early days of the experiment, there was no special control of atmosphere or temperature, but in 1988, air conditioning was installed. The sequence of falling drops allowed to calculate the viscosity of the pitch at room temperature, which is approximately 2.3×10^{11} times that of water. Similar experiments have been also started in other places; e.g., the Dublin pitch drop experiment [8] was set up in Trinity College in 1944.

From detailed experiments on viscosity of a large number of very different materials, it can be concluded that vitrification is much more complex and not at all a purely kinetic phenomenon only. When measuring viscosity or relaxation times as function of temperature over a large time or frequency range, one can observe two limiting behaviors [9–11]: Strong liquids follow an ideal Arrhenius behavior, $\eta \sim \exp(\Delta/T)$, where Δ corresponds to an energy barrier in Kelvin against molecular reorientation or displacement. At the lowest temperatures, viscosity grows toward infinity, but strictly only at $T = 0$ K. Window glasses based on amorphous SiO_2 are prominent examples approximately showing this so-called strong behavior. On the other hand, fragile liquids in an Arrhenius representation, log η versus $1/T$, are strongly curved, displaying so-called super-Arrhenius behavior. The viscosity in these cases can approximately be described by a Vogel–Fulcher–Tammann (VFT) law, $\eta \sim \exp[B/(T - T_{VF})]$, and diverges at the Vogel–Fulcher temperature T_{VF}, indicating some kind of low-temperature (hidden) phase transition at T_{VF}, well below the glass transition temperature T_g, which is defined by a viscosity of 10^{12} Pa s or alternatively by an average relaxation time of ~100 s. Many low molecular weight glass formers, like ortho-terphenyl, follow this fragile temperature dependence of the viscosity.

Fig. 1 University of Queensland pitch drop experiment at 1990 two years after the seventh drop and 10 years before the eight drop fell. At that time, the experimenter was Professor John Mainstone (*Image Source* John Mainstone, University of Queensland)

This notably different relaxation characteristics is indicated in Fig. 2a in an Arrhenius-type representation, where the logarithm of the viscosity is plotted versus the inverse temperature. The Arrhenius behavior yields infinite relaxation time at 0 K, while VFT behavior implies diverging relaxation rates at a finite temperature T_{VF}, which, however, is located well below the glass transition temperature. The Arrhenius law implies a well-defined temperature-independent energy barrier against molecular motion, while a VFT behavior can be ascribed to temperature-dependent energy barriers E(T) (cf. dashed lines in Fig. 2b). The increase of the energy barriers on decreasing temperatures can be explained in terms of growing molecular clusters, which cooperatively reorient, and it seems clear that larger clusters will exhibit larger energy barriers as schematically indicated within the circles at the right of Fig. 2b. This picture can be traced back to the early days of glass physics, where the super-Arrhenius behavior of the viscosity was explained by an increasingly cooperative character of molecular motions, in terms of growing cooperatively rearranging regions [12]. An experimental proof of growing length scales and, hence, an increasing number of correlated particles in approaching the glass transition have been given recently by nonlinear dielectric-susceptibility experiments [13, 14].

To summarize, the Arrhenius law of strong liquids implies a continuous increase in viscosity down to 0 K. In contrast, the VFT behavior signals a critical temperature. From early on, these two rather contradicting aspects, a purely kinetic freezing phenomenon versus the assumption of a hidden phase transition, which, however, are not accessible in real experiments, guided the theoretical description of the glass transition over the last century. For example, some theories describe glasses simply as kinetically constrained liquids [15] with a strongly increasing viscosity, which

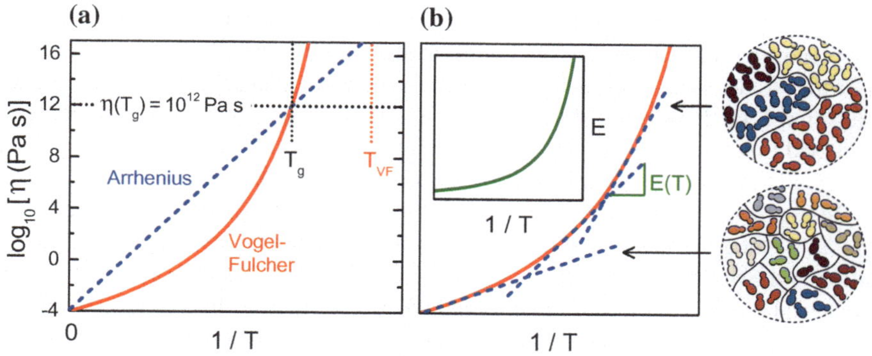

Fig. 2 **a** Temperature dependence of the logarithm of the viscosity of glass formers versus the inverse temperature [14]. Strong liquids follow an Arrhenius behavior, while the temperature dependence of fragile liquids can approximately be described in terms of a Vogel–Fulcher–Tammann law. The two different behaviors are normalized at the glass transition temperature T_g characterized by a viscosity of 10^{12} Pa s. The glass transition temperature and the Vogel–Fulcher temperature T_{VF}, where the relaxation times become infinitely long, are indicated. **b** Vogel–Fulcher–Tammann behavior derived from temperature-dependent energy barriers E(T), which are explained in terms of growing molecular clusters, reorienting cooperatively, as schematically indicated by molecules of same color within the circles at the right [14]. The inset shows the temperature dependence of energy barriers versus the inverse temperature, revealing the dramatic increase in decreasing temperature

makes them seem effectively rigid at low temperatures. Alternatively, other theories assume the existence on an underlying thermodynamic phase transition to a state with frozen-in but well-defined disordered positions [16].

2 Dielectric Spectroscopy

To unravel the mystery of glass formation, the slowing down of molecular motion has to be recorded in the broadest dynamic regime possible. This can be achieved by dielectric spectroscopy, where 20 decades of frequency of the applied ac field can be covered. For this purpose, a vast number of different experimental techniques have to be employed, ranging from time-domain spectroscopy over frequency-response analysis and coaxial reflection/transmission methods far into the optical regime utilizing terahertz and far-infrared techniques [17, 18]. Over the last decades, dielectric spectroscopy has been developed as an extremely useful tool to unravel glassy dynamics [18–23]. These experiments do not only allow to categorize supercooled liquids into strong and fragile, but also demonstrated an enormous complexity of the relaxation dynamics, with the main structural relaxation accompanied by secondary relaxations, fast processes, and microscopic peaks in the optical regime. Dielectric spectroscopy has also been proven to be an ideal tool to study hole burning in supercooled liquids, an experimental proof of heterogeneity of the relaxation dynamics, excluding any homogeneous freezing scenario [24]. It is ideally suited to follow the aging of glasses [25], and, in recent times, it even contributed to the dispute about the existence of a further transition, which is located deep in the glass state, the so-called Gardner transition [26]. Finally, nonlinear dielectric spectroscopy has been developed and provided first hints concerning the cooperative character of glassy freezing [13, 14, 27, 28]. These latter aspects will not be covered by this short review.

To characterize glassy matter, usually the complex dielectric permittivity $\varepsilon^* = \varepsilon' - i\varepsilon''$ is measured, which allows to determine the temperature and frequency evolution of the real part, ε', and the imaginary part, ε'', of the dielectric constant. When a material contains molecular dipoles and when its structure allows these dipoles to reorient, this molecular motion can be followed and mapped out by dielectric spectroscopy. Figure 3a schematically shows the real and imaginary part of the dielectric constant as function of frequency for three different temperatures. On increasing frequencies, the dielectric constant exhibits a step-like decrease from a low-frequency "static" value, ε_s, to a high-frequency value, ε_∞, which is governed by the ionic and electronic polarizability of the material under investigation. A peak in the dielectric loss accompanies this step-like decrease of the real part. In the case of non-interacting relaxing dipoles, the loss peak is fully symmetric and has a width of 1.14 decades in frequency. This is the case of a pure Debye-like relaxation [29]. On increasing temperatures, the step in the real part and the peak in the imaginary part move toward higher frequencies, indicating the speeding up of dipolar reorientation, driven by thermal activation. The peak maximum (ν_p) and the point of inflection of the step-like decrease characterize the relaxation rate $\nu_\alpha = \nu_p$ or the relaxation time τ_α via the relation $\tau_\alpha = 1/(2\pi\nu_\alpha)$ and, hence, are a measure for the molecular mobility. The other important parameter of the Debye relaxation is the dipolar relaxation strength Δ, which is given by the height of the relaxation step or alternatively by the area under the loss peak.

In the vast majority of real materials, dipolar relaxations, however, do not follow this Debye-like behavior. The loss peaks broaden considerably and in many cases have an asymmetric shape. Over the decades, several empirical functions have been developed to fit experimental data. The most common ones are the Cole–Cole (CC) [30], Cole–Davidson (CD) [31], and the Fourier transform of the Kohlrausch–Williams–Watts (KWW) function [32, 33], to name a few. The dielectric loss calculated from these empirical functions is shown in Fig. 3b in addition to the Debye function. It is important to note that the asymmetrically broadened CD and KWW functions follow a pure Debye-like behavior on the low-frequency side of the peak, namely a linear increase of $\varepsilon''(\nu)$. The broadening can be explained assuming a distribution of relaxation times: Each molecule relaxes exponentially in a Debye-like fashion, however, with different relaxation times of molecules at different sites. Then, for the different functions mentioned above, an average relaxation time $\langle\tau_\alpha\rangle$ can be calculated. Usually, via $\langle\tau_\alpha\rangle \approx 1/(2\pi\nu_p)$, it can also be well approximated by deducing the peak frequency. The heterogeneous dynamics of the primary relaxation is treated in an enormous body of experimental and theoretical work, nicely summarized in Refs. [34–36]. However, also other scenarios are possible: For example, a temperature-independent Gaussian distribution of energy barriers results in sym-

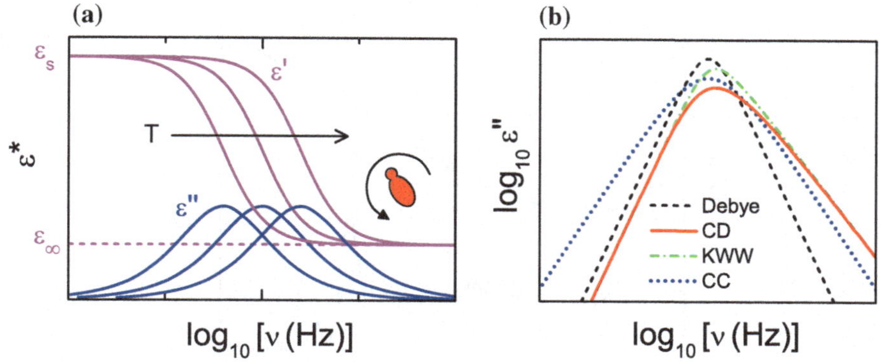

Fig. 3 Schematic view of the characteristics of dielectric spectra of glass formers as revealed by dielectric spectroscopy. **a** Typical relaxation spectra of real and imaginary part of the permittivity, shown for three temperatures. The horizontal arrow indicates increasing temperatures. The dashed line denotes the high-frequency dielectric constant. A molecular electric dipole, which via thermal activation can reorient, is indicated. **b** Model functions of the dielectric loss vs frequency on a double logarithmic plot. These functions are commonly used for the description of loss spectra: The symmetric Debye function, asymmetrically broadened Cole–Davidson (CD), and Kohlrausch–Williams–Watts (KWW) functions, as well as the symmetrically broadened Cole–Cole (CC) function are plotted (see text for details)

metrically broadened loss peaks, which can be described by the CC function. In this latter case, on decreasing temperature, the width of these loss peaks would linearly increase following a 1/T behavior [26, 37].

In some cases, when a material contains mobile charge carriers like electrons or ions, the dielectric response is governed by charge transport. In these cases, it is better to describe the experimental results using a conductivity formalism. It should be noted, however, that there is a one-to-one correspondence between complex permittivity and complex conductivity, which reads $\sigma^* = i\omega\varepsilon^*$, and, thus, in principle, both representations contain the same information. A representative plot of the complex conductivity is shown in Fig. 4: The real part of the conductivity, σ', reveals frequency-independent dc contributions at low frequencies, crossing over into a pure ac conductivity toward higher frequencies. Ac conductivity contributions exhibit a characteristic power-law increase with frequency, $\sigma' \sim \omega^s$, with $s < 1$ [38, 39]. The imaginary part of the conductivity is not sensitive to dc contributions and can be described by a similar power-law behavior with a different prefactor in the complete frequency range. The increase of the conductivity with increasing frequency results from the fact that the charge carriers, which are localized in potential minima with a broad distribution of energy barriers, at high frequencies easily can follow the applied ac electric field by back and forward jumps across small barriers. On the contrary, at low frequencies, when moving over significantly larger distances, large barriers, which cannot easily be crossed, hinder them and long-range diffusion becomes suppressed (see schematic sketch shown as inset of Fig. 4) [40]. Here, it should also be mentioned that sometimes especially for ionically conducting materials, instead of

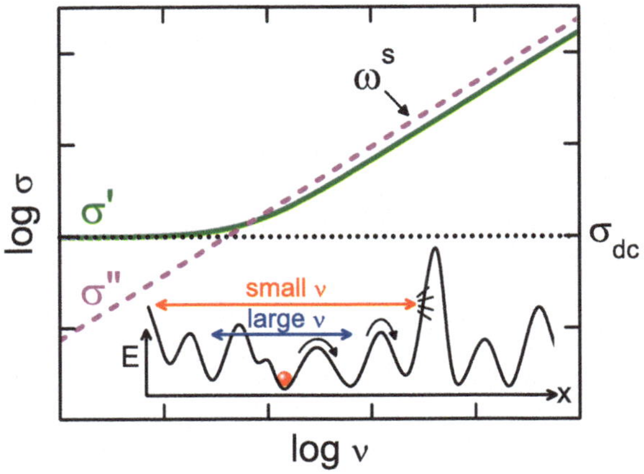

Fig. 4 Frequency dependence of real (σ', full line) and imaginary part (σ'', dashed line) of the complex conductivity, plotted on a double-logarithmic scale. A dotted line indicates the frequency-independent dc conductivity. In the lower part of the figure, a schematic energy surface with a distribution of hindering barriers is indicated, illustrating the likely origin of the commonly observed increase of σ' with frequency

the conductivity representation the dielectric modulus $M^* = 1/\varepsilon^*$ is analyzed to characterize the ionic dynamics [41]. Spectra of the imaginary part $M''(\nu)$ reveal similar peaked behavior as $\varepsilon''(\nu)$ of non-conducting glass formers and can be analyzed in an analogous way. This enables the determination of the so-called conductivity relaxation time τ_σ, which is believed to provide a characteristic timescale for molecular motion in ionic conductors. However, one should be aware that the applicability and interpretation of the electric modulus are quite controversial (see, e.g., [42–45]).

Finally, we would like to come back to the temperature dependence of the average relaxation time of the primary relaxation, as observed in the vast majority of supercooled liquids. The structural or primary relaxation in glass-forming materials, as detected by dielectric spectroscopy, usually is called α-relaxation and, as will be shown later, is strongly coupled to the viscosity. In most glass-forming liquids, this strong coupling is valid for all temperatures above the glass transition temperature and the dynamics of dipolar reorientation strictly maps the viscosity of the material under consideration. Of course, decoupling phenomena can appear below the glass transition temperature, where molecular dipoles still may reorient, even in a rather rigid lattice. For the vast majority of supercooled liquids, with the prominent exception of mono-hydroxy alcohols [46], the α-relaxation is the dominating peak in frequency- and temperature-dependent dielectric spectra. Its tremendous slowing down under cooling in a narrow temperature range signals the increase in viscosity. As a rule of thumb, a viscosity of 10^{12} Pa s corresponds to a relaxation time of the order of 100 s.

As outlined already above and documented in Fig. 2 (in this case, treatment of the temperature dependence of the viscosity), against naive expectations the viscosity and the average relaxation time, $\langle\tau_\alpha\rangle(T)$, in most supercooled liquids do not follow a simple Arrhenius law, $\langle \tau_\alpha \rangle = \tau_0 \exp[E/(k_B T)]$, which arises from purely thermally activated molecular motion over an energy barrier E separating neighboring minima in the potential surface of the system. Instead, for most glass formers, a plot of log $\langle \tau_\alpha \rangle$ versus 1/T exhibits significant curvature as indicated by the solid line in Fig. 2a, which sometimes is called super-Arrhenius behavior. It can be formally fitted by the empirical VFT formula, $\langle \tau_\alpha \rangle = \tau_0 \exp[DT_{VF}/(T - T_{VF})]$ [47–49]. Here, D is the so-called strength parameter [50] and T_{VF} is the Vogel–Fulcher temperature. Small values of D imply strong deviations from Arrhenius behavior. Such glass formers are also termed "fragile", in contrast to so-called "strong" glass formers, whose relaxation time more closely follows the Arrhenius law [50]. Alternatively, a fragility index m was defined, which is given by the slope of the relaxation times plotted versus T_g/T, just at the glass transition temperature [51–53]. The strength parameter is related to the fragility index via $m \approx 16 + 590/D$. Hence, strong, Arrhenius-like behavior corresponds to a fragility index $m = 16$, while an upper limit of fragility was estimated to be close to 200 [54]. However, also other estimates exist (see later).

3 Glassy Dynamics Revealed by Dielectric Spectroscopy

In Figs. 5 and 6, we provide some prototypical dielectric broadband loss spectra of materials exhibiting glassy freezing as measured during the last years in our group in different systems. All the spectra shown in these figures are a combination of the results of different techniques. As will be indicated, most of these spectra have been partly published, but in several cases are extended either in frequency or in temperature. They provide an impressive illustration of the slowing down of the structural relaxation in a variety of glass-forming systems, ranging from water-salt mixtures to various molecular systems, documenting the universal aspects of glassy freezing. The spectra reveal the enormous slowing down of the average relaxation rate, documented by the maximum of the loss peak (α-relaxation), which reaches up to 14 decades in frequency in a relatively narrow temperature regime. It has been proven for a variety of supercooled liquids that the α-relaxation closely follows the temperature evolution of the viscosity (see later). Figures 5 and 6 impressively document the power of dielectric spectroscopy, which allows to follow and detect glassy freezing over more than 20 decades in frequency.

In addition to the dominating α-relaxation, secondary processes appear at the high-frequency flank of the α-relaxation peak. There are numerous examples where secondary relaxations are peaked on the high-frequency flank of the structural relaxation and in these cases often are described as prototypical examples of Johari–Goldstein (JG) relaxations [55]. The occurrence of these secondary processes is thought to be an intrinsic property of glass-forming liquids and to be inherent to the glassy state of matter. They do not, e.g., correspond to relaxations of side-groups of the molecules but may arise from relaxations in less-dense packed regions, so-called islands of mobility [55]. An alternative explanation ascribes JG relaxations to motions within a fine structure of the energy landscape experienced by the molecules [56–58]. Follow-

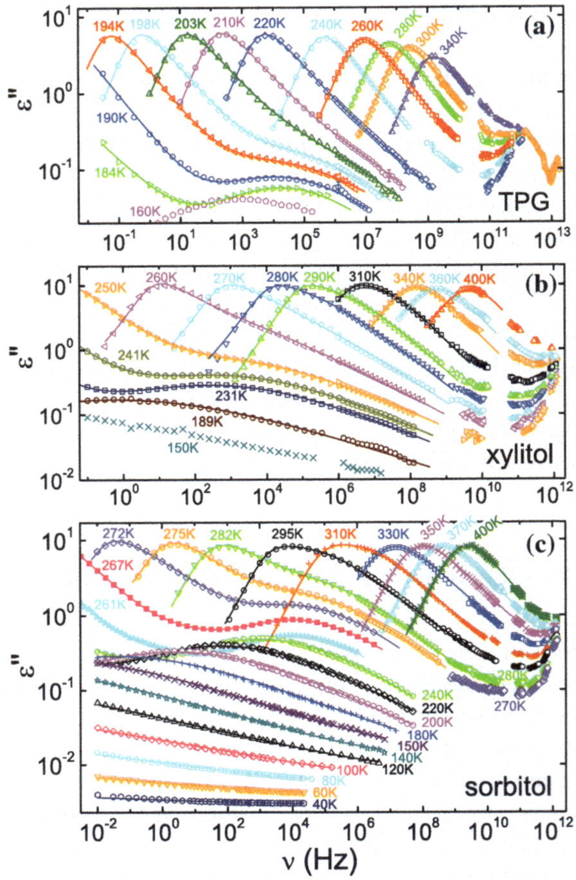

Fig. 5 Broadband dielectric loss spectra of so-called type B glass formers at a series of temperatures ranging from the low-viscosity regime down to the glass transition temperature and below. All spectra are shown on double-logarithmic scales for a series of temperatures. The solid lines represent the sum of model functions for the α-relaxation and the JG secondary ß-process as described in the text. **a** tri-propylene glycol (TPG) [60, 61], **b** xylitol [62], and **c** sorbitol [26, 62]. In sorbitol, the dielectric loss is followed deep into the sub-T_g region

ing the nomenclature introduced by the group of E. Rössler, these systems sometimes are termed type B glass formers [59]. Three examples of broadband spectra of typical type B glass formers are shown in Fig. 5.

The frequency and temperature dependence of tri-propylene glycol (TPG) is shown in Fig. 5a. This spectrum covers more than fifteen decades in frequency. These results up to about 1 THz have been published by Köhler et al. [60]. Here, we have added the high-frequency infrared spectra [61], which cover the boson peak and will be discussed later. This figure provides the main characteristics of the relaxation dynamics governing glassy freezing. At high temperatures, α-relaxation and boson peak are not well separated and almost merge. This fact shows that, at high temperatures in the low-viscosity limit, relaxational and vibronic timescales heavily overlap. On decreasing temperatures, the structural relaxation rapidly slows down, documenting the tremendous separation of timescales of relaxations and vibrations. A gap opens between the boson peak and the α-relaxation. However, the loss in between is strongly enhanced and cannot be described as a simple overlap of the

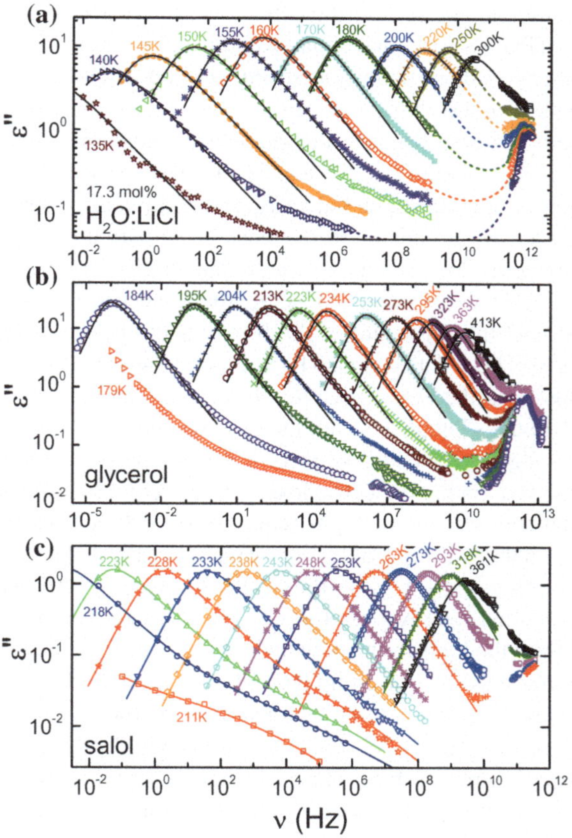

Fig. 6 Broadband dielectric loss spectra of type A glass formers. All spectra are shown on double-logarithmic scales for various temperatures. **a** 17.3 mol% water: LiCl solution [65]: The solid lines represent fits by a HN function for the α-relaxation. The dashed lines are drawn to guide the eye and to find a smooth extrapolation between the structural relaxation or the excess wing and the high-frequency peak, **b** glycerol [20]: The solid lines represent fits with a CD function for the α-relaxation, **c** salol [66]: The solid lines represent the sum of a HN function for the α-relaxation and a CC function for the secondary ß process. The line through the 211 K data is only a guide to the eye

low-frequency flank of the boson peak and the high-frequency flank of the structural relaxation [60]. In between a characteristic minimum appears, which represents a critical spectrum and is treated, e.g., in the framework of the mode-coupling theory (see later). At not too low temperatures (in the case of TPG for T > 220 K), the structural relaxation peaks up to the onset of the minimum can approximately be fitted by a CD function, yielding a constant slope of the high-frequency flank. However, at lower temperatures, a clear peak evolves on this high-frequency flank, which becomes more prominent at low temperatures and is well established below $T_g \sim 194$ K. In Fig. 5a, this sequence of α- and ß-relaxations is fitted by a sum of CD and CC functions, indicated by the solid lines.

Similar spectra with a slightly reduced frequency scale are shown for xylitol [62] and sorbitol [26, 62] in Figs. 5b, c, respectively. The solid lines in Fig. 5b represent fits with a sum of Havriliak–Negami (HN) [63] and CC functions. The fits in sorbitol, indicated by solid lines in Fig. 5c, were also performed utilizing a sum of HN and CC fits. Despite an overall similar behavior of the dielectric loss as function of frequency and temperature, there are significant differences when comparing these three dielectric loss spectra of type B glass formers: The ß-relaxation in TPG has very low dipolar weight and on increasing temperatures seems to become almost hidden under the strong structural relaxation. The secondary processes in xylitol and sorbitol more strongly increase in increasing temperature, and at elevated temperatures, primary and secondary relaxations seem to be almost of equal weight. See, e.g., sorbitol at 295 and 310 K. The spectra of sorbitol are shown down to very low temperatures, far below T_g, and it is clear from these spectra that the ß-relaxation extremely broadens on decreasing temperatures. At the lowest temperatures, the dielectric loss is very low and characterized by a continuous decrease at least up to MHz frequencies. At 40 K, the dielectric response is nearly flat, signaling an almost infinite width of the ß-relaxation, passing over into constant-loss behavior [26]. These data can also be interpreted as signature of the so-called nearly constant loss, ascribed to a separate physical process. A prominent example is the "caged dynamics" treated within the extended coupling model [64].

Figure 6 provides three further examples of glassy freezing of supercooled liquids viewed via the dielectric loss, where secondary relaxations do not show up as well-defined peak on the high-frequency side of the α-relaxation, but rather are indicated as change of slope, only. Figure 6a shows the dielectric loss as measured in H_2O:LiCl solutions [65], Fig. 6b the results in glycerol [20] and Fig. 6c in salol [66]. These so-called type A glass formers are characterized by a second, more shallow power law at the high-frequency flank of the α peak, termed "excess wing", a nomenclature introduced by our group [67, 68]. Again, at high temperatures $T > T_c$, the structural relaxation seems to be characterized by one asymmetric loss peak, approximately described by KWW or CD functions. On decreasing temperatures, $T_g < T < T_c$, the excess wing evolves. This is, e.g., well documented in the dielectric loss of glycerol and salol slightly above T_g: At 195 K and at 10^3 Hz in glycerol and at 228 K and 10^4 Hz in salol, the high-frequency flank of the α-relaxation reveals a clear change of slope. At further decreasing temperatures, e.g., at 179 K in glycerol and at 211 K in salol, well below the glass transition temperature, a clear shoulder-like curvature evolves, which seems to be the remainder of the JG ß-relaxation. Here, it should be noted that these sub-T_g results had to be taken in thermodynamic equilibrium, which was reached after keeping the samples at the measurements temperature for up to five weeks [25, 68, 69]. As for several type A glass formers in this way the excess wing was found to develop into a shoulder at low temperatures, it has been argued that this spectral feature is due to a JG relaxation, which is more closely coupled to the structural relaxation than in type B systems [68]. This was later on corroborated by various other experiments [70–73].

In what follows, we will describe in more detail the phenomenology of glassy relaxation as revealed in Figs. 5 and 6. In glassy matter, the tremendous increase in viscosity on decreasing temperatures is clearly mapped by the slowing down of the dielectric loss peak, shifting through the enormous frequency window in a relatively narrow temperature regime. Structural relaxation and microscopic processes in the THz regime overlap at high temperatures. The microscopic processes remain almost temperature independent, and on cooling, a strongly temperature-dependent minimum occurs at GHz frequencies, which is described, e.g., by mode-coupling theory. At lower temperatures, JG secondary processes appear separating more and more from the structural relaxation approaching the glass transition temperature. This sequence of processes will be described in the following chapters.

3.1 Structural Relaxation

As outlined above, the drastic increase of the viscosity of glass-forming materials is documented in the temperature evolution of the average relaxation rate ν_α or the corresponding relaxation time $\langle\tau_\alpha\rangle = 1/(2\pi\nu_\alpha)$. For temperatures above T_g, the temperature dependence of $\langle\tau_\alpha\rangle$ is strongly coupled to the viscosity and it can be measured by a variety of techniques, including ultrasound experiments, neutron- and light-scattering experiments, as well as by dielectric spectroscopy. For some systems, it has been experimentally checked if these average relaxation times really coincide with the temperature dependence of the viscosity, which of course is the true measure of glassy freezing. Examples are documented in Fig. 7a for glycerol and in Fig. 7b for propylene carbonate [74]. The most complete data set is available for glycerol. In Fig. 7a, we compare our results of the average relaxation time as determined by dielectric spectroscopy in glycerol [75], with average relaxation times derived from shear modulus [76] and ultrasound results [77], as well as from neutron-scattering [78, 79] and light-scattering experiments [80]. These average relaxation times are compared with existing viscosity results [81–83]. It is clear that viscosity results and relaxation times can be freely scaled against each other, however, only by a constant factor. In doing so, we find almost perfect agreement between all measurements in an astonishingly wide temperature and frequency regime. This figure also documents that the glassy dynamics can be studied by dielectric spectroscopy in the widest possible frequency range and with the highest precision. If we take this comparison seriously and accept the used scaling of viscosity and average relaxation times, we find that the glass transition temperature, which corresponds to a viscosity of 10^{13} Poise or 10^{12} Pa s determines an average relaxation time of 1000 s. As a rule of thumb, it is often assumed that the relaxation time at the glass transition temperature equals 100–250 s.

Figure 7b provides a similar comparison for propylene carbonate. Here, the average relaxation times as determined from our dielectric results [75] are compared with those deduced from neutron [84] and light-scattering studies [85]. Again, this temperature dependence of the average relaxation times is compared to the viscosity results [86] as described in detail in the figure caption. We find good agreement, with the exception that the neutron-scattering data [84] seem to be significantly too low. If we accept scaling by a constant factor, we find that 10^{12} Pa s correspond to 10.000 s, which seems somewhat too large. Chen et al. [87] performed a critical comparison of viscosity and dielectric times. The main conclusion of their work was that rotational modes and viscosity are generally coupled, displaying parallel traces in the activation maps, although they can exhibit slight decoupling especially in the millisecond relaxation time range. To conclude about the experimental facts docu-

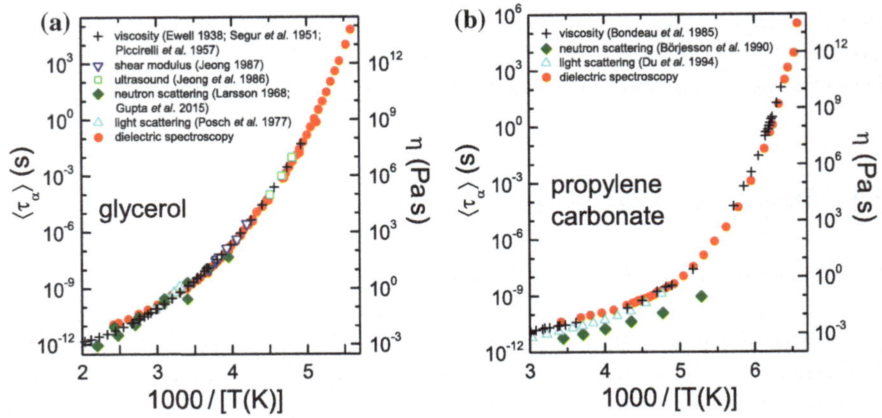

Fig. 7 Comparison of average relaxation times as determined by a variety of experimental techniques (left scale) compared with the temperature dependence of the viscosity (right scale) in **a** glycerol and **b** propylene carbonate [74]. The data for glycerol were taken from Refs. [75–83], those on propylene carbonate from [75, 84–86]. The data are plotted in an Arrhenius-type representation. In both frames, the ranges of the axes of average relaxation times and viscosities were chosen to cover the same total number of decades and, in addition, their start values were adapted to match both data sets as close as possible

mented in Fig. 7a, b, we can state that the average relaxation times determined by dielectric spectroscopy certainly are strongly coupled to the viscosity and broadband dielectric spectroscopy is an ideal and unprecedented tool to study glassy dynamics.

The most relevant parameter that can be obtained from the loss peaks as documented in Figs. 5 and 6 is the average relaxation time. For the supercooled liquids shown in these figures, we collected the temperature dependence of the average relaxation times and added the results of propylene carbonate (PC) and propylene glycol (PC). The results are shown in Fig. 8a. Lunkenheimer et al. [75] have collected most of these temperature dependencies of the average relaxation times. The $\langle\tau_\alpha\rangle$ values of the H_2O:LiCl solutions were taken from Ref. [65]. As indicated by the solid lines, the temperature evolution of the average relaxation times of this variety of different systems—in an extreme dynamic range—can rather nicely be parameterized by the VFT law. However, it has to be clearly stated that the VFT law is a mere parametrization of the temperature dependence of the average relaxation times, respectively, of the viscosity. There are a number of alternative approaches to model the temperature-dependent dynamics in glass-forming liquids. For example, Mauro et al. [88] have proposed an equation avoiding the divergence at a critical temperature inherent to the VFT law. In their formalism, the relaxations times, respectively the viscosity, show a divergence at 0 K. Lunkenheimer et al. [75] provided a critical comparison of these two significantly different approaches, describing glass relaxation with or without the use of a critical temperature. They were able to show that in some cases this new Mauro-type temperature dependence really is superior. However, a number of examples were given, where the VFT law still provides the superior fits.

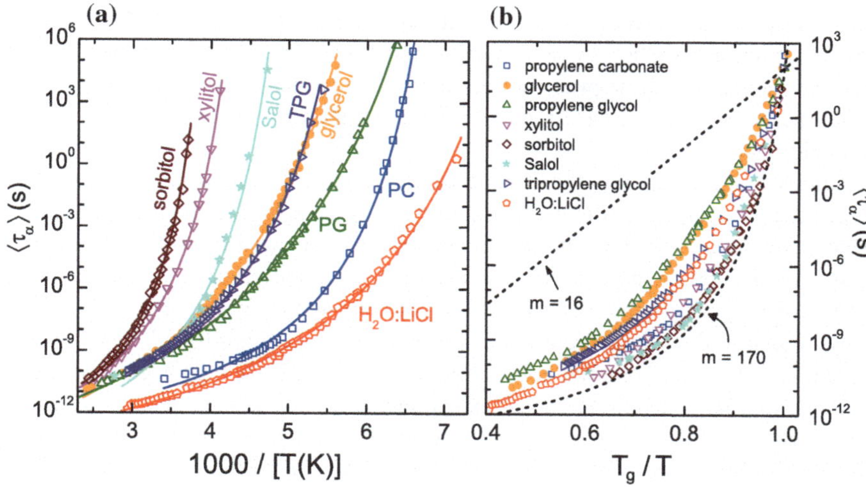

Fig. 8 **a** Temperature dependence of average relaxation times in a variety of glass formers in an Arrhenius-type representation [65, 75]: The logarithm of relaxation times is plotted versus the inverse temperature. The solid lines represent fits with the Vogel–Fulcher–Tammann law (see text). **b** Scaled relaxation times [21] in an Angell plot [89]: All materials shown are scaled to their glass transition temperatures T_g. The dashed lines with m = 16 and m = 170 mark the extremes between strong (Arrhenius) and ultra-fragile behavior, respectively (see text)

For a direct comparison, in Fig. 8b, all average relaxation time results presented in Fig. 8a are normalized to the glass transition temperature and are plotted on a scale T_g/T. In this so-called Angell plot [89], strong and fragile glass formers can easily be identified: The extreme values of strong and fragile are indicated by the dashed lines illustrating fragility indices of m = 16 and m = 170. The value 16 corresponds to Arrhenius behavior, and the value of 170 is indicated as the highest possible fragility index. This value has been adapted from Ref. [90] and is slightly lower than the fragility index m = 200 [53] cited above. The examples shown in Fig. 8b all rather belong to the fragile class of supercooled liquids, with sorbitol and salol being close to the fragile extreme.

3.2 *Johari–Goldstein Relaxation and Excess Wing*

As noted above, characteristic secondary relaxations are observed on the high-frequency flank of the structural-relaxation loss peaks. In some cases, secondary relaxations are caused by intramolecular modes, for example, side-chain motions of polymers. From the viewpoint of glass physics, these relaxations are of minor interest. Here, we are concerned with secondary relaxations, which are thought to be inherent to the glassy state of matter. In their seminal work [55], Johari and Goldstein documented that secondary relaxations are also found in glass-forming systems with rigid molecules, where side-chain motions are unlikely to exist. To discriminate these inherent secondary processes, which seem to be intimately linked to the structural relaxation, these processes are often termed Johari–Goldstein relaxations or, alternatively, slow ß-relaxations to distinguish them from the fast ß-process of the mode-coupling theory.

Figures 5 and 6 already document that these secondary relaxations, in the frequency domain, appear as distinct relaxation peaks (type B glass formers) or just as a change of slope of the high-frequency wing of the structural relaxation, leading to the so-called excess wing [68] (type A glass formers). However, it should be noted that it still is quite controversial if the excess wing is indeed due a JG relaxation, partly hidden in the spectra below the dominating α-relaxation [68, 70, 91], or if it is due to an additional, separate relaxation process [59, 92]. In type B glass formers, these genuine JG relaxations appear as symmetric peaks. This is best documented in Fig. 5c for sorbitol. In type A glass formers, the maximum of the dielectric loss can hardly be observed and only the high-frequency wing of the JG-relaxation peak can be fitted. In most cases, these secondary processes are analyzed in terms of CC-type relaxations. In the temperature dependence of the dielectric loss, a significant contribution from the JG relaxation shows up, too. Figure 9a documents $\varepsilon''(T)$ of sorbitol, a prominent example of a type B glass former, for measurement frequencies between 10 mHz and 20 kHz [26]. The characteristic feature of the α-relaxation is a significant peak close to room temperature. On decreasing temperature, a second peak is observed, arising from the JG relaxation, well separated from the structural relaxation. It shifts to lower temperatures for lower measuring frequencies. It should be noted that in Fig. 9a a change of slope to a weaker temperature dependence of ε'' is found below T_g. This feature is primarily due to the sample falling out of thermodynamic equilibrium at the glass transition and less due to the onset of JG-relaxation contributions to $\varepsilon''(T)$. In fact, especially for the two higher frequencies shown in Fig. 9a, the JG relaxation also significantly contributes to the left flanks of the α-relaxation peaks at $T > T_g$, which can be easily verified by a comparison with the loss spectra in Fig. 5c at temperatures somewhat above T_g (e.g., 275 or 282 K).

In clear distinction to sorbitol, the secondary relaxation in glycerol (Fig. 9b) does not show up as a second peak in $\varepsilon''(T)$. Glycerol belongs to the class of type A glass formers, and its loss spectra (Fig. 6b) show an excess wing on the high-frequency flank of the structural relaxation [68]. In the temperature dependence, the dielectric loss reveals two changes of slope. The first one, occurring at higher temperatures is induced by the onset of significant contributions of the excess wing to the loss, which exists already significantly above T_g. In Fig. 9b, these contributions are indicated by the dashed lines. The second change of slope appears close to the glass transition temperature, where the system falls out of thermodynamic equilibrium. Here, the excess wing is completely dominating the dielectric response. As discussed above, there is no final consensus concerning the microscopic origin of JG relaxations. They could be interpreted as jumps between local minima in a multiwell energy landscape, e.g., local small-angle librations [56–58] or as local processes in less-dense regions of the structural glass, where, due to the frozen-in heterogeneity, islands of mobility exist [55].

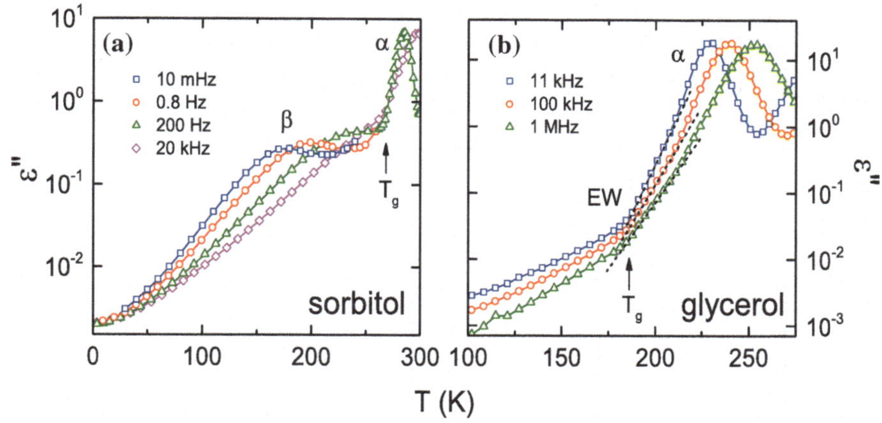

Fig. 9 Temperature dependence of the dielectric loss at selected frequencies in **a** sorbitol [26], a prototypical type B glass former, and in **b** glycerol, a type A glass-forming system. Both figures are shown on semilogarithmic scales. Arrows indicate the glass transition temperatures in both materials. The dashed lines in (**b**) indicate excess wing (EW) contributions at temperatures > T_g

From an evaluation of the frequency-dependent permittivity at different temperatures, it is possible to generate relaxation maps with α- and ß-relaxations in an Arrhenius-type representation. These are presented in Fig. 10a, b for two typical examples. Figure 10a shows the average relaxation times for sorbitol [62], which is a prominent type B glass former. Below T_g, in these materials, the secondary relaxations are well separated from the structural relaxation and strictly follow an Arrhenius type of behavior. It is evident that this relaxation dynamics mostly appears in a frozen glass matrix, which is out of thermodynamic equilibrium. In sorbitol, the JG ß-relaxation decouples from the α-relaxation at temperatures well above the glass transition temperature. It is not an easy task to analyze the temperature dependence of the ß-relaxation times for $T > T_g$, due to the fact that the corresponding features in the dielectric spectra strongly overlap and the resulting average relaxation times bear large experimental uncertainties. However, in literature, there are some reports that τ of the ß-process passes through a shallow minimum before finally approaching the α-process [60, 93, 94]. Dyre and Olsen explained this minimum in terms of a so-called minimal model for ß-relaxations [93], while an "encroachment" of the relaxation time of an additional γ-relaxation with the JG relaxation was proposed in Ref. [95]. A minimum in the temperature dependence of the average relaxation time of the ß-process in sorbitol before merging with the structural relaxation is compatible with the experimental results shown in Fig. 10a [62]. However, the precision of the data analysis does not allow finite conclusions.

The temperature dependence of the secondary relaxation times in type A glass formers is significantly different: The secondary process, viewed via the excess wing, is always much stronger coupled to the structural relaxation. One problem arises because, in most cases, only the high-frequency flank of this secondary process can

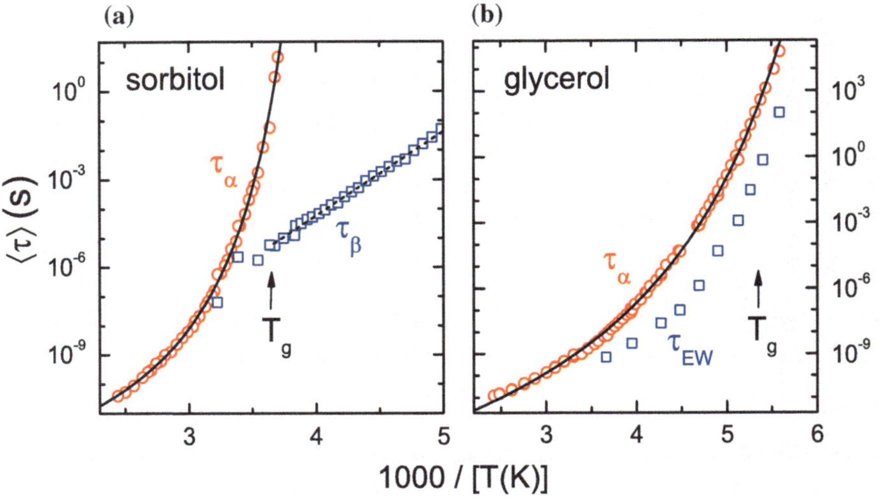

Fig. 10 Temperature dependence of average relaxation times of α- and ß-relaxation in an Arrhenius representation. Relaxation times are plotted versus the inverse temperature for **a** sorbitol [62] and **b** glycerol [75, 91]. The former is a type B glass former, the latter is of type A. The solid lines are fits with the VFT equation. The dashed line in (**a**) corresponds to Arrhenius behavior

be analyzed and the tentative maximum is hidden under the dominating α-relaxation. An example where the secondary relaxation closely follows the primary relaxation in a distinct non-Arrhenius type of behavior is shown in Fig. 10b for glycerol [75, 91]. This figure documents that the JG ß-relaxation in glycerol certainly follows a super-Arrhenius behavior and is closely coupled to the structural relaxation, in clear distinction to type B glass formers. Figure 10b also documents that the excess wing in glycerol can be analyzed well above the glass transition temperature, even if large experimental uncertainties are considered. Similar results, namely that the JG relaxation is non-Arrhenius, were also found, e.g., for propylene carbonate, propylene glycol, and ethanol [60, 91, 96]. Whether the excess wing below the glass transition temperature finally evolves into an Arrhenius behavior, cannot be finally decided. From the observations outlined above, one may conclude that the JG ß-process in type B glass formers, at low temperatures, strictly follows an Arrhenius behavior and at high temperatures probably exhibits a shallow minimum before merging with the structural relaxation. On the contrary, in type A glass formers, the relaxation time of the excess-wing relaxation, also interpreted as inherent JG relaxation, always appears to closely follow the α-relaxation time, exhibiting an explicit super-Arrhenius temperature dependence, and never seems to strongly decouple from the structural relaxation as found in type B systems.

3.3 *The Fast Process*

The mode-coupling theory (MCT) [97–99] belongs to the most prominent theories of the glass transition. In 1984, Bengtzelius et al. [97] and, independently, Leutheusser [98] showed that a specific version of the MCT of liquids exhibits a dynamical singularity, with striking characteristics of the liquid–glass transition. This theory predicts, in addition to a critical slowing down of the structural relaxation, a fast process, sometimes termed fast-ß process, located in the GHz to THz frequency regime. This fast process is thought to be associated with a rattling motion of a particle dynamically caged by its neighbors in the supercooled liquid state. The rattling is responsible for the fast process at high frequencies, while the decay of the transient cage corresponds to the structural relaxation at lower frequencies. Consequently, the generalized susceptibility exhibits a critical spectrum in the regime between the microscopic peak and the structural relaxation, which according to MCT follows specific scaling predictions, all of which are experimentally accessible and, hence, stimulated enormous experimental interest. A review on the variety of experimental tests of MCT predictions was given by Götze [100].

In frequency domain, in the transition region between the microscopic peak and the structural relaxation, the fast ß-process shows up as a significant minimum in the imaginary part of the susceptibility (which corresponds to the loss ε'' in the dielectric case) with enhanced fluctuations. MCT predicts scaling laws for the temperature dependence of this minimum, namely for its frequency ν_{min} and the minimal susceptibility χ''_{min}. Moreover, the frequency of the peak maximum of the structural relaxation ν_α also should exhibit critical behavior with the critical exponents being determined by the shape parameters of the minimum (see below). First experimental records, proving that this critical spectrum with a significant self-similar enhancement above background noise really does exist, have been provided by neutron scattering [101] and light scattering [102]. In due course, the scaling behavior of this critical spectrum and the existence of a critical temperature T_c were verified by dynamic light [103–106] and quasielastic neutron scattering [105]. At that time, dielectric spectroscopy in the GHz to THz frequency regime in supercooled liquids was hardly accessible. Based on dielectric experiments in salol up to 10 GHz, it was argued that there is no minimum in the dielectric spectra and, thus, that a critical slowing down and a critical temperature do not exist and that MCT predictions are invalid for dielectric spectroscopy [107]. However, due to the development of continuous-wave submillimeter and THz spectroscopy as well as advanced transmission and reflection techniques in the microwave range [18], it was later possible to measure the dielectric loss in this frequency regime of the critical spectrum with high precision and to prove the qualitative validity of MCT [108, 109]. It should be noted that, in addition to MCT, there are also other approaches explaining the finding of a shallow susceptibility minimum at high frequencies. The most prominent example is the extended coupling model [64] where a nearly constant-loss contribution is predicted, which is ascribed to "caged dynamics". Indeed, dielectric spectra including the region of the high-frequency minimum have successfully been fitted using this approach [108, 110, 111].

All the data sets determined from the different experimental techniques, like neutron and light scattering, as well as dielectric spectroscopy, finally yielded susceptibilities in qualitative agreement with MCT predictions, however, also with sometimes quantitative differences. A critical comparison of the imaginary part of the dynamical susceptibility of glycerol, as measured by neutron and light scattering as well as by dielectric spectroscopy, is provided in Ref. [19]. It covers the frequency regime of the critical spectrum including structural relaxation and microscopic peak. Similar comparisons have also been made for other glass formers [19, 110]. Of course, it still is unclear to what extent MCT can be applied to real materials with irregularly shaped molecules and with very different microscopic intermolecular potentials. However, it is one of the main predictions of MCT that similar model parameters should arise from all observables, which couple to density fluctuations. In comparing the experimental results of neutron scattering, light scattering, and dielectric spectroscopy, one has to take into consideration that these experimental probes couple to different internal degrees of freedom: Dielectric spectroscopy and light scattering primarily probe the orientational dynamics of the molecules, while neutron scattering primarily couples to density fluctuations. One also should consider that MCT in its original form did not take into account non-spherical molecules and orientational degrees of freedom. Hence, perfect quantitative agreement between the results of these different experimental results cannot be expected. Later theoretical works, generalizing MCT by incorporating orientational degrees of freedom [112–114], have provided at least a qualitative understanding of many of these differences. Moreover, for supercooled propylene carbonate, it was shown that the high-frequency susceptibilities from dielectric-, light-, and neutron-scattering experiments can be consistently described within an extended mode-coupling approach [115, 116].

Nowadays, the detection of glassy dynamics in this high-frequency regime by dielectric spectroscopy is well established and an example of the frequency and temperature dependence of the dielectric loss in the frequency regime of the critical spectrum is presented in Fig. 11a for salol [61, 117]. It shows the evolution of the dielectric loss ε'' in the frequency regime between structural relaxation (<GHz) and microscopic peak (>THz) for temperatures between 263 and 361 K. Please note that the glass transition temperature in salol is close to 220 K and, hence, all these measurements were performed well above T_g. It is well established that the dynamic liquid-to-glass phase transition as predicted by MCT appears at a critical temperature T_c, which is always located significantly above the glass transition temperature. The dielectric loss of salol reveals the well-known susceptibility minimum. It can be well fitted by the MCT prediction, namely the sum of two power laws with an exponent $-b$, accounting for the increase toward the structural relaxation at low frequencies, and with an exponent a for the increase at high frequencies (solid lines in Fig. 11a). We find exponent parameters a and b of 0.352 and 0.75, respectively. Both parameters are directly related to each other and temperature independent; i.e., the minima at different temperatures can be scaled onto each other. On increasing temperature, the minimum significantly shifts to higher frequencies and the minimum value of

ε'' strongly increases. According to mode-coupling theory, both quantities should follow strict scaling behaviors. An analysis of these results in terms of MCT scaling is documented in Fig. 11b–d [117]. The variation of the minimal dielectric loss ε''_{min} (Fig. 11b), the slowing down of the minimum frequency ν_{min} (Fig. 11c), and the predicted temperature dependence of the relaxation rate of the structural relaxation ν_α (Fig. 11d), all can nicely be described using the critical scaling predictions of MCT with critical exponents that are partly directly related to the exponents a and b deduced from the shape of the minimum. The critical behavior of all three quantities shown in Fig. 11b–d consistently points to the same critical temperature of T_c = 256 K. By light scattering, exponent parameters $a = 0.327$ and $b = 0.641$, as well as a critical temperature $T_c = 256$ K, were determined [104], which are well consistent with the present results from dielectric spectroscopy. We feel that these and similar results in other glass formers [19, 21, 60, 108, 109] really document the predictive power of mode-coupling theory and are a proof of the existence of a critical dynamic transition between the supercooled liquid and the structural glass. Lastly, we want to point out that the results of Fig. 11 finally settle the dispute on the presence of the high-frequency minimum and of the critical dynamics in the dielectric response of salol [104, 107].

3.4 Boson Peak

A rather intriguing, universal, and still highly controversially debated feature of disordered matter is the appearance of the so-called boson peak in vibrational spectra and thermodynamic properties. In heat-capacity experiments, it shows up as an excess contribution in the 10 K range. In vibrational spectra, this peak is revealed as an excess contribution to the density of states in the low THz range, which in crystalline materials usually exhibits a Debye-like ν^2 frequency dependence. The boson peak has been predicted to appear in Brillouin-scattering experiments in amorphous solids [118] and later on has been observed in a variety of disordered systems by Raman spectroscopy [119] and in inelastic neutron-scattering experiments [120]. The name boson peak results from the fact that its amplitude exhibits only a weak temperature dependence, which obviously closely follows Bose–Einstein statistics. The boson peak can also be observed in dielectric spectroscopy, and illuminating examples are shown in Fig. 5a for tri-propylene glycol and in Fig. 6b for glycerol.

Numerous models have been formulated to explain the microscopic origin of the boson peak, and it is far out of the scope of this work to discuss and critically evaluate these vastly different models. Theoretical proposals include rather material-specific models [121], the involvement of soft anharmonic potentials [122], dynamics in fractal structures [123], phonon scattering by intrinsic density fluctuations [124], spatially fluctuating force constants [125], random spatial variation of the shear modulus [126], or transverse vibrational modes arising from low-density defect structures [127]. It also has been proposed that the boson peak is a signature of a phase tran-

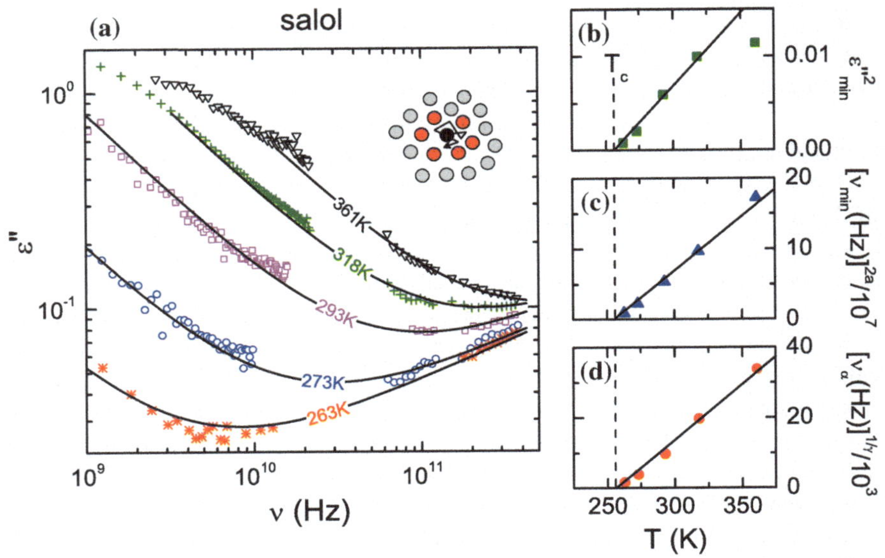

Fig. 11 **a** Dielectric loss of salol in the frequency regime of the fast ß process [61, 117]. The solid lines are fits with the mode-coupling theory (see text). The inset shows a schematic sketch of the rattling process responsible for the frequency and temperature dependence in this regime. Frames **b–d** show an analysis of these results within the framework of the mode-coupling theory: **b** temperature dependence of the minimum dielectric loss, **c** temperature dependence of the minimum frequency, and **d** temperature dependence of the structural relaxation rate [61, 117]. Experimental results (symbols) are compared with theoretical predictions (solid lines). ε''_{min} is the minimum dielectric loss at a given temperature, ν_{min} corresponds to the frequency of the minimum, the exponent a corresponds to the high-frequency exponent of the minimum, ν_α is the relaxation rate of the structural relaxation, and the exponent γ is given by $\gamma = 1/2a + 1/2b$, where b is the low-frequency exponent of the minimum

sition, where the potential energy surface changes from a minima-dominated phase with phonons to a saddle-point dominated phase without phonons [128].

As an example, in Fig. 12a, we show the dielectric loss in glycerol in the frequency regime of the boson peak for a series of temperatures from 363 to 184 K, ranging from the low-viscosity liquid regime to the glass transition temperature [129, 130]. In glycerol, the boson peak is located at high frequencies, $\nu > 1$ THz, and is only weakly temperature dependent. At lower frequencies, a strong temperature dependence appears, partly due to the superposition with the strongly temperature-dependent structural relaxation, but partly also because in this frequency regime the region of the critical dynamics of MCT is entered [19]. Only at the lowest temperature shown in Fig. 12a, at 184 K, the boson peak appears nearly unobscured, extending from about 0.5 to 10 THz. At frequencies above 10 THz, internal molecular vibrations are detected, which become narrow and well defined on decreasing temperatures. Astonishingly, the boson peak in glycerol seems to exhibit a double-peak structure, indicating that a complex density of vibrational states contributes to it.

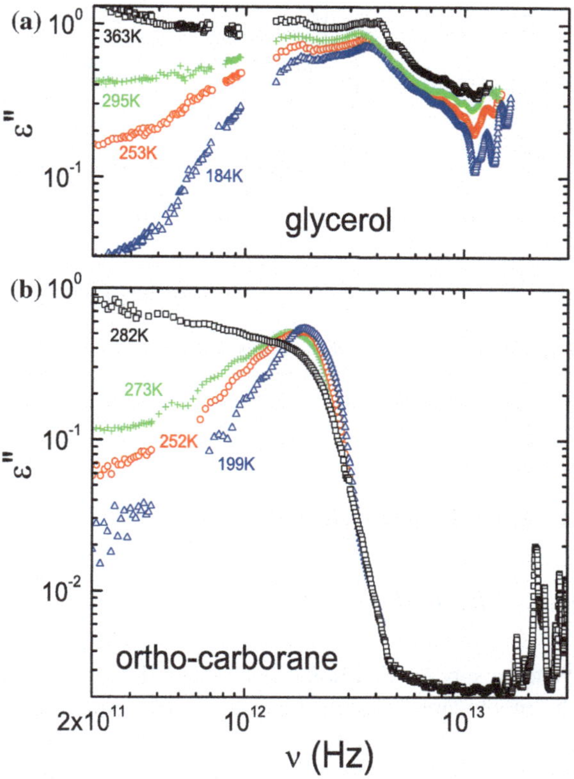

Fig. 12 Boson peak in **a** the supercooled liquid glycerol for temperatures between 184 and 363 K and **b** in the plastic crystal ortho-carborane at a series of temperatures between 199 and 282 K [130]

In Fig. 12, the boson peak in glycerol is contrasted with the corresponding feature as observed in a plastic crystal [130]. Plastic crystals are disordered systems, in which the centers of the molecules form a regular crystalline lattice and obey translational symmetry, while the orientational degrees of freedom are dynamically disordered and in some materials undergo glassy freezing into an orientationally disordered state [131]. Figure 12b shows the boson peak in ortho-carborane for temperatures between 199 and 282 K [130]. Ortho-carborane is a highly symmetrical molecule and, at first sight, can be described as rigid sphere. The rigid molecules form the analogue of a mono-atomic lattice and, hence, should be characterized by acoustic modes only. Intramolecular excitations appear above 20 THz and are well separated from the acoustic modes. In this case, it seems that the boson peak simply maps the acoustic density of states with a cut-off frequency close to 4 THz. In Ref. [130], it was speculated that the coupling to the THz radiation may arise from a disorder-induced breaking of selections rules or a strong hybridization of collective phonon modes with local relaxational excitations. In any case, Fig. 12b provides clear experimental hints that in the glassy, orientationally disordered state, the boson peak results from acoustic modes. On increasing temperatures, these modes become less and less well defined and finally merge with the relaxational mode close to 282 K.

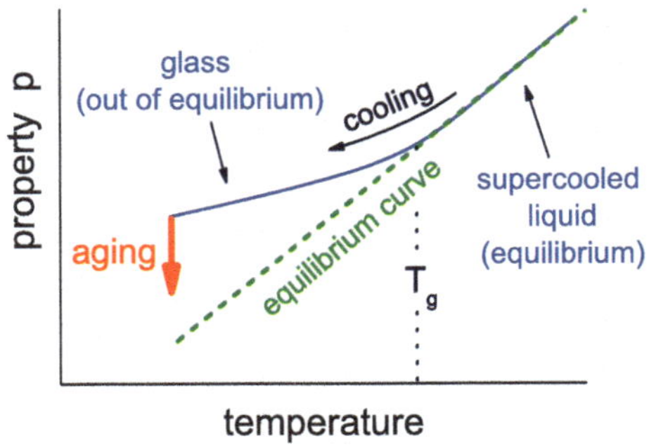

Fig. 13 Schematic plot of the temperature dependence of a material property around the glass temperature (solid line). When crossing T_g, a weaker temperature dependence is found. The dashed line shows the equilibrium curve. The vertical arrow indicates the approach of equilibrium under aging at constant temperature

4 Sub-T_g Behavior of Glasses

4.1 Physical Aging

When a supercooled liquid is cooled below its glass temperature T_g, it falls out of thermodynamic equilibrium leading to the non-ergodic glass state. In the liquid state above T_g, under cooling the molecules (or other glass-forming entities like ions, atoms, polymer segments, etc.) can always adjust their arrangement in reaction to the falling temperature. However, this no longer is possible in the glass state because the relaxation time quickly reaches huge values below T_g (cf. Fig. 2) and, before the entities can adapt their arrangement, the temperature has fallen to even lower values. This results in a frozen glass state whose properties approximately reflect those of an equilibrium state at a higher temperature, termed fictive temperature T_f. Overall, this loss of equilibrium under cooling results in a weaker temperature dependence of various material properties below T_g as schematically indicated in Fig. 13. When, after cooling into the glass state, the material is held at constant temperature, in principle it will approach the equilibrium state. Then, its properties will change with time, a process termed physical aging. This is indicated by the vertical arrow in Fig. 13. However, one should be aware that sufficiently far below T_g, due to the huge relaxation times arising already relatively shortly below T_g, this time dependence will be extremely slow and not detectable within realistic experimental timescales. Only for temperatures not more than several K below T_g, significant effects can be expected.

As an example, Fig. 14a, b shows the time dependence of the dielectric loss of glycerol and xylitol under physical aging [25]. These experiments were performed after quickly cooling the samples from above T_g to temperatures about 6 K (glycerol) or 5 K (xylitol) below the glass transition. Aging at constant temperature lasting up to 5 weeks was necessary to ensure that equilibrium was reached in these experi-

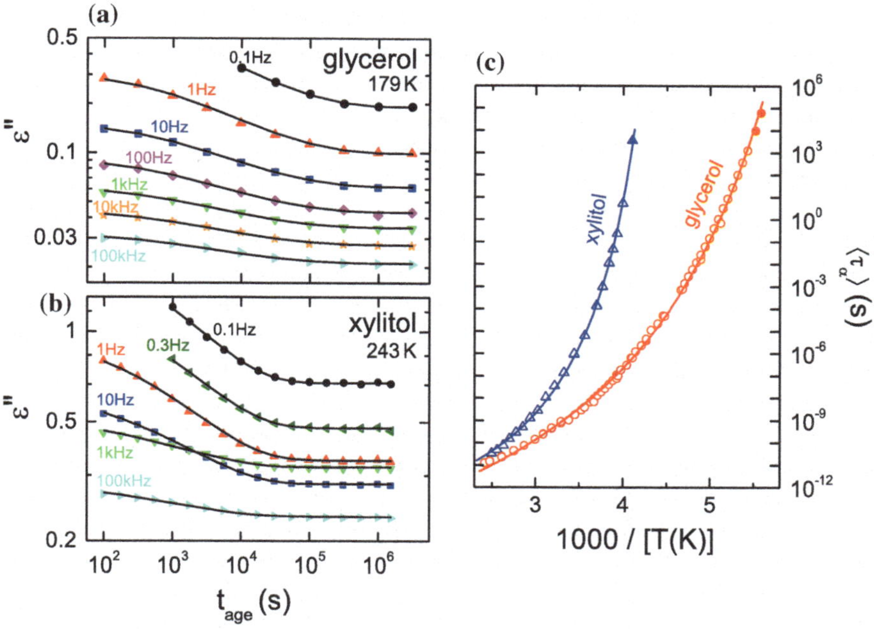

Fig. 14 Time dependence of the dielectric loss of glycerol (**a**) and xylitol (**b**) at temperatures below T_g [25]. The lines are fits with the modified KWW law as promoted in Ref. [25]. Frame **c** shows the average α-relaxation times determined from equilibrium experiments (open symbols) and the relaxation times deduced from the aging experiments, extending the equilibrium results by several decades (closed symbols) [25, 75]. The lines are fits with the VFT function

ments. For both materials, a clear decrease of the dielectric loss is observed for all investigated frequencies. At these temperatures, the α peak is located outside of the investigated frequency range and $\varepsilon''(t)$ in most cases reflects the behavior at the right flank of this peak. As discussed in Ref. [25], obviously the α relaxation shifts to lower frequencies under aging, resulting in the observed decrease of ε''. For xylitol aging at 243 K, at the higher frequencies, the β relaxation partly contributes to the detected loss (cf. Fig. 5b), which explains the partial crossing of the corresponding aging curves in Fig. 14b [25, 132].

The analysis of such data is not straightforward. As realized already long ago, it is clear that, due to the structural rearrangements occurring during aging, the relaxation time of the material itself must be time dependent [133–136]. Therefore, it is quite oversimplifying to fit such time-dependent data with simple stretched-exponential behavior with a fixed aging relaxation time τ_{age} as done, e.g., in Refs. [137–139]. Clearly, the parameters obtained by such an analysis should not be compared with equilibrium results. For a more sophisticated treatment of such data, the Tool—Narayanaswamy–Moynihan formalism [133, 134, 140] can be employed, which traces back the aging-induced variation of properties to the time-dependent variation of the fictive temperature, leading to a time-dependent relaxation time. However, the application of this formalism is not straightforward, and for rather simple aging experiments as in the present case, especially for aging triggered by a single temperature jump, our group has suggested an alternative phenomenological approach [25]. It assumes a stretched-exponential time dependence (also termed KWW behavior), however, modified by introducing a time-dependent relaxation rate which itself also varies with a KWW function. This ansatz can be solved by recursion, and the lines in Fig. 14a, b are fits with this approach, excellently describing the experimental data. All $\varepsilon''(t)$ curves obtained at the different frequencies are simultaneously fitted with the same relaxation rate and width parameter, which is not possible when using a simple KWW function with time-independent τ_{age} [141]. Most importantly, only when employing this approach, the obtained aging parameters are revealed to be consistent with those determined in equilibrium experiments at $T>T_g$ [25, 132, 142]. As an example, Fig. 14c shows the temperature-dependent α-relaxation times τ_α of glycerol and xylitol as determined from equilibrium measurements (open symbols) and, in addition, the relaxation times τ_{age} resulting from the analysis of the aging data (closed symbols) [25, 75]. The latter provide a perfect extrapolation of the equilibrium data, thus extending the range of $\tau(T)$ by up to three decades. The stretching parameter of the KWW function deduced from aging also is consistent with the equilibrium results [25]. From these and similar findings in other glass formers [25, 132], one can conclude that physical aging is governed by the same dynamics as the α relaxation and that $\tau_\alpha = \tau_{age}$.

4.2 Gardner Transition

As outlined above, in the research field of the glassy state of matter, there exists a long-standing controversy about a possible true phase transition and the existence of an ideal glass with infinite viscosity: The question whether there is a hidden phase transition at nonzero temperature, which cannot be reached experimentally as the system falls out of thermodynamic equilibrium at the glass transition temperature, or whether glassy materials just become more and more sluggish on decreasing temperatures has not been finally answered. However, from an experimental point of view, there is growing evidence of increasing length scales of cooperativity when approaching the glass transition [13, 27, 28], favoring a true phase-transition scenario. It came as a big surprise that recent theories suggest a further transition, deep in the glass state, which is thought to be even observable in non-equilibrium conditions. This so-called Gardner transition was originally predicted in mean-field theories of p-state spin glasses [143, 144], but recently was also expected to exist in a variety of model glasses in infinite [145–147] as well as in three dimensions [148–150].The Gardner transition corresponds to a further fractionalization of the energy landscape making the glass even more heterogeneous in the low-temperature phase. However, it also has been argued that the Gardner transition might be strongly affected by finite dimensional fluctuations and doubts have been raised about its relevance for structural glass formers [151, 152].

At the glass transition temperature, usually, it is thought that the system is trapped in one of the many metabasins of the energy landscape and that the α-relaxation correspond to jumps between different metabasins. This scenario is indicated in Fig. 15e (left inset) at high temperatures. In real glasses, the structural relaxation slows down its relaxation time following a VFT-type temperature dependence (solid line in Fig. 15e) and, approximately at the critical temperature T_c of mode-coupling theory, secondary JG relaxation processes decouple from the primary relaxation (dashed line). Within the metabasins, these secondary relaxations can be thought as sub-basins, arising from local molecular motions (Fig. 15e, middle inset). At the hypothetical Gardner transition, a further roughening of the energy landscape occurs (right inset), which could give rise to a possible Gardner relaxation as indicated by the dash-dotted line in Fig. 15e. Hence, in real glasses, the Gardner transition, if existent, has to be accompanied by a fractionalization of the JG sub-basins. As the Gardner transition can be also observed in non-thermodynamic equilibrium, it is clear that the experiment of choice is a detailed temperature-dependent dielectric loss study of secondary relaxations well below the canonical glass transition temperature T_g. Long time ago, this scenario has been proposed by Kirkpatrick and Wolynes [153]: Specifically with reference to the Gardner transition, these authors pointed out that instabilities of the glass state at low temperatures may be associated with secondary relaxations.

A scenario of the temperature evolution of the frequency-dependent dielectric loss including a possible Gardener transition is indicated in Figs. 15a–d. (a) At temperatures above the mode-coupling T_c, in the highly viscous liquid phase, the mode-coupling critical dynamics appears between structural relaxation and boson peak. (b) Roughly at T_c, secondary relaxations decouple from the primary relaxation and shift through the frequency window in the supercooled liquid. (c) Below the glass transition temperature T_g, the α-relaxation is shifted out of the experimentally accessible frequency window. Now all experiments are clearly out of thermodynamic equilibrium, and only JG-type ß-relaxations can be observed. (d) On further cooling, the ß-relaxation considerably broadens and the mean relaxation rate of the secondary relaxation is also shifted out of the experimental frequency window, leaving only its high-frequency tail visible. Any further fractionalization of the energy landscape, the scenario of a possible Gardner transition, then can be assumed to yield additional loss contributions, as indicated in (d). In high-precision dielectric loss experiments, it should be possible to identify this onset of excess loss at low temperatures and to relate it to the appearance of a possible Gardner transition.

In Fig. 5c, we provided already broadband loss spectra of sorbitol in an extremely wide temperature range, including a detailed evolution of the secondary relaxation well below the glass transition temperature [26, 62]. Similar dielectric loss spectra are shown in Fig. 16 for xylitol [26]. In both glass formers, we find a strong shift of the β-relaxation peaks toward low frequencies accompanied by an extreme broadening. At the lowest temperatures, only the high-frequency flank of these secondary loss peaks can be detected. In Ref. [26], these loss peaks have been fitted using a Cole—Cole function, or, in case when the peak was completely shifted out of the frequency window of the experiment, by only fitting a power-law behavior accounting for the

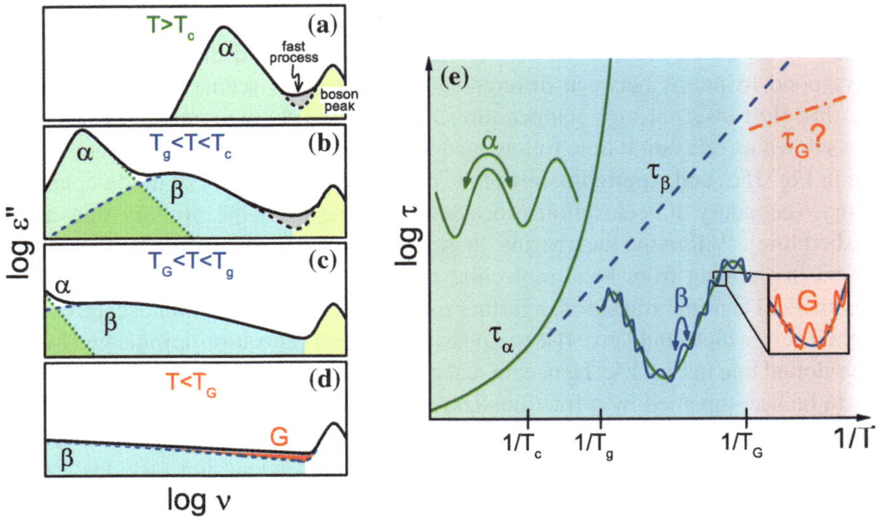

Fig. 15 **a–d** Schematic loss spectra versus frequency of glass-forming materials on double logarithmic scales as function of temperature [26]: **a** The liquid state above T_c. **b** The supercooled liquid for $T_g < T < T_c$ with a well-developed secondary Johari-Goldstein relaxation. **c** The rigid glass below T_g but above the Gardner transition temperature T_G, $T_G < T < T_g$. **d** The glass far below the glass temperature and below the Gardner transition, $T < T_G < T_g$. The different dynamic processes are indicated by color: the α relaxation (green), the JG β-relaxation (blue), the fast process (gray), and the boson peak (yellow). The red area in (**d**) indicates the suggested additional contribution arising from the sub-basins in the energy landscape induced by the Gardner transition. **e** Typical temperature dependence of the α- and β-relaxation times in an Arrhenius plot with a possible additional Gardner relaxation arising below T_G [26]. In the different temperature regions, we provide schematic views of the energy-landscape scenario: α-relaxation, JG β-relaxation, and possible modification of the local β-relaxation basins by the Gardner transition, leading to a fractal roughening of the landscape

high-frequency flank. From these fits, it was possible to deduce the temperature evolution of the width parameter α of the symmetric CC function down to the lowest temperatures [26]. We want to point out that these experiments were performed at sufficiently low temperatures to exclude any effects from physical aging. As demonstrated, e.g., by Fig. 14b, detectable aging effects occur at temperatures much closer to T_g.

The analysis of the dielectric loss results of Fig. 16 reveals a strong increase of the width parameter on decreasing temperatures, nearly reaching unity at 0 K, as indicated in Fig. 17a. One should have in mind that the approach of $\alpha = 1$ on decreasing temperatures signals that the loss peaks become infinitely broad, resulting in a constant-loss behavior at the lowest temperatures. One possible scenario to explain this behavior relies on a model using a fixed distribution of local hindering barriers, namely assuming a temperature-independent Gaussian distribution of barrier heights [26, 37]. This distribution characterizes the energy landscape of the sub-basins responsible for secondary relaxations as indicated in Fig. 15e. That it is temperature independent follows from the fact that, below T_g, the local structure is

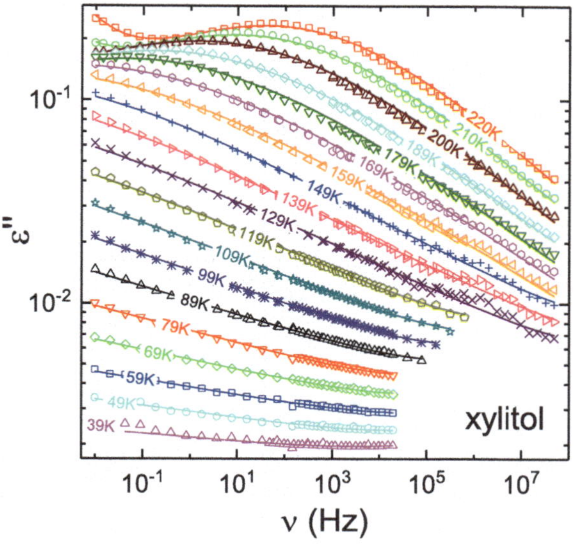

Fig. 16 Sub-T_g spectra of the dielectric loss versus frequency of xylitol for various temperatures between 39 and 220 K [26]. The lines are fits by a combination of power laws and the CC function as described in the text and Ref. [26]

effectively frozen-in. This model leads to symmetrically broadened loss peaks that closely follow CC behavior, justifying its application for the JG β relaxation. Interestingly, such a *temperature-independent* energy barrier distribution leads to a strongly *temperature-dependent* distribution of relaxation times, namely the half-width W_τ of the latter is expected to diverge with 1/T [37, 59]. From the experimentally determined width parameters of the CC distribution, shown in Fig. 17a, the half widths W_τ were calculated and plotted in Fig. 17(b) [26]. Indeed, the overall behavior follows the general trend of a 1/T divergence. However, the line in Fig. 17b, showing the expected $W_\tau \propto 1/T$ behavior, reveals some significant deviations. Here, the proportionality parameter of the 1/T law was adapted to match the high-temperature data. Then, $W_\tau(T)$, determined from α(T) of the loss peaks (circles), at low temperatures becomes even broader compared to the half-width resulting from a constant Gaussian distribution of energy barriers (line). This strong additional low-temperature increase in the half-width becomes even more obvious in the inset of Fig. 17b by plotting $W_\tau \times T$. As shown by the line in Fig. 17a, the 1/T behavior of W_τ leads to an approximately linear increase of α(T). Again, deviations show up and the experimental width parameter approaches $\alpha = 1$ already at higher temperatures.

Geirhos et al. [26] interpreted this and similar behavior found for sorbitol as an indication of a further roughening of the energy landscape of the sub-basins by the Gardner transition. Based on these experiments, one can speculate that a Gardner transition indeed can be observed by a detailed study of secondary relaxations in real glass formers well below the glass transition temperature and that in xylitol T_G is located approximately at 100 K. It is clear that these observed deviations of a continuously broadening JG relaxation may also have other reasons: In first respect, one might argue that a constant Gaussian distribution of energy barriers is a too oversimplified approach and that the real distribution of energy barriers might be

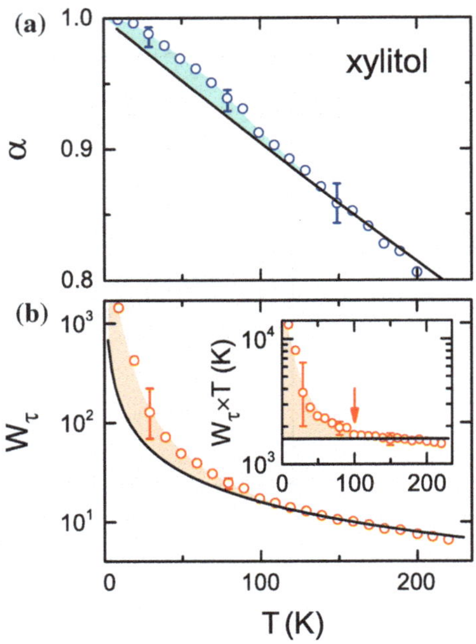

Fig. 17 **a** Circles indicate the temperature dependence of the width parameter α as obtained from fits of the dielectric loss of xylitol as described in the text [26]. The line was calculated from the 1/T dependence of W_τ shown by the line in frame (**b**). **b** The circles show the temperature-dependent half-width W_τ of the distribution of relaxation times, which was calculated from the experimentally determined α(T) assuming a temperature-independent Gaussian distribution of energy barriers. The line in (**b**) shows the expected 1/T divergence of W_τ, adapted to match the experimental data at high temperatures. The inset highlights the strong deviations from this expected behavior below approximately 100 K

much more complex. In addition, constant-loss contributions due to caged dynamics have been predicted within the extended coupling model by Ngai [64]. It could well be that, at low temperatures, these additional constant-loss contributions superimpose with the loss of the high-frequency flank of the JG relaxations and finally become visible below 100 K in the case of xylitol. One has to await further experiments to further verify these speculations about the experimental observation of the Gardner transition in real glasses.

5 Concluding Remarks on the Phenomenology of Glassy Freezing

In the previous chapters, we have discussed the phenomenology of glassy freezing as observed by broadband dielectric spectroscopy. For most supercooled liquids, the relaxation dynamics can be characterized by a dominating structural or α-relaxation

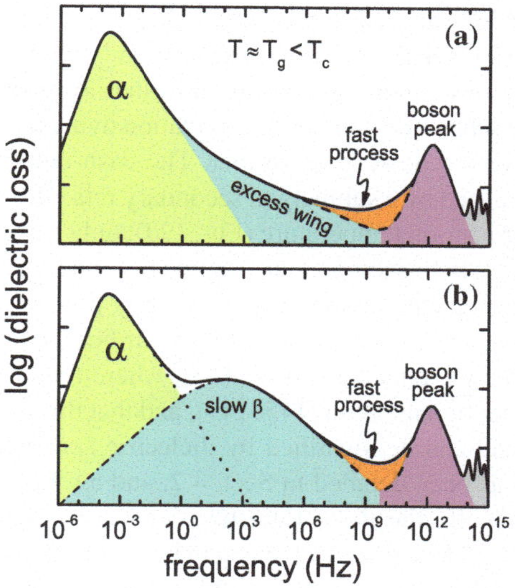

Fig. 18 Frequency dependence of the dielectric loss at a temperature close to the glass transition **a** in a prototypical type A glass former with an excess wing and **b** in a type B glass former with a canonical JG ß relaxation [20]

followed in the spectra by an additional secondary JG-type process on the high-frequency flank of the α-relaxation peak. At GHz frequencies, we enter the frequency domain of the critical dynamics described by the mode-coupling theory, followed by the boson peak at THz frequencies. Intramolecular excitations at even higher frequencies have not been discussed in this review. A schematic overview of this complex relaxation dynamics is plotted in Fig. 18, depicting the situation for a rather low temperature, close to T_g [20]. Figure 18a shows the characteristic features of type A glass formers, where the high-frequency flank of the structural relaxation only exhibits a change of slope. We interpret this excess-wing phenomenon [68] as an inherent JG relaxation, which is more strongly coupled to the structural relaxation as compared to type B glass formers and follows a super-Arrhenius behavior in its temperature evolution. Figure 18b shows the relaxation dynamics at a temperature close to the glass transition for a type B glass former with the characteristic JG ß-relaxation. For temperatures $T < T_g$, the average relaxation time of the secondary relaxation strictly follows an Arrhenius behavior and is strongly decoupled from the structural relaxation.

Finally, Fig. 19 provides the prototypical evolution of a supercooled liquid when cooled from a low-viscosity state at high temperatures to temperatures far below the glass transition [14, 21]. At the highest temperatures (Fig. 19a), the dielectric loss of the low-viscosity liquid shows a dominant structural relaxation, almost merging with the boson peak. On lowering temperatures (Fig. 19b), and below the melting temperature T_m, the structural relaxation in the supercooled liquid state separates from the fast processes creating a characteristic minimum, which is well described by the critical dynamics of the mode-coupling theory. This critical spectrum follows a strict scaling behavior with a critical temperature T_c. Of course, in real materials, this critical spectrum never softens completely and hopping processes play a dominant role. The dash-dotted line in Fig. 19b indicates a possible CC peak responsible for the critical law [154]. Approximately for $T < T_c$, secondary relaxations evolve, which become more prominent on further lowering the temperatures. Figure 19c provides a schematic view of the relaxation dynamics close to the glass transition temperature for a type A glass former. The dash-dotted line indicates how the excess wing is caused by a submerged secondary relaxation peak [68]. Finally, well below the glass transition temperature (Fig. 19d), only secondary relaxations survive. The dielectric loss in this temperature regime becomes almost flat and can be described as constant-loss phenomenon or as secondary process, which becomes infinitely broad. There are recent speculations that a further characteristic temperature appears below T_g, the Gardner transition [143], where the energy potential surface exhibits a further fractionalization [145–149], and that this transition gives rise to additional processes and can be identified by dielectric spectroscopy in sub-T_g experiments [26]. This has been detailed in Sect. 4.2, and a schematic loss spectrum is shown in Fig. 15d. Finally, the glassy freezing as viewed by dielectric spectroscopy via the dielectric loss (Fig. 19a–d) is confronted in Fig. 19d with the dielectric loss observed in a crystalline material. In a crystal with long-range translational and orientational order, phonons are the only relevant absorption processes and the idealized dielectric spectrum at low frequencies is completely loss-free.

In conclusion, we have documented in this short review the power of broadband dielectric spectroscopy to unravel the complexity of glassy freezing in supercooled liquids. The understanding of the temperature evolution of a variety of processes, including structural relaxation, Johari–Goldstein relaxation, the fast ß process of the mode-coupling theory, and the boson peak, which appear in very different frequency windows, is a prerequisite for a deeper understanding of the glass transition. One also should have in mind that, irrespective of the microscopic interpretations, glassy dynamics looks very similar in a variety of systems, spanning small-molecule glass formers, polymers, ionic melts, or plastic crystals. It seems that a universal freezing scenario governs the glass transition in all these very different materials. Finally, we have pointed out the significance of sub-T_g experiments, referring to the importance of aging and to recent experiments possibly indicating a further phase transition, the so-called Gardner transition, deep in the solid glass sate, which can be measured even in non-thermodynamic equilibrium.

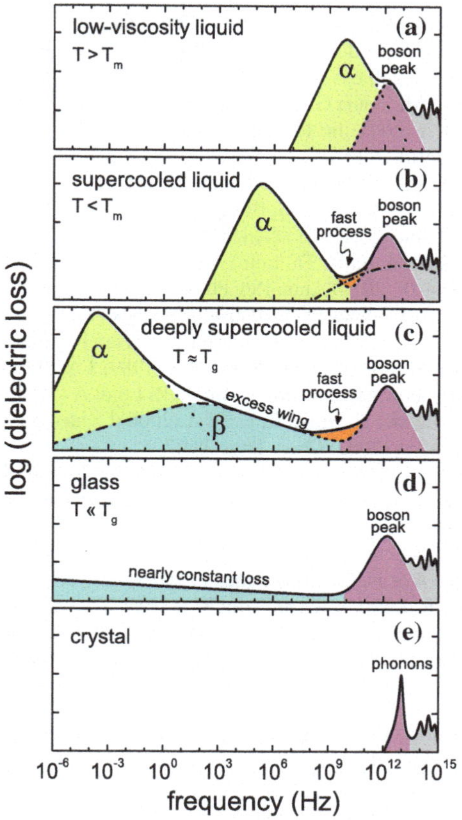

Fig. 19 Pictorial view of the temperature dependence of relaxation processes of supercooled liquids from the low-viscosity regime until deep into the glass phase, compared to the dielectric loss spectrum of an ideal crystal [14, 21]. **a** In the low-viscosity liquid, the structural relaxation almost merges with the boson peak. **b** In the regime of the supercooled liquid well above T_g, the structural relaxation and boson peak separate. In the minimum in between, the mode-coupling fast process emerges. The dash-dotted line indicates the CC peak discussed to be responsible for the critical dynamics [154]. **c** Close to the glass transition temperature, a secondary β relaxation is observed. In the shown case of a type A glass former, it leads to an excess wing. **d** At very low temperatures, in the sub-T_g regime, the structural relaxation is shifted out of the accessible frequency window. A nearly constant loss appears as a remainder of secondary relaxations. **e** Loss spectrum of crystalline matter, with long-range order. Here, only phonons and intramolecular excitations exist

References

1. Parisi G (2000) Phys A 280:115
2. Berthier L, Ediger M (2016) Phys Today 69:41
3. Reiner M (1964) Phys Today 17:62
4. Zanotto ED (1998) Am J Phys 66:392
5. Zanotto ED, Gupta PK (1999) Am J Phys 67:260
6. Pasachoff JM (1998) Am J Phys 66:1021

7. Edgeworth R, Dalton BJ, Parnell T, Eur J Phys **198** (1984)
8. Johnston R, Nature News, 18 July 2014. https://doi.org/10.1038/nature.2013.13418
9. Böhmer R, Ngai KL, Angell CA, Plazek DJ (1993) J Chem Phys 99:4201
10. Angell CA, Ngai KL, McKenna GB, McMillan PF, Martin SW (2000) J Appl Phys 88:3113
11. Böhmer R, Angell CA (2003) In: Kremer F, Schönhals A (eds) Broadband dielectric spectroscopy. Springer, Berlin, p 11
12. Adam G, Gibbs JH (1965) J Chem Phys 43:139
13. Albert S, Bauer Th, Michl M, Biroli G, Bouchaud J-P, Loidl A, Lunkenheimer P, Tourbot R, Wiertel-Gasquet C, Ladieu F (2016) Science 352:1308
14. Lunkenheimer P, Michl M, Bauer Th, Loidl A (2017) Eur Phys J Special Topics 226:3157
15. Chandler D, Garrahan JP (2010) Annu Rev Phys Chem 61:191
16. Dzero M, Schmalian J, Wolynes PG (2012) In: Wolynes PG, Lubchenko V (eds) Structural glasses and supercooled liquids: theory, experiment and applications. Wiley, Hoboken, p 193
17. Böhmer R, Maglione M, Lunkenheimer P, Loidl A (1989) J Appl Phys 65:901
18. Schneider U, Lunkenheimer P, Pimenov A, Brand R, Loidl A (2001) Ferroelectrics 249:89
19. Lunkenheimer P, Schneider U, Brand R, Loidl A (2000) Contemp Phys 41:15
20. Lunkenheimer P, Loidl A (2002) Chem Phys 284:205
21. Lunkenheimer P, Köhler M, Kastner S, Loidl A (2012) In: Wolynes PG, Lubchenko V (eds) Structural glasses and supercooled liquids: theory, experiment and applications. Wiley, Hoboken, p 115
22. Richert R (2015) Adv Chem Phys 156:101
23. Kremer F, Schönhals A (eds) (2003) Broadband dielectric spectroscopy. Springer, Berlin
24. Schiener B, Böhmer R, Loidl A, Chamberlin RV (1996) Science 274:752
25. Lunkenheimer P, Wehn R, Schneider U, Loidl A (2005) Phys Rev Lett 95:055702
26. Geirhos K, Lunkenheimer P, Loidl A (2018) Phys Rev Lett 120:085705
27. Crauste-Thibierge C, Brun C, Ladieu F, L'Hote D, Biroli G, Bouchaud J-P (2010) Phys Rev Lett 104:165703
28. Bauer Th, Lunkenheimer P, Loidl A (2013) Phys Rev Lett 111:225702
29. Debye P (1912) Ann Phys 39:789
30. Cole KS, Cole RH (1941) J Chem Phys 9:341
31. Davidson DW, Cole RH (1950) J Chem Phys 18:1417
32. Kohlrausch R (1854) Ann Phys 167:56
33. Williams G, Watts DC (1970) Trans Faraday Soc 66:80
34. Sillescu H (1999) J Non-Cryst Solids 243:81
35. Ediger MD (2000) Annu Rev Phys Chem 51:99
36. Richert R (2002) J Phys Condens Matter 14:R703
37. Böhmer R (1989) J Chem Phys 91:3111
38. Jonscher AK (1977) Nature (London) 267:673
39. Lunkenheimer P, Loidl A (2003) Phys Rev Lett 91:207601
40. Dyre JC (1988) J Appl Phys 64:2456
41. Macedo PB, Moynihan CT, Bose R (1972) Phys Chem Glasses 13:171
42. Elliott SR (1994) J Non-Cryst Solids 170:97
43. Roling B (1999) J Non-Cryst Solids 244:34
44. Sidebottom DL, Roling B, Funke K (2001) Phys Rev B 63:024301
45. Hodge IM, Ngai KL, Moynihan CT (2005) J Non-Cryst Solids 351:104
46. Böhmer R, Gainaru C, Richert R (2014) Phys Rep 545:125
47. Vogel H (1921) Phys Z 22:645
48. Fulcher GS (1925) J Am Ceram Soc 8:339
49. Tammann G, Hesse W, Anorg Z (1926) Allg Chem 156:245
50. Angell CA (1985) In: Ngai KL, Wright GB (eds) Relaxations in complex systems. Naval Research Laboratory, Washington, DC, p 3
51. Plazek DJ, Ngai KL (1991) Macromolecules 24:1222
52. Böhmer R, Angell CA (1992) Phys Rev B 45:10091
53. Böhmer R (1994) J Non-Cryst Solids 172–174:628

54. Vilgis TA (1993) Phys Rev B 47:2882
55. Johari GP, Goldstein M (1970) J Chem Phys 53:2372
56. Stillinger FH (1995) Science 267:1935
57. Harmon JS, Demetriou MD, Johnson WL, Samwer K (2007) Phys Rev Lett 99:135502
58. Gainaru C, Lips O, Troshagina A, Kaghlau R, Brodin A, Fujara F, Rössler E (2008) J Chem Phys 128:173505
59. Kudlik A, Benkhof S, Blochowicz T, Rössler E (1999) J Mol Structure 479:201
60. Köhler M, Lunkenheimer P, Goncharov Y, Wehn R, Loidl A (2010) J Non-Cryst Solids 356:529
61. Köhler M (2010) Relaxation, rattling, and decoupling. dynamic processes in glassy matter. Mensch und Buch, Berlin
62. Kastner S, Köhler M, Goncharov Y, Lunkenheimer P, Loidl A (2011) J Non-Cryst Solids 357:510
63. Havriliak S, Negami S (1966) J Polym Sci C 14:99
64. Ngai KL (2003) J Phys Condens Matter 15:S1107
65. Lunkenheimer P, Emmert S, Gulich R, Köhler M, Wolf M, Schwab M, Loidl A (2017) Phys Rev E 96:062607
66. Bartoš J, Iskrová M, Köhler M, Wehn R, Šauša O, Lunkenheimer P, Krištiak J, Loidl A (2011) Eur Phys J E 34:104
67. Brand R, Lunkenheimer P, Schneider U, Loidl A (1999) Phys Rev Lett 82:1951
68. Schneider U, Brand R, Lunkenheimer P, Loidl A (2000) Phys Rev Lett 84:5560
69. Lunkenheimer P, Wehn R, Riegger Th, Loidl A (2002) J Non-Cryst Solids 307–310:336
70. Döß A, Paluch M, Sillescu H, Hinze G (2002) Phys Rev Lett 88:095701
71. Hensel-Bielowka S, Pawlus S, Roland CM, Zioło J, Paluch M (2004) Phys Rev E 69:050501(R)
72. Mattson J, Bergman R, Jacobsson P, Börjesson L (2003) Phys Rev Lett 90:075702
73. Blochowicz T, Rössler EA (2004) Phys Rev Lett 92:225701
74. Schneider U (2000) Breitbandige dielektrische Studien der Dynamik struktureller Glasbildner. Books on Demand, Norderstedt. ISBN 3-8311-0921-4
75. Lunkenheimer P, Kastner S, Köhler M, Loidl A (2010) Phys Rev E 81:051504
76. Jeong YH (1987) Phys Rev A 36:766
77. Jeong YH, Nagel SR, Bhattacharya S (1986) Phys Rev A 34:602
78. Larsson KE (1968) Phys Rev 167:171
79. Gupta S, Arend N, Lunkenheimer P, Loidl A, Stingaciu L, Jalarvo N, Mamontov E, Ohl M (2015) Eur Phys J E 38:1
80. Posch HA, Dardy HD, Litovitz TAS (1977) Ber Bunsenges Physik Chemie 88:744
81. Ewell RH (1938) J Appl Phys 9:252
82. Segur JB, Oberstar HE (1951) Ind Eng Chem 43:2117
83. Piccirelli R, Litovic TA (1957) J Acoust Soc Am 29:1009
84. Börjesson L, Elmroth M, Torell LM (1990) Chem Phys 149:209
85. Du WM, Li G, Cummins HZ, Fuchs M, Toulouse J, Knauss LA (1994) Phys Rev E 49:2192
86. Bondeau A, Huck J (1985) J Phys (France) 46:1717
87. Chen Z, Angell CA, Richert R (2012) Eur Phys J E 35:65
88. Mauro JC, Yue Y, Ellison AJ, Gupta PK, Allan DC (2009) Proc Natl Acad Sci USA **106**, 19780
89. Angell CA, Sichina W (1976) Ann NY Acad Sci 279:53
90. Wang L-M, Angell CA, Richert R (2006) J Chem Phys 125:074505
91. Ngai KL, Lunkenheimer P, León C, Schneider U, Brand R, Loidl A (2001) J Chem Phys 115:1405
92. Hensel-Bielowka S, Paluch M (2002) Phys Rev Lett 89:025704
93. Dyre JC, Olsen NB (2003) Phys Rev Lett 91:155703
94. Paluch M, Roland CM, Pawlus S, Zioło J, Ngai KL (2003) Phys Rev Lett 91:115701
95. Ngai KL, Grzybowska K, Grzybowski A, Kaminska E, Kaminski K, Paluch M, Capaccioli S (2008) J Non-Cryst Solids 354:5085

96. Brand R, Lunkenheimer P, Schneider U, Loidl A (2000) Phys Rev B 62:8878
97. Bengtzelius U, Götze W, Sjölander A (1984) J Phys C 17:5915
98. Leutheusser E (1984) Phys Rev A 29:2765
99. Götze W, Sjögren L (1992) Rep Progr Phys 55:241
100. Götze W (1999) J Phys Condens Matter 11:A1
101. Knaak W, Mezei F, Farago B (1988) Europhys Lett 7:529
102. Tao NJ, Li G, Cummins HZ (1991) Phys Rev Lett 66:1334
103. Li G, Du WM, Chen XK, Cummins HZ, Tao NJ (1992) Phys Rev A 45:3867
104. Li G, Du WM, Sakai A, Cummins HZ (1992) Phys Rev A 46:3343
105. Wuttke J, Hernandez J, Li G, Coddens G, Cummins HZ, Fujara F, Petry W, Sillescu H (1993) Phys Rev Lett 72:3052
106. Sokolov AP, Steffen W, Rössler E (1995) Phys Rev E 52:5105
107. Dixon PK, Menon N, Nagel SR (1994) Phys Rev E 50:1717
108. Lunkenheimer P, Pimenov A, Dressel M, Goncharov YuG, Böhmer R, Loidl A (1996) Phys Rev Lett 77:318
109. Lunkenheimer P, Pimenov A, Loidl A (1997) Phys Rev Lett 78:2995
110. Lunkenheimer P, Loidl A (2003) In: Kremer F, Schönhals A (eds) Broadband dielectric spectroscopy. Springer, Berlin, p 131
111. Lunkenheimer P, Pimenov A, Dressel M, Gorshunov B, Schneider U, Schiener B, Loidl A (1997) Am Chem Soc Symp Ser 676:168
112. Götze W, Singh AP, Voigtmann Th (2000) Phys Rev E 61:6934
113. Franosch T, Fuchs M, Götze W, Mayr MR, Singh AP (1997) Phys Rev E 56:5659
114. Schilling R, Scheidsteger T (1997) Phys Rev E 56:2932
115. Wuttke J, Ohl M, Goldammer M, Roth S, Schneider U, Lunkenheimer P, Kahn R, Rufflé B, Lechner R, Berg MA (2000) Phys Rev E 61:2730
116. Götze W, Voigtmann T (2000) Phys Rev E 61:4133
117. Lunkenheimer P, Wehn R, Köhler M, Loidl A (2018) J Non-Cryst Solids 492:63
118. Martin AJ, Brenig W (1974) Phys Status Solidi B 64:163
119. Malinovsky VK, Sokolov AP (1986) Sol Stat Commun 57:757
120. Malinovsky VK, Novikov VN, Parshin PP, Sokolov AP, Zemlyanov MG (1990) Europhys Lett 11:43
121. Buchenau U, Nücker N, Dianoux AJ (1984) Phys Rev Lett 53:2316
122. Buchenau U, Galperin YuM, Gurevich VL, Schober HR (1991) Phys Rev B 43:5039
123. Orbach R (1986) Science 231:814
124. Elliott SR (1992) Europhys Lett 19:201
125. Schirmacher W, Diezemann G, Ganter C (1998) Phys Rev Lett 81:136
126. Schirmacher W, Ruocco G, Scopigno (2007) Phys Rev Lett 98:025501
127. Shintani LH, Tanaka H (2008) Nat Mater 7:870
128. Grigera TS, Martin-Mayor V, Parisi G, Verrocchio P (2003) Nature 422:289
129. Lunkenheimer P, Loidl A (2001) Adv Solid State Phys 41:405
130. Lunkenheimer P, Loidl A (2006) J Non-Cryst Solids 352:4556
131. Brand R, Lunkenheimer P, Loidl A (2002) J Chem Phys 116:10386
132. Wehn R, Lunkenheimer P, Loidl A (2007) J Non-Cryst Solids 353:3862
133. Tool AQ (1946) J Am Ceram Soc 29:240
134. Narayanaswamy OS (1971) J Am Ceram Soc 54:240
135. Scherer GW (1986) Relaxation in glass and composites. Wiley, New York
136. Hodge IM (1994) J Non-Cryst Solids 169:211
137. Leheny RL, Nagel SR (1998) Phys Rev B 57:5154
138. Yue YZ, Jensen SL, Christiansen JC (2002) Appl Phys Lett 81:2983
139. Yardimci H, Leheny RL (2006) J Chem Phys 124:214503
140. Moynihan CT, Macedo PB, Montrose CJ, Gupta PK, DeBolt MA, Dill JF, Dom BE, Drake PW, Easteal AJ, Elterman PB, Moeller RP, Sasabe H, Wilder JA (1976) Ann NY Acad Sci 279:15
141. Lunkenheimer P, Wehn R, Loidl A (2006) J Non-Cryst Solids 352:4941

142. Richert R, Lunkenheimer P, Kastner S, Loidl A (2013) J Phys Chem B 117:12689
143. Gardner E (1985) Nucl Phys B 257:747
144. Gross DJ, Kanter I, Sompolinsky H (1985) Phys Rev Lett 55:304
145. Charbonneau P, Kurchan J, Parisi G, Urbani P, Zamponi F (2014) Nat Commun 5:3725
146. Biroli G, Urbani P (2016) Nat Phys 12:1130
147. Kurchan J, Parisi G, Urbani P, Zamponi F (2013) J Phys Chem B 117:12979
148. Berthier L, Charbonneau P, Jin Y, Parisi G, Seoane B, Zamponi F (2016) Proc Natl Acad Sci USA 113:8397
149. Jin Y, Yoshino H (2017) Nat Commun 8:14935
150. Charbonneau P, Jin Y, Parisi G, Rainone C, Seoane B, Zamponi F (2015) Phys Rev E 92:012316
151. Scalliet C, Berthier L, Zamponi F (2017) Phys Rev Lett 119:205501
152. Hicks CL, Wheatley MJ, Godfrey MJ, Moore MA (2018) Phys Rev Lett (in press)
153. Kirkpatrick TR, Wolynes PG (1987) Phys Rev B 36:8552
154. Sperl M (2006) Phys Rev E 74:011503

Glassy Dynamics as Reflected in the Inter- and Intra-molecular Interactions

Friedrich Kremer, Wilhelm Kossack and Arthur Markus Anton

Abstract The inter- and intra-molecular interactions of low molecular weight and polymeric glass-forming model systems are studied by broadband dielectric (BDS) spectroscopy and Fourier-transform infrared (FTIR) spectroscopy. Analyzing the temperature dependence of specific IR absorption bands, reflecting the intra-molecular potentials of dedicated molecular moieties, enables one to unravel on an intra-molecular scale the process of glass formation and to compare it with the dielectrically determined primarily inter-molecular dynamics. Molecular systems to be studied are typical glass formers as the polyalcohols glycerol, threitol, xylitol, and sorbitol, as salol and three poly(ethylene-glycol) derivatives, namely poly(ethylene-glycol)methyl-ether-acrylate, poly(ethylene-glycol)phenyl-ether-acrylate, and poly(ethylene-glycol)-dibenzoate. Within this experimental framework, a wealth of novel information is obtained proving that the different molecular moieties of a glass former show characteristic features in the course of glassy solidification. This demonstrates the fundamental importance of intra-molecular dynamics for the dynamic glass transition, providing refined insights into the underlying interactions beyond coarse-grained models, approximating, for instance, glass-forming molecules as hard spheres.

1 Introduction

Most theories of glassy materials and the (dynamic and calorimetric) glass transition [1–8] treat glass-forming molecules as hard spheres (s. Chap. 1 in this book). Thus, the *intra*-molecular degrees of freedom are fully neglected and only *inter*-molecular interactions are taken into account. This has in part historical reasons

F. Kremer (✉) · W. Kossack · A. M. Anton
Peter Debye Institute for Soft Matter Physics, University of Leipzig,
Linnéstr. 5, 04103 Leipzig, Germany
e-mail: kremer@physik.uni-leipzig.de

F. Kremer and A. Loidl (eds.), *The Scaling of Relaxation Processes*,
Advances in Dielectrics, https://doi.org/10.1007/978-3-319-72706-6_3

as for a long time the calorimetric glass "transition" was in focus, instead of the full course of glassy solidification. It reflects as well the fact that computer simulations employ usually coarse-grained models in which the intra-molecular potentials of a glass-forming molecule and their temperature dependencies are only partly considered. It is the objective of this chapter to place emphasis on this discrepancy and to demonstrate by combined dielectric and IR experiments how intimately the *inter*- and *intra*-molecular dynamics are interrelated with each other [9–12]. This is demonstrated for a homologous series of the polyalcohols, glycerol, threitol, xylitol, and sorbitol, for the glass-forming liquid salol and for the three oligomeric poly(ethylene-glycol)(PEG)-derivatives, poly(ethylene-glycol)methyl-ether-acrylate (PEG-MEA), poly(ethylene-glycol)-phenyl-ether-acrylate (PEG-PEA), and poly(ethylene-glycol)dibenzoate (PEG-DB).

2 Inter- and Intra-molecular Interactions in Polyalcohols

Polyalcohols (Table 1) among them glycerol, threitol, xylitol, and sorbitol belong to the best analyzed glassy systems. They are characterized by two relaxations, an α-relaxation (ascribed to the dynamic glass transition) and a secondary β-relaxation. With increasing molecular weight their calorimetric glass transition temperature shifts from T_g = 185 K for glycerol to T_g =274 K for D-Sorbitol. The α-relaxation (Fig. 1a) is assigned to fluctuations between structural sub-states of the molecule as a whole, while the β-relaxation (Fig. 1b) reflects small angle librations of the O–H moiety or even a whole segment (CHOH). [13, 14], which decouple from the former with increasing molecular weight. By that the β-relaxation gains dielectric strength and separates from the α-relaxation; for glycerol, it is observed as a wing on the high-frequency side of the α-relaxation, while for sorbitol, xylitol, and threitol it shows up as separate relaxation. Both relaxations have to be considered as coupled [15–19]. In the IR spectrum, characteristic vibrations of glycerol are evident (Fig. 2a–d), such as the deformation vibration $\delta(CH_2)$, the symmetric stretching vibrations ν(CCO), ν(CO) and ν(OH) and the asymmetric $\nu_{as}(CH_2)$ [20].

By determining the mean relaxation rate and the relaxation strength from the dielectric spectra and the vibrational frequencies and the respective oscillator strength from the IR spectra, analogous quantities are obtained enabling one to compare the evolution of the *inter*-molecular dynamics with that of the *intra*-molecular potentials in the course of the dynamic glass transition. This is demonstrated and summarized in (Fig. 3) depending on the rescaled temperature T_g/T. In detail, the following findings can be extracted; (i) the mean relaxation rate of the α-relaxation (Figs. 3a and 6b, watch the rescaled temperatures in Fig. 3a) decreases with increasing molecular weight, whereas the respective relaxation strength Δε (Fig. 3b) decreases; (ii) for the secondary β-relaxation (Fig. 3b) both, the mean relaxation rate and the respective relaxation strength increase with increasing molecular weight; (iii) with decreasing molecular weight the activation energy of the β-relaxation is enhanced undergoing a transition from an Arrhenius-like (for Sorbitol and Xylitol) to a Vogel-Fulcher-

Table 1 Chemical details of the polyalcohols under study

	Molecular formula	Molecular weight [Da]	Glass transition temperature [K]	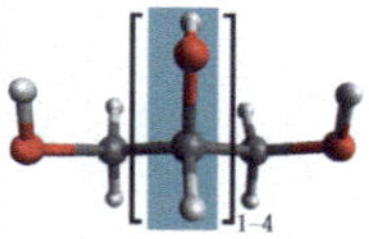
Glycerol	$C_3H_5(OH)_3$	92.1	185	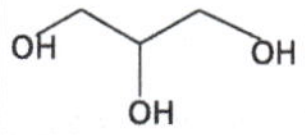
DL-Threitol	$C_4H_6(OH)_4$	122.1	224	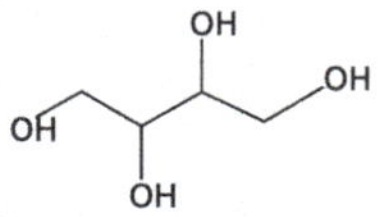
Xylitol	$C_5H_7(OH)_5$	152.2	248	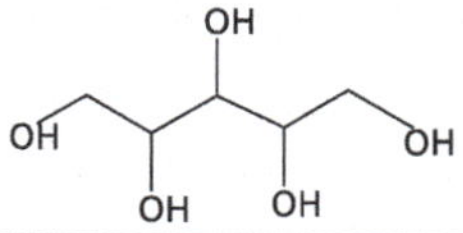
D-Sorbitol	$C_6H_8(OH)_6$	182.2	274	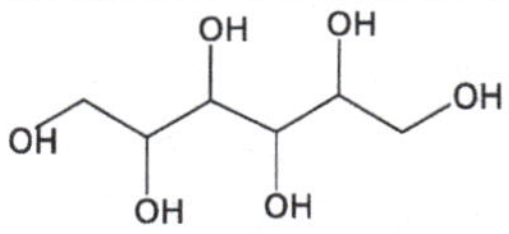

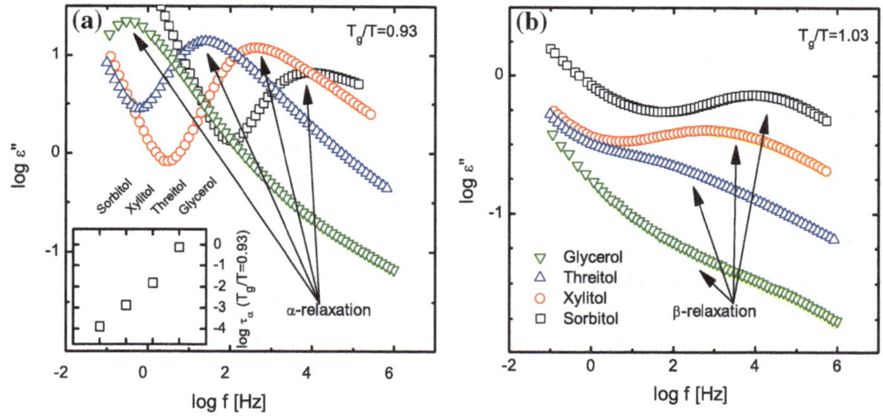

Fig. 1 Dielectric loss spectra at different temperatures showing in **a** the α- and in **b** the β-relaxation for glycerol, threitol, xylitol, and sorbitol as indicated; the inset depicts the position of the maximum at $T_g/T = 0.93$. Error bars are smaller than the size of the symbols

Tammann type of thermal activation (in the case of glycerol); (iv) for the ν(CCO), δ(CH_2), and ν(CO) vibrations, a *red*shift is observed with increasing temperature (Fig. 3c, e, g); only for the δ(CH_2), a kink at T_g is found, while the others show a smooth response at this temperature; (v) for the respective oscillator strengths of the three vibrations, kinks at $T = T_g$ are found (Fig. 3d, f, h) being less pronounced for glycerol than for sorbitol; (vi) for the ν(OH) stretching vibration, a strong *blue*

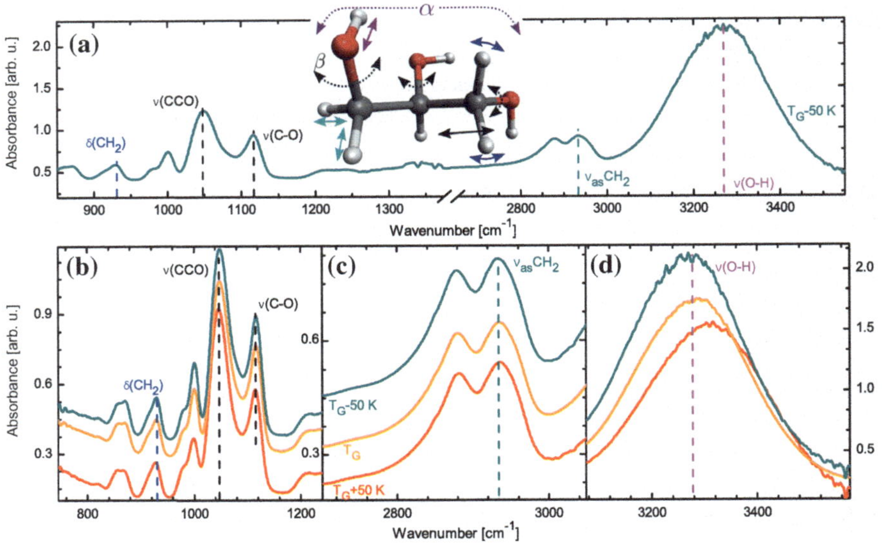

Fig. 2 **a** IR absorbance spectra of glycerol at $T = 135$ K$=T_g$-50 K; the inset depicts the chemical structure of glycerol (red: oxygen, gray: carbon, white: hydrogen), together with arrows indicating the displacement direction during molecular vibrations (solid) and dielectrically active fluctuations (dashed). The lower 3 panels are magnifications of the full spectrum at different temperatures, 135 K$=T_g$-50 K (dark cyan), $T_g = 185$ K (orange), and $T_g + 50$ K$=235$ K (red). Vertical lines mark the frequencies of the indicated vibrations

shift takes place (watch the scale in Fig. 3i) with increasing temperature; (vii) for the respective oscillator strengths, a kink at T_g is found (Fig. 3j) being more developed for the molecules with higher molecular weight.

A quantitative analysis of the above findings would require temperature dependent *atomistic* simulations, which are not available at present. However, it is possible to comprehend qualitatively the aforementioned results. The most striking outcome is the extraordinarily strong *blue* shift of the ν(OH) stretching vibration with increasing temperature (Fig. 3i). It can be semiquantitatively understood as follows: with increasing temperature the intra-molecular (Morse–) potential of O–H narrows, as the inter-molecular hydrogen bond weakens (inset in Fig. 4a). Consequently, the O–H distance x decreases and the ν(OH) stretching vibration is blue-shifted. The relation between the length of the OH bond and the frequency of its stretching vibration ν(OH) as known for molecular crystals [21] is shown in Fig. 4b. Based on the correlation between the *inter*-molecular O–O and the *intra*-molecular O–H distance, a relationship for the frequency of the ν(OH) vibration with the bond length $(x+d)$ is established, assuming the latter scales with $\rho^{(-1/3)}$, ρ being the bulk density. This correlation allows to estimate the temperature dependence of the OH stretching frequency based on density values for glycerol [22]. This estimate reproduces the measured values surprisingly well (Fig. 4a).

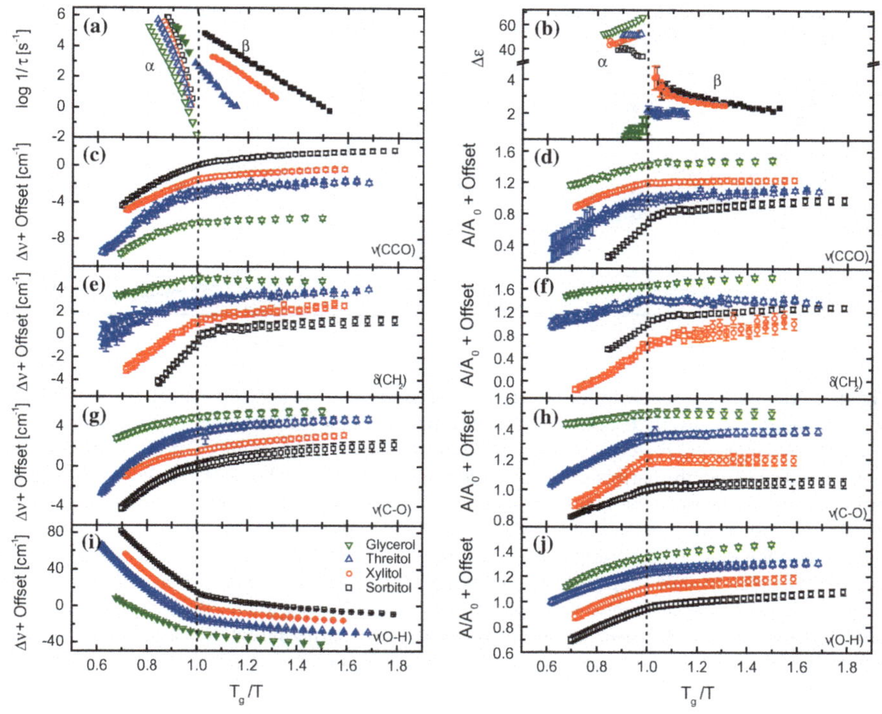

Fig. 3 Inter-molecular relaxations (**a**, **b**) and intra-molecular vibrations (**c**–**j**) as measured by broadband dielectric spectroscopy and FTIR spectroscopy, respectively, in dependence on the inverse temperature scaled by the calorimetric glass transition temperature T_g; panel **a** and **b** display the mean relaxation time (maximum of the Havriliak–Negami (HN) fits and the dielectric strength $\Delta\varepsilon$ of the α (open symbols)- and β (solid symbols)-relaxations, respectively. Panels **c**, **e**, **g**, **i** show the spectral position of the molecular vibrations as indicated within each panel, whereas (**d**, **f**, **h**, **j**) depict the respective integrated absorbance (oscillator strength). Error bars are smaller than the size of the symbols, if not indicated otherwise

With increasing temperature, the ν(CCO), ν(CO), and $\delta(CH_2)$ vibrations show a *red*shift (Fig. 3c, e, g). This can be semiquantitatively comprehended as caused by the temperature dependence of the experimentally determined bulk density $\rho(T)$. By approximating the above vibrations as taking place within a Morse potential and assuming its width a scales according to $\rho^{(-1/3)}$ (Fig. 5a), the experimentally observed *red*shift is obtained (Fig. 5b). It is not surprising that below T_g this approximation deviates from the experimental results. An analysis of the oscillator strengths in (Fig. 3d–j) is not possible within these simplified descriptions using Morse potentials; presumably because it neglects the inter-molecular interactions.

The intimate relationship between intra- and inter-molecular interactions is illustrated in Fig. 6a or b. For all alcohols, the ν(OH) stretching vibration changes its temperature dependence at the calorimetric glass transition temperature $T_{g,}$ with a less pronounced effect for the lower molecular weight polyalcohols. In parallel,

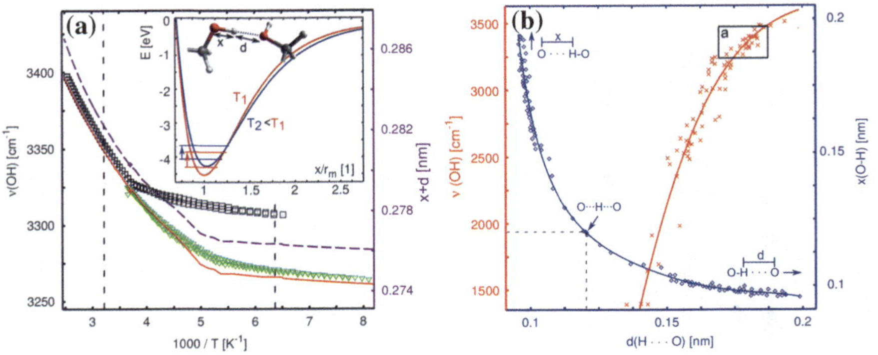

Fig. 4 **a** Vibrational frequencies of ν(OH) for glycerol (green open triangles) and sorbitol (black open squares) versus inverse temperature. On the right-hand axis, the inter-molecular O…O distance ($x + d$, dashed magenta) is shown as determined from scaling the inter-molecular O…O distance at 279 K (deduced from the correlation in panel **b**) with $\rho^{(-1/3)}$ [22]. Above 311 K and below 156 K (marked by vertical dashed lines), the density is extrapolated. The solid red line marks the vibrational frequency as obtained from the scaled distance. **b** Vibrational frequency (red crosses) of the ν(OH) stretching vibration (left abscissa) in dependence on the O…H distance, based on crystallographic data of molecular crystals [21]. Blue dots represent the distance from the hydrogen to the closest neighboring oxygen atom (right abscissa) depending on the distance to the other oxygen atom. The point at 0.124 nm can be considered as the symmetry state during proton transfer. Solid lines are regressions of the correlations. The box marks the temperature and spectral range of the ν(OH) stretching vibration for glycerol and sorbitol shown in (**a**)

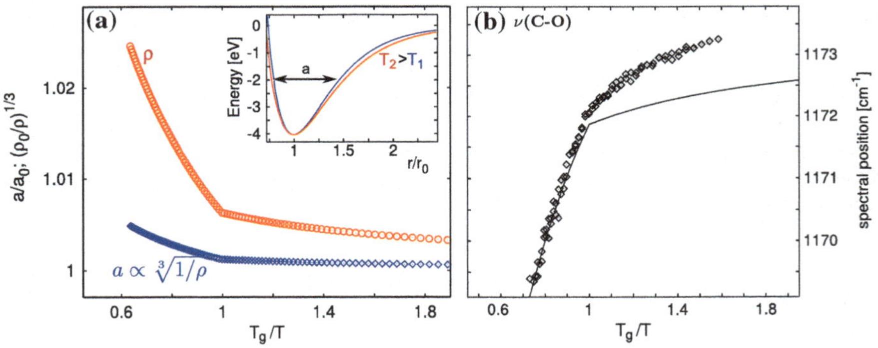

Fig. 5 **a** Scaled inter-molecular distance (a/a_0) resp. $(\rho_0/\rho)^{(1/3)}$ versus rescaled temperature T_g/T based on experimental data of the temperature dependence of the bulk density [22] for glycerol; T_g is the glass transition temperature of glycerol. It is assumed that the width a of a Morse potential scales $\sim\rho^{(-1/3)}$ The inset displays schematically the change of the width a of a Morse potential with temperature. **b** Redshift of the ν(C–O) stretching vibration of glycerol versus rescaled temperature T_g/T. The solid line is a fit assuming that the width a of the Morse potential changes with temperature according to $a \sim \rho^{(-1/3)}$

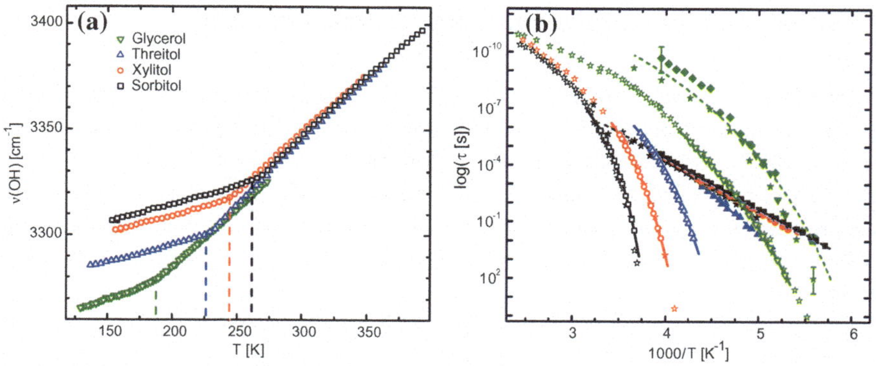

Fig. 6 **a** Vibrational frequency ν(OH) for glycerol, threitol, xylitol, and sorbitol as noted, the calorimetric glass transition temperatures T_g are indicated as dashed lines. **b** Activation plot for the α- (open symbols) and β- (solid symbols) relaxations describing their scaling with respect to the inverse temperature; data represented by stars and squares are taken from [15, 23–25]. Fits according to VFT or Arrhenius dependencies are displayed as solid or dashed lines, respectively

the secondary β-relaxation being assigned to librations of the dipolar COH–moiety turns from a well-decoupled secondary β-relaxation (for sorbitol and xylitol), having nearly identical activation energies, to a still well-separated process with increased activation energy for threitol and finally to a process occurring as high-frequency wing of the α–relaxation in case of glycerol, which can be only formally separated from the latter (Fig. 6b) [15, 26].

This change of the temperature dependence of the relaxation times is attributed to a change of the underlying length and timescales. For larger molecular weights (as in the case of polymers), the α-relaxation is assigned to a structural relaxation of 2–3 segments, [27] confined by the chain backbone. When the number of segments decreases, restrictions due to the chain connectivity are reduced and the α-relaxation turns into a fluctuation of the molecule as a whole. Such a motion involves significantly smaller length- and timescales and is only restricted by inter-molecular interactions, e.g., H-bonds. Consequently, the respective timescales influence each other (as for threitol), and finally couple (as for glycerol). In contrast, when the molecular weight increases, (for xlitol and sorbitol) the length scales of the dielectric relaxations decouple as seen by the increasing resemblance of the β-relaxations (Fig. 6b) and of the vibrational frequencies of ν(OH) (Fig. 6a). Similar scenarios involving merging and coupling of the α- and β-relaxation have also been reported depending on temperature or molecular weight for propylene glycol oligomers, [15, 26] sorbitol [19], and even salol [28].

3 Inter- and Intra-molecular Interactions in Salol

Salol is a glass-forming liquid which was studied with a variety of experimental techniques such as broadband dielectric spectroscopy (BDS) [15–17], photon correlation spectroscopy [29], mechanical spectroscopy [29], Nuclear Magnetic Resonance (NMR) [30, 31], DC- and AC- Calorimetry [32–35], X-ray scattering [36–38], and Infrared and Raman spectroscopy [39–41]. Despite the fact that salol can form H-bonds, it is considered as van der Waals liquid, because it bonds primarily intra-molecularly.

In the IR spectrum (Fig. 7a) of salol well-separated characteristic vibrations are evident, such as the deformation vibration δ(CH) and the stretching vibrations ν(C–O), ν(C–OH), and ν(C=O) which change only weakly within the temperature range of $T_g \pm 50$ K (Fig. 7b–d). It is well known that salol has a strong tendency to crystallize [36–38], which drastically affects the vibrational spectra as demonstrated for the ν(C–O) and ν(C=O) stretching vibration (Fig. 8). The oscillator strength decreases strongly with increasing temperature, especially between T_g +60 K and T_g +70 K. At the latter, salol is considered as fully crystalline. [36, 39].

By comparing the inter-molecular (Fig. 9a or b) with the intra-molecular (Fig. 9c–f) interactions, the following findings can be deduced; (i) The ν(C=O) stretching vibration (Fig. 9c) appears around 1690 cm^{-1}, in agreement with the value for intra-molecularly hydrogen bonded hydroxyl-esters and significantly below the

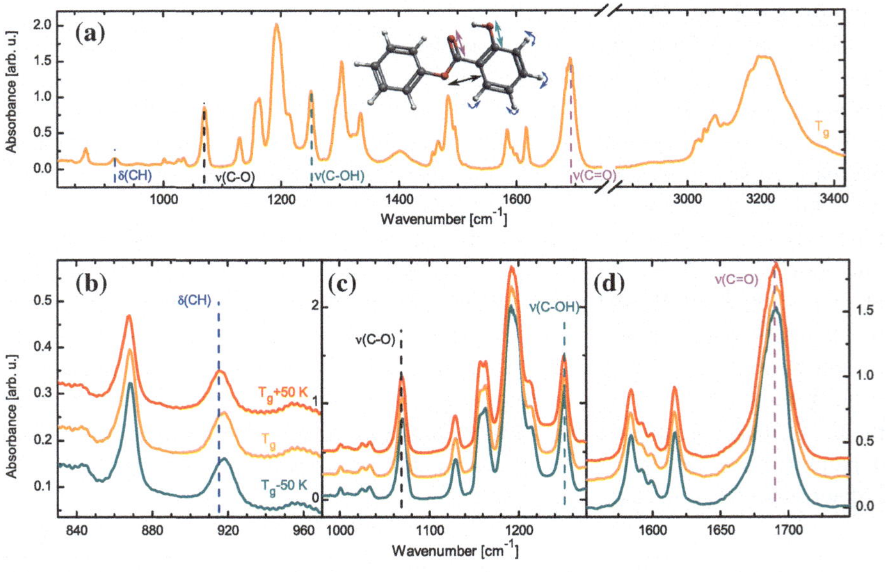

Fig. 7 **a** IR spectrum and vibrational assignments for salol at $T_g = 220$ K; **b–d** enlarged spectra of selected IR vibrations of salol at temperatures as indicated, T_g +50 K (red), T_g (orange) and T_g-50 K (green)

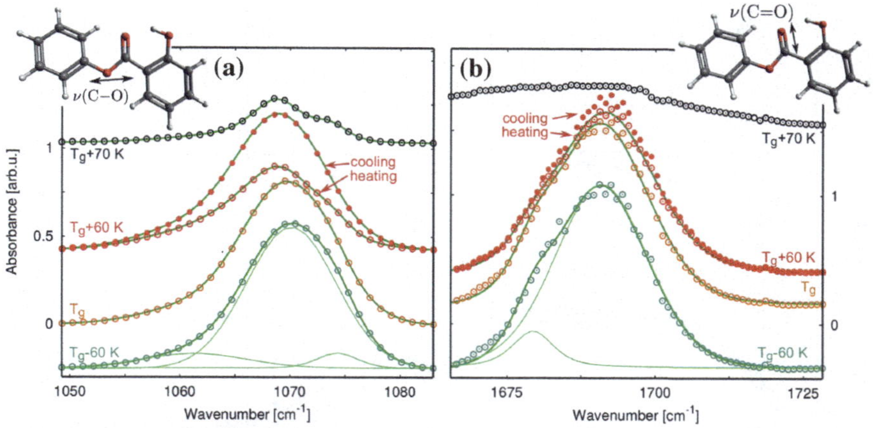

Fig. 8 **a** ν(C–O) and **b** ν(C=O) stretching vibrations fitted by a sum (dark green lines) of Voigt profiles (light green lines) at temperatures as indicated (cooling: solid, heating: open symbols), T_g + 60 K(red), T_g (orange), T_g-60 K (green), and T_g + 70 K (black)

frequency of free phenyl esters (1770 cm^{-1}). [20] The weak blue shift with increasing temperature reflects the negligible changes in the H-bonding pattern and justifies the classification of salol as "van der Waals" liquid. It is remarkable that the blue shift takes place only at $T > T_g$; below the calorimetric glass transition temperature the spectral position is constant within the limits of experimental accuracy; (iii) the oscillator strength of the ν(C=O) stretching vibration (Fig. 9d) does not change at T_g but reduces considerably when the crystallization sets in at T_{c1} and T_{c2}. Thus, the C=O moiety having a dipole moment of 1.85 Debye [42, 43] plays a substantial role in crystallization; (iv) the ν(C–OH) (Fig. 9c) and the ν(C–O) (Fig. 9e) stretching vibrations show a continuous redshift with increasing temperature without any kink or discontinuity at T_g or T_{c1} and T_{c2}. In contrast, the oscillator strength of both vibrations reduces strongly due to crystallization (Fig. 9d, f) with a pronounced hysteresis on heating and cooling. These two observations prove that the vibrations originate within the amorphous phase. (v) the δ(CH) out of plane deformation shows as well a redshift with increasing temperature, but in contrast to the smooth curves of the ν(C–OH) and ν(C–O) vibrations, the deformation vibration displays a kink at the calorimetric glass transition temperature T_g. This reflects the fact that the aromatic C–H bonds at the periphery of the salol molecule are more involved in the inter-molecular interactions than, for instance, the C–OH and C–O units. (vi) The dielectric α- and β-relaxation overlap strongly and follow similar, VFT-like temperature dependencies above T_g (Fig. 9a) [44]. They have to be considered as coupled processes. By carrying out thermally stimulated depolarization current measurements [28], a relaxation process has been determined (Fig. 9a), which bends off from the structural relaxation below 210 K and agrees roughly with the dielectrically determined β-process close to T_g. Assuming, that both are indeed the same process, which changes its temperature dependence at T_g, [18, 19, 26] raises the question

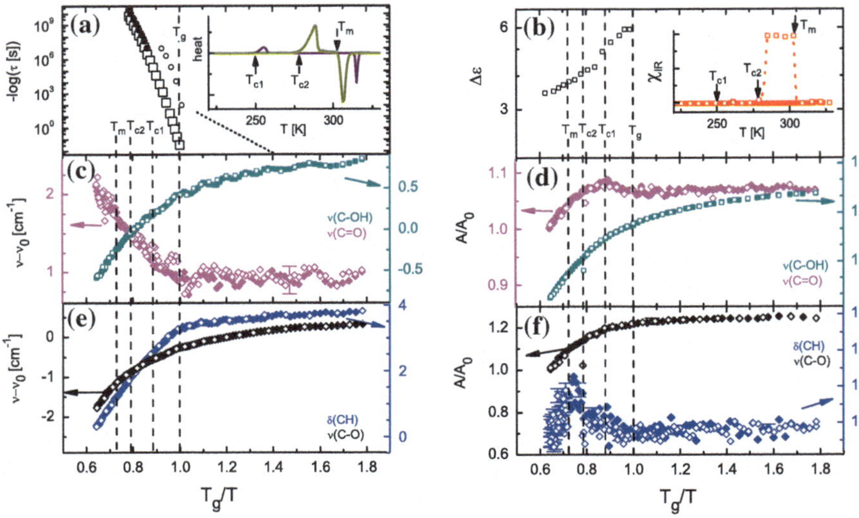

Fig. 9 **a** Mean relaxation rate and **b** relaxation strength $\Delta\varepsilon$ of salol versus rescaled temperature T_g/T for the α-relaxation (open squares) and the β-relaxation (open circles, taken from [44], dotted line taken from thermally stimulated depolarization current [28]). The characteristic rate of the phenyl group rotation (filled circles in **a**) around the molecular long axis is taken from [13, 14]. In **c** and **e**, the spectral positions of the stretching vibrations ν(C–OH), ν(C=O), ν(C–O) and the deformation vibration δ(CH) are shown with the respective oscillator strength's in (**d, f**). Open and full symbols indicate heating and cooling runs, respectively. In the inset in panel **a,** the heat flow during heating with 10 K/min (yellow) and 2 K/min (lilac) is displayed versus temperature [28]. The inset in (**b**) shows the crystallinity determined from IR data as it changes with temperature on heating (open squares) and cooling (solid squares). If not indicated, otherwise the error bars are smaller than the size of the symbols

about its microscopic origin. As salol consists of 2 rigid planar structures [the phenyl group and the intra-molecularly H-bonded hydroxyl-benzoate, see (i)], we assign the β-relaxation to small angle fluctuations of these two units with respect to the molecular long axis. Indeed, well above T_g such a fluctuation has been observed by NMR spectroscopy to be about a factor of 4 faster than the α-process (Fig. 9a) [13, 14]. The tight coupling of the two fluctuations above T_g can be rationalized, as both involve similar length scales: half of the molecule or the whole molecule.

In summary, it is demonstrated how distinctly and specifically the various moieties of salol respond to glassy solidification and crystallization. The structural relaxation and secondary relaxation are strongly coupled above the glass transition temperature T_g and appear to decouple only below, when the inter-molecular structure is essentially frozen. This scenario can be understood, if one assigns the secondary relaxation as a small angle libration of the two rigid units of salol.

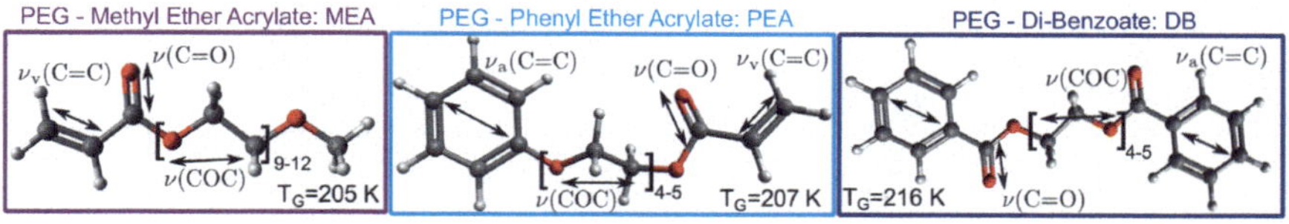

Fig. 10 Chemical structure of the poly(ethylene-glycol) (PEG) derivatives PEG-methyl-ether-acrylate (MEA), PEG-phenyl-ether-acrylate (PEA), and PEG-dibenzoate (DB). The calorimetric glass transition temperatures are indicated; the IR vibrations analyzed in the further discussion are displayed

4 Inter- and Intra-molecular Interactions in Three Oligomeric Poly(Ethylene-Glycol) Derivatives

In the series of (Fig. 10) poly(ethylene-glycol) (PEG) derivatives PEG-methyl-ether-acrylate (MEA), PEG-phenyl-ether-acrylate (PEA), and PEG-dibenzoate (DB), the refined interplay between glassy dynamics and crystallization can be studied in detail. While MEA shows a strong tendency to crystallize [45], PEA and DB are fully amorphous. In the dielectric spectra (Fig. 11) of PEA and DB, a structural α- and secondary β-relaxations are evident [46, 47]; the latter are assigned to librational fluctuations of the C=O and the C–O–C moieties. In contrast for MEA, a strong decline of the α-relaxation is observed with decreasing temperature as a consequence of crystallization. The relaxations of the amorphous and the crystalline systems partly resemble each other at temperatures below the calorimetric glass transition of PEA and DB (s. Fig. 14a or b).

The IR spectra (Fig. 12a or b) of the three PEG derivatives display well-separated vibrational bands reflecting the similarities and differences of their chemical structure. One finds the characteristic vinyl doublet at 1635 cm^{-1}, and the one related to the aromatic structures between 1590 and 1600 cm^{-1} [20, 48]. Similarly, the bands related to the aromatic CH stretching can be found slightly below 3100 cm^{-1}.The stretching of the carboxyl groups ν(C=O) is found around 1720 cm^{-1}. In the backbone stretching region (1000–1250 cm^{-1}), the IR spectra appear very similar. The higher absorbance of MEA arises from scaling the spectra to a constant number of terminal groups. Comparing the bands related to trans COC (1110 cm^{-1}) and gauche COC (1140 cm^{-1}) moieties, a higher absorbance of the latter is observed only in case of PEA, whereas for MEA the absorbance of the trans conformers dominates. [48, 49] This effect is even stronger for DB.

The comparison (Fig. 13a–c) of the temperature dependence of the ν(COC), ν(C=C) and ν(C=O) stretching vibration for DB, PEA, and MEA demonstrates how differently the distinct molecular moieties are involved in glassy solidification and crystallization. While the ν(COC) vibration changes fundamentally in the course of crystallization for MEA (Fig. 13a), the ν(C=O) vibration is only marginally modified. For glassy DB and PEA, the changes are much subtle and require numerical fits

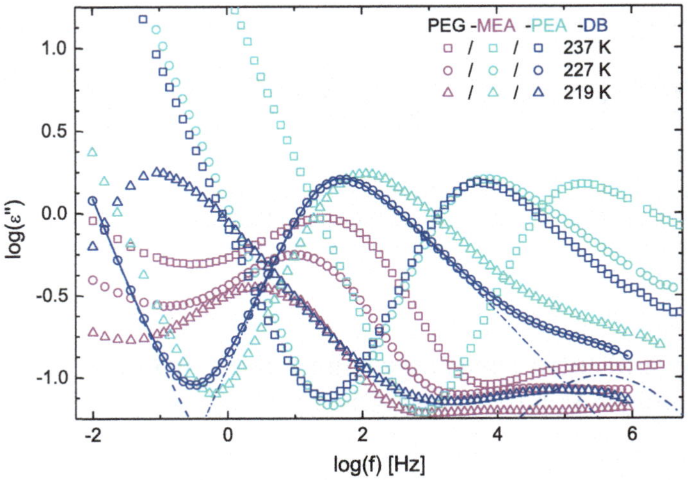

Fig. 11 Dielectric spectra of PEG-methyl-ether-acrylate (MEA), PEG-phenyl-ether-acrylate (PEA), and PEG-dibenzoate (DB) at temperatures as indicated. The solid line in the spectrum of DB at a temperature of 227 K is the superposition of a conductivity contribution (dashed blue line) and two Havriliak–Negami relaxation functions (dashed-dotted blue line) describing the structural α- and a secondary β-relaxation. The experimental uncertainty bars are smaller than the size of the symbols

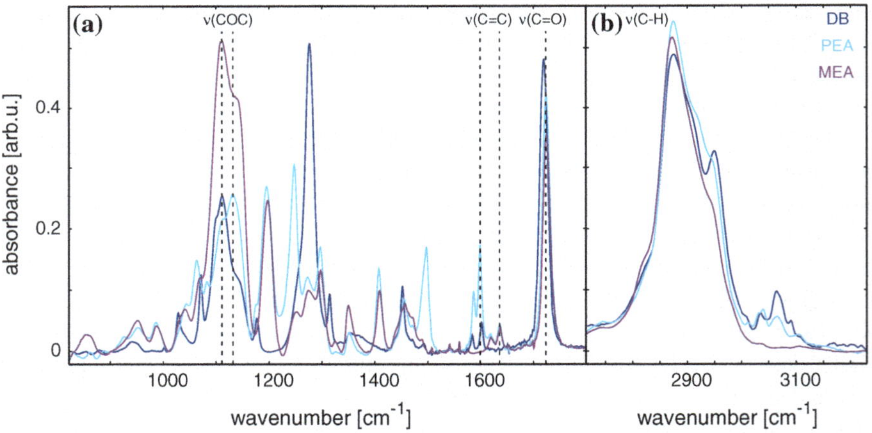

Fig. 12 **a** IR spectra of the three PEG derivatives PEG-methyl-ether-acrylate (MEA), PEG-phenyl-ether-acrylate (PEA), and PEG-dibenzoate (DB) at room temperature. The ν(COC), ν(C=C), and the ν(C=O) stretching vibrations are indicated. **b** Enlargement of the ν(C–H) stretching vibration

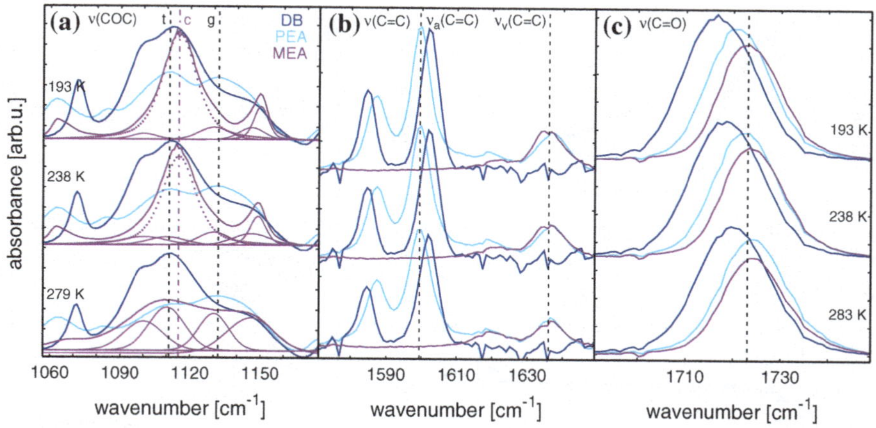

Fig. 13 Temperature dependence of **a** the ν(COC), **b** the ν(C=C), and **c** the ν(C=O) stretching vibration of PEG-methyl-ether-acrylate (MEA), PEG-phenyl-ether-acrylate (PEA), and PEG-dibenzoate (DB). For ν(COC), the decomposition into the trans-(t), the gauche-(g), and the crystal-specific (c, dashed line) vibration is shown

in order to determine the temperature dependencies of the peak frequencies and the oscillator strengths of the various vibrations.

The inter- and intra-molecular interactions in the PEG derivatives under study are dominated by two characteristic temperatures, that of crystallization (in case of MEA) and the calorimetric glass transition temperature (in case of DB and PEA). In detail (Fig. 14a–h), the following findings can be extracted; (i) the glassy dynamics of DB and PEA is similar for both the structural α-relaxation and the secondary β-relaxations. This holds for the relaxation rates (Fig. 14a) and the dielectric strengths (Fig. 14b); (ii) the relaxations of the amorphous and the crystalline systems resemble each other at temperatures below the calorimetric glass transition of PEA and DB (Fig. 14a or b), reflecting the fact that the molecular systems are similar; the slowest relaxation in MEA corresponds to the dynamics of the amorphous regions between the microcrystals and close to their surface [46, 47, 50]; (iii) the spectral positions of the aromatic stretching vibration ν_a(C=C) of DB and PEA (Fig. 14c) are nearly temperature independent; for the respective oscillator strengths, a continuous decline is found (Fig. 14d); (iv) a shift in the spectral position of the vinyl stretching vibration ν_v(C=C) for the crystallizing MEA takes place between T_c and the calorimetric glass transition temperatures of DB and PEA (Fig. 14c); for the glassy PEA, a kink in the oscillator strength of ν_v(C=C) at the calorimetric glass transition temperature is found; (v) the resemblance of the three molecular systems shows up as well in the temperature dependencies of the spectral positions and the oscillator strengths of the ν(C=O) stretching vibration for temperatures below T_g of DB and PEA (Fig. 14e or f). Above this temperature, the two glassy systems are nearly identical while for MEA an increase with a shallow maximum is found proving the involvement of the C=O moiety in the process of crystallization. Above the melting temperature $T_{M,}$ the

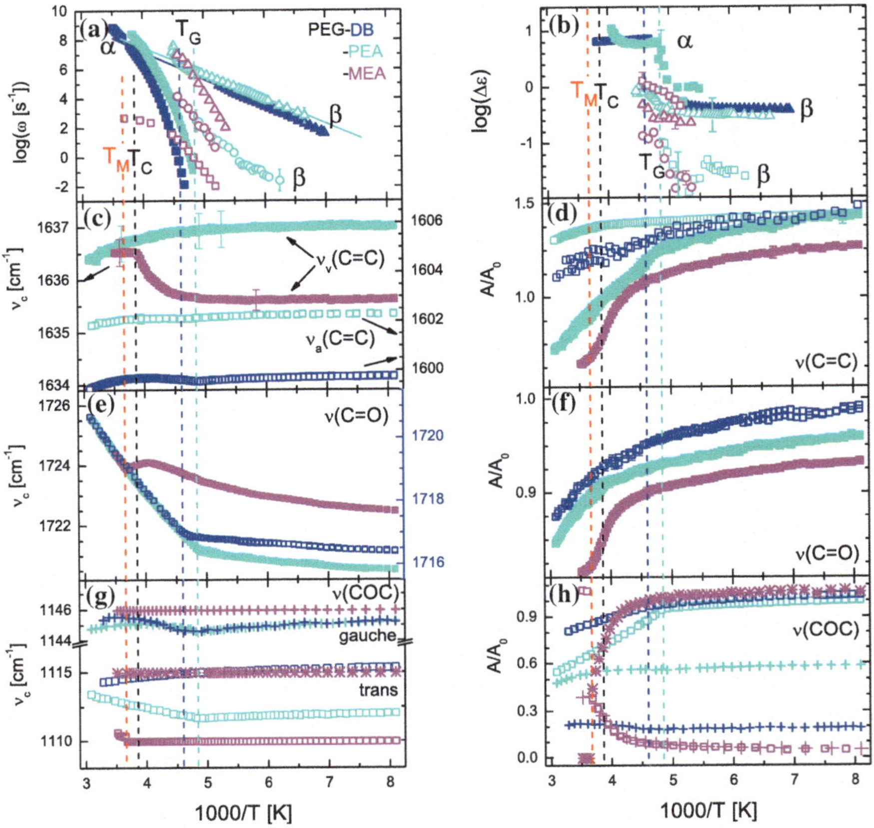

Fig. 14 **a** Mean relaxation rate and **b** relaxation strength Δε of PEG-methyl-ether-acrylate (MEA), PEG-phenyl-ether-acrylate (PEA), and PEG-dibenzoate (DB) versus inverse temperature 1000/T for the α-relaxation (solid squares) and the β-relaxation in case of PEA and DB and for the dielectric relaxation processes in MEA (open magenta symbols) in the crystalline state. The temperature dependencies of the spectral positions of the stretching vibrations ν_V(C=C) of the vinyl and the ν_a(C=C) of the aromatic C=C bonds are shown, respectively in **c**, for the ν(C=O) vibration and the ν(COC) vibration of the gauche and trans conformations the spectral positions are depicted in (**e**, **g**), respectively. The corresponding oscillator strengths are displayed in (**d**, **f**, **h**). The experimental uncertainty is smaller than the size of the symbols if not indicated otherwise. The stars represent a crystal-specific band at 1110 cm^{-1} for MEA

temperature dependencies of the ν(C=O) stretching vibration for all three molecular systems coalesce; (vi) The respective oscillator strengths (Fig. 14f) do not show any discontinuity at T_g of DB and PEA. The effect of crystallization shows up as a strong decline in the case of MEA similarly as for the ν(C=C) vibration; (vii) the COC segments in all three PEG derivatives exist in trans and gauche conformations. For the amorphous DB and PEA, the spectral positions of the related ν(COC) vibrations are nearly temperature independent with a weak change in slope for the trans-specific vibration of PEA and the gauche-specific one of DB (Fig. 14g). Comparing the corresponding oscillator strengths, a strongly different scenario is observed; for the gauche conformation of DB and PEA, nearly no temperature dependence is found, while for the trans conformation beginning with the respective T_g's a decline takes place (Fig. 14h). For MEA, the oscillator strength of the ν(COC) vibration at 1110 cm^{-1} being specific for the crystalline state increases strongly in oscillator strength on the expense of that of the trans *and* gauche conformers.

In summarizing the findings for the three oligomeric PEG derivatives, one has to conclude that the modifications of *one* end group in the chemical structure and slight changes of molecular weight (Fig. 10) suffices to induce or prevent crystallization. For the two glassy PB and PEA, the changes in the intra-molecular potentials are more subtle but by no means less specific.

5 Concluding Remarks

In this chapter, it is demonstrated how diversely and characteristically the different moieties of molecular systems respond to glassy solidification and crystallization and how specific the interrelationship between inter- and intra-molecular interactions is. This has to be considered in any theoretical descriptions having the ambition for a quantitative understanding of the physics beyond.

References

1. Wong J, Angell CA (1976) Glass: structure by spectroscopy. Marcel Dekker, New York
2. Donth EJ (1981) Glasübergang. Akademie Verlag, Berlin
3. Zallen R (1983) The physics of amorphous Solids. Wiley, New York
4. Elliott SR (1990) Physics of amorphous materials. Longman Scientific & Technical, London
5. Donth EJ (1992) Relaxation and thermodynamics in polymers, glass transition. Akademie Verlag, Berlin
6. Donth EJ (2001) The glass transition. Springer Verlag, Berlin
7. Ngai K (2011) Relaxation and diffusion in complex systems. Springer Verlag, Berlin
8. Götze W (2012) Complex dynamics of glass-forming liquids—a mode-coupling theory. Oxford Scientific Publications, Oxford
9. Papadopoulos P, Kossack W, Kremer F (2013) Soft Matter 9:1600–1603
10. Kossack W, Adrjanowicz K, Tarnacka M, Kipnusu WK, Dulski M, Mapesa EU, Kaminski K, Pawlus S, Paluch M, Kremer F (2013) Phys Chem Chem Phys 15:20641–20650
11. Kipnusu WK, Kossack W, Iacob C, Zeigermann P, Jasiurkowska M, Sangoro JR, Valiullin R, Kremer F (2013) Soft Matter 9:4681–4686
12. Kossack W, Kipnusu WK, Dulski M, Adrjanowicz K, Kaminska E, Mapesa EU, Tress M, Kaminski K, Kremer F (2014) Chem Phys 140:215101
13. Döß A, Paluch M, Sillescu H, Hinze G (2002) J Chem Phys 117(14):6582–6589
14. Döß A, Paluch M, Sillescu H, Hinze G (2002) Phys Rev Lett 88:095701
15. Kremer F, Schönhals A (eds) (2003) Broadband dielectric spectroscopy. Springer
16. Adrjanowicz K, Kolodziejczyk K, Kipnusu WK, Tarnacka M, Mapesa EU, Kaminska E, Pawlus S, Kaminski K, Paluch M (2015) J Phys Chem C 119(25):14366
17. Wagner H, Richert R (1999) J Chem Phys 110(23):11660–11663

18. Paluch M, Roland CM, Pawlus S, Zioło J, Ngai KL (2003) Phys Rev Lett 91(11)
19. Böhmer R, Diezemann G, Geil B, Hinze G, Nowaczyk A, Winterlich M (2006) Phys Rev Lett 97(13)
20. Socrates G (2001) Infrared and Raman characteristic group frequencies. Tables and Charts, 3rd edn. Wiley
21. Steiner T (2002) The hydrogen bond in the solid state. Angew Chem Int Ed 41(1):48–76
22. Blazhnov IV, Malomuzh NP, Lishchuk SV (2004) J Chem Phys 121(13):6435–6441
23. Kastner S, Köhler M, Goncharov Y, Lunkenheimer P, Loidl A (2011) J Non-Cryst Solids Bd 357(2):510–514
24. Lunkenheimer P, Kastner S, Köhler M, Loidl A (2010) PRE 81:051504
25. Ngai KL, Lunkenheimer P, León C, Schneider U, Brand R, Loidl A (2001) J Chem Phys 115(3):1405–1413
26. Köhler M, Lunkenheimer P, Goncharov Y, When R Loidl A (2010) J Non-Cryst Solids 356(11–17):529–534
27. Bahar I, Erman B, Kremer F, Fischer EW (1992) Macromolecules 25:816–825
28. Ramos M, Correia NT, Diogo HP (2004) Phys Chem Chem Phys 6(4):793–798
29. Hansen C (1997) Dissertation Universität Mainz
30. Chang, Fujara F, Geil B, Heuberger G (1994) J Non-Cryst Solids 172–174
31. Döß A, Hinze G, Böhmer R, Sillescu H (2000) J Chem Phys 112(13):5884–5892
32. Hanaya M, Hikima T, Hatase M, Oguni M (2002) J Chem Thermodyn 2(34):1173–1193
33. Cheng SZD (ed) (2002) Handbook of thermal analysis and calorimetry. Elsevier Science B.V
34. Paladi F, Oguni M (2003) J Phys: Condens Matter 15(23):3909
35. Hanaya M, Hikima T, Hatase M, Oguni M (2002) J Chem Thermodyn 34(8):1173–1193
36. Eckstein E, Qian J, Hentschke R, Thurn-Albrecht T, Steffen W, Fischer EW (2000) Phys Rev 113(11):4751–4762
37. Hikima Z, Hanaya M, Oguni M (1995) Solid State Commun 93(8):713–717
38. Bilgram JH, Dürig U, Wächter M, Seiler P (1982) J Cryst Growth 57(1):1–5
39. Baran J, Davydova NA (2010) Phys Rev E 81(3):031503
40. Hanuza J (2004) Vib Spectrosc 34(2):253
41. Kalampounias AG, Yannopoulos SN, Steffen W, Kirillova LI, Kirillov SA (2003) J Chem Phys 118(18):8340–8349
42. Neese F (2012) The ORCA program system. Wiley Interdiscip Rev Comput Mol Sci 2(1):73–78
43. Krishnan R, Binkley JS, Seeger R, Pople JA (1980) J Chem Phys 72:650. The value is Calculated based on Density Functional Theory calculations using the Orca program package (v 4) with 6-311 ++G(2d2p) basis sets
44. Bartoš J, Iskrová M, Köhler M, Wehn R, Šauša O, Lunkenheimer P, Krištiak J, Loidl A (2011) Eur Phys J E 34:104
45. Takahashi Y, Tadokoro H (1973) Macromolecules 6:672–675
46. Barroso-Bujans F, Cerveny S, Alegría A, Colmenero J (2013) Macromolecules 46:7932–7939
47. Barroso-Bujans F, Fernandez-Alonso F, Pomposo JA, Cerveny S, Alegría A, Colmenero J (2012) ACS Macro Lett 1(5):550–554
48. Matsuura H, Fukuhara K (1986) J Polym Sci B Polym Phys 24(7):1383–1400
49. Rozenberg M (1998) Spectrochim Acta Part A Mol Biomol Spectrosc 54(12):1819–1826
50. Fragiadakis D, Runt J (2010) Macromolecules 43(2):1028–1034

Universality of Density Scaling

Andrzej Grzybowski and Marian Paluch

Abstract Investigations of the sought after complete and commonly accepted theory of the glass transition and related phenomena have recently gained an essential support from a very promising idea of the *density scaling of molecular dynamics* in viscous liquids. This idea, often known as the *thermodynamic scaling concept*, has been initiated by many phenomenological observations, which have shown that dynamic quantities (e.g., viscosity, structural relaxation time, or segmental relaxation time in case of polymers) measured in different thermodynamic conditions (e.g., along different isobars and isotherms) can be scaled onto one master curve well described by a function of the single variable that is a product of the inverse temperature and the density power with the scaling exponent considered as a material constant independent of thermodynamic conditions. However, a crucial advantage of the phenomenological description has become its theoretical grounds relied on an effective short-range intermolecular potential, which has been derived from the well-known Lennard-Jones potential and satisfactorily verified by computer simulations. A relation suggested between the scaling exponent and the exponent of the dominant repulsive part of the effective intermolecular potential gives a tempting opportunity to study the macroscopic properties of materials by using the underlying intermolecular potential and *vice versa* to determine the intermolecular potential parameters based on measurements of macroscopic quantities. It opens new perspectives for our better understanding of complex physicochemical phenomena occurring near the glass transition. In this chapter, we present the density scaling concept as the idea that bears hallmarks of universality in case of both various materials and different quantities. We show that the density scaling law may concern not only dynamic but also thermodynamic quantities, constituting a convenient tool to explore relationships between molecular dynamics and thermodynamics based on the effective

A. Grzybowski · M. Paluch (✉)
Silesian Center for Education and Interdisciplinary Research,
Institute of Physics, University of Silesia in Katowice,
75 Pułku Piechoty 1, 41-500 Chorzów, Poland
e-mail: marian.paluch@us.edu.pl

A. Grzybowski
e-mail: andrzej.grzybowski@us.edu.pl

F. Kremer and A. Loidl (eds.), *The Scaling of Relaxation Processes*,
Advances in Dielectrics, https://doi.org/10.1007/978-3-319-72706-6_4

short-range intermolecular potential. We demonstrate predictive capabilities of the density scaling law that implies several rules for activation quantities and fragility parameters defined in different thermodynamic conditions, which enable to discover and verify physically well-defined invariants. We also discuss some nontrivial cases of the thermodynamic scaling for which the power density scaling law with a constant scaling exponent is not sufficient, but we can find density or timescale-dependent counterparts of the exponent. The exceptions to the standard power density scaling law delimit further challenges in making progress toward the development of the density scaling idea and its applicability range.

1 Power Density Scaling Law—Its Applicability Range and Theoretical Grounds

In 1995, Philip W. Anderson claimed in Science [1] that "The deepest and most interesting unsolved problem in solid state theory is probably the theory of the nature of glass and the glass transition.", anticipating that "This could be the next breakthrough in the coming decade." However, the problem has not been resolved yet. Despite intensive investigations conducted in the field of the liquid–glass transition physics in the last half-century, no theoretical idea has been acknowledged as the complete and commonly accepted theory.

In the first decade of the twenty-first century, great expectations have been raised that this challenge can be met by a very promising idea of *the thermodynamic scaling* of molecular dynamics of viscous liquids, often known as *the density scaling* idea, which has been initiated by many phenomenological observations [2, 3] and described theoretically in the most interesting way by the theory of isomorphs formulated by Jeppe Dyre's group [4]. In case of various material groups (especially including van der Waals liquids and polymer melts, but also ionic liquids and even metallic glass formers), this important research trend has shown that dynamic quantities such as viscosities η and structural relaxation times τ (either segmental relaxation times in case of polymers or ionic conductivity relaxation times in case of ionic liquids) measured in different isobaric and isothermal conditions in the supercooled liquid state can be scaled onto one master curve well described by a function $f(\Gamma)$ of a single scaling variable $\Gamma = \rho^{\gamma}/T$, where T and ρ denote thermodynamic variables (temperature and density, respectively) and the scaling exponent γ is a material constant independent of thermodynamic conditions.

Although the power density scaling law (the PDS law) has been formulated phenomenologically for different quantities, [2, 3, 5] the greatest impact on the development in the theory of the glass transition and related phenomena is certainly exerted by its version for the timescale of molecular dynamics defined by the structural relaxation time in the following way:

$$\tau = f(\rho^{\gamma}/T) \tag{1}$$

where the density ρ is usually replaced by the inverse specific volume υ^{-1}. It is worth noting that Eq. (1) implies and also results from the criterion for the density scaling with a constant value of γ [6],

$$\rho^{\gamma}/T = const \quad \text{at}\, \tau = const, \tag{2}$$

which enables to reliably verify whether or not the PDS law is obeyed for a given material at least in the considered temperature-density range. Moreover, Eq. (2) underlies a convenient method for determining the scaling exponent γ for a given material as a slope of the linear dependences obtained by finding logarithms of its both sides [6]

$$\log T = \gamma \log \rho + C(\tau) \quad \text{at}\, \tau = const \tag{3}$$

where the intercept $C(\tau)$ depends on the structural relaxation time τ at which temperatures T and densities ρ have been collected. An undoubted advantage of Eq. (3) is its straightforward origin from the density scaling criterion, which is independent of any models of the thermodynamic evolution of the timescale of molecular dynamics. In Fig. 1, the procedure of the density scaling based on the criterion given by Eq. (3) is demonstrated.

Such results of the experimental data analysis have become extremely interesting, when their first commonly accepted theoretical grounds have been suggested [7–9], because the scaling exponent γ has been related to the exponent ($m_{IPL} = 3\gamma$) of the dominant repulsive part $U_{IPL}(r) = A_r r^{-m_{IPL}}$ of the effective short-range potential for intermolecular interactions $U_{eff}(r) = U_{IPL}(r) - A_t$ with a small constant or linear attractive background A_t. It has been shown, also by means of molecular dynamics (MD) simulations [8, 9], that the effective potential

$$U_{eff}(r) = A_r r^{-m_{IPL}} - A_t \tag{4}$$

with appropriately fitted values of its parameters very well approximates the generalized Lennard-Jones (LJ) potential, $U_{LJ}(r) = C_r r^{-m} - C_t r^{-n}$, in case of small intermolecular distances r. For instance, it has been found that $18 \leq m_{IPL} \leq 20$ for the standard LJ potential with $m = 12$ i $n = 6$ [8, 9].

The MD simulations in simple models based on the Lennard-Jones potential have also shown that different quantities (such as structural relaxation times, viscosities, diffusivities, and even configurational isochoric heat capacities) [7–14] can be scaled with the same value of the scaling exponent γ for a given system to a good approximation in a relatively narrow temperature-density range, although the specific scaling functions in the PDS law can be different for different quantities. In contrast to the simple simulation models, in case of real materials that obey the PDS law, the density scaling with the constant value of γ for a given material is usually valid in a wide temperature-density range, and for instance the deviations from the PDS law reflected in the density-dependent scaling exponent are very rarely observed in the pressure range, which can be typically covered up to 2 GPa in measurements of

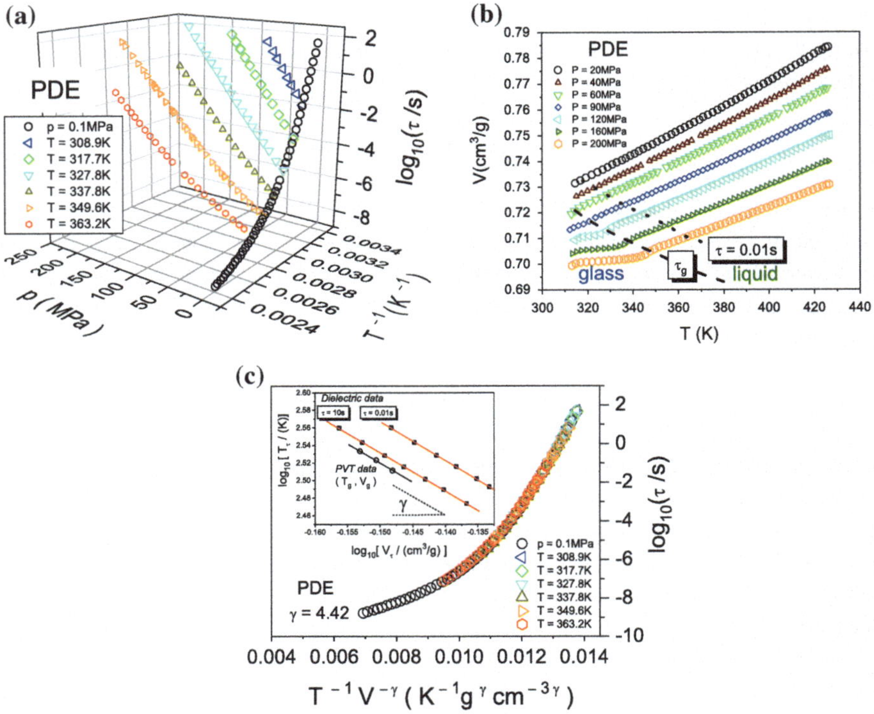

Fig. 1 Illustration of the density scaling procedure based on the density scaling criterion (Eq. 3) applied to dielectric and *pVT* experimental data measured for PDE (phenylphthalein-dimethylether) in the supercooled liquid state. **a** Plot of the structural relaxation times determined in the supercooled liquid state from broadband dielectric measurements as a function of temperature and pressure. **b** Plot of the measured isobaric temperature dependences of volume in the supercooled liquid and glassy states. **c** Plot of the density scaling of the structural relaxation times, which has been performed after transforming the dielectric data from the T-p domain to the T-V one by employing the *pVT* measurement data and finding the scaling exponent γ based on Eq. (3). The inset illustrates the determination of the scaling exponent γ based on Eq. (3) with $\rho = V^{-1}$ by using selected *pVT* experimental data along the glass transition line and dielectric data chosen at selected constant relaxation times

dynamic quantities by using the dielectric and mechanical spectroscopy techniques. The differences between simple model and real supercooled liquids as well as the deviations from the PDS law, which comply or do not comply with the mentioned theory of isomorphs, will be discussed at the end of the chapter. Nevertheless, there are also striking similarities of the results obtained from MD simulations and measurements, which can be explained within the theory of isomorphs. Among them, the isochronal superposition of dielectric loss spectra should be noted.

The invariance of the shapes of the dielectric structural loss peaks in different temperature-pressure conditions along an isochrone was reported for supercooled van der Waals liquids and polymer melts [15, 16]. Later, the invariance of different correlation functions along an isomorph being a curve in the temperature-density diagram (discussed in more detail in the last section of the chapter) has been postulated and confirmed by MD simulations in the simple models based on the Lennard-Jones potential [4]. One of the conditions, which enables to determine the isomorph curve, is the invariance of the structural relaxation time expressed in the special reduced units considered within the isomorph theory, i.e., $\tilde{\tau} = \tau\rho^{1/3}T^{1/2}(k_B/M)^{1/2}$, where k_B is the Boltzmann constant, and M is the average particle mass. It has turned out that different correlation functions in simple simulation models are invariant at $\tilde{\tau} = const$. For instance, this isochronal invariance is satisfied by the incoherent intermediate self-scattering function F_s plotted as a function of time expressed according to the isomorph theory in the following reduced units, $\tilde{t} = t\rho^{1/3}T^{1/2}(k_B/M)^{1/2}$. According to the isomorph theory, the relaxation functions and the relaxation time in the reduced units are in general isomorph invariant. The special dielectric case of the general hypothesis has been thoroughly discussed recently [17]. Based on the isomorph theory, the expressions, $\chi_e(\tilde{\tau})\rho^{\gamma-1}$ and $\chi_e(\tilde{\tau})T/\rho$, involving the electric susceptibility χ_e, have been predicted to be isomorph invariant, which maintain their invariance along an isomorph also in the frequency domain, respectively, in the forms, $\chi_e(\tilde{f})\rho^{\gamma-1}$ and $\chi_e(\tilde{f})T/\rho$, where $\tilde{f} = (2\pi\tilde{t})^{-1}$. Consequently, by applying the latter findings [17] to the Kirkwood-Fröhlich [18, 19] formula via the equivalence of the imaginary parts of the relative permittivity and the electric susceptibility, $\varepsilon_r''(\tilde{f}) = \chi_e''(\tilde{f})$, the product of the local field correction factor and the correlation factor has been found to be isomorph invariant. To verify experimentally the isomorph theory predictions for the dielectric relaxation, the relative permittivity has been expressed by the measured capacitance of a full cell $C(\tilde{f})$ divided by the empty capacitance C_{empty}, i.e., $\varepsilon_r''(\tilde{f}) = -C''(\tilde{f})/C_{empty}$. Thus, the corresponding isomorph invariants, $-(C''(\tilde{f})/C_{empty})\rho^{\gamma-1}$ and $-(C''(\tilde{f})/C_{empty})T/\rho$, have been successfully validated by means of the broadband dielectric measurements. Consequently, interesting predictions about the dielectric loss spectra have become possible. For instance, if we consider two isochronal states (p_1,T_1) and (p_2,T_2) with densities ρ_1 and ρ_2, respectively, the capacitance in the state (p_2,T_2) can be determined from the capacitance in the state (p_1,T_1) via the relation, $-C_2''(\tilde{f}) = -C_1''(\tilde{f})(\rho_1/\rho_2)^{\gamma-1}(C_{empty_2}/C_{empty_1})$, which shows that the frequency-dependent capacitances can be superimposed by shifting in the well-defined manner in the logarithmic plot. Since one can assume that $C_{empty_2} = C_{empty_1}$, a dielectric loss spectrum can be predicted from another one at $\tilde{\tau} = const$ if densities in both the isochronal states and the scaling exponent γ are known. For this purpose, we do not need even the scaling exponent, if temperatures are given in the isochronal states, because alternatively, $-C_2''(\tilde{f}) = -C_1''(\tilde{f})(T_1/T_2)(\rho_2/\rho_1)(C_{empty_2}/C_{empty_1})$. These findings have been confirmed by using experimental data, and the reduced units suggested by the isomorph theory have actually not affected qualitatively the measured isomorph invariants, which maintained their invariance also at $\tau = const$ in the typical frequency domain that is not modified by the reduced units. This is a subsequent example, which shows that the density scaling of the experimental data is usually only slightly improved by using the reduced units of the isomorph theory [20, 21], although these reduced units have been found to be important to successfully scale the simulation data. Thus, the isochronal superposition of measured dielectric loss spectra can be argued within the framework of the isomorph theory independently of the employed units.

The density scaling idea has originally focused on the molecular mobility reflected in the primary relaxation process, which is the characteristic of a given material group (e.g., the structural relaxation in case of supercooled van der Waals liquids). Further investigations have shown that the PDS law with the same scaling exponent γ is valid for the timescales of both the primary relaxation and the secondary relaxation called the Johari-Goldstein (JG) relaxation [22–24]. This secondary process has been most often attributed to an intermolecular origin relied on small-angle reorientations of entire molecules. According to the coupling model (CM) [25] and the extended CM [26] formulated by K. L. Ngai, the JG-secondary process can be identified at least approximately with a primitive relaxation being a precursor of the primary relaxation, and the timescales of the primary and JG-secondary relaxations, τ and τ_{JG}, are connected via the relation, $\tau \approx (t_c^{-n_{CM}} \tau_{JG})^{1/(1-n_{CM})}$, where the coupling parameter, $n_{CM} = 1 - \beta_{KWW}$, depends on the nonexponential stretching parameter β_{KWW} of the Kohlrausch-Williams-Watts (KWW) correlation function, $\Phi(t) = \exp[-(t/\tau)^{\beta_{KWW}}]$, which describes one-body independent relaxation at $\beta_{KWW} = 1$ for $t < t_c$ and many-body relaxation at $0 < \beta_{KWW} < 1$ for $t > t_c$. The onset time t_c of the cooperative many-body relaxation is a material constant equaled to *ca.* 1–2 ps, which can be established from quasielastic neutron scattering experiments and MD simulations [27]. Very recently, after thorough analyses of experimental data, both the coupling parameter n_{CM} and the JG-secondary relaxation time τ_{JG} have been well depicted as functions of the single scaling variable $\Gamma = \rho^\gamma / T$. Thus, the coupling model equation has found its density scaling representation, $\tau \approx [t_c^{-n_{CM}(\rho^\gamma/T)} \tau_{JG}(\rho^\gamma/T)]^{1/(1-n_{CM}(\rho^\gamma/T))}$, which suggests that the PDS law (Eq. 1) for the primary relaxation time can originate from the density scaling of τ_{JG} and n_{CM}.

2 Density Scaling-based Rules for Activation Quantities and Fragility Parameters Defined in Different Thermodynamic Conditions

A significant progress in the high-pressure experimental techniques resulted in many data measured not only in atmospheric conditions but also at elevated pressure, giving new opportunities to study properties of materials subjected to compression. In such investigations, there is a need to consider material properties in different pressure-volume-temperature (pVT) conditions that can be conveniently separated into isobaric, isochoric, and isothermal planes in the thermodynamic space. Therefore, the activation energy and the kinetic fragility, which are the important parameters originally describing the strongly non-Arrhenius behavior of the temperature dependence of the structural relaxation times near the glass transition at ambient pressure, were defined in different thermodynamic conditions already in the 1960s of the twentieth century [28], and since then they have been widely applied to gain better insight into molecular dynamics of supercooled liquids.

It is worth noting that the commonly used definitions of the isobaric activation energy (i.e., activation enthalpy) $E_p^{act} = R(\partial \ln \tau/\partial(1/T))_p$ and the isochoric activation energy $E_V^{act} = R(\partial \ln \tau/\partial(1/T))_V$ are closely related to the activation Arrhenius law, $\tau = \tau_0 \exp(E_X^{act}/RT)$, which is considered at $X = const$ where $X = p$ or $X = V$, while R is the gas constant and τ_0 is a pre-exponential factor. In contrast to the classical Arrhenius law, the activation energy E_X^{act} is not constant and can even strongly depend on temperature when a supercooled liquid is approaching the glassy state. This dependence is called a non-Arrhenius or super-Arrhenius behavior, which is treated as a common characteristic of cooperative and heterogeneous molecular dynamics near the glass transition. Taking into account the isothermal compression effect on molecular dynamics, the Arrhenius law should be extended to the form, $\tau = \tau_0 \exp(E_X^{act}/RT + pV_T^{act}/RT)$, which involves the activation volume that in general also changes with varying thermodynamic conditions, although it has been originally assumed to be constant. Thus, in isothermal conditions, a relevant activation parameter is the activation volume that can be defined as follows: $V_T^{act} = RT(\partial \ln \tau/\partial p)_T$.

The activation isochoric and isobaric energies have been mainly used to investigate the interplay between temperature and density fluctuations, the crucial role of which has been recognized in molecular dynamics of glass-forming liquids, giving evidence that the purely free volume and thermally activated models are its extreme cases [2, 3, 29]. To quantify the relative influence of the fluctuations of temperature and density on molecular dynamics in the vicinity of the glass transition, a few measures have been suggested. Among them, the most popular one is probably the ratio, E_V^{act}/E_p^{act} [2, 3]. The density scaling idea has shed a new light on this measure by finding a relationship between the ratio E_V^{act}/E_p^{act} and the scaling exponent γ. To derive the relationship, we need to invoke an older measure of the relative influence of temperature-density fluctuations on molecular dynamics, which is the ratio of the isochronal and isobaric thermal volume expansivities $\alpha_\tau = (\partial \ln V/\partial T)_\tau$ and $\alpha_p = (\partial \ln V/\partial T)_p$, indicating the purely free volume dynamics when $|\alpha_\tau|/\alpha_p \to \infty$, and the purely thermally activated dynamics when $|\alpha_\tau|/\alpha_p \to 0$ [30]. Since it has been found [31] that $E_V^{act}/E_p^{act} = 1/(1 + \alpha_p/|\alpha_\tau|)$, the ratio of the activation energies is a convenient measure normalized between 0 and 1, where the extreme values of E_V^{act}/E_p^{act} quantify the dominant role of density and temperature fluctuations, respectively. Taking into account that the isochronal expansivities of real materials are negative numbers, i.e., $|\alpha_\tau| = -\alpha_\tau$, we obtain the formula $E_V^{act}/E_p^{act} = 1/(1 - \alpha_p/\alpha_\tau)$, which can be transformed by using the density scaling criterion (Eq. 2) to the following relation [32]

$$\left(\frac{E_V^{act}}{E_p^{act}}\right)_\tau = \frac{1}{1 + \gamma T_\tau \alpha_p(T_\tau)} \tag{5}$$

To arrive at Eq. (5), the density scaling criterion has been differentiated with respect to temperature, $\left(\partial \rho^{\gamma} T^{-1}/\partial T\right)_{\tau} = 0$, yielding $\alpha_{\tau} = -1/\gamma T_{\tau}$ [3]. Although Eq. (5) has been often limited to the glass transition, i.e., at the structural relaxation time $\tau = \tau_g$ at which the glass transition occurs, we prefer its presented general form, which is valid at any constant timescale of molecular dynamics that obeys the power density scaling law (Eq. 1).

The importance of Eq. (5) relies on its relation to the effective short-range intermolecular potential (Eq. 4) via the scaling exponent γ assumed to be equal to $m_{IPL}/3$. By analyzing the relation Eq. (5) between E_V^{act}/E_p^{act} and γ based on experimental data, one can recognize that the scaling exponent γ can also quantify the relative influence of temperature and density fluctuations on molecular dynamics. This is because the product $T_{\tau}\alpha_P(T_{\tau})$ is nearly constant. For instance, it usually ranges between 0.16 and 0.20 for most materials considered at the glass transition at ambient pressure. Thus, if γ tends to 0, then $E_V^{act}/E_p^{act} \to 1$ and molecular dynamics is purely thermally activated, while if $\gamma \to \infty$, then E_V^{act}/E_p^{act} tends to 0 and molecular dynamics is governed only by the free volume. Another analysis, which should not be confused with the previous one, can be performed for a given material at a constant structural relaxation time or along the glass transition line. As have been shown based on experimental data [33, 34], the product $T_{\tau}\alpha_P(T_{\tau})$ slowly decreases with increasing pressure in isochronal conditions. Then, from Eq. (5) with a constant value of γ, one can easily see that the ratio E_V^{act}/E_p^{act} increases with increasing pressure at $\tau = const$. It means that the role of temperature fluctuations in molecular dynamics of the examined system grows under the influence of the system compression. This apparently surprising result can be understood if we realize that the intermolecular distances become smaller and smaller with increasing pressure and the free spaces become reduced. Consequently, their effect on the molecular mobility is diminished.

The latter effect of pressure on molecular dynamics near the glass transition has been often discussed by using an equivalent representation of the measure E_V^{act}/E_p^{act} via the ratio of the isochoric and isobaric fragility parameters [32, 35–41], $m_V = \left(\partial \log_{10}\tau/\partial(T_g/T)\right)_{V,T=T_g}$ and $m_p = \left(\partial \log_{10}\tau/\partial(T_g/T)\right)_{p,T=T_g}$, which yields $m_V/m_p = E_V^{act}/E_p^{act}$ at the glass transition temperature $T_g(p)$. Although T_g increases with increasing pressure p, to a good approximation, one can assume a constant structural relaxation time $\tau_g = \tau(T_g(p),p) = const$ (usually $\tau_g = 100$ s) at the glass transition independently of pressure conditions for a given material. The isobaric fragility parameter has been widely popularized by Angell [42, 43] as a relevant characteristic of glass-forming liquids, which classifies such materials according to the degree of their molecular dynamic sensitivity to changes in temperature on approaching the glassy state. The mentioned isochoric fragility m_V and the isothermal fragility, $m_T = \left(\partial \log_{10}\tau/\partial(V_g/V)\right)_{T,V=V_g}$, have been defined and employed in the glass transition studies to better understand molecular dynamics investigated experimentally near the glass transition not only in the temperature domain in atmospheric conditions, but also in the entire thermodynamic (pVT) space. Within the framework of the density scaling idea, a few general rules have been formulated for the fragility parameters defined in different thermodynamic conditions, which order interrelations between the fragility parameters and predict their behavior in different thermodynamic conditions in a manner consistent with the experimental results [34].

To establish a general relation between m_T and m_V, we can exploit the density scaling criterion (Eq. 2) at $\tau = \tau_g$. Based on the definition of m_T, one can derive the equation, $m_T = -m_V T_g^{-1} \alpha_{\tau_g}^{-1}(T_g)$. Then, if the power density scaling law is obeyed, we obtain the following simple relation,

$$m_T = \gamma m_V, \tag{6}$$

because Eq. (2) yields $\gamma = -T_g^{-1} \alpha_{\tau_g}^{-1}(T_g)$ at $\tau = \tau_g$. As shown in Refs. [3, 6], and [34], in the PDS regime, the density scaling in terms of Eq. (1) can be reduced in case of isotherms to the scaling vs. V_g/V, where V_g is the glass transition volume. Then, all isotherms $\tau(p)$ measured for a tested material can be superimposed onto one master curve given by a function of V_g/V. Thus, from the definition of the isothermal fragility m_T, one can conclude that m_T considered at $\tau_g = const$ is a material invariant if the PDS law is obeyed by this material. Since both m_T and γ are material invariants, Eq. (6) shows the material isochronal invariance of m_V, which means that m_V is independent of thermodynamic conditions at a constant τ_g, if the PDS law is satisfied. To analyze the pressure effect on the isobaric fragility m_p, we can invoke the already-mentioned relation $m_V / m_p = E_V^{act} / E_p^{act}$ and exploit Eq. (5) at $\tau_g = const$ as follows:

$$m_p = m_V \left(1 + \gamma T_g \alpha_p (T_g)\right) \tag{7}$$

Taking into account that m_V and $\gamma$are material invariants in the PDS regime, Eq. (7) shows that the isobaric fragility should decrease with increasing pressure for a given material, because the product $T_g \alpha_P (T_g)$ decreases with increasing pressure in isochronal conditions as mentioned above and any material has not been found yet to exhibit properties which are contradictory to these phenomenological observations. To complete the analysis of the fragility parameters in the PDS regime, it is worth noting that the same conclusion about the pressure behavior of the isobaric fragility can be drawn from the relation, $m_p = m_T \left(\gamma^{-1} + T_g \alpha_p (T_g)\right)$, which can be straightforwardly derived from Eqs. (6) and (7). Therefore, one can claim that the PDS law underlies three rules prospected for the fragility parameters [34]:

(i) Compression brings about the decrease in the isobaric fragility m_p.
(ii) The isothermal fragility m_T is an invariant parameter characteristic for a given liquid.
(iii) The isochoric fragility m_V is also an invariant parameter characteristic for a given liquid.

Based on the rules (ii) and (iii), one can make an interesting observation that the fragility parameters m_T and m_V are material invariants defined in the T-V domain, which enable to investigate separately the effect of the density fluctuations on molecular dynamics in case of the isothermal fragility and the effect of the temperature fluctuations on molecular dynamics in case of the isochoric fragility, whereas both the density and temperature fluctuations influence the isobaric fragility m_p, which is defined in the T-p domain and depends on the thermodynamic conditions. Thus, Eq. (6) supports the interpretation of the scaling exponent γ as a measure of the relative effect of density and temperature fluctuations on molecular dynamics, which has been earlier suggested by analyzing Eq. (5).

From our recent investigations of invariant properties characterizing viscous liquids in the PDS regime [44], we can draw a similar conclusion that the T-V representation is more relevant than the T-p representation to gain a better insight into fundamental properties of molecular dynamics near the glass transition, whereas the measurements of dynamic quantities, e.g., τ and η, respectively, by means of broadband dielectric spectroscopy (BDS) and mechanical spectroscopy (MS), are typically carried out in isobaric and isothermal conditions due to technical limitations. Thus, the dielectric or mechanical measurements in the T-p domain should be supplemented with the pVT measurements providing the temperature-pressure dependences of volume, $V = V(T,p)$, to enable the transformation from the T-p to the T-V representation. However, in case of computer simulations, we can obtain various dynamic quantities (such as diffusion coefficient, τ, and η) as a function of the variables p, V, and T straightforwardly from the performed computations.

Using experimental data, we have shown that the ratio of the dynamic modulus defined in T-V domain by the equation, $M_{V-T} = E_V^{act}/V_T^{act}$, and the thermodynamic modulus defined by the isothermal bulk modulus as the inverse isothermal compressibility, $B_T = -(\partial \ln V/\partial p)_T^{-1}$, is a material constant independent of thermodynamic conditions if the PDS law (Eq. 1) is obeyed by the material, although such an invariance has been earlier improperly suggested [45–47] by employing the dynamic modulus defined in the T-p domain, $M_{p-T} = E_p^{act}/V_T^{act}$. It should be noted that all numerical tests suggested by us [44] to verify the invariance of the ratio B_T/M_{V-T} have been relied on the density scaling criterion (Eq. 2) and not required any additional models of the thermodynamic evolution of the structural relaxation time to calculate needed activation quantities. It was possible because Eq. (2) provides a very useful and accurate method (see Supplementary Information to Ref. [44]) for generating dynamic data (e.g., structural relaxation times) along different isobars, isochores, and isotherms if we know the scaling exponent γ, the dependence $V = V(T,p)$, and for instance the dependence $\tau(T)$ at ambient pressure. The outcome of the numerical tests has found its grounds also in analytical calculations based on Eq. (2). To a known equation [46, 47], $\left(E_p^{act}/V_T^{act}\right)_\tau = T_\tau(\partial p/\partial T)_\tau$, we have applied a generalized form of the equation for the glass transition temperature coefficient dT_g/dp [48], which is valid not only at the glass transition but also at any $\tau = const$, because it is a consequence of the PDS criterion [49],

$$\left(\frac{\partial T}{\partial p}\right)_\tau = \frac{\gamma T_\tau / B_{T_\tau}(p)}{1 + \gamma T_\tau \alpha_p(T_\tau)}. \tag{8}$$

Then, one can find the equation, $\left(E_p^{act}/V_T^{act}\right)_\tau = (1+\gamma T_\tau \alpha_P(T_\tau))B_{T_\tau}(p)/\gamma$, which can be combined with Eq. (5) to arrive at the relationship, $\left(E_V^{act}/V_T^{act}\right)_\tau = B_{T_\tau}(p)/\gamma$. Since its left-hand side constitutes the dynamic modulus M_{V-T} in the T-V domain at a constant structural relaxation time, one can show the following rule:

$$B_T/M_{V-T} = \gamma \quad \text{at any } \tau = const, \tag{9}$$

which means that the ratio of the thermodynamic and dynamic moduli is a material invariant if the PDS law is satisfied by the material and the dynamic modulus is expressed in the T-V representation.

It should be noted that the invariance of the ratio of the thermodynamic and dynamic moduli B_T/M_{V-T} is independent of the considered timescale of molecular dynamics due to the invariance of the scaling exponent γ for a given material in the PDS regime, while the invariances of the isothermal and isochoric fragilities, m_T and m_V, depend on the structural relaxation times. It means that the values of m_T and m_V are invariant along a given isochrone, but they change with varying timescale of molecular dynamics, and only their ratio, $m_T/m_V = \gamma$, remains unchanged at any $\tau = const$. As already mentioned, the scaling exponent γ expressed by the ratio m_T/m_V relates the exponent $m_{IPL}/3$ of the repulsive part of the effective short-range intermolecular potential (Eq. 4) to the measure of the relative influence of the density and temperature fluctuations on molecular dynamics near the glass transition. However, Eq. (9) shows that the scaling exponent γ and also its relation to the IPL exponent $m_{IPL}/3$ gain a robust interpretation as the material invariant that indeed links molecular dynamics and thermodynamics of viscous liquids, providing an additional argument for the concept of "the thermodynamic scaling" of molecular dynamics as sometimes the density scaling has been called [44].

Very recently, the aforementioned method for generating isobars, isochores, and isotherms based on the density scaling criterion (Eq. 2), which was successfully applied [44] to verify the invariance of the ratio B_T/M_{V-T}, has been also used to reveal a breakdown of the Arrhenius law at high temperatures near the melting point in isobaric conditions [50]. This important finding has been inspired by isothermal high-pressure mechanical measurements [51], which identified an inflection point p_{infl} in the isothermal dependences of viscosity on pressure, $\eta(p)$, in the low viscosity range. Analogous dielectric measurements in the limit of sufficiently short structural relaxation times are difficult due to high dc-conductivity σ_{dc}. Therefore, we have measured the dc-conductivity of the materials, for which there is no decoupling observed between σ_{dc} and , and the dc-conductivity can be used to investigate the global molecular dynamics instead of the structural relaxation times, because σ_{dc} satisfactorily mimics τ in this case. As a result, we have found the inflection points p_{infl} in the experimental isothermal dependences $\sigma_{dc}(p)$, which are obviously at odds with the pressure version of the Arrhenius law, $\tau = \tau_0 \exp(pV_T^{act}/RT)$, where $V_T^{act} = const$. However, the dielectric measurements at an elevated pressure as well as those at sufficiently high temperatures to explore the fast molecular dynamics region near the melting point are technically complicated. For this reason, we have performed numerical investigations by using dense sets of points generated along many isobars, isochores, and isotherms of structural relaxation times based on the density scaling criterion (Eq. 2) with the scaling exponent that enabled to scale also the experimental dependences $\sigma_{dc}(p)$ for a given material. It has turned out that the inflection points p_{infl} have been also found in the generated isothermal dependences of the structural relaxation times $\tau(p)$ and their low pressure ranges at $p < p_{infl}$ have been successfully described by the relation dependent on the scaling quantity, $\Gamma = \rho^{\gamma}/T$, as follows:

$$\tau = \tau_0 \exp(A\rho^{\gamma}/T) \tag{10}$$

where the pressure experimental dependences of density have been accurately parametrized by the equation, $\rho(T, p) = \rho(T, p_0)(1 + C_1(p - p_0))^{1/C_2}$, which exploit the density $\rho(T, p_0)$ at ambient pressure p_0 and the fitting parameters C_1 and C_2. Since C_1 is positive and the ratio $\gamma/C_2 < 1$ and also positive for each tested material, Eq. (10) has properly described the isotherms $\tau(p)$ at $p < p_{infl}$ at which these dependences are increasing, but they have a negative curvature. In the next step of the analysis, we have established that both the isothermal and isobaric dependences of the structural relaxation times near the melting point can be very well fitted to Eq. (10) with the same set of the values of the fitting parameters, which enabled to describe the negative curvature of the isothermal dependences $\tau(p)$ at $p < p_{infl}$. What is more, it has not been able in any way to reduce Eq. (10) to the simple temperature version of the Arrhenius law, $\tau = \tau_0 \exp(E_p^{act}/RT)$ with $E_p^{act} = const$ at high temperatures even at ambient pressure, which has been commonly expected until now. It means that the isobaric activation energy has to depend on density, $E_p^{act} \sim \rho^{\gamma}$, which straightforwardly follows from the comparison between the temperature version of the Arrhenius law and Eq. (10) near the melting temperature. Therefore, the deviations from the Arrhenius law occur not only near the glass transition, which is often quite large and called a super-Arrhenius behavior, but also near the melting point, although they are considerably less pronounced in this region due to numerically small contribution of density changes to the isobaric activation energy with changing temperature. Nevertheless, the density dependence of the isobaric activation energy at high temperatures can be additionally rationalized by analyzing the ratio E_V^{act}/E_p^{act}. As we have shown in Refs. [44, 50], the numerically calculated values E_V^{act}/E_p^{act}, which have been determined from isochores and isobars generated from the density scaling criterion (Eq. 2), decrease with increasing T and decreasing ρ^{γ}/T for each tested material. Thus, the role of density fluctuations predominates over the temperature fluctuations in molecular dynamics at high temperatures at any pressure, and the density effect on the isobaric activation energy is reflected in the proportionality $E_p^{act} \sim \rho^{\gamma}$.

3 Density Scaling-based Entropic Models and Density Scaling of Entropy

The meaning of the density scaling idea and the *T-V* representation in recent investigations of glass-forming liquids caused a development of models describing the thermodynamic evolution of the timescale of molecular dynamics near the glass transition, which could intrinsically comply with the PDS law. Among them, the MYEGA and Avramov models are especially interesting due to their origin based on entropic fundamentals.

The MYEGA model originally suggested in the temperature domain [52] has been later extended to a function of the scaling variable $\Gamma = \rho^{\gamma}/T$ [33]. The temperature version of the MYEGA model has been derived from the commonly known Adam-Gibbs model [53], $\tau = \tau_0 \exp(A_{AG}/TS_c)$, and has been relied on the configurational system entropy S_c similarly to its precursor. In the original MYEGA model, the parameter A_{AG} is an effective activation barrier, and the configurational entropy S_c is modeled by using the constraint theory [54] and the analysis of energy landscape [55] on the assumption of topological degrees of freedom per atom in a simple two-state system, where both A_{AG} and S_c depend on the system composition x, and S_c is also temperature-dependent. To formulate the *T-V* version of the MYEGA model, we have assumed that both A_{AG} and S_c additionally depend on density. Then, $A_{AG}(x, \rho) = a(x)(\rho/\rho_r)^{\gamma}$ and $S_c(x, T, \rho) = f(x, T, \rho)Nk_B \ln \Omega$, where γ is the scaling exponent, ρ_r is a reference density, k_B is the Boltzmann constant, and Ω is the number of degenerate configurations per floppy mode, while the topological degrees of freedom in a simple two-state system may depend on both T and ρ in the following way, $f(x, T, \rho) = 3\exp(-H(x, \rho)/k_B T)$, where the energy difference between two states $H(x, \rho) = h(x)(\rho/\rho_r)^{\gamma}$. As a result, we have arrived at the MYEGA model in the *T-V* domain,

$$\tau(x, T, \rho) = \tau_0 \exp\left[D_M(x)\rho^{\gamma} T^{-1} \exp\left(A_M(x)\rho^{\gamma} T^{-1}\right)\right] \tag{11}$$

where the parameters D_M and A_M have their physical meaning given by the expressions, $D_M(x) = a(x)/\rho_r^{\gamma} 3Nk_B \ln \Omega$ and $A_M(x) = h(x)/\rho_r^{\gamma} k_B$, but they are actually fitting parameters like τ_0 and γ. It is worth noting that Eq. (11) provides an analytical equation for the isobaric fragility,

$$m_p = \left(1 + \gamma T_g \alpha_p(T_g)\right) D_M(x) \exp(A_M(x))(1 + A_M(x))/\ln 10 \tag{12}$$

and the isochoric fragility,

$$m_V = D_M(x) \exp(A_M(x))(1 + A_M(x))/\ln 10 \tag{13}$$

which enable to determine the fragility parameters by using the fitting parameters D_M and A_M.

The *T-V* version of the MYEGA model has been successfully verified by using the experimental *T-p* dependences of structural relaxation times of glass-forming liquids analyzed in the *T-V* domain [33]. Our experience in using Eq. (11) and the *T-V* version of the Avramov model discussed later in this section shows that both the models can be alternatively used to describe the dependences τ(T,V) or τ(T,ρ) collected for glass formers in the PDS regime. However, in case of longitudinal relaxation times $\tau_{\|}$ of liquid crystals measured at elevated pressure, which obey the PDS law Eqs. (1) and (11), has turned out [56] to be a better model than the *T-V* version of the Avramov model at least within the fitting capabilities, because the latter resulted in very slowly convergent fits in the narrow range of the timescale $\tau_{\|}$.

In the context of the Adam-Gibbs model variations in the *T-V* domain, an important question arises about the possible density scaling of the configurational entropy, because S_c should satisfy the PDS law (Eq. 1) in case of materials that obey this scaling law if the configurational entropy indeed governed the molecular dynamics as well as the dramatic slowdown in its timescale near the glass transition as often suggested also within the Adam-Gibbs model. Recently, to solve the problem, we have formulated [57] two new *T-V* versions of the Adam-Gibbs model, which involve the temperature-volume dependence of the configurational entropy $S_c(T,V)$ that is not any explicit function of the scaling variable, ρ^γ/T or $1/TV^\gamma$, in contrast to the *T-V* version of the MYEGA model within which $S_c(T,V)$ has been assumed as a function of the single variable ρ^γ/T. These new *T-V* versions of the Adam-Gibbs model differed only in the parameter A_{AG}, which was either constant according to the original assumption made by Adam and Gibbs [53], or proportional to ρ^γ in the Adam-Gibbs equation, $\tau = \tau_0 \exp(A_{AG}/TS_c(T, V))$, considered in the *T-V* representation, where the density-dependent parameter A_{AG} better complied with some earlier suggestions, e.g., made by Johari [58] that the parameter A_{AG} should be pressure dependent. By applying both the T-V versions of the Adam-Gibbs model to describe the experimental dependences τ(T,V) of glass-forming liquids that obey the PDS law (Eq. 1), it has turned out that the model with $A_{AG} = const$ improperly predicted the pressure dependences of the isobaric and isochoric fragilities, which both have been to increase with increasing pressure. However, the Adam-Gibbs model with $A_{AG} \sim \rho^\gamma$ properly reproduced that $m_V = const$ and m_p decrease with increasing *p*. When we thoroughly analyzed the configuration entropy $S_c(T,V)$ parametrized within this model with $A_{AG} \sim \rho^\gamma$, we have found that $S_c(T,V)$ obeys the PDS law (Eq. 1) with the same value of the scaling exponent γ that scales the structural relaxation times τ, but the scaling functions for $\tau$and S_c are different, i.e., $\tau = f_\tau(\rho^\gamma/T)$ and $S_c = f_{S_c}(\rho^\gamma/T)$. It should be stressed that this important finding has been achieved without making any limiting assumptions for the configurational entropy calculations, while some earlier attempts [59–61] to give evidence for the density scaling of both S_c and τ with the same scaling exponent γ suggested an alternative way to calculate S_c, which required determining reference values of S_c along a chosen isochrone. It is also worth noting that our results based on experimental data are in accord with the predictions [62, 63] concluded from the random first-order transition theory and the analyses made [64, 65] by using simulation data in a simple model based on the Lennard-Jones potential. Since both S_c and τ can be scaled with the same value γ, it should be possible to find a single variable function $\tau = g(S_c)$, which have been indeed achieved for the material examined by us in Ref. [57]. In Fig. 2, the comparative analysis of the density scaling of τ and S_c is demonstrated.

We have shown that a decrease in the configurational entropy straightforwardly causes the dramatic slowdown in the molecular dynamics reflected in the rapid increase in its timescale τ near the glass transition. Thus, the configurational entropy S_c seems to be sufficient to govern the structural relaxation of supercooled liquids without any contributions from additional factors. Nevertheless, the configurational entropy S_c should be estimated as the difference between the entropy of the melt and the vibrational contribution to the entropy (rather from the glass than from the crystal) [66]. It makes the configurational entropy impossible to determine from experimental data in any direct way, and we need to exploit some indirect methods to estimate S_c, especially at elevated pressure. Therefore, models based on the total system entropy S (such as different versions of the Avramov model) have not depreciated, giving a better opportunity to verify their predictions about both molecular dynamics and thermodynamics of glass-forming liquids.

The Avramov model has been gradually developed, starting from its temperature version [67–71] via two representations in the temperature-pressure domain (the original one [72] that predicted a constant isobaric fragility in isochronal conditions

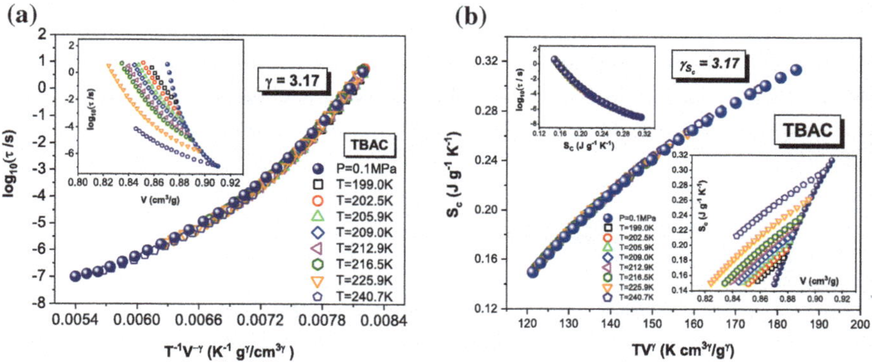

Fig. 2 Illustration of the comparative analysis of the density scaling of the structural relaxation time τ and the configurational entropy S_c for TBAC (tributyl-2-acetylcitrate) in the supercooled liquid state. **a** Plot of the density scaling of the structural relaxation times, which has been performed after transforming the dielectric data from the *T*-*p* domain to the *T*-*V* one and finding the scaling exponent γ based on Eq. (3) with $\rho = V^{-1}$. The isobaric and isothermal dielectric relaxation times determined from broadband dielectric measurements at ambient and elevated pressure, which have been scaled according to Eq. (1), are shown in the inset in the volume domain after employing *pVT* experimental data $V = V(T, p)$. **b** Plot of the density scaling of the corresponding configurational entropies shown in the lower inset as a function of volume, which have been calculated by using the heat capacity measured in the liquid and glassy states at ambient pressure and applying thermodynamic Maxwell's rules possible to be employed due to possessing *pVT* experimental data. The density scaling exponent γ_{S_c} has been determined based on Eq. (3) with $\rho = V^{-1}$ by replacing τ with S_c. Since $\gamma = \gamma_{S_c}$, the plot of the structural relaxation time as a single variable function of the configurational entropy is possible as shown in the upper inset

and the modified one [73] that removed this flaw to reflect the non-invariant pressure behavior of m_p) and ending with its temperature-volume version [74, 75]. In each form of the Avramov model, the total system entropy S has been employed, but its different thermodynamic representations have been implemented. Moreover, in contrast to the MYEGA model, the Avramov model does not invoke the energy landscape concept and the structural relaxation time is assumed to be inversely proportional to an average frequency, $<\nu>\sim\exp(-E_{max}/\sigma)$ of the jump over the energy barrier for the structural relaxation, which is characterized by an energy barrier distribution with its dispersion σ dependent only on the total system entropy S and its maximal energy barrier E_{max} originally [67–74] treated as a material constant independent of thermodynamic conditions. At their fundamentals, the Avramov model describes the structural relaxation time τ as a single variable function of the total system entropy S, $\tau = \tau_0 \exp\left[E\text{max}\sigma_r^{-1} \exp\left(-2(S - S_r)Z^{-1}R^{-1}\right)\right]$, where S_r and σ_r are the total system entropy and the dispersion of the energy barrier distribution in a reference state, Z denotes the number of available pathways for local motions of a molecule or polymer segment, and R is the gas constant.

To formulate the T-V version of the Avramov model, Casalini et al. [74] considered the total system entropy in the known thermodynamic representation expressed in the T-V domain as follows: $S(T, V) = S_r + \int_{T_r}^{T} C_V(T, V_r)d\ln T + \int_{V_r}^{V} (\partial p(T, V)/\partial T)_V dV$, where C_V is the isochoric heat capacity, and T_r and V_r are temperature and volume in a reference state, which is usually assumed at the glass transition at ambient pressure. Using the expression $(\partial p(T, V)/\partial T)_V = \left(C_p - C_V\right)T^{-1}V^{-1}\alpha_p^{-1}$ on the assumption that the isobaric and isochoric heat capacities C_p and C_V as well as the isobaric thermal volume expansivity α_p are constant near the glass transition, the total system entropy can be expressed in a scaling form [20, 74],

$$S(T, V) = S_r + C_V \ln\left(\frac{T V^{\gamma_S}}{T_r V_r^{\gamma_S}}\right) \tag{14}$$

where γ_S is the Grüneisen parameter defined thermodynamically as $\gamma_G \equiv V\alpha_P C_V^{-1}\kappa_T^{-1}$, where κ_T is the isothermal compressibility. We can expect a constant value, $\gamma_S = \gamma_G$, due to another thermodynamic relation, $(\partial p(T, V)/\partial T)_V = \left(C_p - C_V\right)T^{-1}V^{-1}\alpha_p^{-1} = \gamma_G C_V V^{-1}$, which has been assumed at constant values of C_p, C_V, and α_p near the glass transition. Casalini et al. [74] applied Eq. (14) to the general Avramov equation for the structural relaxation time, which involves the difference $S - S_r$ and the maximal energy barrier E_{max} for the structural relaxation, keeping the original assumption that E_{max} is a material constant independent of thermodynamic conditions. Consequently, the authors suggested the following T-V version of the Avramov model, which is expressed below in the T-ρ domain:

$$\tau = \tau_0 \exp[(A\rho^{\gamma}/T)^D] \tag{15}$$

where A and D have their physical meaning given by the expressions, $A = (E\max/\sigma_r)^{1/D} T_r/\rho_r^{\gamma}$ and $D = 2C_V Z^{-1} R^{-1}$, but they are actually fitting parameters like τ_0 and γ.

The T-V version of the Avramov model was the first convenient tool commonly used in case of many glass-forming materials to find their proper value of the scaling exponent γ as a fitting parameter to Eq. (15) [3]. Nevertheless, because of the origin of Eq. (15) based on Eq. (14), this version of the Avramov model has been fraught with a fundamental flaw [20, 75] followed from the relation $\gamma_S = \gamma_G$, which implies that also the scaling exponent valid for structural relaxation times should satisfy the relation $\gamma = \gamma_G$. This problem has been already realized by the authors of Eq. (15), who found that the scaling exponent γ for real materials is considerably larger than the Grüneisen parameter γ_G calculated near the glass transition from its thermodynamic definition (involving V, C_V, α_p, and κ_T) as well as from its vibrational definition [74]. The authors of Eq. (15) even showed that the total system entropy of a van der Waals liquid can be scaled in terms of Eq. (14) with $\gamma_S \approx \gamma_G$ [59], but they tried to remove the discrepancy $\gamma > \gamma_G$ by considering some kind of the configurational version of the Grüneisen parameter [74], which was at odds with the total system entropy basis of the Avramov model. An illustration of the difference in the values of the scaling exponents γ and γ_S is presented in Fig. 3.

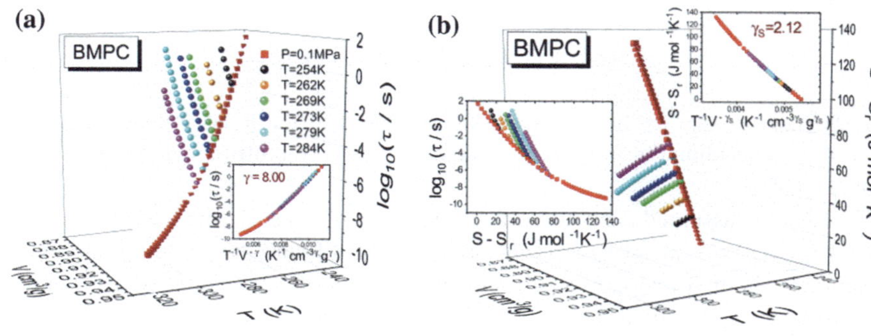

Fig. 3 Illustration of the comparative analysis of the density scaling of the structural relaxation time τ and the total system entropy S for BMPC (1,1′-bis(p-methoxyphenyl)cyclohexane) in the supercooled liquid state. **a** Plot of the isobaric and isothermal structural relaxation times measured by means of broadband dielectric spectroscopy, which have been transformed from the T-p domain to the T-V one by employing the pVT measurement data. The inset shows the density scaling of the structural relaxation times τ according to Eq. (1) with the scaling exponent γ determined based on Eq. (3) with $\rho = V^{-1}$. **b** Plot of the corresponding total system entropies as a function of temperature and volume, which have been calculated by using the heat capacity measured in the liquid state at ambient pressure and applying thermodynamic Maxwell's rules possible to be employed due to possessing pVT experimental data. As the reference state value S_r, the total system entropy at the glass transition at ambient pressure has been assumed. The right inset shows the density scaling of the total system entropy with the density scaling exponent γ_S, which has been determined based on Eq. (3) with $\rho = V^{-1}$ by replacing τ with S. In fact, $\gamma_S \approx \gamma_G$, because $\gamma_G \approx 2.26$ for this material. Since $\gamma > \gamma_S$, the plot of the structural relaxation time as a single variable function of the configurational entropy is impossible as shown in the left inset

We have suggested another solution of the aforementioned problem [75, 76], which is relied on the assumption that the maximal energy barrier E_{max} for the structural relaxation in the Avramov model can be scaled with density, $E\text{max} = A_E(\rho/\rho_r)^{\gamma_E}$, which is reasonable and comparable with the assumptions made later to formulate the aforementioned *T-V* versions of the Adam-Gibbs model. It has been turned out that the density modification of E_{max} in the Avramov model does not change the final *T-V* form of this model given by Eq. (15), and only the physical meaning of its fitting parameter A is changed, because then $A = (A_E/\sigma_r)^{1/D} T_r/\rho_r^{\gamma}$ [75], where $A_E = E\text{max}(\rho_r)$ is the maximal energy barrier in a chosen reference state, and a possible nature of the scaling exponent for the structural relaxation is revealed by the following relation [75, 76],

$$\gamma = \gamma_G + \gamma_E/D \tag{16}$$

In context of Eq. (16), the analytical equations earlier derived [74] from the *T-V* version of the Avramov model for the isobaric and isochoric fragilities, m_p and m_V, should be revised to the following formulae:

$$m_p = \left(1 + \gamma T_g \alpha_p(T_g)\right) D(E\text{max}(\rho_r)/\sigma_r)/\ln 10 \tag{17}$$

and the isochoric fragility:

$$m_V = D(E\text{max}(\rho_r)/\sigma_r)/\ln 10 \tag{18}$$

where $E\text{max}(\rho_r)/\sigma_r = \ln(\tau_r/\tau_0)$ can be estimated at the glass transition, i.e., at $\tau_r = \tau_g$ to determine the fragility parameters by using the fitting parameters D and A of Eq. (15).

Besides the assumed capability of the exponent γ_E to scale the maximal energy barrier $E_{max}(\rho)$ for the structural relaxation, we have found its two important interpretations. The first one is related to a new equation of state (EOS) derived in the PDS regime [77], which is discussed in the next section, and the second one is straightforwardly related to the density scaling of the total system entropy S with a different value of the scaling exponent γ_S from the scaling exponent γ that scales the structural relaxation times γ [20]. In this second analysis, we have shown that the dependence of the structural relaxation time on the total system entropy should be supplemented with a density factor-dependent of the scaling exponents γ and γ_S, i.e., $\tau = h_S\left(\rho^{\Delta\gamma} w(S)\right)$, where $\Delta\gamma = \gamma - \gamma_S$ cannot be reduced to zero, if the PDS laws for the structural relaxation time $\tau = f_\tau(\rho^\gamma/T)$ and the total system entropy $S = f_S(\rho^{\gamma_S}/T)$ are satisfied by different values of the scaling exponents $\gamma \neq \gamma_S$ for a given material. By using dielectric, heat capacity, and volumetric experimental data, we have verified that the total system entropy S can be scaled according to Eq. (14) with the scaling exponent γ_S equals the Grüneisen parameter γ_G to a good approximation ($\gamma_S \approx \gamma_G$) for each tested material and the structural relaxation times τ can be scaled according to Eq. (15) with the scaling exponent γ considerably larger than γ_S. Thus, Eq. (16) is a consequence of the decoupling established between τ

and S, the measure of which can be the difference $\Delta\gamma = \gamma - \gamma_S$, where $\gamma > \gamma_S$. This is because the general dependence, $\tau = h_S\left(\rho^{\Delta\gamma} w(S)\right)$, can be expressed in terms of the T-V version of the Avramov model as $\tau = \tau_0 \exp\left[\left(\rho^{\Delta\gamma} w(S)\right)^D\right]$, where $w(S) = \exp(-(S - S_r)/C_V)$ and $\Delta\gamma = \gamma_E/D$, if we properly remain the different values of the scaling exponents γ and γ_S for a given material. The relation, $\gamma_E = D\Delta\gamma$, has been also verified using experimental data that could be successfully scale according to the function, $\tau = \tau_0 \exp\left[\rho^{\gamma_E} w^D(S)\right]$. As the experimental data analyses yielded $\Delta\gamma > 0$ and $D > 1$ near the glass transition, one can for instance deduce that the structural relaxation time increases with increasing density in isoentropic conditions at which $S = const$ [20]. This is in contrast to the behavior of the timescale of molecular dynamics in the isoentropic conditions at which $S_c = const$, because then also $\tau = const$, and there is no decoupling observed between the structural relaxation time τ and the configurational entropy S_c as thoroughly discussed in the earlier part of this section [57].

It is worth noting that both the MYEGA and Avramov models formulated in the T-V domain can be reduced to Eq. (10), which has been very recently established to be valid near the melting point instead of the typical Arrhenius activation model. That is Eq. (10) is a limiting case of Eq. (11) if the parameter $A_M(x)$ tends to zero as well as Eq. (15) if its parameter $D \to 1$. Recent experimental data analyses have shown [50, 78] that both the MYEGA and the Avramov models are able to reproduce the thermodynamic evolution of the timescale of molecular dynamics from the melting to the glass transition with the same value of the scaling exponent γ, but their other parameters may require some changes in such a wide range of molecular dynamics timescale.

4 Density Scaling-based Equations of State

In the previous sections, we have only briefly mentioned that the density scaling of experimental isobars and isotherms requires pVT measurements, which should be parametrized by an equation of state (EOS) to conveniently and accurately determine the temperature-pressure dependences of volume or density in the supercooled liquid state. Commonly, in such analyses, an old EOS proposed by Tait [79] in 1888 has been used. However, from the cognitive point of view, it was highly expected to formulate an EOS relied on the theoretical grounds underlying the PDS law (Eq. 1). We have responded to this tempting challenge that should provide a linkage between the molecular dynamics and the thermodynamics of glass-forming liquids based on the same effective short-range intermolecular potential (Eq. 4).

In series of papers, we have discovered subsequent properties of the sought after EOS. For simplicity, we have first investigated an isothermal EOS [80], which describes the configurational pressure, $p^{conf} =< W > /V$, calculated in a manner suggested by Bardik and Shakun [81] from the average total system viral $< W >= -(1/3)(\sum_i \mathbf{r}_i \cdot \nabla_i U)$ by applying the Euler theorem on homogeneous functions to the effective potential (Eq. 4) employed in $< W >$, and then considering the first-order Taylor series expansion of $<W>$, which resulted in the approximate proportionality, $< W >\sim \rho^{\gamma_{EOS}}$, where $\gamma_{EOS} = m_{IPL}/3$ is straightforwardly related to the exponent m_{IPL} of the dominant repulsive part of Eq. (4). Then, we have arrived at a working EOS, $p^{conf}(V, T) \approx p^{conf}(V_0, T) + B(T)((V_0/V)^{\gamma_{EOS}} - 1)$, where V_0 is the reference volume at a given T, which should be valid in the low compressibility region. Thus, it could have been expected that $\gamma_{EOS} \approx \gamma$, where the scaling exponent γ could have scaled the structural relaxation times according to Eq. (1). However, our tests exploiting experimental data did not confirm this hypothesis [80]. We found that γ_{EOS} is considerably larger than γ that scales τ following the PDS law. To better explore this intriguing issue, we conducted further investigations. By solving the differential equation, $B_T^{conf} = -\gamma_{EOS}\left(\partial p^{conf}/\partial \ln V^{\gamma_{EOS}}\right)_T$, which originates from the definition of the configurational isothermal bulk modulus $B_T^{conf} \equiv -\left(\partial p^{conf}/\partial \ln V\right)_T$, and considering the first-order Taylor series expansion of the solution about $p^{conf} = p_0^{conf}$, where $p_0^{conf} = p^{conf}(V_0, T)$ at a given T, we have found the physical meaning of the temperature -dependent parameter $B(T)$ of the first EOS, which is given by the expression $B(T) = B_T^{conf}(p_0^{conf})/\gamma_{EOS}$. In this way, the isothermal EOS describing the volume V by the configurational quantities was expressed,

$$(V_0/V)^{\gamma_{EOS}} = 1 + \gamma_{EOS}(p^{conf} - p_0^{conf})/B_T^{conf}(p_0^{conf}) \tag{19}$$

By following a similar procedure based on the first-order Taylor series expansion (about $p = p_0$) of the solution of the differential equation, $B_T = -\gamma_{EOS}(\partial p/\partial \ln V^{\gamma_{EOS}})_T$, which exploits a mathematical trick suggested by Papathanassiou [82] to incorporate the exponent γ_{EOS} into the definition of the isothermal bulk modulus, $B_T \equiv -(\partial p/\partial \ln V)_T$, we arrived at an analogous isothermal EOS describing the volume V by nonconfigurational quantities [77],

$$(V_0/V)^{\gamma_{EOS}} = 1 + \gamma_{EOS}(p - p_0)/B_T(p_0), \tag{20}$$

where the reference volume $V_0 = V(T, p_0)$ is usually chosen at ambient pressure p_0.

We have applied both Eqs. (19) and (20) to successfully describe isothermal *pVT* experimental data, showing that values of the exponents γ_{EOS} obtained from the EOSs are almost the same for a given material. We have confirmed that not only Eq. (19) but also Eq. (20) can be interpreted in terms of the effective short-range intermolecular potential (Eq. 4), expecting that $\gamma_{EOS} = m_{IPL}/3$. Although we have found that γ_{EOS} enables a linear scaling of volumetric data [77], the values of the scaling exponent γ_{EOS} have been considerably larger than those of γ scaling the structural relaxation times according to Eq. (1). Then, we formulated a hypothesis that $\gamma_{EOS} \approx \gamma_E$, where γ_E has been assumed to scale the maximal energy barrier E_{max} for the structural relaxation with density in the Avramov model, which has been discussed in the previous section. This hypothesis has been positively validated for both the measurement and the simulation data by verifying the relation given by Eq. (16). In case of experimental data, we focused on glass-forming liquids, the molecular dynamics of which obeys the PDS law (Eq. 1), and additionally we established the value of the parameter D by fitting the experimental data to Eq. (15) near the glass transition (for instance at $\tau > \tau_B$, where τ_B denotes the timescale of the dynamic crossover suggested by Stickel [83]) [75, 77, 84], but we also investigated representatives of other material groups (such as polymer melts and supercooled ionic liquids) [76, 84]. Thus, the scaling exponent γ_{EOS} was related not only to γ_E, but also to γ_S, which scales the total system entropy of real materials (as argued in the previous section). It should be stressed that a successful result has been surprisingly obtained also from the test of Eq. (16) for the relation $\gamma_E \approx \gamma_{EOS}$. Although $\gamma_{EOS} \approx \gamma$ in case of simple models exploiting the Lennard-Jones potential, which firmly confirms the same grounds based on the effective short-range intermolecular potential (Eq. 4) for the EOSs given by Eqs. (19) and (20) as well as for the PDS law (Eq. 1), the relation expressed by Eq. (16) has been satisfied by simulation data in these models when $\gamma_E = \gamma_{EOS}$ [85].

Due to finding a proper physical interpretation of the scaling exponent γ_{EOS}, it was reasonable to develop the EOSs not limited to isothermal conditions [84], which originated from Eqs. (19) and (20), but enabled a convenient fitting pVT data. Such EOSs have been also widely exploited in the aforementioned and further successful tests of Eq. (16) as more convenient tools to describe the volumetric data than the isothermal EOSs. To formulate them, two temperature parameterization functions have been used, which are presented here in case of EOS based on Eq. (20) for nonconfigurational quantities. At the reference pressure p_0 (usually considered as ambient pressure), we have assumed [84] the exponential temperature function of the isothermal bulk modulus, $B_T(p_0) = B_{T_0}(p_0)\exp(-b_2(T - T_0))$, based on the analytical approximation of the temperature dependence of B_T earlier suggested by Garai and Laugier [86], and the commonly exploited quadratic temperature function of the volume, $V_0 = V(T, p_0) = A_0 + A_1(T - T_0) + A_2(T - T_0)^2$. Then, we have arrived at the following EOS based on Eq. (20),

$$V(T, p) = \frac{A_0 + A_1(T - T_0) + A_2(T - T_0)^2}{\left[1 + \gamma_{EOS}(p - p_0)\exp(b_2(T - T_0)/B_{T_0}(p_0)\right]^{1/\gamma_{EOS}}} \tag{21}$$

with the fitting parameters: $\gamma_{EOS}, A_0 = V_0(T_0) = V(p_0, T_0), A_1 = \partial V(p_0, T)/\partial T|_{T=T_0}$, $A_2 = (1/2)\partial^2 V(p_0, T)/\partial T^2|_{T=T_0}$, $B_{T_0}(p_0)$, and $b_2 = -\partial \ln B_T(p_0, T)/\partial T|_{T=T_0}$, where T_0 is fixed usually at the glass transition temperature at ambient pressure, $T_0 = T_g(p_0)$. Thus, the applicability range of Eq. (21), which was successfully tested [84] in case of supercooled liquids, is limited to the discussed temperature parameterization functions and the low compressibility region assumed to derive Eqs. (19) and (20), which implies the linear pressure dependences of the isothermal bulk modulus considered, respectively, in the configurational and nonconfigurational manner. It means that Eq. (20) and also Eq. (21) satisfy the following dependence of the isothermal bulk modulus on pressure,

$$B_T(p) = B_T(p_0) + \gamma_{EOS}(p - p_0) \tag{22}$$

where $\gamma_{EOS} = \partial B_T(p)/\partial p|_{p=p_0}$. The dependence given by Eq. (22) is valid in the supercooled liquid state and any constant function $B_T(p)$ should not approximate this dependence $B_T(p)$ [87]. Although the EOSs given by Eq. (20) or Eq. (21) have been derived for supercooled liquids, they can be applied to describe volumetric data in other thermodynamic states, for instance in standard liquids [88] and liquid crystals [56], but mostly different values of the fitting parameters and/or the fixed reference temperature and pressure are needed in different thermodynamic phases [84].

By analogy with Eqs. (20) and (21), we have additionally formulated an EOS [89] that describes the *T*-*p* dependences of the activation volume, $V_T^{act} = RT(\partial \ln \tau/\partial p)_T$, earlier discussed in Sect. 2 as the activation parameter, which is a relevant measure of the molecular dynamic sensitivity of a system to changes in pressure near the glass transition in isothermal conditions, but also having a volumetric character. Using the temperature parametrization functions $V_T^{act}(T, p_0) = F_0 + F_1(T - T_0) + F_2(T - T_0)^2$ and $B_T^{act}(T, p_0) = B_T^{act}(T_0, p_0)\exp(-g_2(T - T_0))$ at a reference pressure p_0 for and the activation volume and the isothermal bulk modulus for the activation volume defined as $B_T^{act} \equiv -(\partial p/\ln V_T^{act})_T$, we have derived the following EOS,

$$V_T^{act}(T, p) = \frac{F_0 + F_1(T - T_0) + F_2(T - T_0)^2}{\left[1 + \gamma_{act}(p - p_0)\exp(g_2(T - T_0)/B_T^{act}(T_0, p_0)\right]^{1/\gamma_{act}}} \tag{23}$$

with the fitting parameters: γ_{act}, $F_0 = V_T^{act}(p_0, T_0)$, $F_1 = \partial V_T^{act}(p_0, T)/\partial T|_{T=T_0}$, $F_2 = (1/2)\partial^2 V_T^{act}(p_0, T)/\partial T^2|_{T=T_0}$, $B_T^{act}(T_0, p_0)$, and $g_2 = -\partial \ln B_T^{act}(p_0, T)/\partial T|_{T=T_0}$, where T_0 is fixed usually at the glass transition temperature at ambient pressure, $T_0 = T_g(p_0)$.

Similar to Eq. (22), a consequence of Eq. (23) is the dependence of the isothermal bulk modulus for the activation volume on pressure,

$$B_T^{act}(p) = B_T^{act}(p_0) + \gamma_{act}(p - p_0) \tag{24}$$

where $\gamma_{act} = \partial B_T^{act}(p)/\partial p|_{p=p_0}$. An advantage of Eq. (23) is also that it enables us to investigate not only isothermal, but also isobaric dependences of the activation volume. From this kind of analyses, we have found that the dependences $B_T^{act}(T, p)$ increase with increasing temperature at a constant pressure for real glass formers, whereas the dependences $B_T^{act}(T, p)$ determined from simulation data collected in the prototypical model of supercooled liquids (which is the Kob-Andersen binary Lennard-Jones liquid (KABLJ) [90]) decrease with increasing *T* in isobaric conditions [89].

Since the activation volume typically requires collecting a rich set of experimental data measured at elevated pressure to be directly and accurately determined in the *T*-*p* domain, Eq. (23) has turned out to be a useful tool to investigate this activation parameter, for instance in search of correlations between the dynamic heterogeneity and different activation parameters [91], and even in comparison with the pressure effect on molecular dynamics of glass-forming liquids and liquid crystals [56]. What is more, we have established several interesting relations during the development of Eq. (23). Additionally exploiting the Avramov model (Eq. 15), a counterpart of Eq. (16) has been found [89],

$$\gamma_{act} = \gamma_{EOS}/(\gamma_{EOS} - \gamma D), \tag{25}$$

which has been satisfied by all tested real materials and simulation models that obey the PDS law (Eq. 1). Further analyses have also shown that

$$B_T^{act}(T, p_0) = B_T(p_0)/(\gamma_{EOS} - \gamma D). \tag{26}$$

It means that Eq. (24) enables us to predict the pressure dependence of $B_T^{act}(p)$ without any calculations of the activation volume. In case of all tested systems that obey the PDS law (Eq. 1), the pressure effects on the isothermal bulk moduli B_T and B_T^{act} were opposite, i.e., $B_T(p)$ increased, while $B_{act}(p)$ decreased with increasing pressure in isothermal conditions. It has been rationalized by the negative values $\gamma_{act} < 0$ if we take into account the values of the other exponentsγ, γ_{EOS} and D, including $\gamma > 0$. Based on the analytical derivations, two important relationships, $V_T^{act}(T, p_0)/V_T^{act}(T, p) = (V(T, p_0)/V(T, p))^{\gamma_{EOS}-\gamma D}$ and $V_T^{act}(T, p_0)/V_T^{act}(T, p) = (B_T(p_0)/B_T(p))^{-(\gamma_{EOS}-\gamma D)/\gamma_{EOS}}$, have been exhibited. These findings enable to distinguish the strong volumetric contribution to the activation volume from the dynamic hallmark that is reflected in their exponents, in which the parameter γ plays an important role as the exponent of the PDS law for molecular dynamics near the glass transition.

It should be noted that the relations given by Eqs. (25) and (26) exploit the Avramov model (Eq. 15). For further verification and broader opportunities, their counterparts invoking the MYEGA model (Eq. 11) have been also formulated [56]

$$\gamma_{act} = \frac{\gamma_{EOS}}{\gamma_{EOS} - \gamma[1 + A_M\Gamma_0(2 + A_M\Gamma_0)/(1 + A_M\Gamma_0)]} \tag{27}$$

$$B_T^{act}(p_0) = \frac{B_T(p_0)}{\gamma_{EOS} - \gamma[1 + A_M\Gamma_0(2 + A_M\Gamma_0)/(1 + A_M\Gamma_0)]} \tag{28}$$

where $\Gamma_0 = T^{-1}V^{-\gamma}(T, p_0)$ is constant at $T = const$. Since one can expect similarly to γ_{EOS} that $\gamma_{act} = \left(\partial B_T^{act}(p)/\partial p\right)_T \cong const$ at different temperatures, the value of the parameter Γ_0 should remain unchanged for a given material at least to a good approximation. The validity of Eqs. (27) and (28) compared to Eqs. (25) and (26) has been confirmed in studies of the activation volume of liquid crystals [56].

5 Dynamic Heterogeneity in the Density Scaling Regime

In 1965, G. Adam and J. H. Gibbs postulated [53] the increasing degree of the dynamic heterogeneity of supercooled liquids as a phenomenon closely related to the rapidly increasing timescale of the primary relaxation process (e.g., the structural relaxation in van der Waals liquids) when the liquid systems are approaching the glassy state. The authors assumed that some cooperative rearrangement regions (CRR) form and grow in the liquids on supercooling to formulate one of the most famous models of the thermodynamic evolution of the timescale of molecular dynamics near the liquid–glass transition. After a half-century, this idea has been considerably developed by applying the formalism of the four-point correlation function to study the dynamic heterogeneity of the systems near the glass transition. The four-point dynamic susceptibility χ_4 has been acknowledged as a relevant measure of the dynamic heterogeneity, because it involves both the temporal and the spatial correlations. In particular, the maximum χ_4^{max} of the function χ_4 has been taken into consideration, which is formally interpreted as a characteristic correlation volume for the structural relaxation [N26], and hence can quantify the degree of the system dynamic heterogeneity. Such an application of the formalism of the four-point correlation function has been broadly propagated, starting from the publication by L. Berthier et al. in Science [92], where an increase in the maximum of the four-point dynamic susceptibility χ_4^{max} has been shown, when a liquid system is approaching the glassy state. Moreover, a few estimates have been suggested to evaluate this quantity, because a direct measurement of the four-point dynamic susceptibility χ_4 is a very difficult task that requires detecting a nonlinear response of the examined sample.

In this context, it is worth noting that such a direct determination of the time-dependent four-point dynamic susceptibility function $\chi_4(t)$ is possible by using simulation data, whereas its measurements are extremely rare and still controversial. Until recently, the measurements of the four-point correlation function have been possible by means of the four-dimensional nuclear magnetic resonance (4D NMR) [93, 94]. However, such measurements have been able to carry out at temperatures that are relatively high in comparison with the glass transition temperature. In this situation, a tempting alternative has been provided by the measurements of the three-point dynamic susceptibility χ_3 performed by means of the nonlinear dielectric spectroscopy at ambient [95, 96] and elevated [97] pressure. However, their significance for a reliable direct experimental determination of the four-point measure of the dynamic heterogeneity is reasonably questioned [98], mainly because the four-point dynamic susceptibility function χ_4 is unreachable by means of the dielectric experiments, which only enable to find an approximate number of dynamically correlated molecules N_{corr}, estimated on the basis of the measured frequency spectrum of the three-point dynamic susceptibility $\chi_3(f)$. For the mentioned reasons, until it is possible to reliably determine experimentally the values of the four-point measure of the dynamic heterogeneity near the liquid–glass transition, it is extremely interesting at least to evaluate them by using the suggested estimates [92]. Such an analysis has been first performed [99] by exploiting experimental data measured only at ambient pressure. A great significance of the dynamic heterogeneity idea for studying physical phenomena that occur near the liquid–glass transition and the opportunity for employing the promising formalism of the four-point correlation function in the analysis of the degree of the system dynamic heterogeneity have inclined us to be carried out in a series of the analyses by means of the estimates of the four-point measure $\chi_4^{\max}$ by using experimental data measured in both the ambient and the elevated hydrostatic pressure conditions in the standard dielectric experiment that enables us to detect a linear response of the examined sample, which provides a two-point correlation function in the time domain, for instance, given by the KWW function described earlier in Sect. 1. The $\chi_4^{\max}$—estimate based on the enthalpy fluctuations, $\chi_4^H = (\beta_{KWW}/e)^2(\partial \ln \tau/\partial \ln T)^2(k_B/\Delta c_p)$, involves just the stretching parameter β_{KWW} of the KWW correlation function, the Boltzmann constant k_B, and the specific isobaric heat capacity c_p, which has been suggested to be replaced with the jump in the specific heat Δc_p at the glass transition temperature T_g in case of real materials at the glass transition [92]. Since measurements of the specific heat capacity at elevated pressure are very difficult, in order to investigate $\chi_4^{\max}$ as a function of T and p, we have evaluated Δc_p also as a function of T and p from the known thermodynamic relation, $\Delta c_p(T, p) = \Delta c_p(T, p_0) - T\int_{p_0}^{p}\left(\partial^2 V/\partial T^2\right)dp$, where a reference, $c_p^{glass}(T_0, p_0) = c_p^{glass}\left(T_g(p_0), p_0\right)$, is chosen in the experimental temperature dependence, $\Delta c_p(T, p_0) = c_p^{liquid}(T, p_0) - c_p^{glass}(T_0, p_0)$ at ambient pressure, and the dependence $V(T,p)$ is taken from pVT measurements usually parameterized by an equation of state [100].

Even the first results of the analysis of the high-pressure effect on the system dynamic heterogeneity near the liquid–glass transition have turned out to be extremely interesting. Exploiting experimental dielectric and volumetric data of selected supercooled van der Waals liquids measured in the ambient and elevated hydrostatic pressure conditions as well as the heat capacity data of the materials measured at ambient pressure, we have given evidence [100] that the degree of the change in the estimated values of the four-point measure of the dynamic heterogeneity $\chi_4^{\max}$ in the supercooled liquid region between two dynamic states defined by two different structural relaxation times (choosing $\tau = 1$ μs and $\tau = 100$ s as examples) depends on the thermodynamic path covered between the dynamic states. We have shown that the isobaric cooling of a liquid causes a larger increase in the estimated values of the dynamic heterogeneity measure $\chi_4^{\max}$ than the isothermal compression of the liquid makes. However, the isochoric cooling of the liquid exerts the strongest effect on the increase in its dynamic heterogeneity. These observations have been confirmed by performing the analysis of the thermodynamic paths along which the opposite experiments occur, such as the isochoric heating, the isobaric heating, and the isothermal pressure lowering, which take the system from the dynamic state defined by the structural relaxation time $\tau = 100$ s to the dynamic state at $\tau = 1$ μs. It should be emphasized that the extremely essential and surprising outcome of the analyses [100] is the decrease in the dynamic heterogeneity with increasing pressure found at a constant structural relaxation time. To gain a better insight into this important finding, achieved [100] by applying the estimate χ_4^H derived [92] from fluctuations of the system enthalpy H to evaluate the four-point dynamic susceptibility χ_4, our further investigations of the effect of pressure on the system dynamic heterogeneity near the liquid–glass transition have been focused on a thorough analysis [101] of the effect of temperature and density fluctuations on the system dynamic heterogeneity, distinguishing the contributions from temperature and density fluctuations by exploiting the separate estimates, χ_4^T and χ_4^ρ, based, respectively, on the fluctuations of temperature T and density ρ, which have been earlier suggested also by Berthier et al. [92]. Employing the estimates χ_4^H, χ_4^T, and χ_4^ρ in the study of supercooled van der Waals liquids, measured in ambient and elevated hydrostatic pressure conditions, we have established [101] qualitatively the same variation of the dynamic heterogeneity as that previously obtained [100] by means of the estimate χ_4^H along the different thermodynamic paths (i.e., isothermal, isobaric, and isochoric paths) for both the contributions following separately from temperature and density fluctuations. As a result, we have also found that both the contributions to the dynamic heterogeneity decrease with increasing pressure at a constant structural relaxation time [101]. We have shown that the proper procedure for applying the estimates χ_4^H and χ_4^T to experimental data of real supercooled liquids requires involving the jumps in the heat capacities, Δc_p and Δc_V, which can be considered as estimates of the configurational isobaric and isochoric heat capacities, where the latter is calculated by analogy with the aforementioned Δc_p but at $V = const$. A similar suggestion earlier made by Berthier et al. in Ref. [92] has not been appreciated by the authors in their later analysis of experimental data measure only at ambient pressure [99]. Taking into consideration the experimental data also measured in the elevated hydrostatic pressure conditions, we have been able to clarify [101] that the expected correspondence between the maximal value of the estimate χ_4^H and the sum of the maximal values of the estimates, χ_4^T and χ_4^ρ, i.e., $\left(\chi_4^H\right)^{\max} \approx \left(\chi_4^T\right)^{\max} + \left(\chi_4^\rho\right)^{\max}$, can be achieved only if we exploit Δc_p in χ_4^H and Δc_V in χ_4^T. We have observed that the estimates, χ_4^H and χ_4^T, with the nonconfigurational values of c_p and c_V lead to an unphysical result, which is a considerably larger contribution to the dynamic heterogeneity from the density fluctuations χ_4^ρ (even nearly the same as that obtained from the enthalpy fluctuations χ_4^H) than that originating from the temperature fluctuations χ_4^T. However, the earlier investigations [2, 29] have shown that the supercooled liquids, which belong to van der Waals liquids, are characterized by the ratio of isochoric and isobaric activation energies for the structural relaxation, which takes the value $E_V^{act}/E_p^{act} \approx 0.5$ near the glass transition temperature at ambient pressure. As already mentioned in Sect. 2, the ratio E_V^{act}/E_p^{act} is a well-grounded measure of the relative role of thermal activation and free volume in molecular dynamics near the liquid–glass transition. Therefore, its value, $E_V^{act}/E_p^{act} \approx 0.5$, reflects the comparable effects of changes in temperature and density on the molecular dynamics of real supercooled van der Waals liquids.

The described details of the analysis carried out by using the estimates of the four-point measure of the dynamic heterogeneity should not obscure two fundamental and extremely essential conclusions reported in Refs. [21, 100, 101], which are as follows:

(i) The width of the dielectric loss spectrum, previously often related to the degree of the heterogeneity of the distribution of structural relaxation times τ and often quantified by the value of the stretching parameter β_{KWW}, does not reflect the degree of the dynamic heterogeneity defined by $\chi_4^{\max}$, because the estimated values of $\chi_4^{\max}$ decrease with increasing pressure at a constant structural relaxation time τ in case of the examined typical supercooled liquids, which are characterized [15, 16] by the isochronal superposition of dielectric loss spectra measured at different temperatures and pressures.

(ii) A given timescale of molecular dynamics near the liquid–glass transition, which is defined by a constant structural relaxation time τ, reflects various length scales of the molecular dynamics, related to the correlation volume for the structural relaxation, which is defined by the four-point measure of the degree of the dynamic heterogeneity $\chi_4^{\max}$, because the estimated values of $\chi_4^{\max}$ vary with changing thermodynamic conditions at $\tau = const.$

The particular significance of the latter conclusion has inclined us to its deepened analysis within the framework of the density scaling law [21] in search of the answer to the intriguing question whether a relationship exists between the ideas of the density scaling and the dynamic heterogeneity quantified by using the formalism of the four-point correlation function, and if it is, which form of the relationship is proper.

It should be stressed that the decoupling found [21, 100, 101] between time and length scales of molecular dynamics near the glass transition is a surprising finding that provides a new important example, which shows that huge cognitive benefits follow from the analysis which takes into account the pressure variable besides the typically considered temperature dependence as it has been previously discussed for other issues in Refs. [2, 3, 29]. Berthier et al. [99, 102], exploiting experimental data measured at ambient pressure and results of computer simulations performer in simple force fields, suggested that the structural relaxation time τ can be expressed as a single variable function of $\chi_4^{\max}$, e.g., $\tau = g_\chi\left(\chi_4^{\max}\right)$, and consequently the characteristic timescale of molecular dynamics, defined as the structural relaxation time τ, should be also a single variable function of $\xi_4^{\max}$, which is the characteristic length scale of molecular dynamics, determined from the correlation volume $\chi_4^{\max}$ for the structural relaxation. The already-mentioned theory of isomorphs also implies [4] that the four-point time-dependent dynamic susceptibility function $\chi_4(t)$ should be isomorph invariant. It means that we should observe a superposition of the dynamic susceptibilities $\chi_4(t)$, determined in different thermodynamic conditions at a constant structural relaxation time $\tilde{\tau}$ expressed in the reduced units of the theory of isomorphs. Thus, according to the theory of isomorphs, $\chi_4^{\max} = const$ at $\tilde{\tau} = const$, where the structural relaxation time $\tilde{\tau}$ is expressed in the reduced units suggested by the isomorph theory (see Sect. 1). There are also other models that lead to the conclusion that $\tau = g_\chi\left(\chi_4^{\max}\right)$, which has been discussed in Ref. [21].

The briefly presented state of the art in the investigations of the characteristic scales of molecular dynamics in the domains of time and space shows that the decoupling found by us between τ and $\chi_4^{\max}$ required a theoretical explanation. A satisfactory attempt at formulating such an explanation has been provided based on the density scaling idea [21]. Making the assumption that both the structural relaxation time (and the segmental relaxation time and the conductivity relaxation time in case of polymers and ionic liquids, respectively) and the correlation volume for the given relaxation, defined as the maximum of the four-point dynamic susceptibility $\chi_4^{\max}$, obey the power law density scaling laws, $\tau = f(\rho^\gamma/T)$ and $\chi_4^{\max} = w(\rho^{\gamma_\chi}/T)$, which permits in general different values of the scaling exponents, $\gamma_\chi \neq \gamma$, we have proven that the dependence between τ and $\chi_4^{\max}$ requires supplementing with an additional density-dependent factor $u_\chi(\rho)$ according to the following rule, $\tau = g_\chi\left(u_\chi(\rho)w^{-1}\left(\chi_4^{\max}\right)\right)$, where w^{-1} is the inverse function of the function w. We have also shown that such an assumption implies a power form of the density-dependent factor, $u_\chi(\rho) = \rho^{\Delta\gamma}$, the exponent of which, $\Delta\gamma = \gamma - \gamma_\chi$ defined by the difference between the scaling exponents for the quantities τ and $\chi_4^{\max}$, can be regarded as a measure of the decoupling between these quantities.

The results of the theoretical discussion on the density scaling of τ and $\chi_4^{\max}$ have been confirmed [21] by analyzing ambient and high-pressure experimental data of glass formers that belong to two different material groups (i.e., van der Waals liquids and polymer melts, and additionally ionic liquids unpublished yet), the molecular dynamics of which well satisfies the power law density scaling of the timescale $\tau = f(\rho^\gamma/T)$. We have confirmed the power law density scaling of the degree of the dynamic heterogeneity, $\chi_4^{\max} = w(\rho^{\gamma_\chi}/T)$, employing the estimate of $\chi_4^{\max}$ based on the enthalpy fluctuations and finding a considerable difference in the values of the scaling exponents $\gamma_\chi \neq \gamma$. For the examined materials, we have established that $\gamma_\chi < \gamma$, which explains the decrease in $\chi_4^{\max}$ with increasing pressure at $\tau = const$. Qualitatively the same result has been obtained for the structural (or segmental) relaxation time $\tilde{\tau}$ expressed in the reduced units of the theory of isomorphs. Moreover, we have suggested a phenomenological form of the function, $\tau = g_\chi\left(u_\chi(\rho)w^{-1}\left(\chi_4^{\max}\right)\right)$, which successfully describes the relation between τ and $\chi_4^{\max}$, based on our prior finding that the functions f and w that scale, respectively, the timescale of molecular dynamics of real supercooled liquids and their degree of the dynamic heterogeneity, $\tau = f(\rho^\gamma/T)$ and $\chi_4^{\max} = w(\rho^{\gamma_\chi}/T)$, are morphologically different. The temperature-density version of the Avramov model $\left(\tau = \tau_0 \exp\left[(A\rho^\gamma/T)^D\right]\right.$ as shown in Eq. (15) successfully applied by us to the temperature-density dependences of τ the structural relaxation time has not been able to describe the corresponding dependences of $\chi_4^{\max}$, but they have been satisfactorily fitted to a nonexponential formula,

$$\chi_4^{\max} = (\chi_4^{\max})_0 (A_\chi \rho^{\gamma_\chi}/T)^{D_\chi}, \tag{29}$$

where $(\chi_4^{\max})_0$, A_χ, D_χ, and γ_χ are fitting parameters, the number of which can be diminished by introducing the auxiliary parameter $b_\chi = (\chi_4^{\max})_0^{1/D_\chi} A_\chi$. Then, we have been able to construct the phenomenological forms of the function g_χ and its inverse function g_χ^{-1} as follows:

$$\tau = \tau_0 \exp\left[\left(\rho^{\Delta\gamma}\left(\chi_4^{\max}/(\chi_4^{\max})_0\right)^{1/D_\chi} A/A_\chi\right)^D\right] \quad (30)$$

$$\chi_4^{\max} = (\chi_4^{\max})_0\left(\rho^{-\Delta\gamma}(\ln(\tau/\tau_0))^{1/D} A_\chi/A\right)^{D_\chi} \quad (31)$$

It is worth noting that the assumed form of Eq. (29) has been chosen to extend a logarithmic relation between τ and $\chi_4^{\max}$ earlier suggested [99, 102] by using experimental data at ambient pressure near the glass transition, i.e., Eq. (29) combined with Eq. (15) result in Eq. (31), which is such a logarithmic form supplemented with the additional density-dependent factor $\rho^{-\Delta\gamma}$, where its exponent $\Delta\gamma = \gamma - \gamma_\chi$. We have applied Eq. (29) to fit the degrees of the dynamic heterogeneity evaluated at each (T,p) at which dielectric measurements have been carried out for a given material. As a result, we have found that the fits of $\chi_4^{\max}$ to Eq. (29) are of a very high quality. Then, the scaling $\chi_4^{\max}(\rho^{\gamma_\chi}/T)$ for real glass formers is achieved with $\gamma_\chi < \gamma$. It means that the density factor $\rho^{\Delta\gamma}$ in Eq. 1 cannot be reduced to unity, because $\Delta\gamma > 0$. It should be noted that if we evaluate $\chi_4^{\max}$ and perform the scaling in terms of Eqs. (15) and (29) using τ in reduced units of the isomorph theory, the established positive values of $\Delta\gamma$ remain unchanged. Consequently, Eqs. (30) and (31) are able to properly reproduce the decrease in $\chi_4^{\max}(p)$ at $\tau = const$. The comparative analysis of the density scaling of the structural relaxation time τ and the four-point measure of the dynamic heterogeneity $\chi_4^{\max}$ evaluated by its estimate χ_4^H based on enthalpy fluctuations (described at the beginning of this section) is demonstrated in Fig. 4.

We have also performed molecular dynamics simulations in the NVT statistical ensemble in a relatively wide range of temperatures and densities for the sufficiently large system of 8000 particles in the KABLJ model to determine the four-point measure of the dynamic heterogeneity of the prototypical model of supercooled liquid as the variance of the fluctuations incoherent of the intermediate self-scattering function, $F_s(\vec{k},t) = N^{-1}\sum_i < \cos\vec{k}\cdot[\vec{r}_i(t) - \vec{r}_i(0)] >$, i.e., from the expression, $\chi_4(t) = N[< (N^{-1}\sum_i \cos\vec{k}\cdot[\vec{r}_i(t) - \vec{r}_i(0)])^2 > - < N^{-1}\sum_i \cos\vec{k}\cdot[\vec{r}_i(t) - \vec{r}_i(0)] >^2]$, where the brackets $< >$ denote the ensemble average, N is the number of particles (species), $\vec{r}_i$ is the position vector of the ith particle, and $\vec{k}$ is the wave vector calculated at the wave at the first peak of the AA structure factor between particles of the type A at each simulation state (T,ρ) separately, where A denotes the particle specie that constitutes 80% of the particle content of the binary mixture in the KABLJ model. This is a typical way used [14] to establish $\vec{k}$ in the KABLJ model, and then calculate $F_s(\vec{k},t)$, which is usually exploited to determine the structural relaxation time $\tau = t$ at $F_s(\vec{k},t) = e^{-1}$. In comparison with the findings based on the high-pressure experimental data, the analysis of the simulation data in the KABLJ model [90] has yielded different results [21]. Although we have observed some decoupling between τ and $\chi_4^{\max}$, i.e., a slight increase in $\chi_4^{\max}$ with increasing density at $\tau = const$, when the quantities have been expressed in the typical Lennard-Jones potential units, the application of the reduced units of the isomorph theory to the

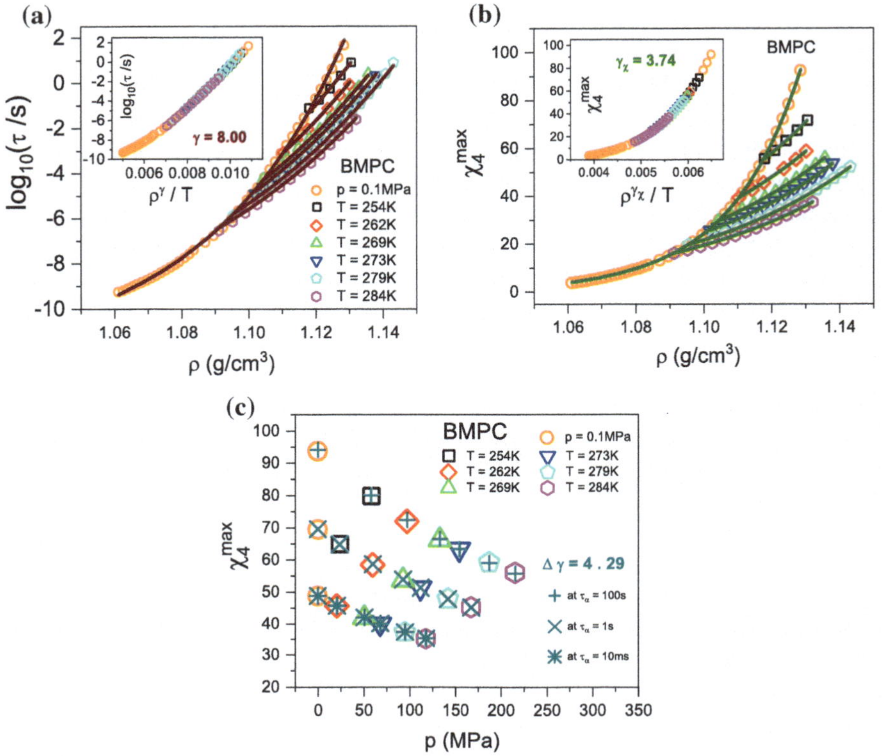

Fig. 4 Illustration of the comparative analysis of the density scaling of the structural relaxation times τ and the estimated values of the four-point measure of the dynamic heterogeneity $\chi_4^{\max}$ (evaluated by its estimate χ_4^H based on enthalpy fluctuations) for BMPC in the supercooled liquid state. **a** Plot of the isobaric and isothermal structural relaxation times as a function of density (the same data shown as a function of temperature and volume in Fig. 3a). The solid lines denote fitting curves to the T-ρ version of the Avramov model, i.e., to Eq. (15). The inset shows the density scaling of the structural relaxation times τ with the scaling exponent γ determined from fitting dielectric data to Eq. (15). **b** Plot of the corresponding estimated values of the four-point measure of the dynamic heterogeneity as a function of density. The solid lines represent fitting curves to Eq. (29). The inset shows the density scaling of the estimated values of the four-point measure of the dynamic heterogeneity $\chi_4^{\max}$ with the scaling exponent γ_χ determined from fitting the estimated values of $\chi_4^{\max}$ to Eq. (29). By comparing the analysis presented in Fig. 3, we can see that $\gamma > \gamma_\chi > \gamma_S$. This pattern of relations between the scaling exponents γ, γ_χ, and γ_S is typical for real materials. **c** Plot of three isochronal pressure dependences of the estimated values of $\chi_4^{\max}$ at selected constant structural relaxation times. The symbols +, ×, * denote the values generated from Eq. (31) at a given structural relaxation time τ. A measure of the isochronal decoupling between τ and $\chi_4^{\max}$ is the difference $\Delta\gamma = \gamma - \gamma_\chi$. For this material, we obtain its quite large value, $\Delta\gamma = 4.29$. Since $\gamma > \gamma_\chi$, the plot of the structural relaxation time as a single variable function of the estimated values of $\chi_4^{\max}$ is also impossible, similarly as it is in case of the total system entropy S

structural relaxation time has actually enabled to find the same values of the scaling exponents $\gamma_\chi = \gamma$, and even morphologically the same functions f and w, which scale $\tilde{\tau}$ and $\chi_4^{\max}$, respectively. It has turned out that Eq. (29) should not have been employed in this case, but we could use Eq. (15) in which $\chi_4^{\max}$ was replaced with $\tilde{\tau}$ as follows: $\chi_4^{\max} = (\chi_4^{\max})_0 \exp[(A_\chi \rho^{\gamma_\chi}/T)^{D_\chi}]$. Then, Eq. (31) should have been transformed to the form, $\chi_4^{\max} = (\chi_4^{\max})_0 \exp\left[\left(\rho^{-\Delta\gamma}\left(\ln(\tau/\tau_0)\right)^{1/D} A_\chi/A\right)^{D_\chi}\right]$.

The comparative analysis of the density scaling of τ and $\chi_4^{\max}$, performed for real and model systems, leads to two important conclusions. Firstly, simple isotropic models with two body interactions as the KABLJ model based on the Lennard-Jones potential can only constitute reference systems, which are not able to reproduce all properties of real supercooled liquids. This conclusion, which seems to be obvious, is often omitted due to a convenient analytical and simulation simplicity of this kind of models. Secondly, various physical quantities can obey the power density scaling law in some range of temperatures and densities; however, the effective short-range potentials for intermolecular interactions, which are responsible for the density scaling of these physical quantities, can be different. Such a hypothesis has been formulated in Ref. [21], where we have argued that one of the reasons for the discussed differences can be different contributions of attractive interactions relevant to various physical quantities to the effective short-range potential for intermolecular interactions, which makes an impact on the value of the exponent of the dominant repulsive term of this potential. This interpretation finds an additional application in the later results of the investigations of the proper relation between the system entropy and the timescale of the system molecular dynamics near the liquid–glass transition, which are discussed in Sect. 3.

6 Beyond the Power Density Scaling Law with the Constant Scaling Exponent

Almost since the beginning of the study on the density scaling idea, the scaling variable, $\Gamma = h(\rho)/T$, has not been limited to the PDS law case; that is, other density scaling functions h than $h(\rho) = \rho^\gamma$ with $\gamma = const$ have been supposed as possible [38]. Such a general approach to the density scaling has found its confirmation in molecular dynamics simulations in simple models of supercooled liquids, which are based on the Lennard-Jones potential, including the prototypical KABLJ model of supercooled liquids. Thorough analyses of molecular dynamics simulation data obtained in the simple models of supercooled liquids led J. Dyre and his group at the Roskilde University to formulate *the theory of isomorphs* [4], which has been mentioned already a few times in this chapter. Results of these theoretical and simulation studies have shown that the nature of the density scaling is most likely more complex than that expected on the basis of the series of earlier phenomenological observations, even in the case of such simple models as that determined by the Lennard-Jones potential. The essence of the observed problem has been the isochoric character (also confirmed by our MD simulations in the KABLJ model) [85] of the strong linear correlation of the instantaneous values of the total system viral W and the total system energy U, which has been postulated for materials, the molecular dynamics of which obeys the power density scaling law, requiring the WU correlation coefficient larger than 0.9. This isochoric character of the correlation WU means that the slope of the linear correlation, suggested as the scaling exponent for dynamic quantities, such as structural relaxation time τ and viscosity η, in terms of the scaling variable, $\Gamma = \rho^{\gamma}/T$, depends on density, i.e., $\gamma(\rho) \neq const$; hence, it is in general no material constant independent of thermodynamic conditions. It is worth noting that there are also further developments of the isomorph theory [103], which suggest that the density dependence of the scaling exponent should be supplemented with temperature dependence, but such an approach seems to diminish unifying benefits of this theory. Thus, we consider herein only the cases in which the scaling exponent γ is well argued to be dependent at most on density for a given system, i.e., when the scaling exponent γ well corresponds to the density-dependent slope of the WU correlation established from MD simulations in the KABLJ model. This result of simulation experiments can be well grounded and generalized within the original theory of isomorphs, which enables to prove [104] that *strongly correlating systems* in general obey the configurational Grüneisen equation of state [105–107], $W = \gamma(\rho)U + \varphi(\rho)$, invoking another postulate of this theory (Appendix A in Ref. [4]), which tells that a system is *strongly correlating* if and only if it has isomorphs to a good approximation in its phase diagram, which are curves of isomorphic state points in the following sense: Two-state points (T_1, ρ_1) and (T_2, ρ_2) are isomorphic if all pairs of their physically relevant microconfigurations $\left(\mathbf{r}_1^{(1)}, \ldots, \mathbf{r}_N^{(1)}\right)$ and $\left(\mathbf{r}_1^{(2)}, \ldots, \mathbf{r}_N^{(2)}\right)$ characterized by identical reduced coordinates $\tilde{\mathbf{r}}_i^{(1)} = \tilde{\mathbf{r}}_i^{(2)}$ (where $\tilde{\mathbf{r}}_i = \rho^{1/3}\mathbf{r}_i$) have proportional configurational *NVT* Boltzmann factors, $\exp\left[-U\left(\mathbf{r}_1^{(1)}, \ldots, \mathbf{r}_N^{(1)}\right)/k_B T_1\right] = C_{12}\exp\left[-U\left(\mathbf{r}_1^{(2)}, \ldots, \mathbf{r}_N^{(2)}\right)/k_B T_2\right]$, where the constant C_{12} depends only on the state points (T_1, ρ_1) and (T_2, ρ_2), not on the microscopic configurations. Within the framework of the theory of isomorphs, which has become a natural generalization about the power density scaling law with the scaling exponent $\gamma = const$, the density dependence of the scaling exponent has been rationalized as $\gamma(\rho) = d\ \ln h/d\ \ln \rho$, i.e., the logarithmic derivative of a scaling function $h(\rho)$, which also depends explicitly only on density and enables to formulate the generalized form of the density scaling, $\tau = f(h(\rho)/T)$, that implies its power law version in the particular case of the scaling function $h(\rho) = \rho^{\gamma}$ with the exponent $\gamma = const$.

This theoretical framework has enabled to figure out an analytic form of the scaling function $h(\rho)$, which is valid for both the Lennard-Jones potential and each potential given by a sum of at least two inverse power laws [108]. Such an IPL term, $U_{IPL}(r) = A_r r^{-m_{IPL}}$, has been already discussed in Sect. 1 in context of the effective short-range intermolecular potential $U_{eff}(r)$ given by Eq. (4). As shown in Ref. [108], the effective short-range intermolecular potential, involving a single power law term and a small attractive background, does not provide any appropriate approximation when we analyze simulation data in a wide range of temperatures and densities, even if we apply the reduced units suggested by the theory of isomorphs to ensure that molecular dynamics is isomorph invariant in the statistical *NVT* and *NVE* ensembles. Then, any constant value of the exponent γ in the repulsive IPL term of the effective short-range intermolecular potential (Eq. 4) does not lead to the density scaling according to the PDS law (Eq. 1) with any constant exponent γ. This theoretical analysis confirmed by computer simulations has naturally required answering the question whether or not it is reflected in the density scaling of real supercooled liquids. The first attempt has been made at exploring this problem also in Ref. [108].

Using experimental data measured by means of broadband dielectric spectroscopy in a wide range of temperatures and pressures (up to $p = 1.8$ GPa), we have performed the analysis of the criterion for the generalized density scaling, which requires to meet the condition $h(\rho)/T = const$ at a constant structural relaxation time $\tilde{\tau} = const$ expressed in the reduced units (see Sect. 1) according to the isomorph theory, where the scaling function can be transformed to the form $h(\rho) = \rho^{\gamma(\rho)}$ with $\gamma(\rho) = d \ln h/d \ln \rho$. Since the scaling function $h(\rho)$, which follows from the potential based on the sum of the inverse power law terms, has not been able to describe properly the nonlinear isochronal dependences $\log_{10} \rho$ on $\log_{10} T$ at various structural relaxation times $\tilde{\tau}$ in case of real supercooled liquids, we have suggested a phenomenological form of the scaling function, $h(\rho) = \exp\left(C_1 \ln \rho + C_2 \ln^2 \rho\right)$ with the fitting parameters C_1 and C_2 for real glass formers. Analyzing experimental data of selected van der Waals liquids, measured in a wide pressure range ($0.1 \text{ MPa} \leq p \leq 1.8 \text{ GPa}$), we have shown [108] that the use of the suggested scaling function $h(\rho)$ results in satisfying the criterion for the generalized density scaling, $h(\rho)/T = const$ at $\tilde{\tau} = const$, which implies that the isochronal dependences $\log_{10} \rho$ on $\log_{10} T$ can be superimposed even if they are nonlinear. Nevertheless, it should be noted that the generalized density scaling (GDS) law, $\tau = f(h(\rho)/T)$, where the scaling function is not limited to $h(\rho) = \rho^{\gamma}$ with $\gamma = const$, requires further investigations, because all other real glass formers tested in terms of the density scaling if they satisfy the density scaling behavior, they obey the PDS law (Eq. 1) in the wide pressure range and the broad timescale range that covers even a dozen decades, besides decahydroisoquinoline (DHIQ) and dibutyl phthalate (DBP) examined in Ref. [108]. The GDS law for real glass formers has also its opponents who are not able to accept the density-dependent scaling exponent. However, their arguments are not sufficiently firm to convincingly show the GDS law invalidity for the real glass formers. In Ref. [109], the density scaling analysis of DHIQ has been performed by using new *pVT* data measured in a very wide pressure range (up to 600 MPa, and the isotherm at $T = 298.2$ K even up to $p = 1.2$ GPa). Using the *pVT* data parameterized by the Tait EOS, the PDS law for DHIQ has been trying to be argued, but the quality of the obtained density scaling according to Eq. (1) is too low (Fig. 2 in Ref. [109]) to undoubtedly approve the found value of the scaling exponent to be invariant for DHIQ, and it can be regarded as an average value valid only to a first approximation. Nonetheless, the way used in Ref. [109] to investigate the density scaling law is certainly able to shed a new light on the PDS and GDS law applications in the real glass formers, because there is still little *pVT* data measured in such wide pressure ranges that cover the pressure domain in which the broadband dielectric spectroscopy measurements are available (i.e., typically up to 2 GPa). This situation results in even large extrapolations of the dependences $V(T,p)$ usually measured only up to $p = 200$ MPa. Therefore, a progress toward a better understanding of the density scaling behavior is possible by performing more volumetric measurements at pressures exceeding the typical pressures applied to the volumetric measurements, which are expected in the near future, but require overcoming various technical difficulties that impede the complex experiments in the high-pressure conditions. It is worth noting that there are even some equations of state already derived in the GDS regime (including the EOSs discussed below) to facilitate the further investigations of such *pVT* data and its impact on the density scaling of dynamic quantities.

Inspired by the findings achieved on the grounds provided by the isomorph theory, we have formulated extended versions of the equations of state [110], which are discussed as Eqs. (19) and (20) in the PDS regime in Sect. 4. In the former case, similarly to the PDS regime, we have shown that it can be derived from the viral representation of the configurational pressure, $p^{conf} = \langle W \rangle / V$, where $\langle W \rangle$ is the average total system viral in a given statistical ensemble. However, in the general case, to determine the average viral $\langle W \rangle$, we have assumed that the effective potential form, $U_{eff}\left((\rho/\rho_0)^{-1/3}\mathbf{R}\right) = h(\rho)U(\mathbf{R}) + v(\rho)$, where ρ_0 denotes density in a reference state, complies with the theory of isomorphs and is characterized by the well-separated configurational and density contributions, $U(\mathbf{R})$ as well as $h(\rho)$ and $v(\rho)$, where $\mathbf{R}$ denotes a position vector of all the system particles, whereas $h(\rho)$ and $v(\rho)$ are some density functions, where the function $h(\rho)$ is expected to be the scaling function in terms of the GDS law, i.e., $\tilde{\tau} = f(h(\rho)/T)$.

To verify the assumptions made to derive the extended equations of state and to confirm that these equations comply with the theory of isomorphs, we have performed additional MD computer simulations [110] in the KABLJ system of 1000 particles in the *NVT* ensemble to increase the temperature-density range two times in comparison with those previously explored [85] to test whether the basic versions of the equations of state comply with the PDS law. The analysis of volumetric simulation data obtained in the wide temperature-density range has shown that both the extended equations of state, which have been formulated [110] assuming the scaling function, $h(\rho) = h_{LJ}(\rho) = c\rho^4 + (1-c)\rho^2$, derived analytically [108] for the Lennard-Jones potential, yield similar values of the parameter c, which lead the KABLJ liquid to the density scaling of its structural relaxation times expressed in the reduced units of the isomorph theory, $\tilde{\tau} = f(h(\rho)/T)$, in the same wide temperature-density range. Since experimental volumetric data measured in the sufficiently wide density range have not been available, we have carried out a comparative analysis of the pressure dependence of the isothermal bulk modulus $B_T(p)$ by means of the equations of state, which follow from the modified definition of the isothermal bulk modulus B_T, derived in the limit of the power law density scaling, i.e., on the assumption that $h(\rho) = \rho^\gamma$, as well as in the generalized density scaling regime, verifying both the scaling functions based on the Lennard-Jones potential, $h(\rho) = h_{LJ}(\rho) = c\rho^4 + (1-c)\rho^2$, and suggested phenomenologically, $h(\rho) = \exp\left(C_1 \ln \rho + C_2 \ln^2 \rho\right)$, which resulted in the better generalized density scaling of real supercooled liquids. In addition, we have considered the standard second-order Taylor series expansion of the pressure function $B_T(p)$ about a reference pressure. Results of the mentioned methods for predicting the pressure dependence $B_T(p)$ have been compared with the recently published experimental data of ultrasonic measurements of the isothermal bulk modulus of propylene carbonate (PC), which were carried out in the wide pressure range from ambient pressure to above 1 GPa [111], Based on the analysis, we have established that the most reliable prediction of the dependence $B_T(p)$ has been made by the extended equation of state with the scaling function, $h(\rho) = \exp\left(C_1 \ln \rho + C_2 \ln^2 \rho\right)$, which has been earlier turned out to be more suitable to perform the generalized density scaling of molecular dynamics of real supercooled liquids than the scaling function, $h_{LJ}(\rho) = c\rho^4 + (1-c)\rho^2$, derived from the Lennard-Jones potential. A high functionality of the originally isothermal GDS equation of state (Eq. 32), which exploits the scaling function (Eq. 33), implying the density-dependent scaling exponent (Eq. 34), has been confirmed by employing this GDS EOS supplemented with two temperature parametrization functions Eqs. (35) and (36) in a novel analysis of complex experimental data of the isobaric thermal volume expansivity α_p of liquid halogenoalkanes measured by means of the scanning transitiometry [88, 112],

$$p = p_0 + \frac{B_T(p_0)}{\gamma(\rho_0)}\left(\frac{h(\rho)}{h(\rho_0)} - 1\right) \tag{32}$$

$$h(\rho) = \exp\left(C_1 \ln \rho + C_2 \ln^2 \rho\right) \tag{33}$$

$$\gamma(\rho) = C_1 + 2C_2 \ln \rho \tag{34}$$

$$\rho_0 = V^{-1}(T, p_0) = \left(\sum_{l=0}^{k} A_l (T - T_0)^l\right)^{-1} \tag{35}$$

$$B_T(p_0) = b_0 \exp(-b_2(T - T_0)). \tag{36}$$

where p_0 and T_0 are, respectively, pressure and temperature in a chosen reference state, and the physical meaning of Eqs. (35) and (36) is the same as the parameters of the temperature parametrizations employed in Eq. (21), but it was useful to expand Eq. (35) up to the cubic term in case of the halgenoalkanes examined by us. From very recent studies comparing the PDS and GDS EOSs [113], one can also conclude that Eq. (36) could give better opportunities for the temperature parameterization of the isothermal bulk modulus in the reference state if their exponential form was replaced with a polynomial representation, because the latter could enable to yield basically nonexponential dependences that could be preferable for some liquids and pVT ranges. However, the problems with the temperature parametrization functions Eqs. (35) and (36) go beyond the density scaling idea, and hence they will not be discussed here in more detail.

It should be noted that the GDS EOS can be reduced to the PDS EOS Eqs. (20) or (21) by taking $C_2 = 0$. Then, $C_1 = \gamma$, which is a constant scaling exponent, and $h(\rho) = \rho^\gamma$. In contrast to the PDS EOS, the applicability range of the GDS

EOS is defined by in general a nonlinear pressure dependence of the isothermal bulk modulus, which involves the density-dependent scaling exponent $\gamma(\rho)$,

$$B_T(p) = \frac{\gamma(\rho)}{\gamma(\rho_0)}\left[B_T(p_0) + \gamma(\rho_0)(p - p_0)\right], \tag{37}$$

which reduces to the linear pressure dependence given by the PDE EOS if we apply Eq. (34) with $C_2 = 0$ to Eq. (37). In our comparative analysis performed by applying the PDS and GDS EOSs to the isobaric volume expansivity data for liquid halogenoalkanes measured in the isothermal scanning transitiometry experiments, it has turned out that the GDS EOS better describes the dependences $\alpha_p(T, p)$, which intersect each other at points satisfying the differential equation, $\left(\partial\alpha_p(T, p)/\partial T\right)_p = 0$. In this way, we have shown that the isotherms $\alpha_p(T, p)$ do not cross in any single intersection point as most often supposed previously [114, 115], but in many different points that constitute a continuous intersection curve, which is in general nonlinear and nonmonotonic in the *T*-*p* diagram. We have established that the advantage of the GDS EOS over the PDS one is related to the significantly wide density range considered for the examined liquid halogenoalkanes despite pressure measurements carried out only up to $p = 200$ MPa. It has revealed that the applicability ranges of the GDS and PDS EOSs should be validated in terms of the density range, which best reflects the combined *T*-*p* changes.

From the viewpoint of the GDS law for dynamic quantities measured in case of real glass formers, the GDS EOS is waiting for a verification that will answer the question whether the scaling function given by Eq. (33) with the values of its fitting parameters found from the *pVT* measurement data is able to scale the structural relaxation times according to the GDS law for a given material. As already mentioned, it is possible for simulation data in the KABLJ model if we use $h(\rho) = h_{LJ}(\rho)$. However, such comparative investigations of the GDS EOS (Eq. 32) with the scaling function $h(\rho)$ expressed by Eq. (33) and the GDS law with the same scaling function $h(\rho)$ require the *pVT* and for instance dielectric measurements in at least approximately the same *pVT* ranges, which still present a challenge to experimentalists.

A separate class of problems is addressed to the density scaling of molecular dynamics in associated liquids. A tendency of the hydrogen-bonded systems to conform the density scaling pattern of behavior is unambiguous. At the beginning of the study on the density scaling, it was shown that sorbitol [32] and glycerol [116, 117] can obey the PDS law, but taking into account the high-pressure dielectric data later measured for glycerol up to $p = 4.5$ GPa [118], one can suppose that the density scaling of glycerol in the extremely wide pressure range would be impossible according to Eq. (1). Although very recently, another H-bonded liquid—a ternidazole (TDZ) drug—has been shown by using experimental data that it conforms to the PDS law with approximately the same value of the scaling exponent $\gamma \approx 2$ for both the structural and JG-secondary relaxations [119], the PDS law is usually not satisfied by molecular dynamics of associated liquids, especially in case of strongly H-bonded systems. The invalidity of the density scaling law for H-bonded systems was suggested by results of MD simulations in simple models [8, 9], but the same

MD simulations [8, 9] also predicted that ionic liquids should not obey the PDS law, which has not been confirmed later by analyzing experimental data [5, 76, 78, 120]. In contrast to ionic liquids, a representative of associated liquids, which does not obey the PDS law, was found a few years after initiating the study on the density scaling. This was dipropylene glycol (DPG) [39, 41], which is classified as a strongly H-bonded system. In case of DPG, the later unpublished tests have shown that the PDS criterion given by Eq. (3) is not met, because the isochronal dependences of $\log_{10} \rho$ on $\log_{10} T$ are nonlinear; hence, $\gamma \neq const$. What is more, these isochronal curves cannot be superimposed. It means that the GDS criterion is also not satisfied by DPG, and consequently it is impossible to find a proper only density-dependent function $\gamma(\rho)$ for this strongly H-bonded supercooled liquid. A similar pattern of behavior has been discovered by us in other strongly and even weakly H-bonded systems (e.g., 2-ethyl-1-hexanol (2E1H) [121] and *N*,*N*-diethyl-meta-toluamide (DEET) [34], respectively). Nevertheless, the PDS law has become invalid for the Debye-like process timescale in monohydroxy alcohol 2E1H only at pressures higher than 0.5 GPa, while the known active ingredient in insect repellents DEET has not obeyed the PDS law for the structural relaxation even at low pressures. Although the latter material has rather a weak tendency to hydrogen bonding, our thorough investigations of their fragility parameters defined in different thermodynamic conditions have shown that all rules prospected for the isobaric, isochoric, and isothermal fragilities in the PDS regime, which are reported in Sect. 2, have been broken in case of DEET, because m_p and m_V increase and m_T decreases with increasing pressure from $p = 0.1$ MPa to $p = 400$ MPa for this material. This finding shows that the rules based on the PDS law for the pressure dependences of the fragility parameters can be used to probe (at least preliminarily) whether or not the PDS law can be valid for a given material in the considered *pVT* range. The observed pressure effect on the isobaric fragility of DEET is in accord with our earlier results [122, 123], which revealed that m_p increases up to about $p = 1$ GPa and then begins to decrease in case of propylene glycol oligomers (di- and tripropylene glycol) as well as rich-in-water mixture of PPG400 (polypropylene glycol of average molar mass equaled to 400 g/mol). This nonmonotonic dependence $m_p(p)$ most likely reflects the opposite compression effects on the hydrogen bonding, depending on changes in temperature in isochronal conditions. Up to 1GPa, the compression densifies the systems, which facilitates the formation of H-bonds. Since the increase in pressure at $\tau = const$ is accompanied with an increase in temperature, when the temperature effect begins to predominate over the compression one in molecular dynamics, H-bonds are more readily broken than formed, and the system begins to behave similarly as a van der Waals liquid. It is reasonable to claim that such a complex behavior of the hydrogen bonding strongly dependent on thermodynamic conditions most likely affects the dependence $m_p(p)$ of the associated liquids. It suggests that changes in the supramolecular structures, which are characteristic of these materials and usually constituted by the strongly H-bonded networks, cannot be excluded in case of weakly H-bonded liquids as DEET and can be regarded as a cause of the invalidity of the PDS law, and even the GDS law for this group of materials.

The cognitive benefits of the density scaling encouraged us to work out a method leading to a density scaling representation for the structural relaxation times of associated liquids that obey neither the PDS law nor its GDS extension. We have suggested that the issue could be figured out on the assumption that the scaling exponent depends on the structural relaxation time, i.e., $\gamma = \gamma(\tau)$ in the PDS law given by Eq. (1). At first glance, it seems to be at odds with the fundamentals of the density scaling. However, the varying supramolecular structure of such materials may provide a rationale for attributing the scaling exponent to the characteristic timescale of molecular dynamics, because one can expect that a supramolecular structure formed at a given timescale is characteristic of the scaling behavior according to Eq. (1) at $\tau = const$, even if various other parameters may change in isochronal conditions. Based on the hypothesis, we have found that a linear function $\gamma(\log_{10}\tau)$ applied to Eq. (1) enables to scale the relaxation times of the Debye-like process in 2E1H measured at $p > 0.5$ GPa (while $\gamma = const$ at lower pressures) [121], and a cubic function $\gamma(\log_{10}\tau)$ incorporated into Eq. (1) leads to the density scaling of the structural relaxation times of DEET in the entire considered *pVT* range, covering ambient and elevated pressures [34]. Nevertheless, further investigations of the density scaling in case of associated liquids are required to verify whether the timescale-dependent scaling exponent $\gamma(\tau)$ can be established for each associated liquid that does not conform to the density scaling, and to understand the predictive capabilities of Eq. (1) with $\gamma = \gamma(\tau)$.

7 Summary and Conclusions

By presenting the undoubted advantages of the density scaling as well as its still only partially discovered aspects, we have aimed at showing the essential contributions of this seminal idea to our better understanding of the glass transition and related phenomena as well as the further challenges that should be taken up to gain a better insight into the generalizations about the power density scaling law and their impact on unifying concepts in the glass transition physics.

We have argued that the power density scaling law valid for various material groups can be satisfied by different dynamic and thermodynamic quantities. The dynamic quantities, such as structural relaxation times, segmental relaxation times, conductivity primary relaxation times, dc-conductivities, viscosities, and diffusivities, which are important characteristics of the primary/global relaxation processes or at least mimic the timescales of such relaxation processes, can be scaled with the same value of the scaling exponent γ for a given material to a good approximation. A similar pattern of the scaling behavior has been observed for the Johari-Goldstein secondary relaxation times, which is reasonable, because this secondary process is regarded as a precursor of the primary/global relaxation. It has turned out also that some thermodynamic quantities and estimates based on fluctuations of thermodynamic variables or quantities can obey the power density scaling law, but it usually requires employing a different value of the scaling exponent from that which enables to scale the aforementioned dynamic quantities. Among the thermodynamic quantities, only the configurational entropy S_c has been found to satisfy the power density scaling law with the same value of the scaling exponent γ that scales the structural relaxation times τ, which implies that it is possible to describe τ as a single variable function of S_c, i.e., $\tau = g(S_c)$. However, the total system entropy S can be scaled according to the power density scaling law with the scaling exponent considerably less than that which scales the structural relaxation times τ, i.e., $\gamma_S < \gamma$, and then there is no single variable function, i.e., $\tau \neq w(S)$. Based on classical thermodynamics, it has been established that the scaling exponent γ_S for the total system entropy can be evaluated as the Grüneisen parameter γ_G defined thermodynamically and calculated in the supercooled liquid state near or at the glass transition. What is more, the PDS laws satisfied by τ and S with different scaling exponents $(\gamma_S \neq \gamma)$ have shown that the established decoupling between τ and S, which is well identified in isochronal or isoentropic conditions, can be quantified by the difference between the scaling exponents, $\Delta\gamma = \gamma - \gamma_S$, because the dependence of τ on S should be supplemented with the density-dependent factor as follows: $\tau = h_S\big(\rho^{\Delta\gamma} w(S)\big)$. Similarly, the estimate χ_4^H based on enthalpy fluctuations for the four-point measure of the dynamic heterogeneity $\chi_4^{\max}$ can be scaled neither with the scaling exponent valid for the structural relaxation time τ nor that for the total system entropy S, but it satisfies the power density scaling law with another scaling exponent γ_χ, which ranges between γ_S and γ for real glass-forming liquids. The PDS laws obeyed for τ, $\chi_4^{\max}$, and S with mutually different values of the scaling exponents γ, γ_χ, and γ_S show that one can observe the decoupling phenomena between τ, $\chi_4^{\max}$, and S, which can be quantified by the differences between scaling exponents, $\gamma - \gamma_\chi$, $\gamma - \gamma_S$, and $\gamma_\chi - \gamma_S$. Such decoupling measures are also useful to represent the case of no decoupling behavior, because then they tend to *0* as it has been established, for instance, between the structural relaxation time and the four-point measure of the dynamic heterogeneity straightforwardly determined from MD simulation data in simple models of supercooled liquids based on the Lennard-Jones potential, for which $\gamma_\chi \approx \gamma$.

In search of the linkage between the relaxation dynamics, thermodynamics, and intermolecular interactions, a class of the equations of state has been formulated in the both PDS and GDS regimes. All the EOSs have been derived or at least well interpreted by using the effective intermolecular potentials that have been suggested to be responsible for the density scaling of the dynamic quantities according to either PDS or GDS laws. Due to limited sets of data measured in the sufficiently wide pressure or density ranges, only the case invoking the PDS law has been thoroughly verified by using experimental data. As a result, different values of the scaling exponent related to the exponent of the dominant repulsive part to the effective short-range intermolecular potential [expressed in this chapter by Eq. (4)] have been established from the density scaling of measured dynamic quantities and from the fitting of measured *pVT* data to the PDS-based EOS, i.e., $\gamma_{EOS} > \gamma$.

The discrepancies between the scaling exponents for dynamic, thermodynamic, fluctuation, and volumetric quantities for a given material can find their rationale in different effective cutoff distances for interactions or different molecular modes reflected in various quantities. One can realize this by considering that the effective potential can be differently modeled, depending on the assumed contribution of attractive interactions, which varies with changing the cutoff distance for interactions, resulting in different values of the exponent of the repulsive part that is dominant in the effective short-range intermolecular potential underlying the PDS law. This hypothesis requires further investigations by using computer simulations in more realistic models than the isotropic models based on the Lennard-Jones potential. On the other hand, thorough experimental verifications of the generalized density scaling law with the scaling exponent dependent on density and its applications to real glass-forming liquids still demand experimental measurements of dynamic quantities and *pVT* data in the same very broad *pVT* range. Additionally, the density scaling of associated liquids should be subject to further theoretical studies.

Nevertheless, a few questions that arose during the study on the density scaling cannot obscure the achievements of the power density scaling law, the very broad applicability range of which is accompanied with its valuable prediction capabilities based on the effective intermolecular potential. Thus, the possible universal applications of the different cases of density scaling, which offer very tempting opportunities to study macroscopic properties of materials by using the relevant intermolecular potential and *vice versa* to determine parameters of this intermolecular potential based on measurements of macroscopic quantities, become both the main advantage and the challenge of further investigations in this field.

Acknowledgements The authors are deeply thankful for receiving the research project within the program MAESTRO 2 financed by the Polish National Science Center, based on Decision No. DEC-2012/04/A/ST3/00337.

References

1. Anderson PW (1995) Science 267:1615
2. Roland CM, Hensel-Bielowka S, Paluch M, Casalini R (2005) Rep Prog Phys 68:1405
3. Floudas G, Paluch M, Grzybowski, Ngai KL (2011) In: Kremer F (ed) Molecular dynamics of glass-forming systems: effects of pressure. Advances in dielectrics, chap. 2, Springer, Berlin
4. Gnan N, Schrøder TB, Pedersen UR, Bailey NP, Dyre JC (2009) J Chem Phys 131:234504
5. Wojnarowska Z, Paluch M (2016) High-pressure dielectric spectroscopy for studying the charge transfer in ionic liquids and solids, chap. 4. In: Paluch M (ed) Dielectric properties of ionic liquids. Advances in dielectrics, Kremer F (ed), Springer, Berlin
6. Paluch M, Grzybowska K Grzybowski A (2007) J Phys Condens Matter 19: 205117
7. Pedersen UR, Bailey NP, Schrøder TB, Dyre JC (2008) Phys Rev Lett 100:015701
8. Bailey NP, Pedersen UR, Gnan N, Schrøder TB, Dyre JC (2008) J Chem Phys 129:184507
9. Bailey NP, Pedersen UR, Gnan N, Schrøder TB, Dyre JC (2009) J Chem Phys 130:039902
10. Bailey NP, Pedersen UR, Gnan N, Schrøder TB, Dyre JC (2008) J Chem Phys 129:184508
11. Coslovich D, Roland CM (2008) J Phys Chem B 112:1329
12. Coslovich D, Roland CM (2009) J Chem Phys 130:014508
13. Schrøder TB, Pedersen UR, Bailey NP, Toxvaerd S, Dyre JC (2009) Phys Rev E 80:041502
14. Pedersen UR, Schrøder TB, Dyre JC (2010) Phys Rev Lett 105:157801
15. Ngai KL, Casalini R, Capaccioli S, Paluch M, Roland CM (2005) J Phys Chem B 109:17356
16. Roland CM, Casalini R, Paluch M (2003) Chem Phys Lett 367:259
17. Xiao W, Tofteskov J, Christensen TV, Dyre JC, Niss K (2015) J Non-Cryst Solids 407:190

18. Froehlich H (1987) Theory of dielectrics, dielectric constant and dielectric loss, 2nd edn. Oxford University Press, Oxford
19. Kremer F, Schonhals A (2003) Broadband dielectric spectroscopy. Springer, Berlin
20. Grzybowska K, Grzybowski A, Pawlus S, Pionteck J, Paluch M (2015) Phys Rev E 91:062305
21. Grzybowski A, Koperwas K, Kolodziejczyk K, Grzybowska K, Paluch M (2013) J Phys Chem Lett 4:4273
22. Ngai KL, Habasaki J, Prevosto D, Capaccioli S, Paluch M (2012) J Chem Phys 137:034511
23. Ngai KL, Habasaki J, Prevosto D, Capaccioli S, Paluch M (2014) J Chem Phys 140:019901
24. Ngai KL, Paluch M (2017) J Non-Cryst Solids J Non-Cryst Solids 478:1
25. Tsang KY, Ngai KL (1996) Phys Rev E 54:R3067
26. Ngai KL (2003) J Phys: Condens Matter 15:S1107
27. Ngai KL (2011) Relaxation and diffusion in complex systems. Springer, New York
28. Williams G (1964) Trans Faraday Soc 60:1548
29. Floudas G, Paluch M, Grzybowski A, Ngai KL (2011) In: Kremer F (ed) Molecular dynamics of glass-forming systems: effects of pressure. Advances in dielectrics, chap. 1, Springer, Berlin
30. Ferrer ML et al (1998) J Chem Phys 109:8010
31. Casalini R, Roland CM (2003) J Chem Phys 119:4052
32. Casalini R, Roland CM (2004) Phys Rev E 69:062501
33. Masiewicz E, Grzybowski A, Sokolov AP, Paluch M (2012) J Phys Chem Lett 3:2643
34. Paluch M, Masiewicz E, Grzybowski A, Pawlus S, Pionteck J, Wojnarowska Z (2014) J Chem Phys 141:134507
35. Tarjus G, Kivelson D, Mossa S, Alba-Simionesco C (2004) J Chem Phys 120:6135
36. Casalini R, Roland CM (2005) Phys Rev B 71:014210
37. Casalini R, Roland CM (2005) Phys Rev E 72:031503
38. Alba-Simionesco C, Tarjus G (2006) J Non-Cryst Solids 352:4888
39. Grzybowski A, Grzybowska K, Zioło J, Paluch M (2006) Phys Rev E 74:041503
40. Casalini R, Roland CM (2007) Phys Rev E 76:013501
41. Grzybowski A, Grzybowska K, Zioło J, Paluch M (2007) Phys Rev E 76:013502
42. Angell CA (1995) Science 267:1924
43. Angell CA (1995) Proc Natl Acad Sci USA 92:6675
44. Jedrzejowska A, Grzybowski A, Paluch M (2017) Phys Chem Chem Phys 19:18348
45. Ingram MD, Imrie CT, Stoeva Z, Pas SJ, Funke K, Chandler HW (2005) J Phys Chem B 109:16567
46. Ingram MD, Imrie CT, Ledru J, Hutchinson J (2008) J Phys Chem B 112:859
47. Ingram MD, Imrie CT (2011) Solid State Ionics 196:9
48. Koperwas K, Grzybowski A, Grzybowska K, Wojnarowska Z, Pionteck J, Sokolov AP, Paluch M (2012) Phys Rev E 86:041502
49. Koperwas K, Grzybowski A, Tripathy SN, Masiewicz E, Paluch M (2015) Sci Rep 5:17782
50. Thoms E, Grzybowski A, Pawlus S, Paluch M (2018) J Phys Chem Lett 9:1783
51. Casalini R, Bair S (2008) J Chem Phys 128:084511
52. Mauro JC, Yue YZ, Ellison AJ, Gupta PK, Allan DC (2009) Proc Natl Acad Sci USA 106:19780
53. Adam G, Gibbs JH (1965) J Chem Phys 43:139
54. Gupta PK, Mauro JC (2009) J Chem Phys 130:094503
55. Naumis GG (2006) J Non-Cryst Solids 352:4865
56. Grzybowski A, Urban S, Mroz S, Paluch M (2017) Sci Rep 7:42174
57. Masiewicz E, Grzybowski A, Grzybowska K, Pawlus S, Pionteck J, Paluch M (2015) Sci Rep 5:13998
58. Johari GP (2003) J Chem Phys 119:635
59. Casalini R, Roland CM (2007) Phil Mag 87:459
60. Casalini R, Roland CM (2007) J Phys: Condens Matter 19:205118
61. Casalini R, Roland CM (2014) Phys Rev Lett 113:085701
62. Xia X, Wolynes PG (2000) Proc Natl Acad Sci USA 97:2990
63. Xia X, Wolynes PG (2001) Phys Rev Lett 86:5526

64. Hall RW, Wolynes PG (2008) J Phys Chem B 112:301
65. Sengupta S, Schrøder TB, Sastry S (2013) Eur Phys J E 36:141
66. Johari GPJ (2000) Chem Phys 112:7518
67. Avramov I, Milchev A (1988) J Non-Cryst Solids 104:253
68. Avramov I (1991) J Chem Phys 95:4439
69. Avramov I (1996) Therm Acta 280/281:363
70. Avramov I (1997) J Mater Sci Lett 13:1367
71. Avramov I (1998) J Non-Cryst Solids 238:6
72. Avramov I (2000) J Non-Cryst Solids 262:258
73. Avramov I, Grzybowski A, Paluch M (2009) J Non-Cryst Solids 355:733
74. Casalini R, Mohanty U, Roland CM (2006) J Chem Phys 125:014505
75. Grzybowski A, Paluch M, Grzybowska K, Haracz S (2010) J Chem Phys 133:161101
76. Paluch M, Haracz S, Grzybowski A, Mierzwa M, Pionteck J, Rivera-Calzada A, Leon C (2010) J Phys Chem Lett 1:987
77. Grzybowski A, Haracz S, Paluch M, Grzybowska K (2010) J Phys Chem B 114:11544
78. Paluch M, Wojnarowska Z, Goodrich P, Jacquemin J, Pionteck J, Hensel-Bielowka S (2015) Soft Matter 11:6520
79. Tait PG (1888) Physics and chemistry of the voyage of H. M. S. challenger, vol 2, Part 4. HMSO, London
80. Grzybowski A, Paluch M, Grzybowska K (2009) J Phys Chem B 113:7419
81. Bardik VY, Shakun KS (2005) Ukr J Phys (Paris) 50:404
82. Papathanassiou AN (2009) Phys Rev E 79:032501
83. Stickel F, Fischer EW, Richert R (1995) J Chem Phys 102:6251
84. Grzybowski A, Grzybowska K, Paluch M, Swiety A, Koperwas K (2011) Phys Rev E 83:041505
85. Grzybowski A, Koperwas K, Paluch M (2012) Phys Rev E 86:031501
86. Garai J, Laugier AJ (2007) Appl Phys 101:023514
87. Grzybowski A, Paluch M, Grzybowska K (2010) Phys Rev E 82:013501
88. Chorążewski M, Grzybowski A, Paluch M (2014) Phys Chem Chem Phys 16:19900
89. Grzybowski A, Koperwas K, Swiety-Pospiech A, Grzybowska K, Paluch M (2013) Phys Rev B 87:054105
90. Kob W, Andersen HC (1994) Phys Rev Lett 73:1376
91. Koperwas K, Grzybowski A, Grzybowska K, Wojnarowska Z, Paluch M (2015) J Non-Cryst Solids 407:196
92. Berthier L, Biroli G, Bouchaud JP, Cipelletti L, El Masri D, L'Hôte D, Ladieu F, Pierno M (2005) Science 310:1797
93. Tracht U, Wilhelm M, Heuer A, Feng H, Schmidt-Rohr K, Spiess HW (1998) Phys Rev Lett 81:2727
94. Qiu XH, Ediger MD (2003) J Phys Chem B 107:459
95. Crauste-Thibierge C, Brun C, Ladieu F, L'Hôte D, Biroli G, Bouchaud JP (2010) Phys Rev Lett 104:165703
96. Bauer T, Lunkenheimer P, Loidl A (2013) Phys Rev Lett 111:225702
97. Casalini R, Fragiadakis D, Roland CM (2015) J Chem Phys 142:064504
98. Samanta S, Richert R (2014) J Chem Phys 140:054503
99. Dalle-Ferrier C, Thibierge C, Alba-Simionesco C, Berthier L, Biroli G, Bouchaud JP, Ladieu F, L'Hôte D, Tarjus G (2007) Phys Rev E 76:041510
100. Grzybowski A, Kolodziejczyk K, Koperwas K, Grzybowska K, Paluch M (2012) Phys Rev B 85:220201
101. Koperwas K, Grzybowski A, Grzybowska K, Wojnarowska Z, Sokolov AP, Paluch M (2013) Phys Rev Lett 111:125701
102. Berthier L (2011) Physics 4:42
103. Schrøder TB, Dyre JC (2014) J Chem Phys 141:204502
104. Ingebrigtsen TS, Bøhling L, Schrøder TB, Dyre JC (2012) J Chem Phys 136:061102

105. Born M, Huang K (1954) Dynamical theory of crystal lattices. Oxford University Press, Oxford
106. Ross M, Young DA (1993) Ann Rev Phys Chem 44:61
107. Burakovsky L, Preston DL (2004) J Phys Chem Solids 65:1581
108. Bøhling L, Ingebrigtsen TS, Grzybowski A, Paluch M, Dyre JC, Schrøder TB (2012) New J Phys 14:113035
109. Casalini R, Roland CM (2016) J Chem Phys 144:024502
110. Grzybowski A, Koperwas K, Paluch M (2014) J Chem Phys 140:044502
111. Kondrin MV, Gromnitskaya EL, Pronin AA, Lyapin AG, Brazhkin VV, Volkov AA (2012) J Chem Phys 137:084502
112. Chorążewski M, Grzybowski A, Paluch M (2015) Ind Eng Chem Res 54:6400
113. López ER, Fandiño O, Cabaleiro D, Lugo L, Fernández J (2018) Phys Chem Chem Phys 20:3531
114. Randzio SL, Deiters UK (1995) Ber Bunsenges Phys Chem 99:1179
115. Deiters UK, Randzio SL (1995) Fluid Phase Equilib 103:199
116. Dreyfus C, Le Grand A, Gapinski J, Steffen W, Patkowski A (2004) Eur Phys J B 42:309
117. Win KZ Menon N (2006) Phys Rev E 73:040501(R)
118. Pronin AA, Kondrin MV, Lyapin AG, Brazhkin VV, Volkov AA, Lunkenheimer P, Loidl A (2010) JETP Lett 92:479
119. Romanini M, Barrio M, Macovez R, Ruiz-Martin MD, Capaccioli S, Tamarit JL (2017) Sci Rep 7:1346
120. Wojnarowska Z, Jarosz G, Grzybowski A, Pionteck J, Jacquemin J, Paluch M (2014) Phys Chem Chem Phys 16:20444
121. Pawlus S, Paluch M, Grzybowski A (2011) J Chem Phys 134:041103
122. Grzybowska K, Paluch M, Grzybowski A, Pawlus S, Ancherbak S, Prevosto D, Capaccioli S (2010) J Phys Chem Lett 1:1170
123. Grzybowska K, Pawlus S, Mierzwa M, Paluch M, Ngai KL (2006) J Chem Phys 125:144507

Scaling of Suprastructure and Dynamics in Pure and Mixed Debye Liquids

Sebastian Peter Bierwirth, Jennifer Bolle, Stefan Bauer, Christian Sternemann, Catalin Gainaru, Metin Tolan and Roland Böhmer

Abstract Supramolecular relaxations of the Debye or near-Debye type are featured by monohydroxy alcohols, water, and several other liquids. Mainly focusing on results from broadband dielectric spectroscopy, shear rheology, X-ray diffraction, and near-infrared absorption, scaling properties of chain-forming and ring-forming monohydroxy alcohols are examined. Deviations from ideal-mixing behavior in binary solutions involving these liquids in their supercooled state are given particular attention. The present survey is selective rather than comprehensive with a focus on exciting recent developments in this scientific area. Although most of the research summarized in this chapter is based on experiments and analyses carried out under linear-response and ambient-pressure conditions, phenomena emerging beyond these regimes are briefly touched upon as well. Finally, aiming at a faithful representation of the molecular dynamics taking place in these liquids at the microscopic level, overarching aspects arising from the complementary application of experimental techniques as well as perspectives for future developments are discussed.

1 Introduction

Supercooled liquids display a number of generic features as described by the three non's [1] which refer to their nonexponential, non-Arrhenius, and nonlinear response. Are these features related to each other? Is the degree of nonexponentiality pivotal to the response of complex materials [2]? Is their super-Arrhenius behavior adequately captured by, e.g., the kinetic fragility [3]? What is the role played by the internal clock that effectively "linearizes" the time dependence of physical aging in glass formers [4]? Often enough, the answers given to these and many other questions are governed by the set of properties one is interested in. There is a large number of overarching similarities in the relaxation of metallic glass formers, inorganic network systems,

S. P. Bierwirth · J. Bolle · S. Bauer · C. Sternemann · C. Gainaru
M. Tolan · R. Böhmer (✉)
Technische Universität Dortmund, Dortmund, Germany
e-mail: roland.bohmer@tu-dortmund.de

F. Kremer and A. Loidl (eds.), *The Scaling of Relaxation Processes*, Advances in Dielectrics, https://doi.org/10.1007/978-3-319-72706-6_5

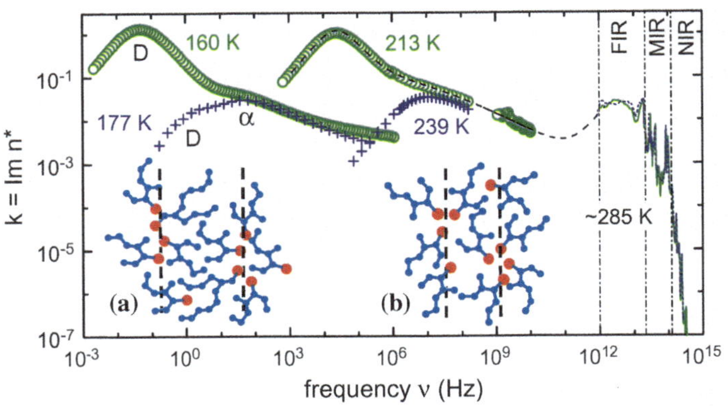

Fig. 1 Imaginary part, k, of the complex refractive index as measured for 2-ethyl-1-hexanol (circles and green solid lines), and 4-methyl-3-heptanol (plusses and blue dashed lines) using broadband dielectric as well as far-, mid, and near-infrared spectroscopies. Adapted from Ref. [8]. The illustration of **a** chain-like and **b** ring-like supramolecular clustering in the inset is taken with permission from Ref. [10]

van der Waals liquids, and so on. Yet, the exceedingly rich phenomenology of each class of glass forming materials has not led to a universally accepted microscopic understanding of their structural relaxation so far.

In this situation, it may be worthwhile to study families of glass formers with interesting non-generic properties. Among them are monohydroxy alcohols (MAs), the materials on which the present chapter focuses. Their dominant Debye-like, i.e., typically single-exponential relaxation shows up prominently in broadband dielectric spectroscopy. This feature is named after Debye since its shape reflects the polarization behavior predicted in his classic 1913 paper [5]. For decades the Debye process in MAs seemed to evade detection by essentially all other experimental methods. Based on the recognition that this usually dominant relaxation is *not* of structural origin [6] and the recent application of a range of different methods for their study [7], the expectation is fueled that an understanding of the Debye process is eventually within reach.

There appears to be a widespread notion that this process is due to changes in the hydrogen-bond-mediated suprastructural organization of the MA liquids. The degree and nature of this organization can be affected, e.g., by temperature and pressure as well as by the molecular structure of the MAs and by suitable mixing. As an example for the impact that a change of molecular structure can have on the supramolecular relaxation in MAs, Fig. 1 shows the comparison of broadband dielectric functions of two isomeric alcohols in terms of the imaginary part, $k = [\frac{1}{2}(\varepsilon'^2 + \varepsilon''^2)^{1/2} - \frac{1}{2}\varepsilon']^{1/2}$, of the complex index of refraction n^* [8]. One of the MAs included in Fig. 1, 2-ethyl-1-hexanol (2E1H), displays a prominent Debye-like response, while for the other, 4-methyl-3-heptanol (4M3H), this supramolecular feature is rather weak.

In a simplified fashion, the insets in Fig. 1 sketch the differences of the envisioned ring-like (less compact [9]) and chain-like (more compact) association patterns [10]

Fig. 2 Sketch of various effects that can occur in binary mixtures of liquids that exhibit strong, weak, or no Debye processes. Examples for some of these features will be presented later in this chapter. The acronyms used here for various substances are defined in Table 1

prevailing in these two MAs (a similar illustration can be found in Ref. [11]). It is obvious from Fig. 1 that the structural (or α-) relaxation is almost unaffected by the change in molecular structure. As a valuable local probe, also of the supramolecular organization, infrared (IR) spectroscopy reveals interesting differences among the two isomers as well [8]. These differences, highlighted in detail further below, are, however, barely visible on the 18-decade frequency scale of Fig. 1.

With some properties of (mostly) pure MAs recently reviewed [7], this chapter focuses on a few issues of current interest. In particular, the addressed questions include the following: How is the supramolecular organization altered when exchanging the hydroxyl group by other molecular sub-units that may or may not sustain intermolecular association? What can happen in mixtures [12]; does the relaxation pattern evolve continuously in an ideal-mixing type of scenario when changing the composition from one end to the other? Or, on the contrary, will strongly nonlinear concentration effects arise? In a simplified way, Fig. 2 briefly summarizes some of the observed phenomena mostly along with corresponding examples of binary mixtures.

Until recently the notion prevailed that "the most intriguing feature of the Debye relaxation found in monohydroxy alcohols is that this process is experimentally observed only by means of dielectric spectroscopy. Other experimental methods do not provide any evidence of the existence of this process" [13]. In the meantime, it has become clear that supramolecular relaxation in MAs can be detected successfully by other techniques as well (some of them reviewed recently [7, 14]). Progress toward a better understanding of material properties has always benefitted from applying several experimental methods in parallel and from striving for a rationalization of all pending data from a unified perspective. In this vein, we will combine results from complementary experimental techniques wherever indicated. Thus, in the present chapter we will not only discuss dielectric spectroscopy of MAs and binary mixtures [15] involving MAs, but include results mostly from infrared absorption, X-ray diffraction, and shear rheology as well. Other experimental techniques useful in studies of MAs will be given less attention. Detailed discussions of pressure application and theoretical modeling are also beyond the scope of the present article. Material for illustration is mostly drawn from work done in the authors' laboratory; this article is selective also in the sense that mainly topics of our own scientific interest are covered.

2 Homologous Series, Functional Groups, and Supramolecular Association

To explore the properties of associating liquids such as MAs that are enormously useful and versatile solvents, it is often rewarding to investigate homologous series of them and of related substances in a systematic fashion. The exploitation of this approach has a long and successful tradition not only in the field of dielectric spectroscopy [16–18]. Examples of homologous MA series that have been studied well and of which, below, we can discuss only a few, include the following:

A. Normal (n-) alcohols of various alkyl chain lengths [19–21]
B. Isomeric MAs for which the position of the OH group is changed while the number of carbon atoms is kept constant [22, 23]
C. MAs with double and triple carbon-carbon bonds in the aliphatic chain [24]
D. MAs with heteroatom substitution in the alkyl main chain such as in 3-methylthio-1-hexanol (α-process is enhanced [25]) or in alkoxy alcohols (no evidence for a Debye process [26])
E. Octanols of the type j-methyl-3-heptanol (jM3H) with $j = 2$, 3, 4, 5, and 6 in which the position of only the methyl group is varied [18, 27–29]; see also Fig. 53 of Ref. [7]
F. Other octanol-based (or also heptanol-based [30]) series with hydroxyl groups at different positions such as 8-, 4-, and 2-methyl-1-heptanol [31] or 2- and 3-methyl-4-heptanol [32]
G. Homologous neo-alcohols of the type H-$(CH_2)_j$-$C(CH_3)_2$-CH_2OH with $j = 1$–8 [33]
H. Branched alcohols of the type $CH_3(CH_2)_{j+2}(CH_2CHOH)(CH_2)_jCH_3$ with j = 1 corresponding to 2E1H [10, 11, 34]
I. MAs like 1-phenyl-1-propanol and 1-phenyl-2-propanol in which the OH group is sterically screened to such an extent that the existence of a Debye process is easily detected [14, 35, 36]

J. Molecules in which the OH moiety is substituted by other functional groups that may (or may not) support intermolecular association [37]

This incomplete list suggests that the hydroxyl (OH) moiety is the most prominent example of a functional group that facilitates hydrogen bonding. But regarding the last item (group J) also NH and, to a lesser extent, SH moieties come to mind. Furthermore, the amine (NH_2), aldehyde (COH), or carboxyl (COOH) groups and ones supporting halogen bonds [38] share the potential to facilitate intermolecular association. Formation of supramolecular structures may be enabled, for instance, also by plain Coulombic interaction or metal-ligand coordination [39].

2.1 Linear and Nonlinear Rheology

For polymeric materials, rheological measurements are indispensable, not the least to study the viscoelastic behavior that relates to their supramolecular modes (normal modes) [40, 41]. Since the mid-1950s MAs have been supposed to show no indication of a supramolecular relaxation in their ultrasonic or other mechanical responses [42]. Much later, work by Kaatze and coworkers removed this misconception [43, 44]. In addition to an ultrasonic absorption peak, these authors identified excess low-frequency viscoelastic contributions in several MAs.

Near room temperature the time scale separation of the two processes was found to be about 150 for 2E1H [43], one of the MAs mentioned above. Even larger separations were found in work of this substance carried out near the glass transition temperature [45]. In Fig. 3a the frequency-dependent shear modulus $G'(\nu)$ of 2E1H is depicted in the form of a master curve. At high frequencies a modulus $G_\infty = 1.1$ GPa is reached while the low-frequency limit $G'(\nu \to 0) \to 0$ is approached in two steps. Moving toward lower frequencies the step near $\nu \approx \nu_{max}$ corresponds to the loss of rigidity caused by the structural relaxation that is characteristic of all liquids. Then, below ν_{max} a regime with a relatively mild power-law variation in $G'(\nu)$ commences until the residual rigidity, here sustained by hydrogen bonds, gets lost in a second step. This behavior resembles that of short-chain polymers [45] for which a rubbery plateau is not fully developed.

The terminal mode that appears at the lowest (scaled) frequencies in Fig. 3 is recognized from its $G'(\nu) \propto \nu^2$ behavior to indicate compliance with the Maxwell model. This approach models a viscoelastic body simply as an equivalent circuit of a spring in series with a dashpot. For 2E1H the modulus at the second step is some three decades smaller than G_∞, suggesting that the impact of the hydrogen bonds might be only minor and this is also what one would naïvely infer from the shear loss modulus shown in Fig. 3b.

However, the impression gained from the frequency-dependent viscosity $\eta'(\nu)$, cf. Fig. 3c, is different. In the representation chosen for this figure, the viscosity associated with the structural relaxation, $\eta'(\nu \approx \nu_{max})$, of 2E1H is seen to amount to about a tenth of the zero-shear viscosity, $\eta'(\nu \to 0)$. To be able to assess this behavior

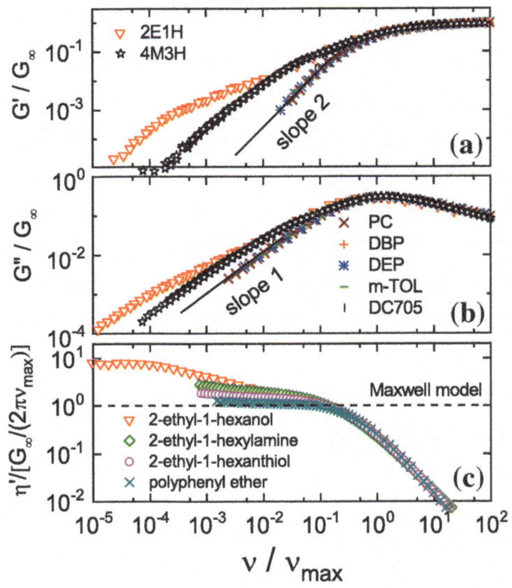

Fig. 3 **a** Real part and **b** imaginary part of the complex shear modulus of 2E1H and 4M3H, adapted from Ref. [45] where also the acronyms of the simple liquids listed in frame (**b**) are defined. **c** Comparison of scaled shear viscosities of three 2-ethyl-1-hexyl derivatives with results from polyphenyl ether, another simple liquid. Adapted from Ref. [37]

in relation to that of a "normal" liquid devoid of associating functional groups, data for the van der Waals fluid polyphenyl ether (5PPE) are included in Fig. 3c [37]. For 5PPE it is seen that $\eta'(\nu < 0.1\nu_{\max})$ is roughly constant. Intermediate behavior is displayed by 2-ethyl-1-hexylamine and by 2-ethyl-1-hexanethiol [37]. These data imply that the amine group enhances the zero-shear viscosity significantly, while corresponding effects are only weakly developed for the substance featuring the thiol group. Furthermore, shear rheology data raise the possibility that in halogen-substituted 2-ethyl-1-hexyl derivatives a faint halogen-bond-related low-frequency feature may be present [46].

So far, this section dealt with *shear relaxation* studies of Debye mode and structural relaxation. Importantly, the Debye process can be detected via the *bulk relaxation* as well, implying that this process couples to density fluctuations [47]. This finding is compatible with observations that changes in the local association patterns occur in terms of fluctuations from predominantly chain-like (more compact [9]) to ring-like (less compact) supramolecular structures and vice versa.

Returning to shear mechanical spectroscopy, it is useful to explore the domain not only of small-amplitude perturbations but also of large-amplitude perturbations, in other words of nonlinear excitation. For polymers such studies have a long history, see, e.g., Ref. [48], and constitute an active field producing interesting recent developments [49, 50].

Corresponding methods were barely applied to MAs; Fig. 4 represents recent activities [51]. Here the linear shear mechanical loss spectrum of 2E1H from Fig. 3b is compared with experimental results obtained under conditions of small- and medium-amplitude oscillatory shear, i.e., for various mechanical stress amplitudes γ. Akin to data for other supramolecular systems [49, 52] the third-order nonlinearity parameter ${}^3Q_0(\nu)$ of 2E1H displays a maximum on the low-frequency flank of the shear

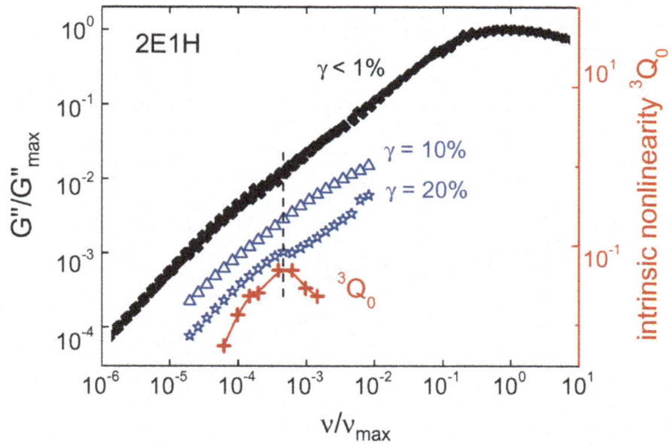

Fig. 4 Results of linear and nonlinear shear mechanical measurements carried out for 2E1H. The small- and medium-amplitude oscillatory shear results are represented as filled and open symbols, respectively, and refer to the scale on the left-hand side. The third-order nonlinearity parameter $^3Q_0(\nu)$ (scale on the right-hand side) is represented by the crosses and the vertical line highlights its frequency-dependent maximum

mechanical spectrum. Experiments carried out at $\gamma = 20\%$ indicate peculiar behavior near the peak in $^3Q_0(\nu)$ as well.

2.2 *Dielectric Spectroscopy*

Dielectric spectroscopy and its optical (solvation-dynamics) variant [53] as well as nonlinear *electrical* excitation, the latter not dealt with here, have been summarized in detail elsewhere [54, 55]. Among the homologous series listed above, in the present section we will focus on group H, i.e., on branched alcohols of the type $CH_3(CH_2)_{j+2}(CH_2CHOH)(CH_2)_jCH_3$ which contain $\ell = 2j + 6$ carbon atoms. Figure 5 depicts dielectric loss peaks for 2E1H ($\ell = 8$, blue dots), 2B1O ($\ell = 12$, black triangles), 2H1D ($\ell = 16$, green diamonds), and 2D1T ($\ell = 24$, red stars) as well as sketches of some of these molecules. The acronyms of these and a few other substances relevant for the present work are defined in Table 1 along with their glass transition temperatures [56–61].

Relative to the losses corresponding to the structural relaxation, in Fig. 5 it is 2E1H which displays the largest Debye-type low-frequency peak. The temperatures for the spectra of all of the MAs shown in Fig. 5 are chosen such that the mainpeak frequencies are near 2 Hz. Since the molecular dipole moments of MAs, $\mu \approx 1.7$ D, vary little from substance to substance [18, 62], the major loss amplitude variation is governed by mutually differing orientational correlations of the molecular electric dipole moments. In other words, the Kirkwood factor [63]

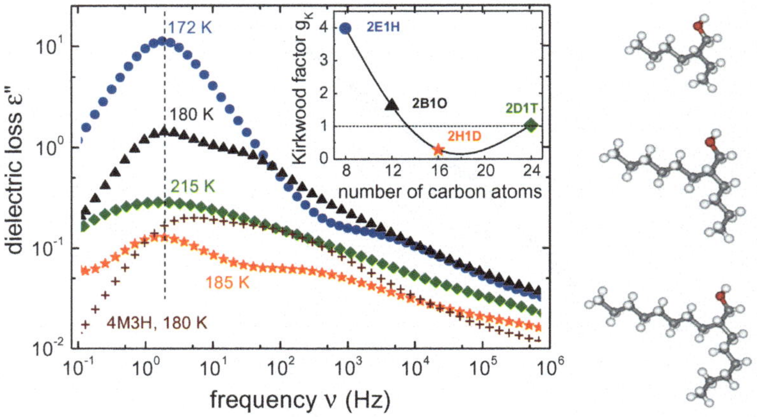

Fig. 5 Dielectric loss spectra of several branched MAs. From top to bottom the structures of 2E1H, 2B1O, and 2H1D are schematically represented on the right-hand side of the figure. The frequency position of the mainpeak is highlighted by the vertical line. For reference purposes, data on 4M3H are included as crosses. For the homologous series the inset shows g_K as function of the number of carbon atoms. The solid line guides the eye and the dashed line marks g_K=1. Adapted from Ref. [10]

$$g_K = \frac{9\varepsilon_0 k_B T}{n\mu^2}\frac{(\varepsilon_s - \varepsilon_\infty)(2\varepsilon_s + \varepsilon_\infty)}{\varepsilon_s(\varepsilon_\infty + 2)^2}, \tag{1}$$

a suitable measure of such correlations, is strongly ℓ dependent; see the inset of Fig. 5 [10]. In Eq. (1) ε_s and ε_∞ are the static and high-frequency permittivities, respectively, and n denotes the number density. Thus, for 2E1H the large g_K indicates chain-like molecular association, while for 2H1D the small g_K implies a near cancelation of the effective dipole moment. The latter scenario can be rationalized if ring-like structures prevail. Incidentally, for 4M3H g_K ($\approx$0.2) is even smaller than for 2H1D. The spectra of all substances shown in Fig. 5, except for 2D1T, exhibit two more or less well resolved relaxation peaks: the slower Debye process and the faster α-relaxation. The molecule with the longest alkyl chain, 2D1T, exhibits a single-peak structure. Here, the absence of a Debye process as well as a Kirkwood factor close to 1 is both indicative for "normal" supercooled-liquid behavior. Obviously the large intramolecular dilution of the hydroxyl groups in 2D1T suppresses suprastructure formation very efficiently.

Let us now discuss the substituted 2-ethyl-1-hexyl derivatives which are representatives of group J according to the classification defined at the beginning of Sect. 2 and for which shear mechanical moduli were presented in Fig. 3c. The only 2-ethyl-1-hexyl derivative, apart from 2E1H, that displays an immediately discernible low-frequency dielectric loss feature is 2-ethyl-1-hexylamine [37, 64]. For 2-ethyl-1-hexanethiol no clear spectral signatures of a Debye-like process could be discerned [37].

Table 1 Acronyms of some MAs and related substances for which data are shown in the present chapter

ℓ	Name	Acronym	$T_{g,cal}$/K	$T_{g,\alpha}$/K	Reference
3	1-propanol	PrOH	96	97	[56]
4	1-butanol	BuOH	113.4	113.6	[57]
	1-bromobutane	BuBr	103.3	99.6	[58]
5	2-methyl-1-butanol	2M1B	123.6	120.0	[59]
6	2-ethyl-1-butanol	2E1B	131.0	129.4	[59]
8	2-ethyl-1-hexanol	2E1H	145.9	144.0	[59]
	4-methyl-3-heptanol	4M3H	162.2	160.8	[59]
	2-ethyl-1-hexylbromide	2E1Br	–	131.0	[60]
10	3,7-dimethyl-1-octanol	3,7D1O	–	152.8	[61]
12	2-butyl-1-octanol	2B1O	160.5	159.0	[11]
16	2-hexyl-1-decanol	2H1D	172.1	171.0	[11]

The number ℓ of carbon atoms and, where available, the glass transition temperatures determined using calorimetry ($T_{g,cal}$) as well as using dielectric spectroscopy ($T_{g,\alpha} = T(T_\alpha = 100s)$) are listed as well

However, when judging weak dielectric low-frequency features caution has to be exercised. This is because owing to its enormous sensitivity dielectric spectroscopy is able to detect the response even of trace amounts of (dipolarly active or charged) impurities. Unwanted charge carriers are an issue in many studies of dielectric properties since the vast majority of dielectric laboratories (including that of the authors) are not regularly taking special measures to purify the liquids or sweep out the ions [65], but use them as provided by the supplier. Thus, almost inevitably dc electrical conductivity, σ_{dc}, is present in the samples, leading to a low-frequency loss $\varepsilon''(\nu) \propto \sigma_{dc}\, \nu^{-y}$ of a magnitude that depends on the number of (unspecified impurity) charge carriers. If only dc conductivity is present, the exponent should be $y = 1$. Yet, typically somewhat smaller values are observed that can be interpreted in terms of constant-phase elements. Consequently, in addition to effects of electrode polarization [66, 67] conductivity effects can "spill over" to the real part of the dielectric function, then leading to an increase of $\varepsilon'(\nu)$ toward frequencies $\nu \rightarrow 0$. Such an upturn usually commences for almost any liquid only a few decades below the dispersion step that is related to the structural relaxation. However, most of the considered 2-ethyl-1-hexyl derivatives display a very long plateau in $\varepsilon'(\nu)$.

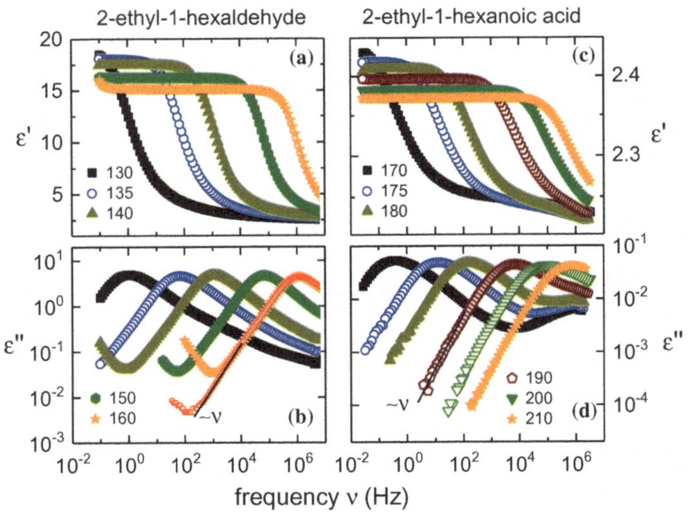

Fig. 6 Complex dielectric function of 2-ethyl-1-hexaldehyde is shown in frames (**a**) and (**b**) and that of 2-ethyl-1-hexanoic acid is shown in frames (**c**) and (**d**) [68]. In frame (**b**) the dielectric loss curve extending to lower ε'' values was obtained from a Kramers–Kronig analysis of the $\varepsilon'(\nu)$ data [65]

As an example Fig. 6 depicts dielectric data for 2-ethyl-1-hexaldehyde and 2-ethyl-1-hexanoic acid. The low-frequency plateau of both materials extends over a range of six or more decades. Nevertheless, for 2-ethyl-1-hexaldehyde conductivity effects preclude the observation of a low-frequency feature in the imaginary part of the dielectric function. Also a Kramers–Kronig analysis [65] of the real part does not reveal a clear signature of an additional process. For 2-ethyl-1-hexanoic acid the electrical conductivity is very small and at the lowest frequencies the dielectric loss obeys an $\propto\nu^{+1}$ dependence. This behavior changes at frequencies much lower than those characterizing the α-peak, cf. Fig. 6d, and signals weak contributions of additional slow dynamics. At this time, it remains open whether this feature is related to the Debye-type processes observed for MAs or whether it reflects a local excitation. A conformational change of a COOH group is held responsible for a weak Debye-like process in the pharmaceutical ibuprofen [69].

In any case, several of our dielectric experiments suggest that the conductivity relaxation time, e.g., defined as $\tau_\sigma = \varepsilon_s\varepsilon_0/\sigma_{dc}$ [70], may be shorter in the presence of supramolecular association and longer in its absence.

2.3 *Infrared Absorption*

An entirely different perspective on supramolecular association is provided by infrared spectroscopy of those absorption bands that change when the degree of association is altered, e.g., by temperature or pressure variation. Excellent reviews exist on MAs with a focus on ultrafast two-dimensional (2D) chemical exchange spectroscopy [71], 2D correlation analysis [72], and other aspects [73]. Despite exciting recent developments in the field of ultrafast spectroscopies, this topic is beyond the scope of this review. Rather than citing pertinent recent work, let us make reference to the extensive but necessarily incomplete literature surveys in Refs. [7, 74]. All of the various IR regimes including the far-, mid, and near-infrared spectral ranges are interesting to study because they can yield complementary information on the supramolecular hydrogen bonding in MAs and related materials.

In the present section we will focus on the near-infrared (NIR) range and as an example Fig. 7 presents NIR spectra of various 2-ethyl-1-hexyl derivatives. In this figure the vertical dashed lines subsection the covered range of wavelengths λ (or wavenumbers $\bar{\nu}$) according to more general considerations [75, 76]: Below about $\lambda < 1350$ nm second overtones of CH stretching vibrations occur. Indeed, for all of the 2-ethyl-1-hexyl derivatives for which we collected spectra the so-called 3ν(CH) band absorbs near 1200 nm. For 1350 nm$< \lambda <$1660 nm the first overtones of CH- or NH-containing functional groups appear. Then, in the range from about 1660 to 1900 nm the first overtones, 2ν(CH), of the CH group stretching vibrations as well as other molecule-specific excitations contribute significantly. The brominated species display a band near 1670 nm that originates from a combined deformation and symmetric stretching vibration of the CH moiety next to the halogen atom [60]. For 2-ethyl-1-hexanoic acid a prominent first overtone of the C=O bond is expected to occur in the mid-infrared at 1725 cm^{-1}, i.e., outside the present range. Its second overtone is detected near 1900 nm, but it is only weakly pronounced and overlaid by broad spectral features originating from dimer structures that typically form in carboxylic acids [77] and thus presumably in 2-ethyl-1-hexanoic acid as well. Apparently, the infrared absorption of 2-ethyl-1-hexanethiol has not been measured so far and information on other thiols also seems to be scarce [78].

The MAs show significant excess intensity in the $2\nu(X\mathrm{H})$ (with $X = $O,N) regime, cf. Fig. 7: The most prominent features are a narrow, so-called monomer (α) band at about 1410 nm and a so-called polymer (δ) band near 1580 nm. The inset of Fig. 8 summarizes the assignment of these bands in terms of their local OH-bonding patterns and their acronyms. The temperature dependence of the various bands has been discussed previously [8, 74, 79]: Most MAs display common features in the 1350–1660 nm regime. As an example, Fig. 8 presents NIR spectra of 2H1D. These data add to the collection published for several other MAs [7, 8, 45, 61, 80]. Regarding intensity and wavelength position, in Fig. 8 the α- and δ-bands are seen to undergo major changes with temperature. Largely temperature independent are the peak wavelengths of the so-called terminal (γ) band as well as that of the feature near 1540 nm which has been assigned to oligomeric (o) species [74].

In the 1350 nm$< \lambda <$1660 nm range which is of prime interest in the present context, the most intense absorption is observed for 2-ethyl-1-hexylamine. Here, first overtone bands due to symmetric, $2\nu_s(NH_2)$, and asymmetric, $2\nu_a(NH_2)$, NH stretching vibrations appear at 1525 and 1485 nm, respectively. The weakest features (on the broad background generally observed for many aliphatic compounds [81])

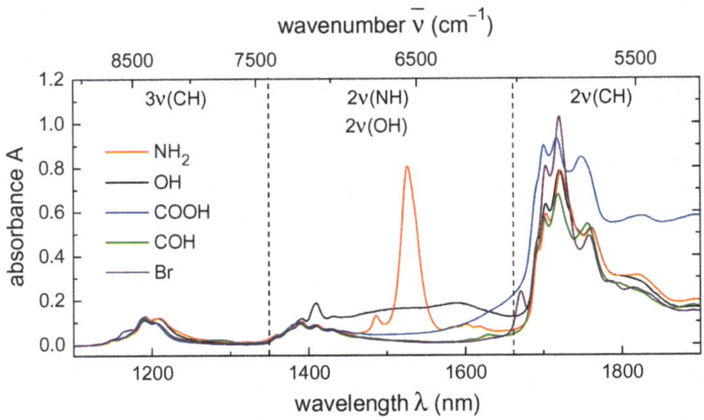

Fig. 7 NIR spectra of several 2-ethyl-1-hexyl derivatives measured at 300 K. The functional groups include the amino (NH_2), the hydroxyl (OH), the carboxyl (COOH), and the aldehyde (COH) moiety. A spectrum for the brominated compound is also shown. Vertical lines separate the various 3ν(CH), 2ν(NH) and 2ν(OH), as well as 2ν(CH) overtone regimes discussed in the text

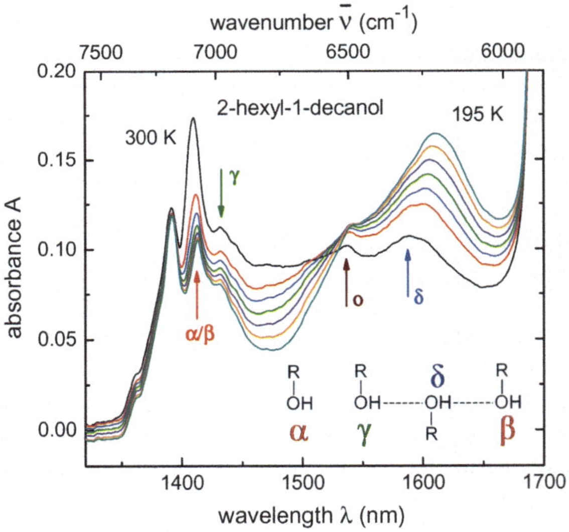

Fig. 8 Temperature-dependent NIR spectra of 2H1D. The inset illustrates the assignment of various bands, and the arrows indicate their spectral positions. The designation α/β-band is used here because in this type of experiment the contributions originating from the β-band are often hard to resolve from those of the α-band; see, however, Ref. [8]

are seen for 2-ethyl-1-hexaldehyde with a barely resolved group-specific absorption band near 1620 nm.

Expectedly and in accordance with the experimental observations, as the temperature decreases, more and more molecules become part of supramolecular structures so that the monomer band decreases and the polymer band increases in intensity. For 2H1D Fig. 9 reveals these trends in a quantitative fashion. In frame (a) the peak absorbances of the various bands are summarized. All band intensities except those relating to the polymeric species display slope changes near 250 K, an observation

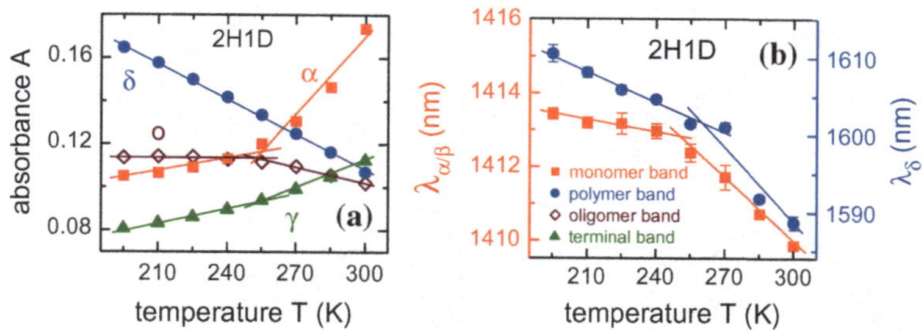

Fig. 9 **a** Peak amplitudes and **b** wavelength positions of various absorbance bands determined from the NIR spectra of 2H1D presented in Fig. 8. Lines are drawn to guide the eye and suggest a slope change to occur near 250 K

previously called the "250 K anomaly" [61, 80]. Frame (b) reveals that the wavelength position of the monomer *and* polymer bands, i.e., of $\lambda_{\alpha/\beta}$ and λ_δ, respectively, both reflect this phenomenon. Related observations were made for a number of MAs not only using IR experiments but also using other techniques such as Raman scattering [82], proton magnetic resonance, and dielectric spectroscopy [79].

In Fig. 9b the peak wavelengths of the various bands are seen to differ with respect to their temperature dependence and they do so for different reasons. First, one recognizes that $\lambda_{\alpha/\beta}$ varies by less than 4 cm^{-1} over a temperature range of more than 100 K. This change can be rationalized by noting that upon cooling the unresolved β-band (with $\lambda_\beta > \lambda_\alpha$ [83]) becomes more intense and the α-band less important. Conversely, the red-shift of the δ-band, that evolves upon cooling, reflects a concomitant overall strengthening of hydrogen bonding, so that the covalent OH bond which is probed in the NIR range is slightly weakened and its stretching frequency decreases.

Although vibrational spectroscopy provides an exquisite local probe, phenomena that are believed to be collective, including the glass transition, can be probed by this technique as well [84, 85]. Recent MA examples include 4M3H, 2E1H, 5-methyl-2-hexanol (5M2H), and 3,7-dimethyl-1-octanol (3,7D1O) [61]. Corresponding data are collected in Fig. 10 where near T_g the δ-bands of all these glass formers show a kink. This means that the readjustment of the hydrogen-bond equilibrium among the "polymer" and the monomer (or small-associates) states which drives the increase of the δ-band with decreasing temperature, freezes in at the glass transition. Thus, at T_g equilibration comes to a halt and the band intensity, $A(\lambda_\delta,T)$, ceases to increase on the time scale set by the experiment.

Akin to studies of physical aging, after temperature jumps a glass former may reach equilibrium in the course of prolonged waiting periods. Depending on the details, including the concentration of the molecular species participating in such an equilibration process, the establishment of equilibrium can either (i) proceed on the time scale of the structural relaxation [86] or (ii) take much longer than that; see, e.g., Fig. 2.16 of Ref. [87]. Case (i) applies here since the coincidence of the slope change in $A(\lambda_\delta,T)$ (i.e., of the absorbance T_g) with the calorimetric

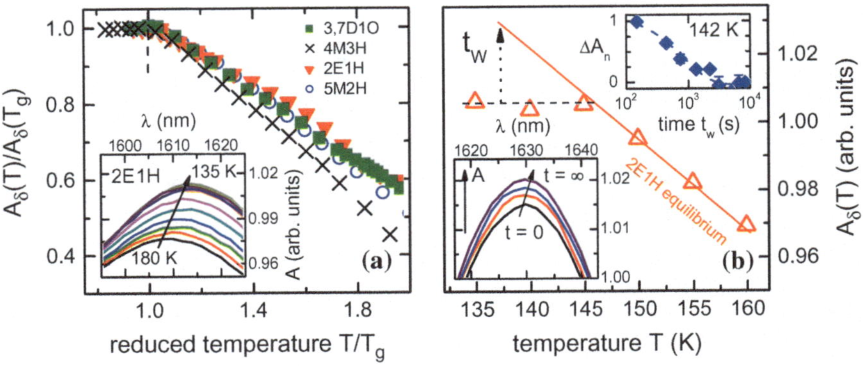

Fig. 10 **a** NIR detection of the glass transition via the maximum absorbance of the polymer band, $A_\delta(T) = A(\lambda_\delta, T)$, in 4M3H, 3,7D1O, 5M2H, and 2E1H. Typically the effective cooling rate in these experiments was −0.5 K/min. The dashed line marks the glass transition temperature T_g. The inset depicts the temperature-dependent band maximum which below T_g ceases to increase. **b** Enlarged view of the temperature-dependent polymer-band absorbance $A_\delta(T)$ near the glass transition of 2E1H. The lines are drawn to guide the eye, and the arrow schematically illustrates the evolution of the absorbance during a waiting time t_w elapsing subsequent to a down-jump in temperature. The lower inset in panel (**b**) shows how the polymer band evolves with t_w. The upper inset in panel (**b**) presents suitably normalized peak values, $\Delta A_n(t_w)$, and the line represents an exponential function, $\exp(-t_w/\tau_{ex})$, with a time constant $\tau_{ex} \approx \tau_\alpha$. Adapted from [Ref. 61, 79]

T_g indicates that in 2E1H the hydrogen-bond equilibration time is essentially given by the structural relaxation time. Stepping the temperature to below T_g, the time-dependent absorbance intensities shown in the inset of Fig. 10 confirm this finding directly. Here, the results of aging experiments are shown that were carried out for 2E1H at a base temperature of 142 K, about 4 K below its calorimetric T_g [79]. Experiments of the type summarized in Fig. 10 have found widespread applications also in other contexts [88].

2.4 X-Ray Diffraction

In view of the survey on diffraction studies of pure MAs given in Ref. [7], a brief introduction may suffice here. For MAs prepeak intensities detected via neutron diffraction, are typically much smaller [7, 89–91] than those from X-ray diffraction and will not be considered in the following. As an example of an X-ray diffraction pattern, Fig. 11a shows data of 2E1H from which a prepeak (at a momentum transfer of $q_{pre} \approx 5.4$ nm^{-1}) and a mainpeak (at $q_{main} \approx 13.7$ nm^{-1}) can be discerned. Numerical simulations are most useful to gain a detailed understanding of the correlations which give rise to specific parts of such diffraction patterns; see, e.g., Refs. [19, 92–94]. With simulation results not at hand for the MAs for which diffraction data are shown in this chapter, a simple decomposition of the scattering intensity,

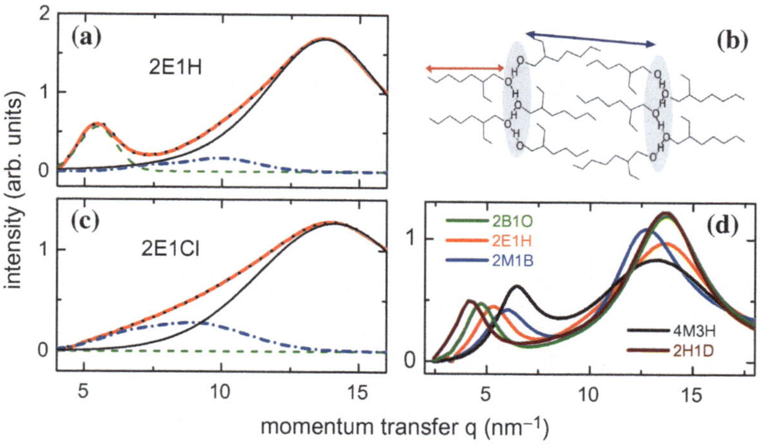

Fig. 11 X-ray diffraction patterns **a** for 2E1H and **c** for 2E1Cl (red solid lines) and their simple decomposition in terms of a prepeak (green dashed lines), a mainpeak (black solid lines), and additional contributions (blue dash-dotted lines); see the text for details. The sum of all contributions is represented by the dots. **b** Sketch of envisioned structure formation in 2E1H. The double arrows define typical molecular length scales that may be relevant in connection with the prepeak. The longer blue double-arrow marks a distance of OH-bonded supramolecular chains and the shorter red double arrow symbolizes a length relating to the effective number of carbon atoms. Adapted from Ref. [46]. **d** Compilation of diffraction patterns of 4M3H, 2M1B, 2E1H, 2B1O, and 2H1D, i.e., of MAs that cover an appreciable range of effective chain lengths [46, 96]

$I(q)$, was performed. Apart, from the prepeak and mainpeak, additional contributions occur at intermediate momentum transfers. The latter can be ascribed to alkyl–alkyl correlations as well as, in the respective samples, to scattering originating from the presence of the halide atoms [19, 46]. In order to estimate the relative intensities in the various q-regimes, the prepeak and mainpeak intensities were described by Pearson functions [95], while for the intermediate intensity a superposition of two Pearson functions was chosen. The results of the corresponding fits for 2E1H, cf. Fig. 11a, capture the measured intensity in the shown q-window very well. This decomposition procedure is not entirely unambiguous, and a comparison with different analyses [46, 96] allows one to assess the uncertainty of the partial scattering contributions.

The determination of the prepeak and mainpeak position is rather insensitive to such considerations. Hence, in simple terms via $d_{\mathrm{pre}}=2\pi/q_{\mathrm{pre}}$ a supramolecular length scale of about 1.2 nm can be associated with the prepeak which essentially reflects intermolecular O–O correlations; see the sketch in Fig. 11b. Conversely, correlations among the carbon atoms are encoded in the mainpeak. When varying the alkyl chain length (or the number ℓ of carbon atoms) the mainpeak thus changes very little (for $\ell \geq 4$), cf. Figure 12 [19, 97]. However, as Fig. 11d reflects, the prepeak positions for 2H1D [10] and for 2B1O [96] appear at momentum transfers q that are smaller than those for 2E1H. Prepeak q-values larger than for 2E1H are observed

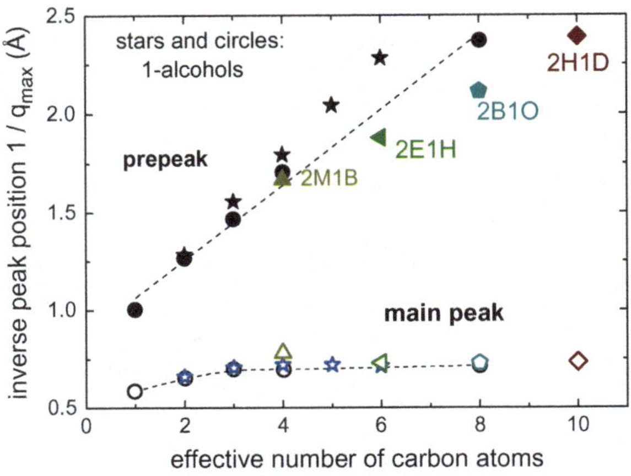

Fig. 12 Inverse momentum-transfer position, $1/q_{max}$, of the peaks in the diffraction patterns as a function of the (effective) number ℓ_{eff} of carbon atoms which (only) for the normal (or 1-) alcohols coincides with the total number of carbon atoms ℓ. The circles reflect data from Ref. [97], the stars from Ref. [19], and the other (colored) symbols were assessed from the patterns for those MAs shown in Fig. 11d that have two alkyl branches. Lines are drawn to guide the eye

for 2M1B and for 4M3H. In alkyl halides a prepeak is not discernible [46, 98, 99] as confirmed in Fig. 11c for 2-ethyl-1-hexylchloride (2E1Cl). Yet, for 2E1Cl significant and broad intensity contributions appear at $q \approx 8\ \text{nm}^{-1}$, see again Fig. 11c, that are much more intense than the corresponding contributions detected for 2E1H near these momentum transfers. This observation shows that for 2E1Cl the intermediate feature is dominated by halogen-related scattering.

The positions $1/q_{main}$ at which the mainpeaks in the diffraction patterns appear show little variation with the length of the alkyl chain, see Fig. 12, thus essentially reflecting the similar densities of the considered alcohols. While it may be difficult to rationalize the prepeak positions generally [100], the normal alcohols display systematic trends [16, 19, 97] as a function of the number ℓ of carbon atoms. The data for the branched MAs compiled in Fig. 12 are represented in terms of a differently defined "effective" number ℓ_{eff} of carbon atoms. This quantity is taken here to reflect the separation of the OH group from the most distant CH_3 group; see the red arrow in Fig. 11b. Other measures for effective lengths of branched MAs can be defined as well [10, 101].

3 Monohydroxy Alcohols Mixed with Alkyl Halides

Countless articles deal with alcohol mixtures. Yet, in the field of solution chemistry the focus is typically on temperatures near ambient. In spite of even attempting to give a survey, here the focus will be on mixtures of MAs with aprotic alkyl halides.

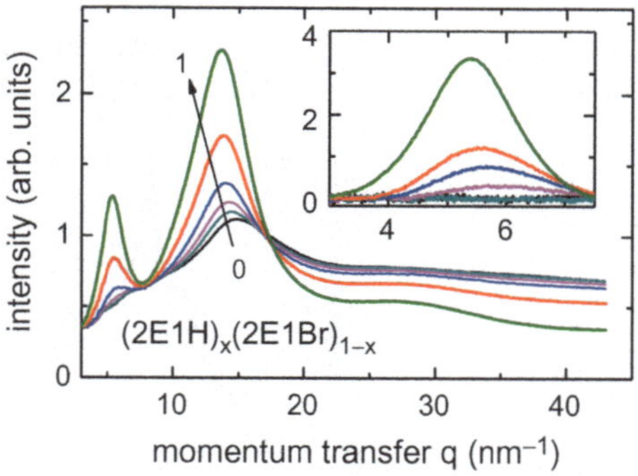

Fig. 13 Synchrotron-based X-ray diffraction intensities of $(2E1H)_x(2E1Br)_{1-x}$. The patterns in the mainframe, with concentrations $x = 0$, 0.23, 0.38, 0.60, 0.83, and 1.0, were obtained at an incident photon energy of 27 keV. The data in the inset refer to an incident photon energy of 10 keV and here the concentrations $x = 0$, 0.20, 0.40, 0.60, 0.80, and 1.0 are shown. Adapted from Ref. [46]

One of the first pertinent low-temperature investigations, on a mixture of iso-amyl bromide and 1-propanol [102], has appeared almost 60 years ago. While numerous studies on mixtures involving alkyl chlorides and bromides followed, we are not aware of ones involving alkyl fluorides or iodides [103]. The present section focuses on recent work involving alkyl halides and will not deal with other systems such as MA mixed, e.g., with water or with alkanes. Thus, solutions of, e.g., PrOH [104], pentanol [105], or 5-methyl-2-hexanol [106] with 2-methyl-pentane or other aprotic solvents [107] do not fit the chosen constraint.

3.1 *Mixing Molecules with Equally Long Alkyl Chains*

Building on previous work on alcohol/alkyl halide mixtures involving partners with different alkyl chain lengths, see Sect. 3.2, our first investigation in the field dealt with dielectric properties of BuOH (μ = 1.66 D) mixed with BuBr (μ = 2.08 D) [108]. A most surprising finding of this work was the appearance of a so-called critical concentration, underscored also in subsequent studies [94, 109]. Next, we will summarize X-ray, NIR, dielectric, and shear mechanical work on related binaries such as $(2E1H)_x(2E1Br)_{1-x}$ [60] and $(2E1H)_x(2E1Cl)_{1-x}$ [46].

First, in Fig. 13 we show diffraction patterns obtained for $(2E1H)_x(2E1Br)_{1-x}$ over the entire range of concentrations [46]. In separate runs the scattering intensities $I(q)$ were obtained in different momentum-transfer ranges and made it possible to assess the positions, widths, and intensities in the regimes of the prepeak and the mainpeak. Below, we will focus on the concentration dependence of just their intensities as obtained from the simple analysis described in Sect. 2.4. More elaborate analyses of these data can be found in Ref. [46].

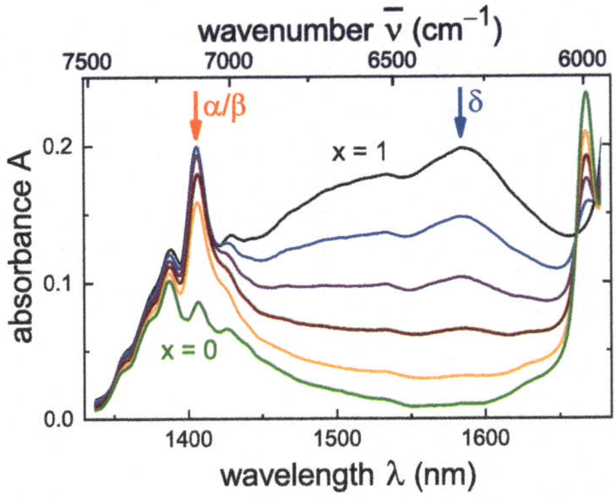

Fig. 14 NIR spectra of $(2E1H)_x(2E1Br)_{1-x}$ with x in steps of 0.2. The arrows highlight the approximate positions of the maxima corresponding to the monomer band and the polymer band. The peak near 6000 cm^{-1} reflects an absorbance band involving the CH_2 group closest to the Br atom. Adapted from Ref. [60]

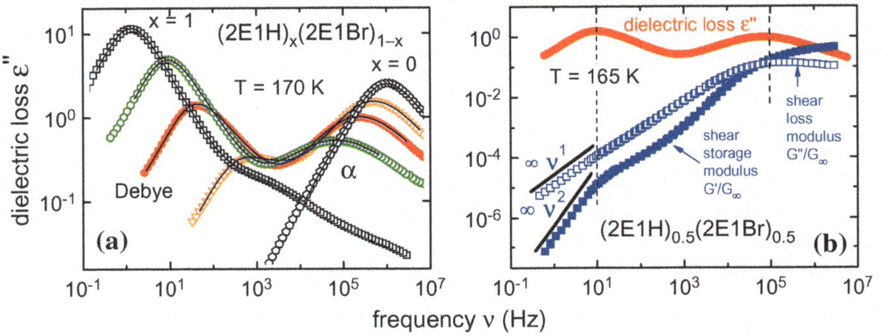

Fig. 15 **a** Dielectric loss spectra of $(2E1H)_x(2E1Br)_{1-x}$ for alcohol concentrations of $x = 0$, 0.31, 0.50, 0.69, and 1.00. For 2E1H the Debye response prevails and for $x = 0$ only the α-relaxation can clearly be recognized. **b** Comparison of dielectric loss spectrum and master curves based on shear mechanical data for $(2E1H)_{0.5}(2E1Br)_{0.5}$. Adapted from Refs. [74] and [60]

The $(2E1H)_x(2E1Br)_{1-x}$ system was investigated also using NIR spectroscopy [60]: From Fig. 14 one recognizes that the overtone intensities of the monomer and polymer bands both increase continuously as the 2E1H concentration increases and simultaneously the Br-related band loses intensity. Before comparing these results with those from X-ray diffraction, let us look at what the dynamic probes, dielectric spectroscopy and rheology, yield for $(2E1H)_x(2E1Br)_{1-x}$. Figure 15a shows how the dielectric loss peaks of this mixture series evolve: For $x = 1$ a dominant Debye peak is seen which becomes weaker as 2E1Br is added. Concomitantly the α-process, in the neat alcohol only displaying a small amplitude, gains considerable strength for $x \to 0$. The latter trend is dominated by the circumstance that the amplitude of the structural relaxation is anomalously small in 2E1H and that 2E1Br has a larger electrical dipole moment than 2E1H does. Thus, with $\upsilon_{1,2}$ denoting the volume fractions of the two components, the average dipole moment $\langle\mu\rangle$ should interpolate between the moments μ_1 and μ_2 of each end member according to [110]

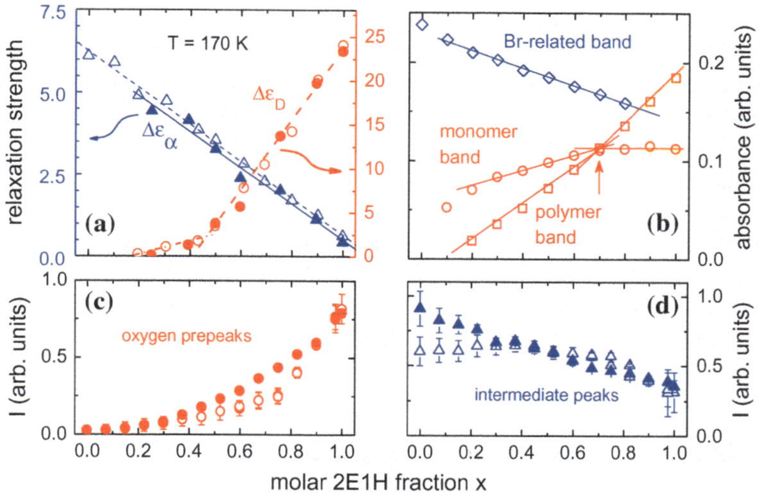

Fig. 16 Composition dependence of various quantities measured for $(2E1H)_x(2E1Br)_{1-x}$ (open symbols) and $(2E1H)_x$ $(2E1Cl)_{1-x}$ (filled symbols) from Refs. [60] and [46]. **a** Dielectric relaxation strengths of α- and Debye process. **b** Intensities of the Br-related band as well as of the monomer and polymer NIR absorption bands. Like in Ref. [60] the intensities of the latter two bands are plotted as differential absorbance, $\Delta A(\lambda,x) = A(\lambda,x) - A(\lambda,0)$. **c** Prepeak intensities and **d** integrated intensities of the intermediate peaks from X-ray diffraction [46] (but analyzed as described in Sect. 2.4). All data except those in panel **a** refer to a temperature of 300 K

$$\langle \mu \rangle = \sqrt{\upsilon_1 \mu_1^2 + \upsilon_2 \mu_2^2}. \tag{2}$$

An additional effect is expected to arise in alcohol-rich structures. Here, akin to the situation in certain types of polymers [111, 112], essentially only the dipole moment components that are oriented perpendicular to the OH-bonded "backbone" contour are expected to contribute to the amplitude of the dielectrically detected α-process.

Figure 15b emphasizes another aspect for an equimolar mixture of 2E1H and 2E1Br: Here, the separation of the two dielectric relaxation processes, that amounts to about four orders of magnitude, is similarly resolved by the shear modulus. This dynamical separation (or decoupling of processes) is composition dependent as will be discussed further in Sect. 5.2. In the present context we are most interested in the x dependence of the peak amplitudes.

Figure 16 summarizes the intensity information available from the diffraction peaks, the infrared absorption bands, and the dielectric amplitudes for $(2E1H)_x(2E1Br)_{1-x}$ and, where available, also for $(2E1H)_x(2E1Cl)_{1-x}$. Quantities reflecting properties of the *alcohol* show deviations from a simple linear behavior. It is remarkable that all different experimental probes yield evidence for a critical concentration. Conversely, properties dominated by the *alkyl halide* do exhibit a linear behavior (referred to as ideal mixing) to a very good approximation, except for the X-ray data of $(2E1H)_x(2E1Br)_{1-x}$ with $x < 0.3$, cf. Fig. 16d.

Due to a lack of sufficient viscoelastic data for the $(2E1H)_x(2E1Br)_{1-x}$ or $(2E1H)_x(2E1Cl)_{1-x}$ systems, at this point it is not possible to test whether, e.g., the scaled zero-shear viscosity z_0 (to be introduced in Sect. 4.4) would conform to the trends seen in Fig. 16 for the quantities related to the Debye process. In view of the finding that "the slower viscoelastic relaxation of the alcohols is assigned to the relaxation of the heterogeneous structure described by the prepeak" [113], such a conformity can be expected.

In the series of alkyl halide mixtures discussed here, the pure alcohol, 2E1H, displays a large Debye process. In mixtures of 2E1Br with MAs featuring a small Debye process (like 4M3H [114] or certain branched MAs) interesting new phenomena occur that will be discussed elsewhere [115].

3.2 Mixtures of Short and Long Alkyl Chains

Earlier dielectric investigations of corresponding systems found remarkable non-monotonic variations of glass transition temperatures $T_g(x)$ with concentration, notably in trichloroethylene, chloroform, and other halides diluted in various normal alcohols [116]. This type of non-ideal mixing is not unique to solutions with halides but is displayed also when other solvents such as, e.g., triethylamine [117] or 2-ethylhexylamine [118] are chosen as the second component. A large body of work on alkanols mixed with CCl_4 is summarized in Refs. [119, 120]. An approach useful to investigate the impact of molecular size variation is to compare mixtures in which the glass transition temperature of the alkyl halide is either (i) larger or (ii) smaller than that of the MA, with isoamylbromide plus 1-propanol (PrOH) [121] or plus 5-methyl-2-hexanol [122], respectively, as examples. Regarding case (ii) additional studies on $(4M3H)_{0.89}(BuBr)_{0.11}$ [123] and on $(2E1H)_x(BuBr)_{1-x}$ [124, 125] are available to which the interested reader is referred.

Regarding case (i) for which the glass transition temperature of the alkyl halide is much higher than that of the MA, Fig. 17 shows dielectric loss spectra for $(PrOH)_x(2E1Br)_{1-x}$. In frame (a) one recognizes how, starting from the pure alkyl halide, upon moderate addition of PrOH a minor low-frequency Debye-like feature appears. The Debye process becomes dominant in the pure MA, while the relaxation strength of the α-process remains more or less the same. As a function of temperature, cf. Fig. 17b, an even more interesting observation can be made for $(PrOH)_{0.15}(2E1Br)_{0.85}$: At the highest T the two processes are clearly discernible; their time scales, however, approach each other upon cooling so that toward lower temperatures only a single relaxation process is visible.

An impressive view of this "approach" of processes is gained from Fig. 18a, a doubly scaled plot. It is clearly seen how the Debye-like feature originating from propanol becomes submerged under the dominant α-peak of the halide-rich mixture. Similar phenomena are known for 2E1H/squalane mixtures [126] and for a number of polymers [127]. As an example, the inset of Fig. 18 shows data for oligomeric propylene glycol [128]. For polymers, the temperature-driven approach of the segmental

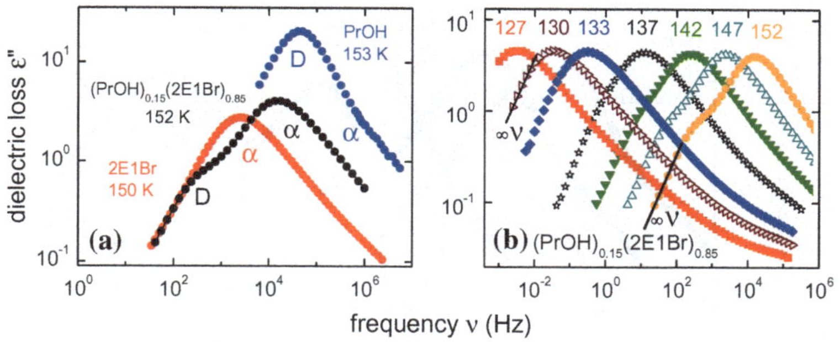

Fig. 17 **a** Dielectric loss spectra of $(PrOH)_x(2E1Br)_{1-x}$ acquired near 150 K for several x. **b** Temperature-dependent dielectric loss spectra of $(PrOH)_{0.15}(2E1Br)_{0.85}$

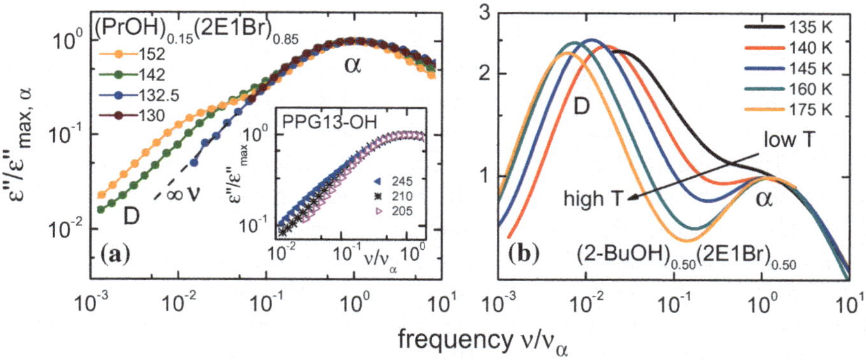

Fig. 18 Plot of doubly scaled dielectric loss versus frequency as generated from the data shown in Fig. 17a. Temperatures are given in Kelvin. For comparison the inset presents analogously scaled dielectric losses of a 13-mer of OH-terminated poly(propylene glycol), abbreviated here as PPG13-OH. The data in the inset are taken from Ref. [130]. **b** Dielectric loss of $(2\text{-}BuOH)_{0.5}(2E1Br)_{0.5}$ scaled with respect to the maximum values corresponding to the α-process, $\varepsilon''(\nu)/\varepsilon''_{max,\alpha}$

modes (that lead to α-peak loss maxima) and the low-frequency normal modes was rationalized within various contexts [129]. Based on a study of poly(propylene glycol) for which chain lengths were systematically varied, it was suggested to exploit the mode coalescence as a molecular ruler [128]. Applying these ideas to MAs, the merging, taking place near 130 K, see Fig. 18a, indicates that the cooperativity length of the structural fluctuations roughly equals the average size of the supramolecular structures giving rise to the Debye mode. Investigations under spatial confinement could provide additional information on this merging phenomenon which seems to be general for liquids/melts supporting suprastructural relaxation dynamics. Finally, for polymers this approach of modes can also be traced using rheology [130] which is particularly useful for polymers devoid of a dipole moment along their backbone. In MAs associated OH groups do form the "backbone" so that a supramolecular dipole moment should always exist.

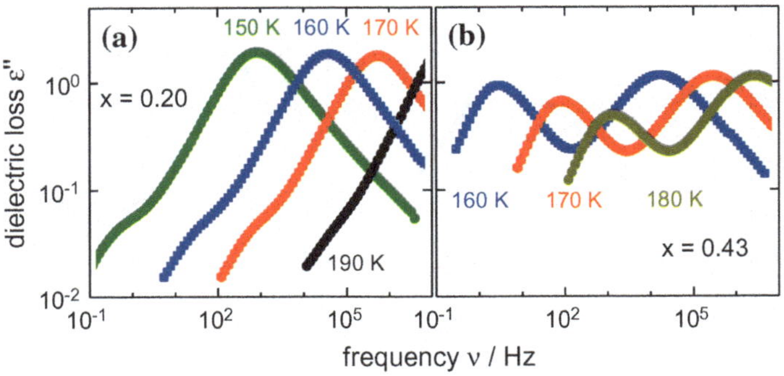

Fig. 19 Dielectric loss spectra of $(2E1H)_x(2E1Br)_{1-x}$ for alcohol concentrations of **a** $x = 0.2$ and **b** $x = 0.43$ [131]

This kind of behavior seems common to mixtures in which T_g of the MA is lower than for the halide: From the data in Fig. 18b one recognizes the tendency of merging toward low temperatures for a mixture of 2E1Br with 2-butanol (2-BuOH). The approach of modes looks, however, completely different from that in Fig. 17. This is because in $(PrOH)_{0.15}(2E1Br)_{0.85}$ the α-process dominates while in $(2\text{-}BuOH)_{0.5}(2E1Br)_{0.5}$ the Debye relaxation is the much stronger one.

In several pure and mixed MAs not a merging but an increasing *apparent separation* of α- and Debye process seems to occur as the temperature is lowered. Since we are not aware of mixtures of short and long alkyl chains revealing this kind of behavior, in Fig. 19a data for $(2E1H)_{0.2}(2E1Br)_{0.8}$ are presented. Two scenarios come to mind to rationalize the apparent separation of the processes: Either the Debye relaxation is more temperature dependent than the α-relaxation or, alternatively, the Debye process gains amplitude upon cooling. Solely based on the spectra shown in Fig. 19a, it is hard to distinguish these two options. In this situation it is helpful to compare these loss spectra with those of $(2E1H)_x(2E1Br)_{1-x}$ for $x > 0.2$ [60]; see the example in Fig. 19b. The Debye process is rapidly gaining amplitude as the temperature is lowered, an observation clearly in favor of the second explanation. The normalized spectra in Fig. 18b, however, hint to the first scenario. Thus, the origin of the apparent separation of processes appears to be system specific.

4 Monohydroxy Alcohols Mixed with Other Monohydroxy Alcohols

Despite some early work [132, 133] and the typically good miscibility of the components many of the low-temperature studies involving mixtures of two MAs seem to be more recent. A rough classification of mixing effects, cf. Fig. 2, may be achieved by choosing mixing partners that display either strong, intermediate, or weak Debye (-like) processes. Some studies start from liquids with pronounced Debye peaks

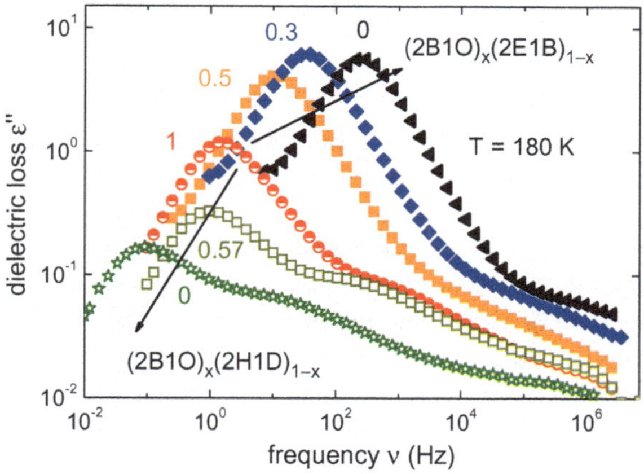

Fig. 20 Dielectric loss spectra of $(2B1O)_x(2E1B)_{1-x}$ (closed symbols) and $(2B1O)_x(2H1D)_{1-x}$ (open symbols). The loss spectrum of neat 2B1O is represented by half-filled symbols

such as ethanol (with hexanol added) [134] or PrOH (with 3,7D1O [12] or 4M3H [135] added), or 4-methyl-2-pentanol [136] mixed with 2E1B. For pure 2E1B several dielectric studies were performed [61, 136, 137]; not all yield the same relaxation strength [138].

Starting from 2E1H, mixtures with 2M1B [12], 2H1D [11], 4M3H [139], or methanol [140] were investigated, where 2H1D and 4M3H themselves display very small Debye-like features. Before turning to mixtures involving 4M3H in Sects. 4.1–4.4, we will first consider 2B1O with its intermediately strong Debye process (cf. Fig. 5) to illustrate mixing with liquids featuring either stronger or weaker Debye processes; see Fig. 20 for data recorded at 180 K. Starting from 2E1B, an MA with a large Debye process, admixture of a sufficient amount of 2B1O leads to a major reduction of the dominant loss peak. When going from 2B1O to 2H1D a further strong decline of the strength of the Debye peak occurs, consistent with what is expected from the Kirkwood factors of these MAs, cf. Fig. 5. Overall the dielectric loss frequencies shift to lower frequencies as one progresses from 2E1B to 2B1O to 2H1D, consistent with the concomitant increase of the glass transition temperatures of these three MAs. Furthermore, it is interesting to observe that the amplitude of the α-process shows little variation across the two series of mixtures.

A close inspection of the data for $(2B1O)_x(2E1B)_{1-x}$ reveals that the dielectric loss peak for the $x = 0.3$ composition is slightly larger than that for pure 2E1B. This remarkable manifestation of non-ideal mixing is also exhibited by $(\text{4-methyl-2-pentanol})_x(2E1B)_{1-x}$ with a slight maximum in the relaxation strength appearing for $x = 0.75$ [136]. For $(2E1H)_x(2M1B)_{1-x}$ mixtures a similar maximum appears for $x = 0.25$ [136] and none was reported to occur for 3,7D1O mixed with PrOH [12].

4.1 Compositional Evolution of Dielectric Debye and α-Process

A maximum in the susceptibility as a function of the concentration is exhibited by several mixtures containing 4M3H as well. Figure 21 summarizes dielectric loss spectra of 4M3H with MAs for which the number ℓ of carbon atoms ranges from 6 to 16. For each mixture series loss spectra of the pure substances as well as for an intermediate concentration are shown, all measured at 180 K. Let us first look at the maximum loss of the mixing partners of 4M3H: The peak height ε''_{max} of 2E1B is similar to that of 2E1H [138], ε''_{max} of 2B1O is much larger than ε''_{max} of 4M3H, and for 4M3H and 2H1D the peak losses are of a similar magnitude. All of the mixtures (with 4M3H fractions near 0.5) compiled in Fig. 21 display a larger Debye (-like) process than either of the pure components. The most dramatic effects are exhibited by $(4M3H)_x(2H1D)_{1-x}$ where the peak maximum of the $x = 0.5$ mixture is about 10 times larger than in the pure MAs. This finding sharply contrasts with a picture of ideal mixing: Since the molecular dipole moments of either of the neat MAs vary very little, $\mu_{1,2} \approx 1.7$ D [18, 62], the average dipole moment $\langle\mu\rangle$ should be largely independent of concentration. Thus, one may suspect that the emergence of special supramolecular association patterns is responsible for the observation of x-dependent susceptibility maxima. Indeed, aided by results from X-ray diffraction experiments, where the x dependence of the mainpeak and the prepeak in the static structure factor was followed, cf. Sect. 4.3, this "susceptibility enhancement" can be rationalized in terms of ring-opening effects [10].

A good overview of the composition dependence for the various 4M3H mixtures can be gained from Fig. 22 where all data refer to temperatures close to 180 K. In frame (a) the loss peak amplitude of the Debye peak, $\varepsilon''_{peak,D}$, is shown. One recognizes a systematic evolution of the maximum "susceptibility amplification" [10] as the mixing partner of 4M3H is varied. This effect is at the edge of being absent for $(4M3H)_x(2E1H)_{1-x}$: While at 185 K it was found that $\varepsilon''_{peak,D}(x)$ is roughly constant for $x < 0.8$ [139], Fig. 21b shows that at 185 K the Debye peak amplitude is largest at intermediate concentrations. The mixtures with 3M3H (displaying only a faint Debye-like peak in the pure substance, not shown) and that with 2E1Br (not an MA, but included here to present a "homologous" liquid devoid of a Debye process) do not feature a mixing induced susceptibility maximum at all.

The x dependence of the *time constants* relative to that of 4M3H, $\tau_D/\tau_{D,4M3H}$, exhibit related trends; see Fig. 22b. Admixture of more and more 2E1Br, obviously a hydrogen-bond breaker, shifts the Debye-like process in 4M3H to shorter and shorter times, while 3M3H addition leaves the time constants of the corresponding peak almost invariant. The $(4M3H)_x(2E1B)_{1-x}$ and $(4M3H)_x(2E1H)_{1-x}$ systems display pronounced maxima near $x = 0.8$ in the time constants, τ_D, characterizing the Debye process. Since $T_{g,2B1O} > T_{g,2E1H}$, for $(4M3H)_x(2B1O)_{1-x}$ the maximum in τ_D is less pronounced and, continuing the trend, absent for $(4M3H)_x(2H1D)_{1-x}$ since 2H1D shows the largest T_g of the 4M3H mixing partners considered here.

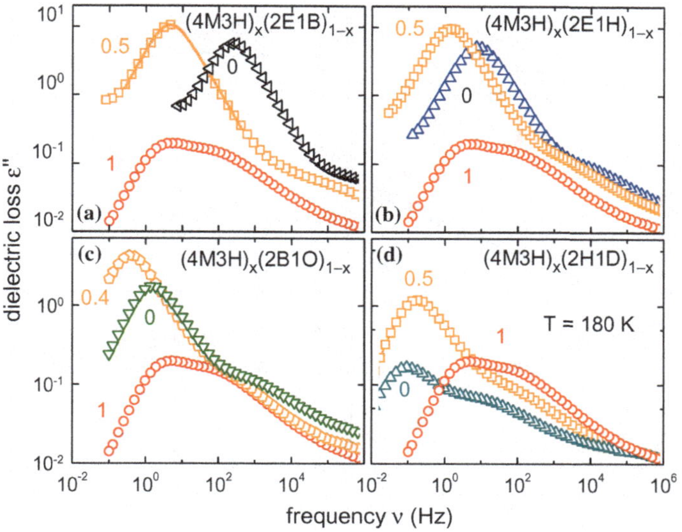

Fig. 21 Dielectric loss spectra of 4M3H mixed with **a** 2E1B (this work), **b** 2E1H [139], **c** 2B1O (this work), and **d** 2H1D [10], all recorded at 180 K. For $(4M3H)_{0.5}(2E1B)_{0.5}$ a line interpolating a Debye process is added where loss data are missing

More properly, when arguing in terms of T_g, reference should be made to the α-process. Figure 22d compiles the composition dependence of the α-time constants, $\tau_\alpha/\tau_{\alpha,4M3H}$, again scaled relative to that of 4M3H. Here the trend in the homologous 2H1D, 2B1O, and 2E1H series is qualitatively the same as that obvious from Fig. 22b. However, for these three MAs the α relaxations vary stronger than the Debye time constants, thereby giving rise to characteristic spectral separations of the associated processes, Sect. 5.2. Finally, from Fig. 22c one recognizes that all binary alcohol mixtures display very similar α-relaxation amplitudes, yet within the homologous 2E1H, 2B1O, and 2H1D series the α-peak amplitudes decrease since the dipole moments become more and more diluted as the alkyl tails become longer. The behavior of $(4M3H)_x(2E1Br)_{1-x}$ deviates strongly from that of the other series since the molecular dipole moment of 2E1Br is larger than for the MAs.

So far, we discussed mixtures of 4M3H with MAs that contain $\ell \geq 6$ carbon atoms. Now let us consider also data for $(4M3H)_x(PrOH)_{1-x}$ (presumably segregating for $x \leq 0.8$) [135], thus with a mixing partner characterized by (i) a larger dipole density and (ii) a much lower glass transition temperature than the other MAs represented in Fig. 22. For $(4M3H)_x(PrOH)_{1-x}$ frame (a) reveals that owing to item (i) a few percent of PrOH added to 4M3H generate a precipitous increase in the strength of the Debye-type process. Most remarkable is the observation that addition of about 5–10% of the low-T_g compound (with its thus relatively fast dynamics) does not speed up, but slows down the supramolecular dynamics with respect to that of pure 4M3H.

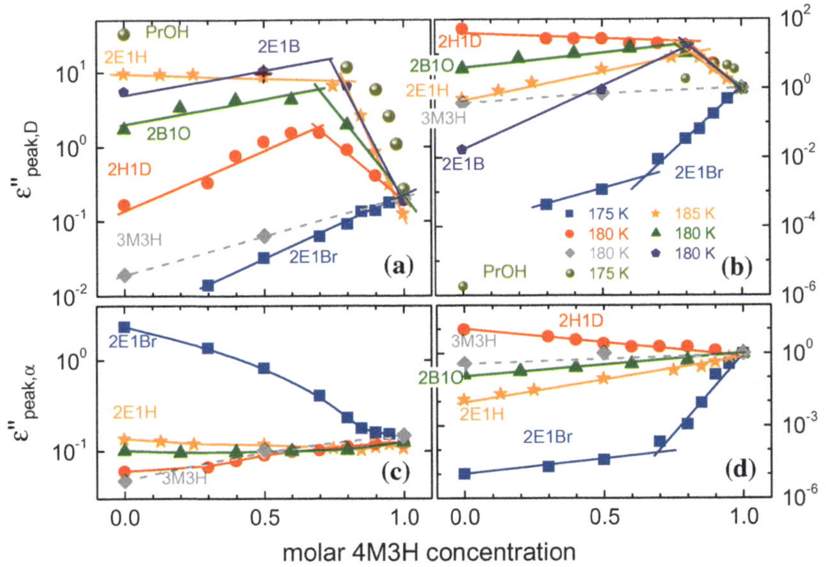

Fig. 22 **a** Amplitudes and **b** time scales of Debye-type process as well as **c** amplitudes and **d** time scales of the structural relaxation in 4M3H mixed with 2E1H [139], 2H1D [10], and PrOH [135] as well as with 2E1B, 2B1O, and 3M3H (all from this work). The Debye loss peak maximum of $(4M3H)_{0.5}(2E1H)_{0.5}$ determined in this work at 180 K is shown as a cross. The other data sets refer to the temperatures given in frame (**b**). For comparison results for $(4M3H)_x(2E1Br)_{1-x}$ are added [60], where the alkyl halide may be taken to represent the limit of a liquid with a vanishing Debye relaxation

4.2 *Stretching Overtones of Monomer and Polymer Bands*

While in Sect. 2.3 NIR spectra of several neat MAs are discussed, here the focus is on mixtures involving 4M3H. Figure 23 compiles spectra of $(4M3H)_x(2E1H)_{1-x}$ and of $(4M3H)_x(2H1D)_{1-x}$ recorded at room temperature over the full concentration ranges. Frame (a) reveals how the strong monomer peak for the ring former 4M3H ($x = 1$) becomes smaller upon addition of 2E1H while simultaneously the absorption of the polymer band, $A(\lambda_\delta)$, increases. By contrast, in $(4M3H)_x(2H1D)_{1-x}$ the intensities of all bands decrease as 2H1D is added to 4M3H (Fig. 23c). The latter observation simply reflects that the density of the OH groups in 2H1D is smaller than in 4M3H. However, as Fig. 23d shows, this does not imply that the band intensities follow a straightforward linear interpolation: A clear break in the concentration dependences of $A(\lambda_{\alpha/\beta})$ and $A(\lambda_\delta)$ is observed for $x \approx 0.7$, consistent with the expectations from the dielectric results, cf. Fig. 22. Furthermore, a close inspection of the wavelength at which the δ-band in $(4M3H)_x(2E1H)_{1-x}$ is maximum, cf. Fig. 23c, reveals its relatively minor concentration dependence on the 2H1D-rich side. Above the critical concentration, i.e., for $x \geq 0.7$, λ_δ is shifted more and more toward shorter wavelengths.

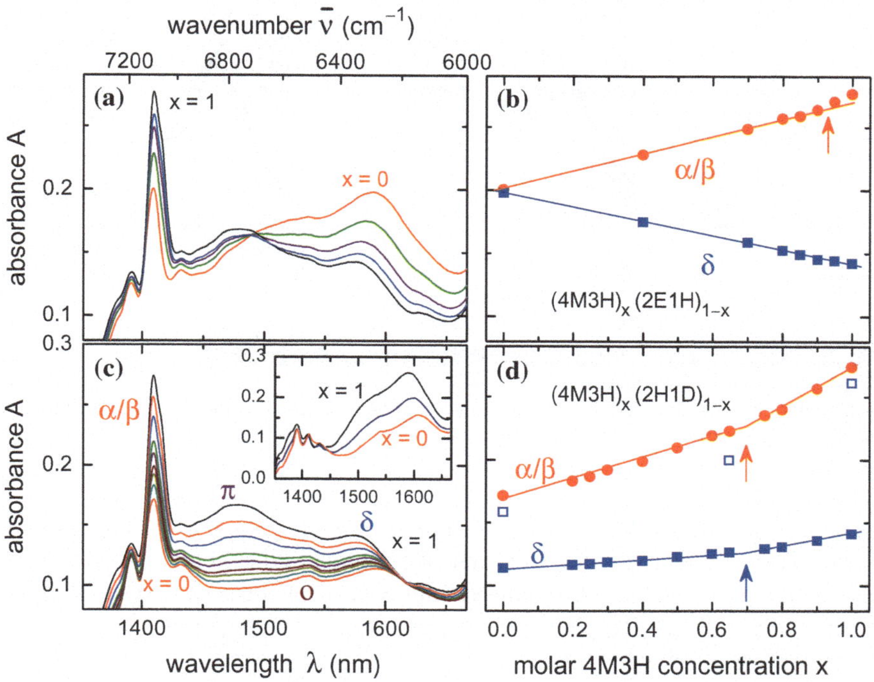

Fig. 23 Panels **a** and **b** present concentration-dependent NIR absorbance spectra (lines for x = 0, 0.4, 0.7, 0.85, and 1) and band intensities (closed symbols), respectively, of $(4M3H)_x(2E1H)_{1-x}$ obtained at 300 K. Panels **c** and **d** provide the analogous information for $(4M3H)_x(2H1D)_{1-x}$, with x = 0, 0.2, 0.3, 0.4, 0.5, 0.6, 0.8, 0.9, and 1. The inset in panel (**c**) features spectra of $(4M3H)_x(2H1D)_{1-x}$ that were recorded at 220 K and in panel (**d**) the α/β- and δ-band intensities referring to 220 K are added as open symbols. The solid lines in panels (**b**) and (**d**) are guides to the eye

Furthermore, as the inset in Fig. 23c shows, for $(4M3H)_x(2H1D)_{1-x}$ a few additional spectra were recorded at 220 K. These spectra emphasize that upon cooling the polymer band grows more strongly for 4M3H than it does for 2H1D. However, a more extensive set of low-temperature spectra is not available for this and related mixtures of MAs. This precludes more detailed analyses regarding the underlying hydrogen-bond equilibria and effects of vibrational anharmonicities that have variously been carried out for neat MAs [7, 8, 74].

4.3 *Prepeaks and Mainpeaks*

For $(4M3H)_x(2H1D)_{1-x}$ diffraction patterns were reported for the entire concentration range [10]. Similar results were obtained for $(4M3H)_x(2B1O)_{1-x}$; see Fig. 24a. Compared to the data for $(2E1H)_x(2E1Br)_{1-x}$ that are presented in Fig. 13, the 4M3H-

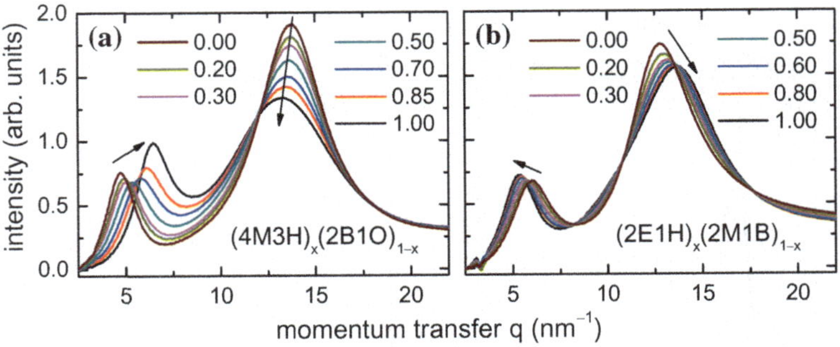

Fig. 24 X-ray diffraction patterns of (**a**) $(4M3H)_x(2B1O)_{1-x}$ and (b) $(2E1H)_x(2M1B)_{1-x}$ recorded at room temperature for several concentrations x. The compositional evolution of the prepeaks as well as of the mainpeaks of the two mixture series is different. The arrows highlight how the prepeak and mainpeak maxima evolve for increasing concentrations x

based mixtures of two MAs display an entirely different pattern as the composition is varied. Most dramatic is the x-dependent evolution of the prepeak. As may have been anticipated from the momentum-transfer-dependent intensities, $I(q)$, of the pure liquids, cf. Fig. 11b, in Fig. 24a it is clearly seen that the prepeak positions shift with x, and that the amplitude of the prepeak is reduced at intermediate concentrations. For $(2E1H)_x(2M1B)_{1-x}$ this feature is almost absent, cf. Fig. 24b.

A hint regarding the microscopic origin of the effects in the 4M3H-based mixtures comes from the observation that supramolecular structures with a low electrical dipole moment "take a higher volume per molecule than the multimers with a high dipole moment" [9]; see also Ref. [141]. The macroscopic densities $\rho_{\text{mac}}(x)$ of the $(4M3H)_x(2B1O)_{1-x}$ mixtures, summarized in Fig. 25a, indicate slightly *positive* deviations from an ideal-mixing behavior at intermediate concentrations. Thus, based on the above quote, one may suspect that the tendency to ring formation which (at room temperature) characterizes 4M3H as well as 2B1O (see Sect. 5.1) is less prevalent in the corresponding mixtures. In Fig. 25c the relative deviations from a linear interpolation of the densities, $\rho_{\text{lin}}(x)$, between the pure ends are shown as $\Delta\rho_{\text{mac}}(x)/|\Delta\rho_0| = [\rho_{\text{mac}}(x) - \rho_{\text{lin}}(x)]/|\Delta\rho_0|$, where $|\Delta\rho_0|$ refers to the density difference of the two pure ends. Thus, very similar to findings for $(4M3H)_x(2H1D)_{1-x}$ [10] a maximum relative deviation of about 20% from simple linear behavior is obtained for $(4M3H)_x(2B1O)_{1-x}$. Changes of the "microscopic" density can be assessed from the compositional evolution of the cubed mainpeak positions, q^3_{main}. As also demonstrated in Fig. 25c, the analogously defined relative deviations of this quantity, $\Delta q^3_{\text{main}}(x)/\left|\Delta q^3_{\text{main},0}\right| = [q^3_{\text{main}}(x) - q^3_{\text{main,lin}}(x)]/\left|\Delta q^3_{\text{main},0}\right|$, follow similar trends, again resembling those for $(4M3H)_x(2H1D)_{1-x}$ [142]. The macroscopic (ρ_{mac}) and microscopic ($\propto q^3_{\text{main}}$) densities both suggest that ring-like structures which are present in both pure MAs tend to become more open upon mixing. Thus, for the corresponding samples, an enhancement of the dielectric strengths and of their dielectric losses, $\varepsilon''_{\text{peak,D}}$, is expected, in harmony with the results for $(4M3H)_x(2B1O)_{1-x}$ and $(4M3H)_x(2H1D)_{1-x}$ that are summarized in Fig. 22b. A pictorial presentation of the

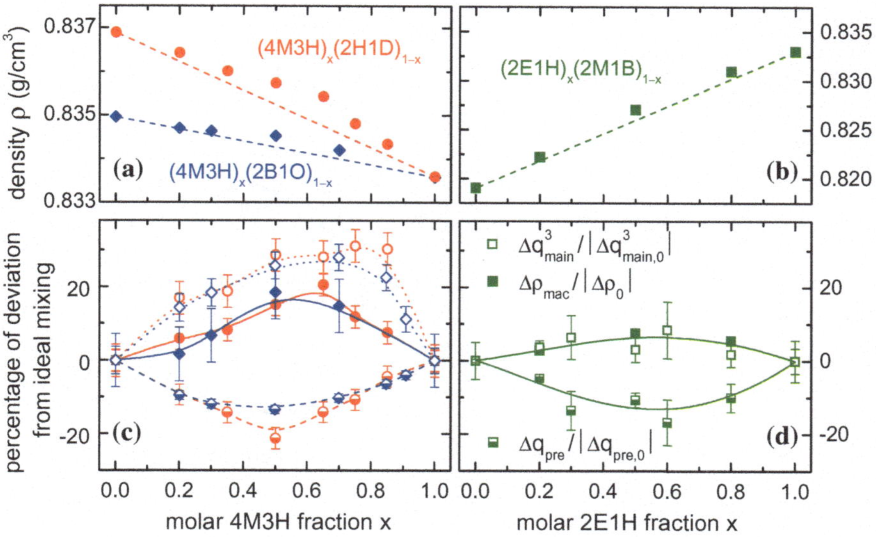

Fig. 25 Density and X-ray data for $(4M3H)_x(2H1D)_{1-x}$ (circles), $(4M3H)_x(2B1O)_{1-x}$ (diamonds), and $(2E1H)_x(2M1B)_{1-x}$ (squares). Macroscopic densities ρ_{mac} are shown in panels (**a**) and (**b**); error bars are smaller than the size of the symbols. The dashed line, representing $\rho_{lin}(x)$, connects the values for the neat alcohols as expected for ideal mixing. Panels (**c**) and (**d**) present relative deviations from ideal mixing for density ($\Delta\rho_{mac}/|\Delta\rho_0|$, filled symbols), cubed mainpeak position ($\Delta q^3_{main}/|\Delta q^3_{main,0}|$, open symbols), and prepeak position ($\Delta q_{pre}/|\Delta q_{pre,0}|$, half-filled symbols); lines are drawn to guide the eye. All measurements refer to room temperature

changes in the molecular arrangement capturing ring-opening effects may be inferred from Fig. 4 in Ref. [10]; see also the inset of Fig. 1. In $(2E1H)_x(2M1B)_{1-x}$ the relative compositional changes of the macroscopic and microscopic densities, see Fig. 25c, d, are much smaller than for the mixtures containing 4M3H. This observation is compatible with the notion that phenomena akin to ring-opening effects are not expected to occur in binary systems composed of predominantly chain-forming MAs.

On this background, remarkable observations can be made for the x dependence of the prepeak positions, q_{pre}. Opposite to the trend for the density-related C–C correlations, the O–O correlations as probed by the prepeak deviate *negatively* from the naïvely expected (linear in x) behavior; see Fig. 25c, d. For this plot, with $|\Delta q_{pre,0}|$ referring to the difference between the two pure ends, we use the definition $\Delta q_{pre}(x)/|\Delta q_{pre,0}| = [q_{pre}(x) - q_{pre,lin}(x)]/|\Delta q_{pre,0}|$ for the relative deviations. In Fig. 25 these are seen to be 15–20% for all binary systems, that is, even for the $(2E1H)_x(2M1B)_{1-x}$ mixtures for which ring-opening effects are not expected. And indeed, also these mixtures display a maximum in their concentration-dependent dielectric relaxation strength $\Delta\varepsilon$ [136] (an observation not possible on the basis of the x-dependent tangent of the maximum loss angle [12] which essentially scales out large $\Delta\varepsilon$).

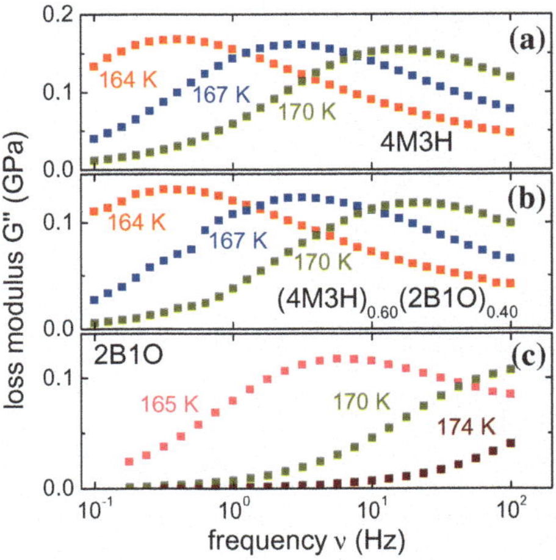

Fig. 26 Frequency-dependent shear loss moduli of $(4M3H)_x(2B1O)_{1-x}$ for **a** $x = 0$, **b** $x = 0.6$, and **c** $x = 1$

Hence, at intermediate concentrations all mixtures represented in Fig. 25 tend to form supramolecular conglomerates with (i) a larger average dipole moment than the pure MAs and (ii) O–O distances larger than expected from linear mixing behavior. Hence, the combined tracking of C–C and O–O correlations provides useful insights into the presence or absence of composition-dependent changes in the local association patterns. Thus, unique information can be gained using X-ray diffraction experiments which is not accessible via macroscopic density measurements alone.

4.4 Terminal Shear Modes and Viscosity Enhancement

Rheological data for several neat MAs are shown as master curves in Sect. 2.1. Here, in Fig. 26 we present raw shear loss modulus spectra of $(4M3H)_x(2B1O)_{1-x}$ with $x = 0$, 0.6, and 1. In contrast to the double-logarithmic representation of such data, cf. Fig. 3b, the semi-logarithmic plot chosen for Fig. 26 de-emphasizes the Debye process. In any case, the temperature-dependent peak frequencies ν_{max} relate to that of the structural relaxation.

To gain insights into the compositional trends also of the Debye-like low-frequency features, the scaled viscosity, $\eta'(\nu)/[G_\infty/(2\pi\nu_{max})]$, as introduced in Sect. 2.1, was evaluated from the data in Fig. 26. The results for $(4M3H)_x(2B1O)_{1-x}$, are shown in Fig. 27 in the form $z(\nu) \equiv \eta'(\nu)/[G_\infty/(2\pi\nu_{max})]$ versus ν/ν_{max}. On the linear ordinate scale of this figure the data, in particular those for $x = 0.6$, appear somewhat more scattered than in double-logarithmic master plots.

In Fig. 27 one recognizes that the scaled zero-shear viscosity, $z_0 = z(\nu \to 0)$, of pure 4M3H and that of pure 2B1O are both larger than that expected for a simple liquid, as exemplified using data for the van der Waals glass former ααβ-tris-naphthylbenzene (TNB) [143, 144]. However, most importantly, at the intermediate

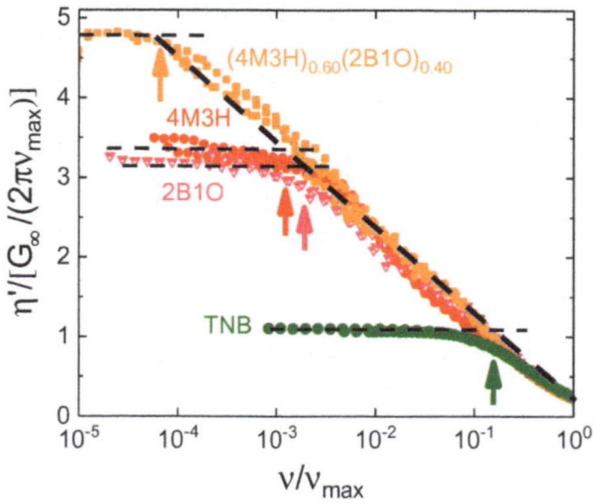

Fig. 27 Scaled shear viscosity master curves versus scaled frequency as calculated from the shear moduli presented in Fig. 26. Additionally, data for a simple supercooled liquid (TNB, from Ref. [143]) are shown. The horizontal lines define the scaled zero-shear viscosity $z_0 = \eta'(\nu \to 0)/[G_\infty/(2\pi\nu_{max})]$. The intersections with the other line (that guides the eye) defines the onset frequency ν_z of the viscosity plateaus and are highlighted by the arrows

concentration, z_0 is maximum, thus signaling a zero-shear viscosity enhancement relative to that of the neat alcohols. This effect is related to the dielectric features seen for $(4M3H)_x(2B1O)_{1-x}$ and other binary systems in Fig. 22b.

Already the mechanical master curves, cf. Fig. 3, suggest that a common low-frequency envelope is approached for frequencies between that of the shear loss peak, ν_{max}, and the onset of the terminal mode. Thus, there should be a correlation of the frequency ν_z / ν_{max} and the scaled modulus at which these deviations start to emerge. Likewise, from the scaled viscosity plot, cf. Fig. 27, a similar pattern is obvious: For $(4M3H)_x(2B1O)_{1-x}$ the arrows in this figure highlight that lower ν_z/ν_{max} correspond to larger scaled zero-shear viscosities z_0. The extent to which this correlation of viscosity enhancement and Debye process-related plateau onset can be generalized to other substances is examined in Sect. 5.2.

It is instructive to track also the compositional variation of the scaled zero-shear viscosities. Figure 28a compiles $z_0(x)$ for several alcohol mixtures and shows that z_0 decreases as the alkyl chains of the mixing partners of 4M3H get longer. Furthermore, the available data suggest that a viscosity enhancement appears for several mixtures. While the present effects are much less pronounced than the dielectric ones, presented in Fig. 22b, deviations from ideal mixing are nevertheless clearly resolved for the zero-shear viscosities, in particular for the mixtures characterized by a large contrast in z_0.

For comparison, Fig. 28b displays the trends regarding the structural relaxation time $\tau_\alpha = 1/(2\pi\nu_{max})$ at a temperature of 170 K. Trends more or less linear in $\log_{10}\tau_\alpha(x)$ are seen which translate to corresponding $T_g(x)$ dependences that are compatible with the behavior deduced from dielectric spectroscopy, cf. Fig. 22d.

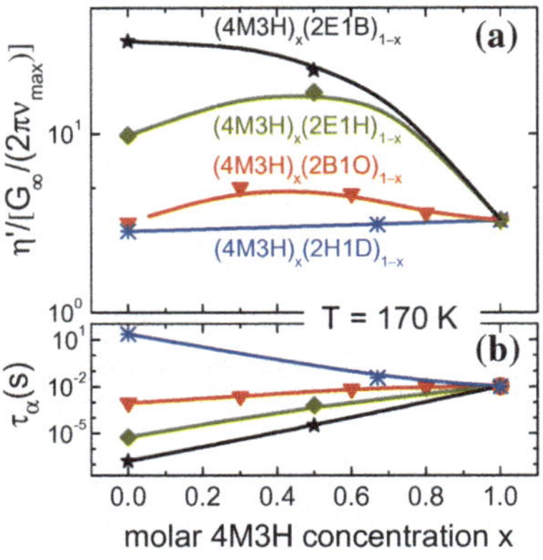

Fig. 28 Concentration dependences **a** of the scaled zero-shear viscosity z_0 and **b** of the structural relaxation times, defined here as $\tau_\alpha = 1/(2\pi\nu_{max})$ for several mixture series involving 4M3H. The quantity $z_0(x)$, which relates to the supramolecular low-frequency viscoelastic response, displays significant deviations from ideal-mixing behavior, but not the structural relaxation time τ_α. Lines are drawn to guide the eye

5 Trends and Perspectives

5.1 Ring-Chain Equilibria and Kirkwood Factors

One of the features of MAs which distinguishes them from most other small-molecule glass formers is their ability to form rather well-defined supramolecular structures with either more chain-like or more ring-like association patterns. The dynamic equilibrium among, e.g., monomeric and transiently polymeric structures can be affected by changes in various external parameters such as temperature [18, 145], pressure [30, 137, 141, 146, 147], as well as sufficiently large electrical [28] (and presumably also mechanical) fields, and in mixtures not the least by the chemical composition.

Typically, the mutual orientational alignment of the MA molecules can be evaluated via Kirkwood's correlation factor g_K; see Eq. (1). Figure 29 collects literature data for g_K for several homologous series of pure MAs including (G) neo-alcohols [33] and (E) the *j*M3H alcohols [135], where the bracketed letters refer to the classification given in Sect. 2. Then, Fig. 29c comprises data for $(4M3H)_{1-x}(2E1H)_x$ [139] and Fig. 29d for other 4M3H-based mixtures as well as for 2B1O and 2H1D. These plots demonstrate common trends: One can distinguish MAs exhibiting a relatively low g_K value across the entire temperature range, others which approach large g_K ($\geq$4) at low T, and some liquids that exhibit a crossing of a correlation factor $g_K = 1$.

Usually the Kirkwood factor is assessed from the full dielectric relaxation strength. Alternatively, it has been suggested to replace ε_∞ in Eq. (1) by $\varepsilon_{inf} = \varepsilon_\infty + \Delta\varepsilon_\alpha + \Delta\varepsilon_\beta$ [135], so that only the orientational correlations due to the Debye process are

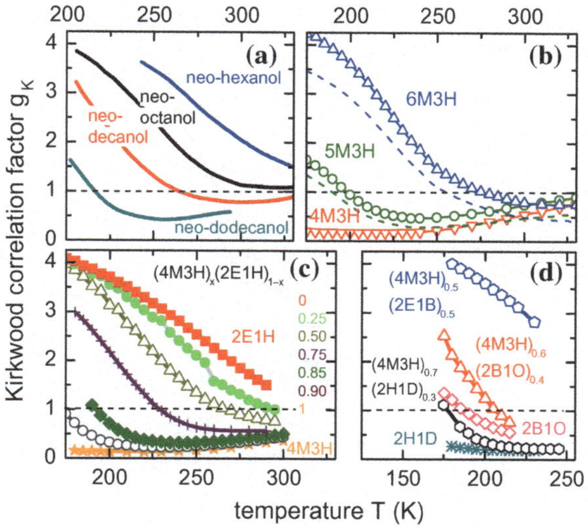

Fig. 29 Kirkwood factor g_K of pure MAs and mixtures: **a** Data for neo-alcohols, scanned from Ref. [33] are represented as solid lines. **b** Correlation factors for *j*M3H with symbols and solid lines obtained by setting $\varepsilon_{inf} = \varepsilon_\infty$ and the dashed curves using $\varepsilon_{inf} = \varepsilon_\infty + \Delta\varepsilon_\alpha + \Delta\varepsilon_\beta$ [135]. **c** Data for the mixture series $(4M3H)_{1-x}(2E1H)_x$ [139]. In panel **d** from the largest to the smallest g_K the data are for $(4M3H)_{0.5}(2E1B)_{0.5}$, $(4M3H)_{0.6}(2B1O)_{0.4}$, 2B1O, $(4M3H)_{0.7}(2H1D)_{0.3}$, and 2H1D [138]. The horizontal dashed lines mark a Kirkwood factor of $g_K = 1$

taken to contribute to g_K. The symbols and dashed lines in Fig. 29b emphasizes that for liquids with small g_K the difference between the two ways of evaluation is of minor importance.

Formally, a Kirkwood factor of $g_K = 1$ reflects an absence of mutual orientational correlations, a case that in dipolar liquids can arise for various scenarios [53] including a situation in which a heterogeneous mixture of associates some with $g_K > 1$ and others with $g_K < 1$ are simultaneously present. Under such conditions, the crossing of $g_K = 1$ is an interesting feature, since it hints at a delicate balance of competing supramolecular structures which can thus be perturbed effectively by external stimuli. This circumstance was already exploited for 5M3H using high-field electrical impedance spectroscopy [148]. Typical magnitudes of maximum slopes, $-dg_K/dT|_{g_{K=1}}$, that may be inferred from Fig. 29 are of the order of 0.04–0.05 K^{-1}.

Extensive compilations of composition-dependent Kirkwood factors for mixtures of MAs with CCl_4, some of them displaying a crossing of $g_K = 1$ as the MA concentration is varied, can be found in Refs. [119, 120]. Furthermore, $g_K = 1$ crossings were observed in pressure studies under isothermal conditions as well, e.g., for 2-methyl-2-hexanol [141], 4M3H [137], and 5M3H [149]. A behavior similar to the latter liquid was inferred for 2-methyl-3-heptanol [145]. It seems that so far, no X-ray, infrared, or other structure-sensitive studies were carried out at the crossing temperatures to investigate the underlying phenomena in detail.

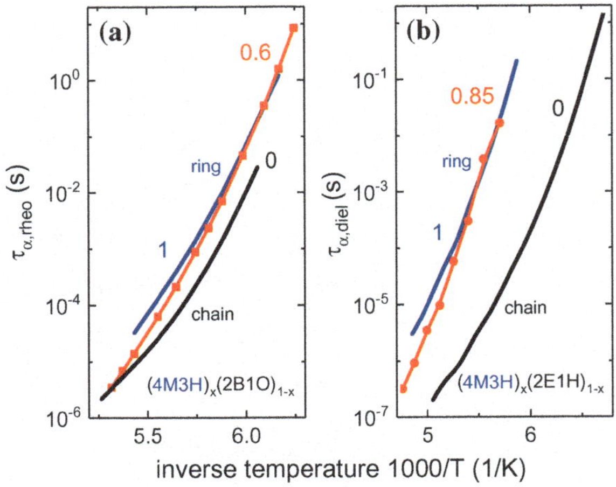

Fig. 30 Arrhenius plot of structural time scales **a** from rheology for $(4M3H)_{0.85}(2E1H)_{0.15}$ and (**b**) from dielectric spectroscopy for $(4M3H)_{0.4}(2B1O)_{0.4}$. For the sake of representation all data are plotted as lines and for the mixtures additionally as symbols

A continuous transition between different supramolecular populations is reflected also in the temperature dependence of the microscopic *mobility*. For $(4M3H)_x(2B1O)_{1-x}$ this statement is illustrated for the rheologically detected structural relaxation times in Fig. 30a. Figure 30b shows similar results for $(4M3H)_x(2E1H)_{1-x}$ obtained by means of dielectric spectroscopy. For both systems the temperature dependence of the time constants is stronger in the mixtures as compared to the pure MAs. This finding indicates that the static intermolecular correlations which are governing the strong T-dependence of the Kirkwood factor (see Fig. 29) have a dynamical counterpart. Put in simple terms, the bulkier the supramolecular structures, the more effective are they in slowing down the local dynamics.

5.2 Correlations and Decoupling of Time Scales

Following the considerations in Sect. 4.4, let us now address the question regarding a more general correlation of the viscosity enhancement, parameterized by the scaled zero-shear viscosity, $z_0 = \eta'(\nu \to 0)/[G_\infty/(2\pi\nu_{max})]$, with the time scale separation. The latter can be estimated from the scaled onset frequencies ν_z/ν_{max} of the plateaus obvious from (master) plots of the shear viscosity, for an illustration see Fig. 27. In Fig. 31 we collect $z_0(\nu_z/\nu_{max})$ data for a number of pure and mixed MAs. Based on these shear mechanical results a broad, but clear correlation of relative strength and time scale separation of supramolecular (Debye-type) and structural relaxation is found. More specifically, the relationship $z_0 \propto (\nu_z/\nu_{max})^{-\beta}$ appears to hold where the exponent is in the range $\beta_z = 0.25$–0.4, cf. the lines in Fig. 31. These lines are drawn to intersect in the lower right of the figure which corresponds to the coordinates expected for typical simple liquids.

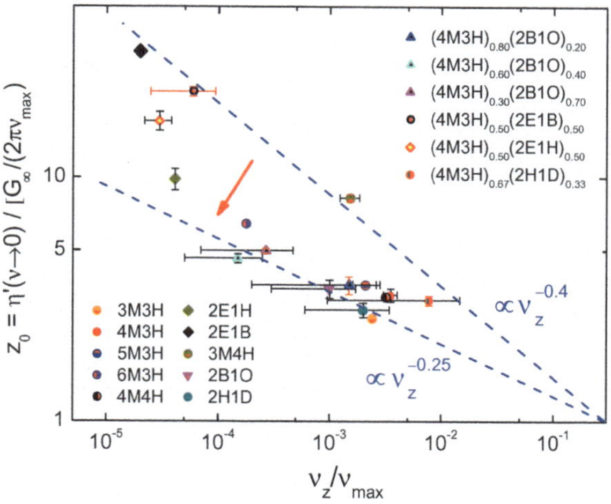

Fig. 31 Scaled viscosity enhancement z_0 is plotted versus the scaled onset frequency $\nu_z/\nu_{\max}$ of the viscosity plateaus for several pure and mixed MAs. All data are based on rheological measurements carried out in the authors' laboratory. The arrow is meant to indicate that, if the form of the *supramolecular* contribution to the viscosity, $\eta'(\nu)$, is describable by a distribution function, e.g., of the Cole-Davidson type, then the expected exponent β_z is smaller than in the absence of such a distribution

Slopes of the order indicated for β_z in Fig. 31 can be rationalized in terms of bead-and-spring models which are often applied for polymers [40]. Within this framework the shear modulus is expected to obey a power law, $G'' \propto \nu^{\beta}$, for frequencies just above the range in which the terminal mode (with $G'' \propto \nu^1$) prevails. For the Rouse model one has $\beta = 1/2$ [150], while including hydrodynamic interactions leads to the Zimm exponent, $\beta = 2/3$ [40]. Since $\eta' \propto G''/\nu$ one expects that $z_0 \propto \nu_z^{\beta-1}$, in the latter case giving $z_0 \propto \nu_z^{-1/3}$. Indeed, within the scatter obvious from Fig. 31, this exponent guides the trend seen for the shear mechanical data of all supramolecular liquids included in this figure.

Previously, correlations of time scale separation, on the one hand, and relative strength of Debye and structural relaxation, on the other, have been sought on the basis of *dielectric* experiments [29, 148]. However, while trends for certain homologous series appeared encouraging, a clear-cut one-to-one correspondence of relative dielectric strengths and dynamic separations could not be substantiated. This shortcoming can now be understood better when taking together the clues provided by the rheology-based correlation seen in Fig. 31 with the results for the Kirkwood factors presented in Sect. 5.1. Obviously, the most important factor that scrambles a dielectric-based correlation (analogous to that in Fig. 31) is that the orientational correlations differ vastly among different MAs and their mixtures. Above all their sometimes pronounced temperature and composition dependences, that are reflected in the dielectric but not in the mechanical data, preclude the establishment of a

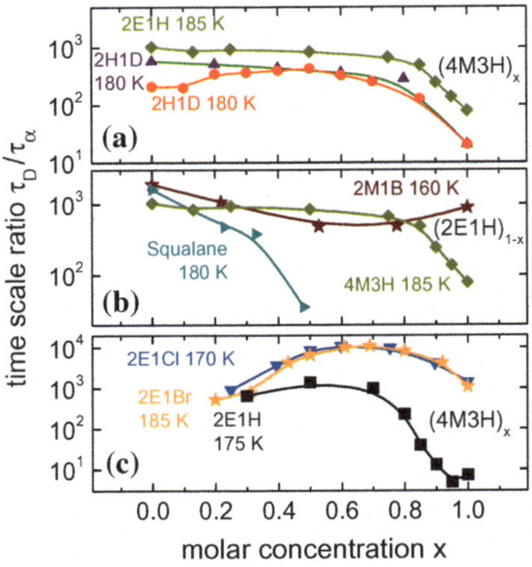

Fig. 32 Time scale ratios, τ_D/τ_α, evaluated from dielectric measurements for binary systems with **a** 4M3H as mixing partner of 2E1H [139], 2B1O, and 2H1D, **b** 2E1H as mixing partner of 4M3H, 2M1B [12], and squalane [105], as well as **c** 2-ethyl-1-hexyl-halides as mixing partners of 4M3H [114] and 2E1H [60, 139]

dielectric spectroscopy based plot similar to Fig. 31 when considering a large array of liquids.

When focusing on results from dielectric spectroscopy alone, it is therefore not advisable to consider relaxation strengths. Rather, next we will focus on exploring whether the time scale ratio, τ_D/τ_α, displays systematic compositional trends. Corresponding data for binary mixtures of alcohols with alkyl halides or with other MAs are summarized in Fig. 32. Time scale ratios τ_D/τ_α of 4M3H-containing mixtures are shown in frame (a), and all share a common trend. Admixture of about 20% of another alcohol to 4M3H increases the separation of α- and Debye process. Adding more of the mixing partners 2E1H, 2B1O, and 2H1D does not alter the decoupling much further, a phenomenon that is obviously not at all affected by the strongly differing relaxation strengths, $\Delta\varepsilon_D$, characterizing these latter MAs. In spite of the common trend evident from Fig. 32a the deviations from ideal mixing of the Debye time, τ_D, alone, cf. Fig. 22b, differ considerably for the 2E1H-, 2B1O-, and 2H1D-containing mixtures. Differences in the glass transition temperatures also play an important role as is evident, e.g., from the concentration dependence of $\tau_D(x)$ for the $(4M3H)_x(PrOH)_{1-x}$ mixture that is included in Fig. 22b.

$(2E1H)_x(2M1B)_{1-x}$, where both MAs possess large Debye processes [151], is one of the binary systems for which ideal mixing with respect to $\tau_D(x)$ has been reported [12] and the same applies for mixtures of 2E1H with methanol [140]. Using τ_D/τ_α as an indicator, slight deviations from a linear behavior are observed for $(2E1H)_x(2M1B)_{1-x}$, cf. Fig. 32b, while 2E1H mixed with the low-T_g liquid squalane [105] displays a strong approach of time scales as the MA is diluted. A dilution-induced *separation* of time scale arises, however, by adding 2E1Br to 4M3H; see Fig. 32c. The largest time scale separation currently known is for mixtures of 2-ethyl-1-hexyl-halides with 2E1H.

5.3 Dynamic Scattering and Electro-Optics

Frequency-dependent calorimetry [6], high-pressure dielectric measurements [125], and dielectric spectroscopy combined with neutron scattering, the latter for a simultaneous study of dynamical processes on both channels [152], are certainly among the experimental approaches that deserve to be applied in future studies also of *mixtures* including MAs. Furthermore, it appears that in order to understand the microscopic dynamics of MAs better, dynamic scattering methods should be exploited more intensively. This class of methods includes dynamic light scattering [14, 153, 154] (with X-ray-based photon correlation spectroscopy barely applied to neat hydrogen-bonded liquids at all [155]), neutron spin-echo spectroscopy [80], or quasielastic neutron scattering in combination with spin polarization analysis [156], as well as synchrotron-based quasielastic nuclear resonant scattering of γ-rays [157], to name a few.

An idea worth pursuing is to try and switch the degree of intermolecular association by means of an external light source. While optically triggered cis-trans isomerization of probe molecules was already utilized to study dynamics in BuOH [158], light-induced conformational changes, for instance, using 2-hydroxystilbene or related dyes are conceivable [159]. In such an approach one would optically switch the hydroxyl group from an orientation which, within the molecular frame, is pointing more "outward" to another one that is pointing more "inward," thereby altering the hydrogen-bond-based association propensity of an MA. Other techniques combine optical and electrical channels in different manners. The time-dependent electro-optical Kerr effect [160, 161], for instance, involves application of dc electrical fields (or ac fields [162, 163]) in conjunction with optical detection.

The latter combination is exploited for electro-absorption spectroscopies as well. In the following, we summarize very briefly some electrically induced optical effects and present initial steps necessary to apply this method to MAs in the NIR spectral range. Variations of external electrical fields $\vec{E}$ can (i) change molecular conformations or ring-chain equilibria, (ii) modulate orientational anisotropies, and (iii) provoke electronic responses. Let us denote $\delta\vec{\mu}$ and $\delta\boldsymbol{\alpha}$ as the field-induced changes in the transition dipole moment and those in the polarizability tensor, respectively, Then, the changes in the peak absorbance frequency ($\delta\bar{\nu}$, vibrational Stark effect) and in the absorbance (A) of a given band can be written as [164]

$$\delta\bar{\nu} = -\delta\vec{\mu}\cdot\vec{E} - \tfrac{1}{2}\vec{E}\cdot\delta\boldsymbol{\alpha}\cdot\vec{E} \tag{3a}$$

and

$$\delta A(\bar{\nu}) = f^2\vec{E}^2\left(aA(\bar{\nu}) + b\frac{dA(\bar{\nu})}{d\bar{\nu}} + c\frac{dA(\bar{\nu})}{d\bar{\nu}}\right), \tag{3b}$$

respectively. Here, the numerical factor $f \sim 1$ accounts for local-field corrections and the coefficients a, b, and c reflect molecule-specific parameters. As far as we know,

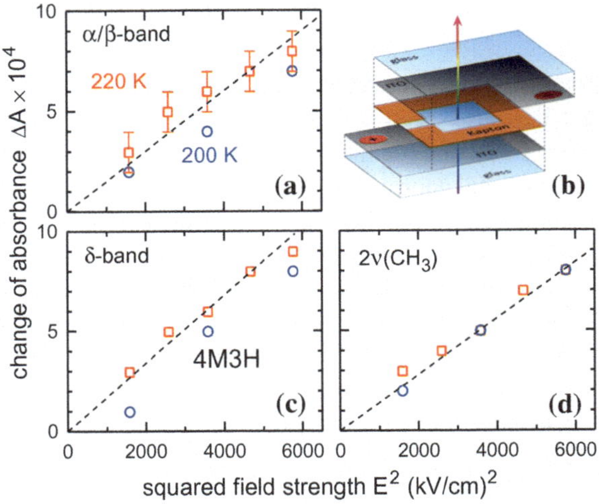

Fig. 33 Field-induced absorbance change as measured for **a** the α/β-band, **c** the δ-band, and **d** the 2ν(CH_3) stretching overtone. Panel **b** sketches the layout of the cell used for these experiments. The red dots indicate the spots at which the voltage source is connected to the indium tin oxide electrodes. The dimensions of the glass plates are $30 \times 24 \times 0.2$ mm^3

field-induced Stark shifts, $\delta\bar{\nu}$, were only occasionally studied, e.g., for the stretching vibrations of OH [165] or NH [166] groups in hydrogen-bonded liquids and we are aware of only a few infrared electro-absorption investigations [165, 167].

Our cell for the electro-optical measurements consists of a 125 μm thick polyimide spacer sandwiched between two glass plates covered with 100 nm thin films of indium tin oxide each having a resistance of 100–200 Ω. This assembly, sketched in Fig. 33b, displayed an only slightly wavelength-dependent absorbance in the regime of the first OH overtone. Measurements on 4M3H for applied voltages up to 950 V are shown in Fig. 33a, c for the α/β- and δ-bands, respectively. It is seen that the differential absorbance, $\Delta A = A(\bar{\nu}, E) - A(\bar{\nu}, 0)$, depends linearly on E^2. For both bands an analysis of these data yields a slope of $\Delta A/E^2 \approx 0.16$ (kV/mm)$^{-2}$ for the studied temperatures. For the 2ν(CH_3) overtone appearing near a wavelength of 1700 nm [75], which reflects the nonpolar methyl groups, an about 10% smaller slope is found, cf. Fig. 33d.

A next step is to devise a state-dependent, i.e., band-specific dielectric spectroscopy where the electrical field is applied in a step-wise fashion [168] or using a sinusoidal excitation to obtain a complex electro-optical susceptibility. This would allow one to resolve and separate the responses of various molecular groups, thus augmenting the repertoire of standard dielectric spectroscopy which integrates over the responses of all polar groups. Furthermore, since not only the change of the electrical dipole moment but also of the polarizability enters, cf. Eq. (3a), the dynamics of nonpolar groups is accessible using this approach as well.

5.4 En Route to Understanding the Relaxation Processes in Alcohols

The secondary (Johari-Goldstein [169, 170]) processes that occur in MAs are presumably to be viewed similar to those of small-molecule glass formers in general [61, 135, 171]. By contrast, the primary relaxation of MAs deserves special attention. Despite the circumstance that this process is often called α- or structural relaxation, dielectric spectroscopy suggests that in MAs this process is unusual in at least two important respects:

(1) While process I, the Debye process, is typically either *larger or smaller* than expected on the basis of the molecular dipole moment (in other words $g_K \neq 1$, cf. Table 1; Sect. 5.1), process II, the α-relaxation, is almost *always smaller* than expected from the molecular dipole moment, at least for sufficiently low temperatures at which supramolecular structures have formed [172]. This latter observation presumably reflects the *temporarily anisotropic character* of the α-relaxation. Such a kind of anisotropy is in harmony with the transient chain model [173] and related approaches [174] so that only the component of the molecular dipole moment that is oriented perpendicular to the supramolecular hydroxyl chain contributes to process II.
(2) Then, on the basis of dielectric results this process shows an unusual *symmetric shape*. This is well established at least for 2E1B [61] or MA mixtures for which the low-frequency flank of process II is clearly resolved. Several arguments come to mind to rationalize this latter behavior. These include (for mixtures) the appearance of nanophase separation [115, 121, 122] or an occurrence of submerged slow hydroxyl-related modes. These modes take place on the time scale identified by nuclear magnetic resonance and neutron scattering [80, 172]. In this respect it should be rewarding to exploit the spectral separation of the α- and the Debye processes in mixtures involving MAs for their selective study and to check whether intermediate modes are present using, e.g., specific-heat spectroscopy.

It is worthwhile to keep in mind that techniques appearing to de-emphasize the contributions of the Debye process in their susceptibility or modulus spectra such as dynamic light scattering [153] or mechanical spectroscopy [175], see, e.g., Fig. 3, detect the familiar *asymmetric* broadening for the structural relaxation. More work seems to be required to fully understand the detailed nature of the dielectrically resolved structural relaxation in MAs.

In Debye liquids such as MAs the most important question is of course that concerning the microscopic nature of the *Debye* process. In addition to the numerous suggestions aiming at rationalizing it as discussed, e.g., in Ref. [7], several new results should be considered. A recent combined dielectric and ac calorimetry study provides evidence for the occurrence of a Debye process not only in ethanol's supercooled *liquid* but also in its plastic *crystalline* state [176]. The crystal matrix represents a self-confining environment which near the glass transition should effectively inhibit molecular transport. Other types of soft and hard confining matrices providing nanoscopic motional constraints [177–179] as well as the mesoscopic confinement emerging during crystallization [151, 180] were previously examined for MAs. Many of these studies yield evidence for dielectric Debye-type relaxations even under conditions of severe spatial confinement which challenges the current understanding of the Debye peak in MAs.

Is the occurrence of narrow (close to) Debye-like relaxation peaks simply related to large dielectric relaxation strengths? A comparative study of 88 glass formers (essentially governed by van der Waals interactions) showed just this [181]: The liquids with the largest dielectric relaxation strengths near the glass transition, $\Delta\varepsilon(T_g)$, obviously show the closest-to-exponential structural relaxation. It is certainly worthwhile to pursue these ideas further also for hydrogen-bonded systems and to take Debye (-type) peaks into account. A step in this direction is the recognition [11] that for some MAs Debye peaks with small relaxation strengths, $\Delta\varepsilon_D$, can display significant deviations from exponentiality as recently scrutinized in detail [34]. Figure 5 shows that 2H1D is an example for this kind of behavior. On the one hand, it is clear that for different MAs $\Delta\varepsilon_D(T_g)$ can differ widely, despite their widths being close to the Debye limit. On the other hand, the nature of very strong Debye-like peaks such as those in secondary amides [182, 183] could possibly be elucidated this way.

Amorphous solids featuring unusual properties, so-called ultrastable glasses, have recently become the focus of exciting scientific activities [184]. Such glasses can be produced by physical vapor deposition at substrate temperatures typically about 15% below the conventional glass transition temperature, T_g. During production, the molecules can move much faster along the free surface than in the bulk, so that energetically low-lying amorphous states become accessible that otherwise would require aging on time scales estimated to be hundreds to many thousands of years [183].

However, several vapor-deposited alcohol glasses including PrOH and 2E1H are special in the sense that they show little or no enhanced kinetic stability when deposited under conditions otherwise employed in such studies [185, 186]. This finding was rationalized by noting that "hydrogen-bonding systems restructure at the free surface to preserve a large fraction of the hydrogen bonds, and so surface mobility in such systems is not as high" [183] as for substances from which highly stable glasses can be formed.

Furthermore, glasses of 4M3H deposited at $0.6T_g$ revealed more than ten times larger dielectric losses with respect to 4M3H glasses produced by deposition near T_g or by conventional cooling of the liquid [187]. This difference was interpreted to indicate that for this MA the formation of ring-like supramolecular structures is inhibited during low-temperature deposition. Most remarkably, the dielectric response of such samples remains much higher and slower upon heating than for ordinary 4M3H, even for temperatures of up to 40 K above the conventional T_g. A similar inheritance of liquid-state properties acquired during vapor deposition of the glass, i.e., a more intense and a slower response for temperatures up to 20% above the conventional T_g was reported also for glycerol [188]. The nature of these latter observations seems to be different [189] from those prevailing in low-temperature-deposited 4M3H.

Much of the present discussion is concerned with the Debye process in *mono*hydroxy alcohols, although the question regarding an appearance of slower-than-structural supramolecular relaxations was raised already in 1961 by Davidson who stated that "in view of the evidence already given for the ability of each OH group in 1,5-pentanediol to freely associate with other molecules, relaxation behavior similar to that of the simple monohydroxyl alcohols might well be expected" [190]. Yet, a depolarized light scattering study did not conclusively resolve the issue for this diol [191] which certainly is a candidate to be revisited using dedicated equipment [153]. Support for Davidson's expectation, albeit for another diol, comes from a recent study of polydimethylsiloxane, a polymer with a flexible chain that is terminated by OH groups on both ends [192]. For two oligomers, with 7 and 14 repeat units, a Debye process was found using dielectric and mechanical spectroscopy suggesting that at this separation of OH groups, they can be viewed as mutually independent. For an oligomer with 56 repeat units, a Debye process could not be detected. The latter is presumably due to an intramolecular OH group dilution as analogously discussed in Sect. 2.2 for 2D1T, a long-chain MA [10]. Obviously, in order to observe supramolecular Debye-type relaxations assigned to "mutually independent" OH groups (although they belong to the same molecule) these groups should be located not too far from and not too close to each other. Also tripropylene glycol with its 11-membered carbon- and oxygen-containing chain that is terminated by OH groups on both ends appears to suit this constraint: This substance develops a Debye-like low-frequency feature, however, only at elevated pressures [193]. Whether and how the recently detected low-frequency rheological signature of glycerol [194] fits into this picture is a question that remains open at present.

6 Concluding Remarks

Supramolecular MA liquids display a rich variety of structural and dynamical effects that mainly emerge as a consequence of differences in their ability to form and sustain hydrogen bonds. With the goal to elucidate the mechanisms driving the intermolecular association and the corresponding relaxation properties of MAs, it is of utmost importance to exploit the various perspectives offered by using different experimental techniques. In the present survey, we mainly included results from broadband dielectric spectroscopy, high-resolution shear rheology, X-ray diffraction, and near-infrared absorption. Within the scope of this review, only brief reference was made to insights provided by other experimental techniques. Due to the variability of the supramolecular arrangements in pure MAs and the often delicate balance of different association patterns that can be generated by suitable mixing, a number of interesting features are revealed as a function of temperature, pressure (not explicitly considered in the present article), composition, molecular size, positioning of the OH group within the molecule, and so on.

By examining scaling properties using different experimental techniques, the identification of critical concentrations and of deviations from ideal mixing appears as recurring important topics that aid in advancing our understanding of the relaxational behavior in MAs from their fluid to their highly viscous state. Our multi-technique approach confirms and underscores the special role played by broadband dielectric spectroscopy for the study of MAs: This experimental method reveals the Debye process (process I) as an often dominant feature and via the Kirkwood factor provides access to the underlying orientational correlations which in many cases evolve in a surprising manner. While rheology monitors the structural relaxation (process II) as a feature with the same characteristics as that of "normal" supercooled liquids, in dielectric studies the α-process is special with respect to its spectral shape and relaxation amplitude.

In spite of the recent steps ahead to an improved understanding of the particular properties of MAs on the experimental side and the quest to refine existing modeling approaches, it is easy to predict that the formulation of fresh theoretical ideas, the application of new experimental techniques, and the development of novel preparation routes will keep the scientific research in the field of hydrogen-bonded liquids exciting and fruitful also in the foreseeable future.

Acknowledgements The support provided by the Deutsche Forschungsgemeinschaft under Grants No. BO1301/8-1, BO1301/8-2, and BO1301/14-1 is highly appreciated. Tina Hecksher kindly made available data for 2-ethyl-1-hexylamine and -thiol. Among our former students we thank in particular Thomas Büning, Kevin Moch, Marcel Preuß, and Hendrik Wittkamp for their contributions. We acknowledge DELTA synchrotron radiation source for allocating beamtime at beamline BL9, Michael Paulus for help with the X-ray diffraction experiments, as well as Patrick Degen and Heinz Rehage for enabling the density measurements.

References

1. Dyre JC (2006) Colloquium: the glass transition and elastic models of glass-forming liquids. Rev Mod Phys 78:953
2. Ngai KL (2011) Relaxation and diffusion in complex systems. Springer, Berlin
3. Böhmer R, Ngai KL, Angell CA, Plazek DJ (1993) Nonexponential relaxations in strong and fragile glass formers. J Chem Phys 99:4201
4. Hecksher T, Olsen NB, Niss K, Dyre JC (2010) Physical aging of molecular glasses studied by a device allowing for rapid thermal equilibration. J Chem Phys 133:174514
5. Debije P (1913) Zur Theorie der anomalen Dispersion im Gebiete der langwelligen elektrischen Strahlung. Verh d Dt Phys Ges 15:777
6. Huth H, Wang L-M, Schick C, Richert R (2007) Comparing calorimetric and dielectric polarization modes in viscous 2-ethyl-1-hexanol. J Chem Phys 126:104503
7. Böhmer R, Gainaru C, Richert R (2014) Structure and dynamics of monohydroxy alcohols milestones towards their microscopic understanding, 100 years after Debye. Phys Rep 545:125
8. Bauer S, Stern J, Böhm F, Gainaru C, Havenith M, Loerting T, Böhmer R (2015) Vibrational study of anharmonicity, supramolecular structure, and hydrogen bonding in two octanol isomers. Vib Spectrosc 79:59
9. Böttcher CJF (1973) Theory of electric polarization. volume 1, Elsevier, New York, §40
10. Bierwirth SP, Büning T, Gainaru C, Sternemann C, Tolan M, Böhmer R (2014) Supramolecular x-ray signature of susceptibility amplification in hydrogen-bonded liquids. Phys Rev E 90:052807

11. Gao Y, Tu W, Chen Z, Tian Y, Liu R, Wang L-M (2013) Dielectric relaxation of long-chain glass-forming monohydroxy alcohols. J Chem Phys 139:164504
12. Wang L-M, Richert R (2005) Ideal mixing behavior of the Debye process in supercooled monohydroxy alcohols. J Phys Chem B 109:8767
13. Pawlus S, Klotz S, Paluch M (2013) Effect of compression on the relationship between viscosity and dielectric relaxation time in hydrogen-bonded primary alcohols. Phys Rev Lett 110:173004
14. Gabriel J, Pabst F, Helbling A, Böhmer T, Blochowicz T (2018) Dynamic light scattering and dielectric spectroscopy: two perspectives on molecular reorientation in supercooled liquids. In: Loidl A, Kremer F (eds) The scaling of relaxation processes. Springer, this book
15. Various aspects of binary mixtures of glass formers are reviewed by Körber T, Rössler EA (2018) Dynamic heterogeneity in binary glass formers. In: Loidl A, Kremer F (eds) The scaling of relaxation processes. Springer, this book
16. Steward GW, Morrow RM (1927) X-ray diffraction in liquids: primary normal alcohols. Phys Rev 30:232
17. Davidson, DW, Cole RH (1951) Dielectric relaxation in glycerol, propylene glycol, and normal-propanol. J Chem Phys 19:1484 varied the number of molecular OH groups
18. Dannhauser W (1968) Dielectric study of intermolecular association in isomeric octyl alcohols. J Chem Phys 48:1911
19. Tomšič M, Jamnik A, Fritz-Popovski G, Glatter O, Vlček L (2007) Structural properties of pure simple alcohols from ethanol, propanol, butanol, pentanol to hexanol: comparing Monte Carlo simulations with experimental SAXS data. J Phys Chem B 111:1738
20. Stickel F (1995) Dissertation, Universität Mainz
21. Kaatze U, Behrends R, Pottel R (2002) Hydrogen network fluctuations and dielectric spectrometry of liquids. J Non-Cryst Solids 305:19
22. Yomogida Y, Sato Y, Yamakawa K, Nozaki R, Mishina T, Nakahara J (2010) Comparative dielectric study of pentanol isomers with terahertz time-domain spectroscopy. J Mol Struct 970:171
23. Kaatze U, Behrends R, von Roden K (2010) Structural aspects in the dielectric properties of pentyl alcohols. J Chem Phys 133:094508
24. Dannhauser W, Guerin R, Flueckinger AF (1970) Dielectric properties of liquid, 2-Propen-l-ol and 2-Propyn-l-ol. J Chem Phys 52:6446
25. Gao Y, Bi D, Li X, Liu R, Tian Y, Wang L-M (2013) Debye-type dielectric relaxation in glass-forming 3-methylthio-1-hexanol. J Chem Phys 139:024503
26. Gao Y, Chen Z, Tu W, Li X, Tian Y, Liu R, Wang L-M (2015) Anomaly in dielectric relaxation dispersion of glass-forming alkoxy alcohols. J Chem Phys 142:214505
27. Sartor G, Hofer K, Johari GP (1996) Structural relaxation and H bonding in isomeric octanols and their LiCl solutions by calorimetry. J Phys Chem 100:6801
28. Singh LP, Richert R (2012) Watching hydrogen-bonded structures in an alcohol convert from rings to chains. Phys Rev Lett 109:167802
29. Hecksher T, Jakobsen B (2014) Communication: supramolecular structures in monohydroxy alcohols: insights from shear-mechanical studies of a systematic series of octanol structural isomers. J Chem Phys 141:101104
30. Wikarek M, Pawlus S, Tripathy SN, Szulc A, Paluch M (2016) How different molecular architectures influence the dynamics of H-bonded structures in glass-forming monohydroxy alcohols. J Phys Chem B 120:5744
31. Dannhauser W, Flueckinger AF (1970) Liquid structure and dielectric relaxation of some isomeric methylheptanols. Phys Chem Liq 2:37
32. Johari GP, Dannhauser W (1968) Dielectric study of intermolecular association in sterically hindered octanol isomers. J Phys Chem 72:3273
33. Dannhauser W, Bahe LW, Lin RY, Flueckinger AF (1965) Dielectric constant of hydrogen bonded liquids. IV. Equilibrium and relaxation studies of homologous neo alcohols. J Chem Phys 43:257

34. Arrese-Igor S, Alegria A, Colmenero J (2017) On the non-exponentiality of the dielectric debye-like relaxation of monoalcohols. J Chem Phys 146:114502
35. Kalinovskaya OE, Vij JK, Johari GP (2001) Mechanism of the major orientation polarization in alcohols, and the effects of steric hindrance-, and dilution-induced decrease on H-bonding. J Phys Chem A 105:5061
36. Johari GP, Kalinovskaya OE, Vij JK (2001) Effects of induced steric hindrance on the dielectric behavior and H bonding in the supercooled liquid and vitreous alcohol. J Chem Phys 114:4634
37. Adrjanowicz K, Jakobsen B, Hecksher T, Kaminski K, Dulski M, Paluch M, Niss K (2015) Communication: slow supramolecular mode in amine and thiol derivatives of 2-ethyl-1-hexanol revealed by combined dielectric and shear-mechanical studies. J Chem Phys 143:181102
38. Metrangolo P, Neukirch H, Pilati T, Resnati G (2005) Halogen bonding based recognition processes: a world parallel to hydrogen bonding. Acc Chem Res 38:386
39. Li S-L, Xiao T, Lin C, Wang L (2012) Advanced supramolecular polymers constructed by orthogonal self-assembly. Chem Soc Rev 41:5950
40. Ferry JD (1980) Viscoelastic properties of polymers. Wiley, New York
41. Roland CM (2011) Viscoelastic behavior of rubbery materials. Oxford University Press, Oxford
42. Lyon T, Litovitz TA (1956) Ultrasonic relaxation in normal propyl alcohol. J Appl Phys 27:179
43. Behrends R, Kaatze U (2001) Hydrogen bonding and chain conformational isomerization of alcohols probed by ultrasonic absorption and shear impedance spectrometry. J Phys Chem A 105:5829
44. Kaatze U, Behrends R (2011) Hydrogen bond fluctuations and dispersive interactions of alcohol/alkane mixtures. An ultrasonic relaxation study. Chem Phys Lett 510:67
45. Gainaru C, Figuli R, Hecksher T, Jakobsen B, Dyre JC, Wilhelm M, Böhmer R (2014) Shear-modulus investigations of monohydroxy alcohols: evidence for a short-chain-polymer rheological response. Phys Rev Lett 112:098301
46. Büning T, Lueg J, Bolle J, Sternemann C, Gainaru C, Tolan M, Böhmer R (2017) Connecting structural and dynamical signatures of supramolecular Debye liquids. J Chem Phys 147:234501
47. Hecksher T (2016) Communication: linking the dielectric Debye process in monoalcohols to density fluctuations. J Chem Phys 144:161103
48. Shamim N, McKenna GB (2014) Mechanical spectral hole burning in polymer solutions: comparison with large amplitude oscillatory shear fingerprinting. J Rheology 58:43 and references cited therein
49. Cziep MA, Abbasi M, Heck M, Arens L, Wilhelm M (2016) Effect of molecular weight, polydispersity and monomer of linear homopolymer melts on the intrinsic mechanical nonlinearity $^3Q_0(\omega)$ in MAOS. Macromolecules 49:3566
50. Wilhelm M, Hyun K (2018) Non-linear shear mechanical responses. In: Richert R (ed) Non-linear dielectric spectroscopy. Springer
51. Bierwirth SP, Gainaru C, Böhmer R (2018) Debye liquids with tiny supramolecular signatures in their dielectric responses: insights from linear and non-linear shear studies (unpublished)
52. Seyboldt R, Merger D, Coupette F, Siebenbürger M, Ballauff M, Wilhelm M, Fuchs M (2016) Divergence of the third harmonic stress response to oscillatory strain approaching the glass transition. Soft Matter 12:8825
53. Richert R (2014) Supercooled liquids and glasses by dielectric relaxation spectroscopy. Adv Chem Phys 156:101
54. Richert R (2017) Nonlinear dielectric effects in liquids: a guided tour. J Phys: Condens Matter 29:363001
55. Richert R (ed) (2018) Nonlinear Dielectric Spectroscopy. Springer
56. Takahara S, Yamamuro O, Suga H (1994) Heat capacities and glass transitions of 1-propanol and 3-methylpentane under pressure. New evidence for the entropy theory. J Non-Cryst Solids 171:259

57. Murthy SSN, Nayak SK (1993) Experimental study of the nature of the glass transition process in monohydroxy alcohols. J Chem Phys 99:5362
58. Murthy SSN, Sobhanadri J, Gangasharan (1994) The origin of β relaxation in organic glasses. J Chem Phys 100:4601
59. Wang L-M, Tian Y, Liu R, Richert R (2008) Calorimetric versus kinetic glass transitions in viscous monohydroxy alcohols. J Chem Phys 128:084503
60. Preuß M, Gainaru C, Hecksher T, Bauer S, Dyre JC, Richert R, Böhmer R (2012) Experimental studies of Debye-like process and structural relaxation in mixtures of 2-ethyl-1-hexanol and 2-ethyl-1-hexyl bromide. J Chem Phys 137:144502
61. Bauer S, Burlafinger K, Gainaru C, Lunkenheimer P, Hiller W, Loidl A, Böhmer R (2013) Debye relaxation and 250 K anomaly in glass forming monohydroxy alcohols. J Chem Phys 138:094505
62. CRC Handbook of Chemistry and Physics (1998) Lide DR (ed) Chemical Rubber, 79th edn. Cleveland, pp 9–43
63. Kirkwood JG (1939) The dielectric polarization of polar liquids. J Chem Phys 7:911
64. Wang L-M, Tian Y, Liu R, Ngai KL (2011) Anomalous component dynamics of a binary mixture of associating glass-forming liquids. J Phys Chem B 115:719
65. Gainaru C, Böhmer R, Williams G (2010) Ion-sweeping in conducting dielectric materials. Eur Phys J B 75:209
66. Ishai PB, Talary MS, Caduff A, Levy E, Feldman Y (2013) Electrode polarization in dielectric measurements: a review. Meas Sci Technol 24:102001
67. Emmert S, Wolf M, Gulich R, Krohns S, Kastner S, Lunkenheimer P, Loidl A (2011) Electrode polarization effects in broadband dielectric spectroscopy. Eur Phys J B 83:157
68. Wittkamp H (2012) Master Thesis, TU Dortmund University
69. Adrjanowicz K, Kaminski K, Dulski M, Wlodarczyk P, Bartkowiak G, Popenda L, Jurga S, Kujawski J, Kruk J, Bernard MK, Paluch M (2013) Communication: synperiplanar to antiperiplanar conformation changes as underlying the mechanism of Debye process in supercooled Ibuprofen. J Chem Phys 139:111103
70. Moynihan CT, Boesch LP, Laberge NL (1973) Decay function for electric-field relaxation in vitreous ionic conductors. Phys Chem Glasses 14:122
71. Fayer MD (2011) Watching ultrafast molecular motions with 2D IR chemical exchange spectroscopy. World Scientific, Bangalore
72. Czarnecki MA (2011) Two-dimensional correlation analysis of hydrogen-bonded systems: basic molecules. Appl Spectrosc Rev 46:67
73. Suhm MA (2009) Hydrogen bond dynamics in alcohol clusters. Adv Chem Phys 142:1
74. Bauer S, Moch K, Münzner P, Schildmann S, Gainaru C, Böhmer R (2015) Mixed Debye-type liquids studied by dielectric, shear mechanical, nuclear magnetic resonance, and near-infrared spectroscopy. J Non-Cryst Solids 407:384
75. Workman J, Weyer L (2008) Practical guide to interpretive near-infrared spectroscopy. CRC Press, Boca Raton, Florida, USA
76. Miller CE (2002) Chemical principles of near-infrared technology. Chalmers JM, Griffiths PR (eds) Handbook of vibrational spectroscopy, vol 1, Chapter 2. Wiley, Chichester
77. Adrjanowicz K, Kaminski K, Wojnarowska Z, Dulski M, Hawelek L, Paulus S, Paluch M, Sawicki W (2010) Dielectric relaxation and crystallization kinetics of ibuprofen at ambient and elevated pressure. J Phys Chem B 114:6579 and references cited therein
78. Bicca de Alencastro R, Sandorfy C (1972) A low temperature infrared study of self-association in thiols. Can J Chem 50:3594
79. Gainaru C, Kastner S, Mayr F, Lunkenheimer P, Schildmann S, Weber HJ, Hiller W, Loidl A, Böhmer R (2011) Hydrogen-bond equilibria and life times in a monohydroxy alcohol. Phys Rev Lett 107:118304
80. Sillrén P, Matic A, Karlsson M, Koza M, Maccarini M, Fouquet P, Götz M, Bauer T, Gulich R, Lunkenheimer P, Loidl A, Mattsson J, Gainaru C, Vynokur E, Schildmann S, Bauer S, Böhmer R (2014) Liquid 1-propanol studied by neutron scattering, near-infrared, and dielectric spectroscopy. J Chem Phys 140:124501

81. Luck WAP, Ditter W (1968) Zur Bestimmung der Wasserstoffbrückenbindung im Oberschwingungsgebiet. J Mol Struct 1:261
82. Hédoux A, Guinet Y, Paccou L, Derollez P, Danède F (2013) Vibrational and structural properties of amorphous n-butanol: A complementary Raman spectroscopy and X-ray diffraction study. J Chem Phys 138:214506
83. Graener H, Ye TQ, Laubereau A (1989) Ultrafast dynamics of hydrogen bonds directly observed by time resolved infrared spectroscopy. J Chem Phys 90:3413
84. Wong J, Angell CA (1972) Spectroscopic detection of the glass transition: the $\pi \rightarrow \pi^*$ transition in nitrate melts. J Non-Cryst Solids 7:109
85. Barkatt A, Angell CA (1978) Use of structural probe ions for relaxation studies in glasses. 2. Temperature-jump and temperature-ramp studies of cobalt(II) in nitrate glasses. J Phys Chem 82:1972
86. Böhmer R, Sanchez E, Angell CA (1992) AC technique for simultaneous study of local and global linear responses near the glass transition: the case of doped $Ca^{2+}/K^+/NO_3^-$. J Phys Chem 96:9089
87. Böhmer R, Angell CA (1994) Global and local relaxations in glass-forming materials. In: Richert R, Blumen A (eds) Disorder effects on relaxational processes. Springer, Heidelberg. pp 11–54
88. Kremer F, Kossack W, Anton M (2018) The dynamic glass transition as reflected in its intermolecular dynamics and intra-molecular mobility. In: Loidl A, Kremer F (eds) The scaling of relaxation processes. Springer, this book
89. Sillrén P, Swenson J, Mattsson J, Bowron D, Matic A (2013) The temperature dependent structure of liquid 1-propanol as studied by neutron diffraction and EPSR simulations. J Chem Phys 138:214501
90. Singh LP, Raihane A, Alba-Simionesco C, Richert R (2015) Dopant effects on 2-ethyl-1-hexanol: a dual-channel impedance spectroscopy and neutron scattering study. J Chem Phys 142:014501
91. Ghoufi A, Hureau I, Lefort R, Morineau D (2011) Hydrogen-bond-induced supermolecular assemblies in a nanoconfined tertiary alcohol. J Phys Chem C 115:17761
92. Perera A (2017) Charge ordering and scattering pre-peaks in ionic liquids and alcohols. Phys Chem Chem Phys 19:1062
93. Mariani A, Ballirano P, Angiolari F, Caminiti R, Gontrani L (2016) Does high pressure induce structural reorganization in linear alcohols? A computational answer. Chem Phys Chem 17:3023
94. Hennous L, Abdel Hamid AR, Lefort R, Morineau D, Malfreyt P, Ghoufi A (2014) Crossover in structure and dynamics of a primary alcohol induced by hydrogen-bonds. J Chem Phys 141:204503
95. Hall MM Jr, Veeraraghavan VG, Rubin H, Winchell PG (1977) The approximation of symmetric x-ray peaks by Pearson type VII distributions. J Appl Crystallogr 10:66
96. Büning T (2016) Supramolecular structure of monohydroxy alcohols, Ph.D. Thesis. TU Dortmund University
97. Vahvaselkä KS, Serimaa R, Torkkeli M (1995) Determination of liquid structures of the primary alcohols methanol, ethanol, 1-propanol, 1-butanol and 1-octanol by X-ray scattering. J Appl Cryst 28:189
98. Bertagnolli H, Schulz G (1986) An x-ray study of specific bromine interaction in liquid ethylbromine and its interpretation by geometrical and potential models. Ber Bunsenges Phys Chem 90:816
99. Zetterström P, Dahlborg U, Wannberg A (1994) Structural studies of liquid 2-bromopropane and 2-chloropropane by neutron diffraction. Mol Phys 83:971
100. Stewart GK (1930) The cybotactic (molecular group) condition in liquids; the nature of the association of octyl alcohol molecules. Phys Rev 35:726
101. Pierce WC, macMillan DP (1938) X-ray studies on liquids: The inner peak for alcohols and acids. J Am Chem Soc 60:779

102. Denney DJ (1959) Dielectric relaxation in some two-component systems. J Chem Phys 30:1019
103. Mopsik FI, Cole RH (1966) Dielectric relaxation in liquid n-octyl iodide. J Chem Phys 44:1015
104. Denney DJ, Ring JW (1963) Dielectric relaxation in the 1-propanol-2-methylpentane system. J Chem Phys 39:1268
105. Wang L-M, Shahriari S, Richert R (2005) Diluent effects on the Debye-type dielectric relaxation in viscous monohydroxy alcohols. J Phys Chem B 109:23255
106. Power G, Nagaraj M, Vij JK, Johari GP (2011) Debye process and dielectric state of an alcohol in a nonpolar solvent. J Chem Phys 134:044525
107. Abdel Hamid AR, Lefort R, Lechaux Y, Moréac A, Ghoufi A, Alba-Simionesco C, Morineau D (2013) Solvation effects on self-association and segregation processes in tert-butanol—aprotic solvent binary mixtures. J Phys Chem B 117:10221
108. El Goresy T, Böhmer R (2008) Diluting the hydrogen bonds in mixtures of n-butanol with n-bromobutane: a dielectric study. J Chem Phys 128:154520
109. Lederle C, Hiller W, Gainaru C, Böhmer R (2011) Diluting the hydrogen bonds in viscous solutions of n-butanol with n-bromobutane. II. A comparison of rotational and translational motions. J Chem Phys 134:064512
110. Li X, Chen Z, Li Z, Gao Y, Tu W, Li X, Zhang Y, Liu YD, Wang L-M (2014) Comparative study of dynamics in glass forming mixtures of Debye-type N-ethylacetamide with water, alcohol, and amine. J Chem Phys 141:104506
111. Adachi K, Kotaka T (1993) Dielectric normal-mode relaxation. Prog Polym Sci 18:585
112. Williams G (2009) Chain dynamics in solid polymers and polymerizing systems as revealed by broadband dielectric spectroscopy. Macromol Symp 286:1
113. Yamaguchi T (2017) Viscoelastic relaxations of high alcohols and alkanes: Effects of heterogeneous structure and translation-orientation coupling. J Chem Phys 146:094511
114. Bierwirth SP, Münzner P, Knapp TA, Gainaru C, Böhmer R (2017) Communication: nonadditive dielectric susceptibility spectra of associating liquids. J Chem Phys 146:101101
115. Bierwirth SP, Gainaru C, Böhmer R (2018) Coexistence of two structural relaxation processes in monohydroxy alcohol–alkyl halogen mixtures. J Chem Phys (submitted)
116. Lesikar AV (1975) On the glass transition in organic halide–alcohol mixtures. J Chem Phys 63:2297
117. Lesikar AV (1977) On the glass transition in mixtures between the normal alcohols and various Lewis bases. J Chem Phys 66:4263
118. Wang L-M, Richert R (2008) Glass transitions in viscous monohydroxy alcohols: calorimetry versus dielectric relaxation. Int J Thermophys 29:2055
119. Weingärtner H, Nadolny H, Oleinikova A, Ludwig R (2004) Collective contributions to the dielectric relaxation of hydrogen-bonded liquids. J Chem Phys 120:11692
120. Vasiltsova T, Heintz A, Nadolny H, Weingärtner H (2009) Application of a new statistical mechanical model for calculating Kirkwood factors in self associating liquid systems to alkanol + CCl_4 mixtures. Phys Chem Chem Phys 11:2408
121. Power G, Vij JK, Johari GP (2007) Dielectric relaxation and crystallization of nanophase separated 1-propanol-isoamylbromide mixture. J Chem Phys 127:094507
122. Power G, Vij JK, Johari GP (2007) Relaxations and nano-phase-separation in ultraviscous heptanol-alkyl halide mixture. J Chem Phys 126:034512
123. Murthy SSN, Tyagi M (2002) Experimental study of the high frequency relaxation process in monohydroxy alcohols. J Chem Phys 117:3837
124. Murthy SSN, Tyagi M (2002) Dielectric study of the miscibility of binary liquids, one being an alcohol. J Sol Chem 31:33
125. Pawlus S, Paluch M, Dzida M (2011) Molecular dynamics changes induced by solvent in 2-ethyl-1-hexanol. Phys Rev E 84:031503
126. See the temperature dependent loss angle shown in Ref [105]
127. See, e.g., Schönhals A, Schlosser E (1993) Relation between main- and normal-mode relaxation A dielectric study on poly(propyleneoxide). Prog Coll Polym Sci 91:158; Mierzwa M, Floudas G, Dorgan J, Knauss D, Wegner J (2002) Local and global dynamics of polylactides:

a dielectric spectroscopy study. J Non-Cryst Solids 307–310:296; Casalini R, Roland CM (2005) Temperature and density effects on the local segmental and global chain dynamics of poly(oxybutylene). Macromolecules 38:1779
128. Gainaru C, Böhmer R (2009) Oligomer-to-polymer transition of poly-(propylene glycol) revealed by dielectric normal modes. Macromolecules 42:7616; Gainaru C, Hiller W, Böhmer R (2010) A dielectric study of oligo- and poly(propylene glycol). Macromolecules 43:1907
129. Ngai KL, Schönhals A, Schlosser E (1992) An explanation of anomalous dielectric relaxation properties of polypropylene glycol. Macromolecules 25:4915; Ilan B, Loring RF (1999) Local vitrification model for melt dynamics. Macromolecules 32:949
130. Gainaru C, Hecksher T, Fan F, Xing K, Cetinkaya B, Olsen NB, Dyre JC, Sokolov AP, Böhmer R (2017) Simple-liquid dynamics emerging in the mechanical shear spectra of poly(propylene glycol). Colloid Polym Sci B 295:2433
131. Preuß M (2010) Master Thesis. TU Dortmund University
132. Denney DJ, Cole RH (1955) Dielectric properties of methanol and methanol-1-propanol solutions. J Chem Phys 23:1767
133. Bordewijk P, Gransch F, Böttcher CJF (1969) Dielectric behavior of mixtures of 1-heptanol and 4-heptanol and the fluid structure of the monoalcohols. J Phys Chem 73:3255
134. Petong P, Pottel R, Kaatze U (1999) Dielectric relaxation of H-bonded liquids mixtures of ethanol and n-hexanol at different compositions and temperatures. J Phys Chem A 103:6114
135. Singh LP, Alba-Simionesco C, Richert R (2013) Dynamics of glass-forming liquids. XVII. Dielectric relaxation and intermolecular association in a series of isomeric octyl alcohols. J Chem Phys 139:144503
136. Gong H, Chen Z, Bi D, Sun M, Tian Y, Wang L-M (2012) Unusual dielectric strength of debye relaxation in monohydroxy alcohols upon mixing. J Phys Chem B 116:11482
137. Pawlus S, Wikarek M, Gainaru C, Paluch M, Böhmer R (2013) How do high pressures change the Debye process of 4-methyl-3-heptanol? J Chem Phys 139:064501 and references cited therein
138. Dielectric data for pure 2E1B are available from Refs. [61,136,137] and from the present work. The present results agree with Ref. [136] and in both cases chemicals from Sigma Aldrich were used. In Refs. [61,137] larger relaxation strengths were found for substances obtained from Alfa Aesar. The reasons for these discrepancies are not clear at present. In the following we use the data obtained in the present work
139. Bauer S, Wittkamp H, Schildmann S, Frey M, Hiller W, Hecksher T, Olsen NB, Gainaru C, Böhmer R (2013) Broadband dynamics in neat 4-methyl-3-heptanol and in mixtures with 2-ethyl-1-hexanol. J Chem Phys 139:134503
140. Li X, Chen Z, Gao Y, Tu W, Wang L-M (2015) Probing the Debye dielectric relaxation in supercooled methanol. Front Mater 2:41; see also Ref [90]
141. Vij JK, Scaife WG, Calderwood JH (1981) The pressure and temperature dependence of the complex permittivity of heptanol isomers. J Phys D Appl Phys 14:733
142. The $\Delta q^3_{\mathrm{main,max}}$ data given in Fig. 2(b) of Ref. [10] were mislabeled. The correct values are shown here in Fig. 25(c)
143. Bierwirth SP, Böhmer R, Gainaru C (2017) Generic primary mechanical response of viscous liquids. Phys Rev Lett 119:248001
144. Bierwirth SP, Gainaru C, Böhmer R (2018) Communication: correlation of terminal relaxation rate and viscosity enhancement in supramolecular small-molecule liquids. J Chem Phys 148:221102
145. Dannhauser W (1968) Dielectric relaxation in isomeric octyl alcohols. J Chem Phys 48:1918
146. Johari GP, Dannhauser W (1969) Effect of pressure on dielectric relaxation in isomeric octanols. J Chem Phys 50:1862
147. Fragiadakis D, Roland CM, Casalini R (2010) Insights on the origin of the Debye process in monoalcohols from dielectric spectroscopy under extreme pressure conditions. J Chem Phys 132:144505
148. Young-Gonzales AR, Richert R (2016) Field induced changes in the ring/chain equilibrium of hydrogen bonded structures: 5-methyl-3-heptanol. J Chem Phys 145:074503

149. Gainaru C, Wikarek M, Pawlus S, Paluch M, Figuli R, Wilhelm M, Hecksher T, Jakobsen B, Dyre JC, Böhmer R (2014) Oscillatory shear and high-pressure dielectric study of 5-methyl-3-heptanol. Colloid Polym Sci 292:1913
150. Zimm BH (1956) Dynamics of polymer molecules in dilute solution: viscoelasticity, flow birefringence and dielectric loss. J Chem Phys 24:269
151. Wang L-M, Richert R (2004) Dynamics of glass-forming liquids. IX. Structural versus dielectric relaxation in monohydroxy alcohols. J Chem Phys 121:11170
152. Sanz A, Nogales A, Puente-Orench I, Jiménez-Ruiz M, Ezquerra TA (2011) Detection of early stage precursor during formation of plastic crystal ethanol from the supercooled liquid state: a simultaneous dielectric spectroscopy with neutron diffraction study. Phys Rev Lett 107:025502
153. Wang Y, Griffin PJ, Holt A, Fan F, Sokolov AP (2014) Observation of the slow, Debye-like relaxation in hydrogen-bonded liquids by dynamic light scattering. J Chem Phys 140:104510
154. Gabriel J, Pabst F, Blochowicz T (2017) Debye process and β-relaxation in 1-propanol probed by dielectric spectroscopy and depolarized dynamic light scattering. J Phys Chem B 121:8847
155. Perakis F, Amann-Winkel K, Lehmkühler F, Sprung M, Pettersson D, Sellberg JA, Pathak H, Späh A, Cavalca F, Schlesinger D, Ricci A, Jain A, Massani B, Aubree F, Benmore CJ, Loerting T, Grübel G, Pettersson LGM, Nilsson A (2017) Diffusive dynamics during the high-to-low density transition in amorphous ice. Proc Natl Acad Sci USA 114:8193
156. Bertrand CE, Self JL, Copley JRD, Faraone A (2017) Nanoscopic length scale dependence of hydrogen bonded molecular associates' dynamics in methanol. J Chem Phys 146:194501
157. Yamaguchi T, Saito M, Yoshida K, Yamaguchi T, Yoda Y, Seto M (2018) Structural relaxation and viscoelasticity of a higher alcohol with mesoscopic structure. J Phys Chem Lett 9:298
158. Grebenkin SY, Syutkin VM (2007) Dynamical heterogeneity in glassy n-butanol. Phys Rev B 76:054202
159. Murohoshi T, Kaneda K, Ikegami M, Arai T (2003) Photoisomerization and isomer-specific addition of water in hydroxystilbenes. Photochem Photobiol Sci 2:1247
160. Coelho R, Khac Manh D (1967) Utilisation de la Biréfringence Électro-optique pour l'Étude de la Relaxation Dipolaire dans les Liquides Polaires Faiblement Conducteurs. C R Acad Sci Paris C 264:641
161. Crossley J, Williams G (1977) Relaxation in hydrogen-bonded liquids studied by dielectric and Kerr-effect techniques. J Chem Soc Faraday Trans 273:1906
162. Ookubo N (1991) New frequency-domain electric birefringence spectrometer using an advanced digital lock-in technique. Rev Sci Instrum 62:948
163. Arenas-Guerrero P, Iglesias GR, Delgado ÁV, Jiménez ML (2016) Electric birefringence spectroscopy of montmorillonite particles. Soft Matter 12:4923
164. Andrews SS, Boxer SG (2000) Vibrational Stark Effects of Nitriles. I. Methods and Experimental Results. J Phys Chem A 104:11853
165. Handler P, Aspnes DE (1967) Electric-Field-Induced Spectral Shifts of the OH Vibrational Absorption Line in Alcohols. J Chem Phys 47:473
166. Hiramatsu H, Hamaguchi H (2002) Association structures of N-methylacetamide in solution studied by infrared electroabsorption spectroscopy. Chem Phys Lett 361:457
167. Hiramatsu H, Kato C, Hamaguchi H (2001) Development of infrared electroabsorption spectroscopy for liquids. Chem Phys Lett 347:403
168. A related approach was exploited by Shilov SV, Müller M, Krüerke D, Heppke G, Skupin H, Kremer F (2002) Molecular arrangements and reorientation behavior in a dibenzopyrene-derivative ferroelectric columnar liquid crystal as studied by time-resolved Fourier-transform IR spectroscopy. Phys Rev E 65:021707
169. Johari GP, Goldstein M (1970) Viscous liquids and the glass transition. II. Secondary relaxations in glasses of rigid molecules. J Chem Phys 53:2372
170. Ngai KL, Paluch M (2004) Classification of secondary relaxation in glass-formers based on dynamic properties. J Chem Phys 120:857
171. Johari GP, Goldstein M (1971) Viscous liquids and glass transition. III. Secondary relaxation in aliphatic alcohols and other nonrigid molecules. J Chem Phys 55:4245

172. For superheated MAs expectedly the $g_K \rightarrow 1$ limit is approached, see Dannhauser W, Bahe LW (1964) Dielectric constant of hydrogen bonded liquids. III. Superheated alcohols. J Chem Phys 40:3058
173. Gainaru C, Meier R, Schildmann S, Lederle C, Hiller W, Rössler EA, Böhmer R (2010) Nuclear-magnetic-resonance measurements reveal the origin of the Debye process in monohydroxy alcohols. Phys Rev Lett 105:258303
174. Wieth P, Vogel M (2014) Dynamical and structural properties of monohydroxy alcohols exhibiting a Debye process. J Chem Phys 140:144507
175. Based on Ref. [143] the structural mechanical response of MAs is expected to display the generic shape
176. Chua YZ, Young-Gonzales AR, Richert R, Ediger MD, Schick C (2017) Dynamics of supercooled liquid and plastic crystalline ethanol: Dielectric relaxation and AC nanocalorimetry distinguish structural α- and Debye relaxation processes. J Chem Phys 147:014502
177. Jansson H, Swenson J (2011) The slow dielectric Debye relaxation of monoalcohols in confined geometries. J Chem Phys 134:104504
178. Gainaru C, Schildmann S, Böhmer R (2011) Surface and confinement effects on the dielectric relaxation of a monohydroxy alcohol. J Chem Phys 135:174510
179. Kipnusu WK, Elsayed M, Kossack W, Pawlus S, Adrjanowicz K, Tress M, Mapesa EU, Krause-Rehberg R, Kaminski K, Kremer F (2015) Confinement for more space: a larger free volume and enhanced glassy dynamics of 2-ethyl-1-hexanol in nanopores. J Phys Chem Lett 6:3708
180. Jensen MH, Alba-Simionesco C, Niss K, Hecksher T (2015) A systematic study of the isothermal crystallization of the mono-alcohol n-butanol monitored by dielectric spectroscopy. J Chem Phys 143:134501
181. Paluch M, Knapik J, Wojnarowska Z, Grzybowski A, Ngai KL (2016) Universal behavior of dielectric responses of glass formers: role of dipole-dipole interactions. Phys Rev Lett 116:025702
182. Wang L-M, Richert R (2005) Identification of dielectric and structural relaxations in glass-forming secondary amides. J Chem Phys 123:054516
183. Gainaru C, Bauer S, Vynokur E, Wittkamp H, Hiller W, Richert R, Böhmer R (2015) Dynamics in supercooled secondary amide mixtures: dielectric and hydrogen bond specific spectroscopies. J Phys Chem B 119:15769 and references cited therein
184. Ediger MD (2017) Perspective: highly stable vapor-deposited glasses. J Chem Phys 147:210901
185. Tylinski M, Chua YZ, Beasley MS, Schick C, Ediger MD (2016) Vapor deposited alcohol glasses reveal a wide range of kinetic stability. J Chem Phys 145:174506
186. Tylinski M, Beasley MS, Chua YZ, Schick C, Ediger MD (2017) Limited surface mobility inhibits stable glass formation for 2-ethyl-1-hexanol. J Chem Phys 146:203317
187. Young-Gonzales AR, Guiseppi-Elie A, Ediger MD, Richert R (2017) Modifying hydrogen-bonded structures by physical vapor deposition: 4-methyl-3-heptanol. J Chem Phys 147:194504
188. Capponi S, Napolitano S, Wübbenhorst M (2012) Supercooled liquids with enhanced orientational order. Nat Commun 3:1233
189. Kasina A, Putzeys T, Wübbenhorst M (2016) Dielectric and specific heat relaxations in vapor deposited glycerol. J Chem Phys 143:244504
190. Davidson DW (1961) Dielectric relaxation in liquids II. Isomeric pentanediols. Can J Chem 39:21
191. Fytas G, Dorfmüller T (1981) Depolarized light scattering studies of liquids: 1,5- and 2,4-pentanediol. J Chem Phys 75:5232

192. Xing K, Chatterjee S, Saito T, Gainaru C, Sokolov AP (2016) Impact of hydrogen bonding on dynamics of hydroxyl-terminated polydimethylsiloxane. Macromolecules 49:3138
193. Prevosto D, Capaccioli S, Lucchesi M, Rolla PA, Paluch M, Pawlus S, Ziolo J (2005) Emergence of a new feature in the high pressure–high temperature relaxation spectrum of tripropylene glycol. J Chem Phys 122:061102
194. Jensen MH, Gainaru C, Alba-Simionesco C, Hecksher T, Niss K (2018) Slow rheological mode in glycerol and glycerol–water mixtures. Phys Chem Chem Phys 20:1716

Dynamic Heterogeneities in Binary Glass-Forming Systems

D. Bock, Th. Körber, F. Mohamed, B. Pötzschner and E. A. Rössler

Abstract Starting with an overview of major results of the main (α) and secondary (β) relaxation in neat glass formers as compiled by dielectric and nuclear magnetic resonance (NMR) spectroscopy as well as by light scattering, the contribution deals with elucidating the component dynamics in binary glass formers. Dynamically asymmetric mixtures with high-T_g contrast of their components are in focus. In addition to polymers, specially synthesized non-polymeric glass formers are considered as high-T_g component and mixed with a low-T_g simple liquid. While the high-T_g component in the mixtures shows relaxation features similar to that of neat glass formers, the low-T_g component displays significantly faster dynamics and pronounced dynamic heterogeneities, i.e., an extremely broad distribution of correlation times $G(\ln\tau)$, which may lead to quasi-logarithmic correlation functions. Two glass transition temperatures with non-trivial concentration dependences are identified. The dynamic heterogeneities are transient in nature as proven by 2D exchange NMR. Thus, liquid-like (isotropic) reorientation of the low-T_g additive as well as exchange within its distribution $G(\ln\tau)$ is observed in an essentially rigid high-T_g matrix. The results show similarity with those collected for glass formers in confining geometries, suggesting that in asymmetric binary glass formers (intrinsic) confinement effects may control the dynamics either. We also investigate the β-process in the mixed glasses introduced by the low-T_g additive. It is rediscovered for all concentrations with virtually unchanged time constants. NMR identifies the β-relaxations as being similar to those of neat glasses. A spatially highly restricted motion with an angular displacement below $\pm 10°$ encompasses all molecules. Very similar spectral features are observed for the high-T_g component in NMR. Apparently, the (small) additive molecules "enslave" the large molecules to perform a common hindered reorientation. At lowest additive concentrations, one finds indications that the β-process starts to disintegrate. We conclude that the β-process is a cooperative process.

D. Bock · Th. Körber · F. Mohamed · B. Pötzschner · E. A. Rössler (✉)
Universität Bayreuth, Experimentalphysik II, 95440 Bayreuth, Germany
e-mail: ernst.roessler@uni-bayreuth.de

F. Kremer and A. Loidl (eds.), *The Scaling of Relaxation Processes*, Advances in Dielectrics, https://doi.org/10.1007/978-3-319-72706-6_6

1 Introduction

1.1 Main Relaxation in Pure Glass Formers

In the past two–three decades, pure glass formers were extensively investigated. In particular, the evolution of the dynamic susceptibilities reflecting, for instance, density or reorientational fluctuations were studied by various techniques covering times scales from picoseconds up to minutes [1–12]. This established a clear-cut phenomenology; yet, a full understanding of the glass transition is still missing. Among its prominent features are the super-Arrhenius temperature dependence of the structural relaxation time $\tau_\alpha(T)$ covering about 15 decades, and the two-step character of the correlation functions with stretching of its long-time decay. Usually, the extent of stretching does not change with temperature; this constitutes time-temperature (or frequency-temperature) superposition (TTS or FTS). Figure 1 presents an example [12]. The generic two-step nature of the correlation functions can be attributed to a temporary caging of molecules in dense liquids by their neighbors. While the first decay results from the "rattling" of the molecules within the cage, the second decay is due to the escape from the cage and, hence, reflects the structural or α-relaxation which virtually does not change in shape. Such "glassy dynamics" are found even above the melting point and lead to a single-step, essentially exponential correlation function only close to the boiling point [13]. A further feature of dense liquids is the presence of dynamic heterogeneities [14–17]. That means mobile and less mobile molecules can be distinguished. According to computer simulations [18–21], they are clustered in space and the correlated regions appear to grow in size upon cooling, thus reflecting an increasing dynamical length scale. The latter may be

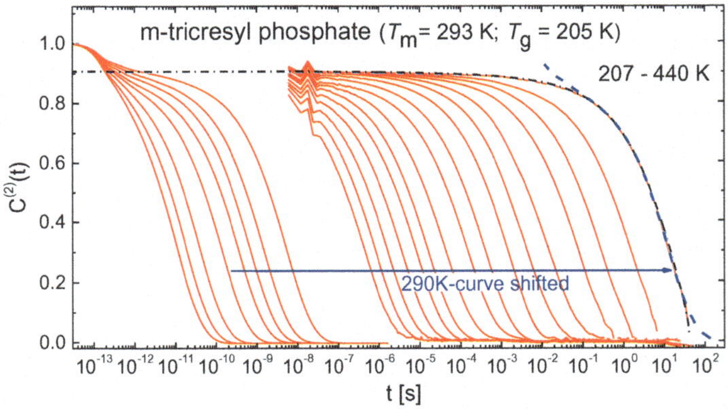

Fig. 1 Two-step reorientational correlation function of m-tricresyl phosphate as obtained from combining several techniques of depolarized light scattering [12]. The stretching of the long-time decay (α-relaxation) does not change with temperature; i.e., time-temperature superposition applies

a possible cause for the increase of activation energy of $\tau_\alpha(T)$ observed upon super-cooling a liquid. Recently, investigating higher-order susceptibilities, the number of correlated molecules has become directly accessible, indeed, demonstrating an increase of cooperativity [22, 23].

1.2 The Next Challenge: Binary Glass Formers

The next challenge is to characterize and to understand the dynamics of binary glass-forming systems. While technologically relevant polymer-additive (or polymer-plasticizer) systems have been the subject of intensive research [24–32] and are also the starting point of the present contribution, less is known about non-polymeric mixed glasses [10, 32–39], in particular about such systems with a high-T_g contrast ΔT_g of their components as it is usually the case in polymer-plasticizer systems. They may be taken as model systems for binary glass formers with significant size disparity of their components. Moreover, with up-to-date methods, the dynamics not only in the super-cooled regime but also in the stable liquid state can be accessed and this research encompasses the important field of dynamics in multi-component liquids. We note that non-polymeric high-T_g glass formers may also be of technological relevance [40]. The outstanding properties of such two-component systems are pronounced dynamical heterogeneities which show similarity to those of glass formers embedded in confining geometries, like liquids in nano-porous systems [41–50]. Actually, while the extent of dynamic heterogeneities, represented by the measured distribution of correlation times $G(\ln\tau)$, is rather small in pure glass formers and sophisticated experimental techniques have to be applied to identify them [51–55], strong experimental effects are observed in binary mixtures. Moreover, it is well established that such systems may exhibit two glass transition temperatures albeit they are fully miscible [56–61], a phenomenon also observed in polymer blends [62,63].

The effects strongly depend on the difference of the T_g values of the pure components, i.e., on the T_g contrast ΔT_g. As many polymers show high-T_g values, such dynamically "asymmetric" binary glass formers are most conveniently prepared by blending a polymer with a low-molecular additive. Yet, extending our survey to non-polymeric asymmetric mixtures we describe the dynamics of especially synthesized high-T_g components mixed with low-T_g components (additives). As will be shown, the pronounced dynamic heterogeneities in such mixed systems are well documented by nuclear magnetic resonance spectroscopy (NMR), in particular with the help of two-dimensional (2D) NMR which also allows to identify their transient nature [29, 32, 64–66]. As the distribution $G(\ln\tau)$ in asymmetric glass formers is much broader than in the case of pure systems it is experimentally quite easy to select a sub-ensemble of molecules and to monitor its evolution in time. Moreover, the selectivity of NMR using different probe nuclei allows to investigate the dynamics of each component separately. Together with dielectric spectroscopy (DS) such experiments provide a detailed picture of the rather complex dynamics in asymmetric binary glass formers [67–70]. The results are described in Sects. 2.1 and 2.2.

1.3 Secondary Relaxations in Glass-Forming Systems

Upon cooling, further relaxation processes at frequencies above those of the main (α-) relaxation appear. These secondary or β-relaxations provide an ubiquitous phenomenon of glass-forming liquids. Since the pioneering work of Johari and coworkers [71], which identified such relaxation processes in glasses made of "rigid" molecules displaying no internal degrees of freedom, they are usually taken as an intermolecular phenomenon and thus intrinsic to the glass formation. As probed in most cases by DS [72–79], in a few cases also by mechanical relaxation [72, 80], the β-process controls the response of the glass former at temperatures below T_g. Above T_g both α-relaxation and β-relaxation merge in many cases, a situation which often renders it difficult to separate both processes above T_g. Still, there appear to exist also non-generic β-relaxations, i.e., such secondary relaxation supposedly reflecting structural particularities of the glass-forming molecule. We mention in passing, that further "secondary" relaxations occur at cryogenic temperatures [81], which are connected to the low-temperature anomalies of glasses [82]. We also note that a well-resolved β-process is not always found [75]. Instead, an excess wing on the high-frequency side of the α-peak is observed (so-called type A glass formers in contrast to type B glass formers, which show a distinguishable β-relaxation [75]). In some cases, even both an excess wing and β-process are observed [75]. It was suggested to take the pressure dependence of the β-process as a classification scheme to distinguish generic and non-generic β-processes [83–85]. In the first case, the ratio of τ_β/τ_α for a given τ_α is pressure independent. This may be taken as a fingerprint that this β-process is of intermolecular (generic) nature and linked to the α-process. We note that a similar idea is behind the reports that for some sort of β-processes, its activation energy in the glass ($T < T_g$) follows the relationship $E_\beta/R \cong 24\ T_g$ [75]. Of course, in a dense liquid or even glass any type of molecular reorientation except that controlled by molecular symmetry (e.g., methyl group rotation) [86] has to involve intermolecular rearrangements to accommodate the free volume needed for molecular reorientation.

In any case, unraveling the nature or more precisely the molecular mechanism of the β-process is difficult being restricted to dielectric studies alone. Here, NMR as well as neutron scattering (NS) [87], the latter, however, being usually restricted to temperatures above T_g, provided deeper insight. Applying solid-state ^{2}H and recently ^{31}P NMR measurements on neat glass formers such as toluene, decaline, tripropyl phosphate [88–94] as well as glassy crystals like ethanol [91] and cyano cyclohexane [95], the molecular mechanism of the β-process was thoroughly investigated. Due to the small-angle sensitivity of the NMR echo technique [92, 96], it was shown that essentially all molecules take part in spatially highly restricted reorientations, a conclusion also drawn from solvation studies [97]. Figure 2 summarizes the typical phenomena for toluene. In all glass formers investigated so far, very similar features

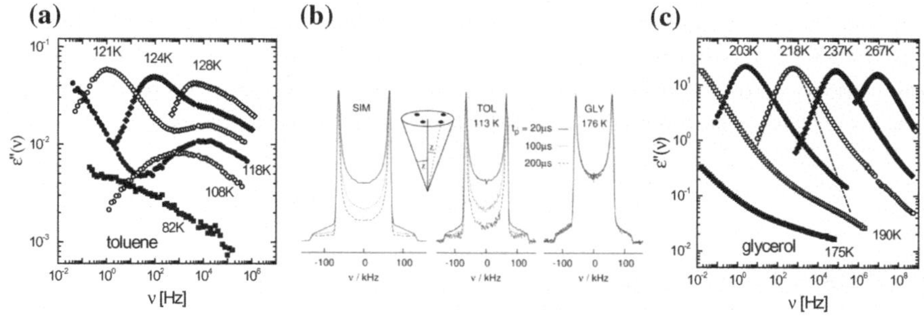

Fig. 2 **a** Dielectric spectra of glassy toluene (TOL) showing α- and β-relaxation and **c** glycerol (GLY) without a β-relaxation as well as **b** affiliated ^{2}H NMR solid-echo spectra measured for different pulse delays t_p. The spectral changes with increasing t_p observed in the case of toluene reflect spatially highly hindered motion of essentially all molecules as demonstrated by random walk simulations (SIM) [75, 91]

are observed, and a wobbling-on-a-cone model [98] with an opening angle smaller than 10° reproduces the salient features of the echo spectra ($T < T_g$) [90]. These findings are in contrast to results from recent MD simulations where rather large reorientation angles were found [99]. The model derived from the NMR experiments is close to the ideas forwarded by Williams and coworkers quite a long time ago [100]. Implicitly, it was assumed that an isotropic, small-angle process not much depending on the particular structure of the glass-forming molecule is involved.

The case of the molecule toluene is of particular interest as there exist contradicting reports on the extent of anisotropy of the motion involved in the β-process in its glass state. A ^{2}H spin-lattice relaxation study came to the conclusion that fluctuations around the pseudo C_2-axis dominate [101]. Recently, we confirmed this picture by measuring partially deuterated toluene-d_3 [102]. Yet, investigating toluene solely deuterated in the para-position clearly demonstrated that indeed the C_2-axis itself fluctuates in the course of the β-process [103] as was already concluded in the previous studies on toluene-d_5 [89, 90]. As the molecular dipole moment is directed along the C_2-axis the dielectric activity of the β-process has to involve fluctuations of this axis. Our previous experiments on toluene-d_3 were spoiled by the presence of the methyl group rotation. The only still possible extent of anisotropy, if at all, may involve an in-plane reorientation of the molecule, a situation recently discussed for structurally planar glass formers [104].

In the case of a polymer, namely poly(ethylene-propylene) (PEP), we found that the β-process is not only anisotropic but may also be dynamically inhomogeneous [102], i.e., not all polymer segments participate below T_g, a finding also reported for other polymers [105, 106]. It appears that the concept of "islands of mobility" first discussed by Johari and Goldstein [71, 107] to explain the origin of the β-process might reappear, at least in the case of polymers and in binary glass formers (see below) [70, 93,94]. More generally, this brings back the question whether a glass, and in particular a mixed glass, is structurally homogeneous. Considering silicate glasses, for example, ideas like percolation channels through a partly broken network of corner-sharing SiO_4 tetrahedra are prominently discussed [108]. Here, it is worthwhile to mention recent MD simulations on a metallic glass where the β-process was associated with cooperative string-like motions still present in the glass ($T < T_g$) [109].

2 Binary Glass Formers

2.1 Polymer-Plasticizer Systems—Main Relaxations

A relaxation process in a (super-cooled) liquid is denoted as a main (or α-) relaxation, if it is described by a correlation function decaying to zero. In terms of molecular reorientation as probed by NMR or DS, the molecular orientation assumes all loci on the unit sphere in the course of time—as expected for a liquid. In contrast, a β-process is spatially restricted leaving a residual correlation for long times; the latter may then relax by the α-process at $T > T_g$. We note that while in molecular systems DS probes a rank-one reorientational correlation function $C_1(t)$ in first approximation, NMR monitors a rank-two function $C_2(t)$. Yet, qualitatively, similar correlation decays are expected in most cases. In the following, all concentrations of the investigated mixtures are given by mass percent of the additive (low-T_g component).

We start with discussing the main features of the polymer-plasticizer system tripropyl phosphate (TPP)/polystyrene (PS) which is investigated by ^{2}H NMR (deuterated PS), ^{31}P NMR (TPP), and DS [67, 68]. Comparing the dielectric response of pure low-molar mass PS and TPP as depicted in Fig. 3a, the latter exceeds the former by three decades in amplitude. Thus, DS probes predominately the dynamics of TPP. In both glass formers, the α-relaxation is well visible, which shifts to higher frequencies with increasing temperature. In addition, a pronounced β-relaxation is observed in the case of TPP (type B glass former). In PS, the high-frequency flank of the α-peak crosses over from one power law behavior to another one, typical of a type A glass former, exhibiting an excess wing without discernible β-relaxation peak [75]. The corresponding time constants $\tau_\alpha(T)$ are included in Fig. 12, compared to the results collected for the mixtures, and discussed below (Sect. 2.3). We emphasize that FTS applies for the spectra; i.e., regarding the α-relaxation, all spectra can be collapsed to a single master curve, a feature characteristic of glassy dynamics in neat systems.

Figure 3b shows the dielectric spectra of 45% TPP/PS. Clearly, the α-peak is significantly broader. In particular, the low-frequency side of the α-peaks follows a power law with an exponent a significantly lower than that in neat TPP ($a = 1$). Due to the anti-plasticizer effect caused by the presence of PS, the relaxation is shifted to higher temperatures compared to pure TPP. The low-temperature spectra displaying the full β-relaxation have been omitted. Figure 3c shows dielectric data for a 18% TPP/PS mixture. In contrast to the 45% mixture now three relaxation phenomena are

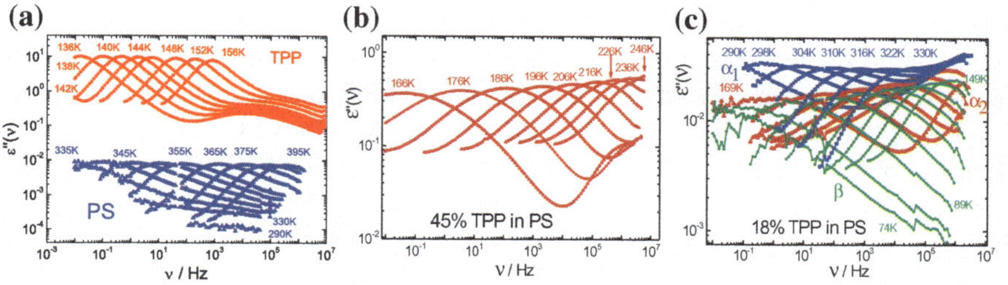

Fig. 3 **a** Dielectric spectra of the pure components, TPP ($T_g = 134$ K) and polystyrene (PS, $T_g =$ 335 K and M = 2250 g/mol). **b** Spectra of the TPP/PS mixtures: $c_{\text{TPP}} = 45\%$ and **c** $c_{\text{TPP}} = 18\%$ sample [67]

observed. Regarding high temperatures (blue squares), a relaxation peak with a low-frequency flank following a Debye behavior, i.e., $\varepsilon'' \sim \nu^1$, is observed, and FTS holds in good approximation. At lower temperatures, another broad relaxation peak (red triangles) is recognized, which exhibits a strong flattening on the low-frequency flank qualitatively similar to that observed for TPP in the 45% mixture. Moreover, FTS does not apply [67]. For reason which will become clear soon, the two relaxations are called α_1-relaxation (observed at higher temperatures) and α_2-relaxation (at lower temperatures). At lowest temperatures a β-relaxation is recognized (green circles).

In order to describe the spectra, a Kohlrausch function is fitted to the α_1-relaxation while a Havriliak-Negami (HN) function is used to interpolate the α_2-relaxation; however, in the case of the HN function, the time constants are obtained by "peak picking"—the results are included in Fig. 12. While $\tau_{\alpha 1}(T)$ shows a strong non-Arrhenius temperature dependence similar to that of neat PS, yet shifted to lower temperature, $\tau_{\alpha 2}(T)$ follows an Arrhenius law at low additive concentration. As discussed further below ^{2}H NMR experiments probing the PS dynamics yield correlation times in good agreement with $\tau_{\alpha 1}(T)$. Thus, the α_1-relaxation probed by DS is connected to the dynamics of PS. Comparing the spectral shapes and amplitudes of the relaxations, the α_1-peak in the 18% mixture is broader and by a factor of about four higher than the α-peak of neat PS. We have to conclude that also the (dielectric) α_1-relaxation is predominantly caused by the TPP molecules. In other words, a fraction of TPP molecules is associated with the PS dynamics, and therefore part of its α-relaxation, a conclusion also drawn by Blochowicz and coworkers investigating methyl tetrahydrofuran/PS mixtures [33]. The nature of the α_2-relaxation remains still to be clarified, though, given the two T_g reported by DSC in the mixtures (see below), it is natural to assume that it belongs to a second population of TPP molecules reorienting much faster than the PS component.

Another phenomenon can be recognized in Fig. 3c. While the amplitude of the α_1-relaxation decreases, that of the α_2-relaxation increases with temperature. It appears that, when temperature rises, more and more TPP molecules are freed from being associated with the polymer dynamics. A quantitative analysis of the relaxation strengths suggests that at some (high) temperature the α_1-relaxation originating from TPP molecules disappears. Interpreting the α_2-relaxation as a population of fast molecules (TPP) in a matrix of slow molecules (PS and a sub-ensemble of TPP), a redistribution of TPP molecules among the two populations takes place, which finally leads to the disappearance of the slowly moving TPP fraction at some temperature. Similar results were reported by Blochowicz and coworkers [33, 38], and as these authors pointed out, such a behavior might be interpreted in the framework of mode coupling theory (MCT) as a type A glass transition, where a dynamic decoupling of large and small particles is anticipated, if the components differ sufficiently in size [110]. In that case, the smaller particles retain mobility below the glass temperature of the larger ones until they undergo a localization transition in the confinement of a frozen matrix.

The TPP/PS mixtures were also studied by NMR [68]. Figure 4 shows a direct comparison of the reorientational correlation function $C_2(t)$ of (deuterated) PS probed by ^{2}H NMR (left) and of TPP monitored by ^{31}NMR (right) as obtained by the stimulated echo method [64, 111, 112]. While the decay curves of PS can be interpolated by a slightly more stretched Kohlrausch function when compared to that of pure PS, the decays of TPP become more and more stretched, the lower the additive (TPP) concentration. Finally, the correlation function assumes a quasi-logarithmic decay at lowest TPP concentrations, which renders it difficult to define a clear-cut mean correlation time. These findings directly correlate with the broad dielectric α_2-relaxation. Yet, the ^{31}P NMR correlation functions of TPP do not display a bimodal decay as is expected from the dielectric spectra which, in addition to the α_2-relaxation, show the α_1-relaxation assumed to originate from a fraction of TPP molecules associated with the motion of PS (α_2-process). However, the ^{31}P NMR decays in Fig. 4 can only be reproduced by transforming the fit of both dielectric α_1- *and* α_2-relaxation into the time domain. Figure 4 directly demonstrates the decoupling of the dynamics of TPP (α_2) and PS (α_1): Covering the same time window, the correlation of PS is monitored at much higher temperatures as compared to that of TPP.

We can summarize the findings collected so far. While the character of the reorientational correlation function of the high-T_g component PS (α_1) does not significantly change in the mixtures compared to that of neat PS, a broad distribution $G(\ln\tau_{\alpha 2})$ is found for the much faster α_2-process of the low-T_g component TPP, yielding an extremely stretched correlation function. The distribution $G(\ln\tau_{\alpha 2})$ has a strong contribution on the long-$\tau_{\alpha 2}$ flank in contrast to what is observed in neat glass formers for which a Kohlrausch or a Cole-Davidson (CD) distribution interpolates the spectra. Note that the CD distribution has a cut-off on the long-τ flank. Such dynamic heterogeneities are directly reflected in the ^{31}P NMR spectra—see Fig. 5a. The TPP spectra are dominated by the chemical shift anisotropy (CSA) interaction yielding an asymmetric solid-state spectrum at low temperatures, while at high temperatures a Lorentzian central line is observed due to fast isotropic (liquid-like) dynamics. At intermediate temperatures "two-phase" spectra are found [29, 111, 113–115]. They can be described by a sum of a Lorentzian line (reflecting liquid-like molecules) and a solid-state sub-spectrum (reflecting slowly reorienting molecules). Such two-phase spectra are a direct fingerprint of strong dynamic heterogeneities, and their explanation is sketched in Fig. 5b. Due to a broad distribution $G(\ln\tau_{\alpha 2})$, only the limiting spectra of fast (isotropic) and slow (isotropic) reorientation are observed;

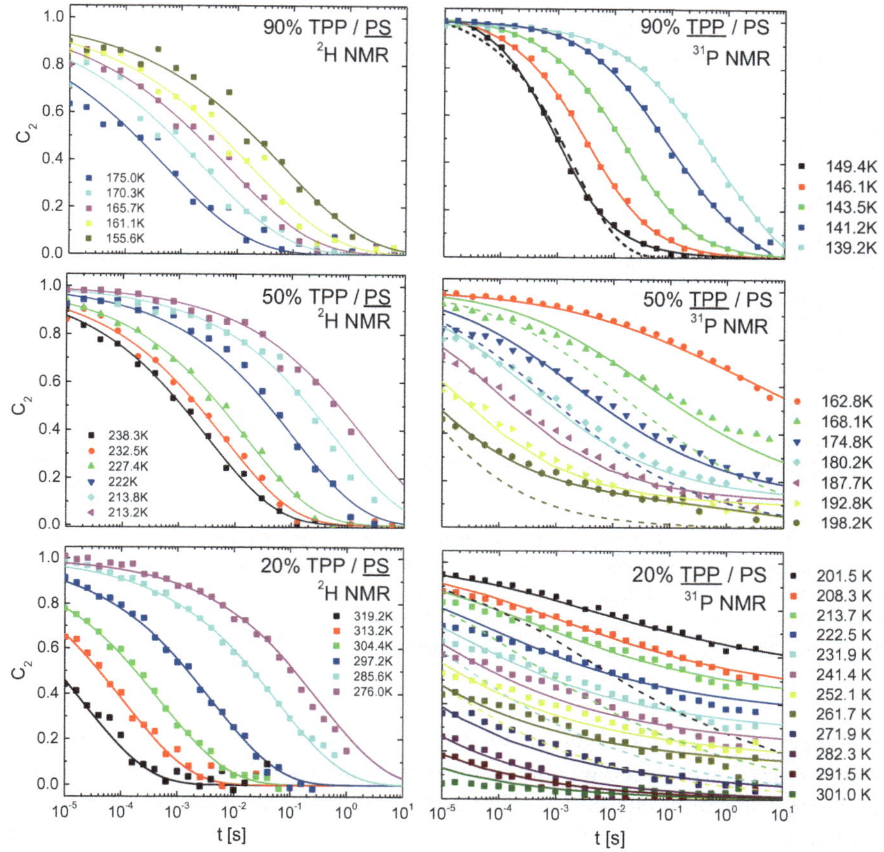

Fig. 4 Reorientational correlation functions $C_2(t)$, measured by the NMR-stimulated echo technique for different (mass) fractions of TPP in PS [68]; (left) $C_2(t)$ of PS by ^{2}H NMR, solid lines: Kohlrausch fits; (right) $C_2(t)$ of TPP by ^{31}P NMR, solid lines for $c_{\mathrm{TPP}} = 90\%$: Havriliak-Negami fits, dashed line: Kohlrausch fits with $\beta_K = 0.53$; solid lines for $c_{\mathrm{TPP}} \leq 50\%$: interpolation based on DS data, taking also α_1-relaxation contribution into account; dashed lines: HN fits accounting solely for the α_2-relaxation, they do not fit

so-called intermediate NMR spectra as observed in the case of neat glass formers can be ignored (cf. Fig. 10), since the fraction of molecules with time constants on the order of the reciprocal NMR coupling constant δ is small in a broad distribution. We emphasize that the appearance of a narrow central Lorentzian line clearly indicates isotropic reorientation of the TPP molecules in a more or less immobilized PS matrix; i.e., the dielectric α_2-relaxation reflects a liquid-like motion of the TPP molecules.

Determined by the quadrupolar interaction, the symmetric ^{2}H NMR spectra of (deuterated) PS are included in Fig. 5a (left). Compared to pure PS [67], the spectral collapse is shifted to lower temperatures reflecting accelerated dynamics due to the

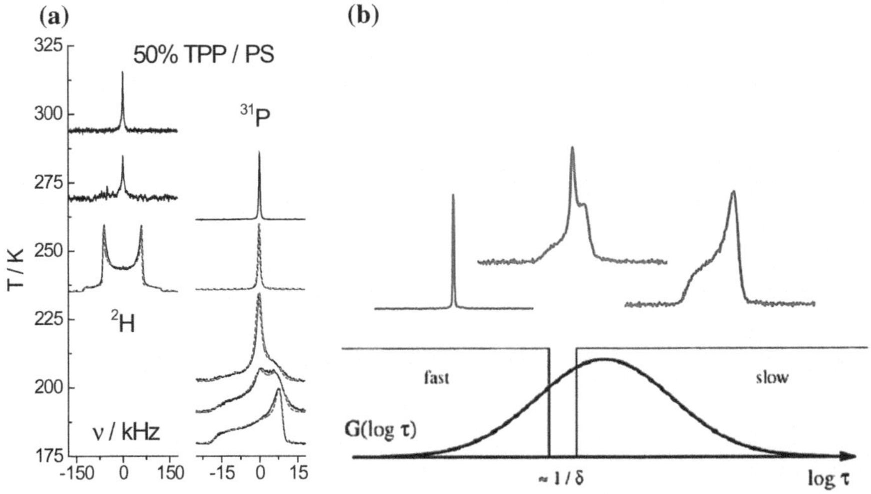

Fig. 5 **a** ^{2}H (left) and ^{31}P NMR (right) 1D spectra of a 50% TPP/PS mixture, baseline of spectra marks temperature on vertical axis [68]. Blue dashed lines are spectral fits by the two-phase model. **b** Two-phase ^{31}P NMR spectra, explained by the presence of a broad distribution G(lnτ). A superposition (center) of a central Lorentzian line (left, fast isotropic dynamics) and a broad solid-state spectrum (right, slow motion) is found while the distribution shifts through the NMR time window

plasticizer effect. Besides a temperature shift no qualitative change of the ^{2}H spectra is observed: The spectral collapse from the solid-state (Pake) to the Lorentzian liquid-state spectrum takes place within a narrow temperature interval at any concentration of the mixtures. Comparing the ^{31}P (Fig. 5a, right) and the ^{2}H NMR spectra, one clearly recognizes the decoupling of the dynamics of the components: At T ≅ 236 K a broad solid-state spectrum characteristic of slow motion is observed for PS, whereas the ^{31}P spectra of TPP still reflect fast isotropic motion (Lorentzian line). Moreover, in contrast to TPP, the high-T_g component PS does not show strong dynamic heterogeneities as no two-phase NMR spectra are observed.

Two-dimensional (2D) NMR exchange spectra directly visualize the evolution of the reorientational process [64, 92, 111]. The 2D spectrum, $S(\omega_1,\omega_2;t_m)$, measures the joint probability, $P(\omega_1,0;\omega_2, t_m)$, to find molecules with a frequency ω_2 at the mixing time t_m and a frequency ω_1 at time $t = 0$. This probability can be expressed as a product of an a priori probability $P(\omega_1)$ that a molecule assumes a NMR frequency ω_1 before the mixing time t_m and the conditional probability $P(\omega_1,0|\omega_2,t_m)$, that a molecule has the NMR frequency ω_2 after the mixing time if it had the NMR frequency ω_1 before. As the NMR frequency ω is connected to the orientation of the interaction tensor (chemical shift anisotropy for ^{31}P, or quadrupolar interaction for ^{2}H), which is fixed in a molecular frame, the spread of intensity out from the diagonal spectrum into the ω_1-ω_2 plane monitors the progress of the reorientational process. Usually, 2D spectra are collected for some fixed values of the mixing time, thus monitoring the progress of reorientation in time.

Such 2D spectra for the $c_{TPP} = 20\%$ TPP/PS sample are depicted in Fig. 6a (left side) for the case of a low temperature [68]. No isotropic motion faster than $1/\delta_{CSA}$ is detectable in the 1D echo spectra (given by the diagonal spectrum), as no central Lorentzian line is visible. The molecules may be slow ($\tau > 1/\delta_{CSA}$) or immobile. For $t_m = 0.05$ ms essentially the entire intensity is found on the diagonal; i.e., no reorientation occurs on this time scale. For $t_m = 500$ ms the signal in the full ω_1-ω_2 plane grows, clearly indicating spatially unrestricted (isotropic) reorientation, and for $t_m = 5000$ ms the off-diagonal signal further gains intensity while diagonal intensity decreases. However, even at longest times some intensity along the diagonal is still recognizable indicating that not all molecules have moved so far. This is not surprising as the distribution $G(\ln\tau_{\alpha 2})$ is extremely broad. Moreover, TPP molecules associated with the still slower PS dynamics (α_1-process) lead to a residual diagonal spectrum. Nevertheless, we again conclude that most of the TPP perform a liquid-like motion in an essentially rigid matrix of PS chains, and the 2D spectra can be reproduced by a broad distribution of isotropic reorientation, i.e., by a weighted sum of the two 2D sub-spectra belonging to not-reorienting and fully isotropically reorienting molecules, respectively (cf. Fig. 6a, right side).

The dynamic heterogeneities reflected by a broad $G(\ln\tau_{\alpha 2})$ are actually of transient nature. This is demonstrated in Fig. 6b, where 2D spectra of the $c_{TPP} = 18\%$ sample at some higher temperature are shown. In this case the 1D spectrum is a superposition of a central Lorentzian line (reflecting fast TPP molecules) and a diagonal spectrum (reflecting slow molecules) similar to those of Fig. 5a. A typical cross-like pattern occurs in the 2D spectra for long t_m, which indicates that fast molecules and slowly reorienting molecules mutually exchange. A detailed analysis [68] shows that isotropic reorientation and exchange occur both on the same timescale as found in neat glass formers [53].

2.2 *Non-polymeric Binary Glass Formers—Main Relaxations*

Extending our activities on binary glass formers, we synthesized several non-polymeric high-T_g glass formers based on spirobichroman derivatives (cf. Fig. 7) which were mixed with a low-T_g component. In a first study, the non-deuterated high-T_g compound "DH" ($T_g = 382$ K) together with m-tricresyl phosphate (m-TCP; $T_g = 206$ K) as additive were employed yielding a T_g contrast $\Delta T_g = 176$ K [69]. In a second work, deuterated "SBC" ($T_g = 356$ K) mixed with TPP achieved $\Delta T_g = 222$ K [70]. As will be demonstrated, very similar dielectric as well as NMR features are observed as in polymer-plasticizer systems. The corresponding time constants $\tau_\alpha(T, c_{\text{add}})$ as well as $T_g(c_{\text{add}})$ will be discussed together with the results from the TPP/PS mixtures (cf. Sect. 2.1) in Sect. 2.3.

Figure 8a displays the dielectric spectra of a 80% m-TCP/DH sample after subtracting the dc conductivity contribution; for comparison one spectrum of neat

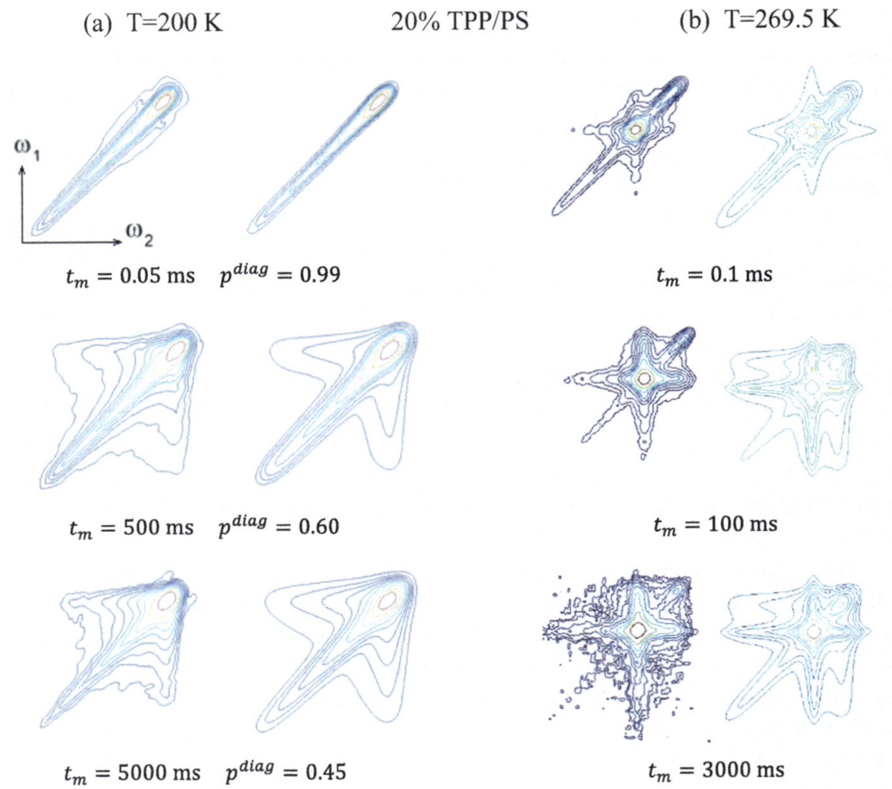

Fig. 6 **a** Experimental 2D ^{31}P NMR exchange spectra of a 20% TPP/PS mixture at $T = 200$ K with mixing times t_m indicated (left) and spectra calculated assuming a broad distribution of isotropic reorientation (right). **b** Corresponding 2D spectra at T = 269.5 K demonstrating exchange among liquid-like and solid-like TPP molecules [68]

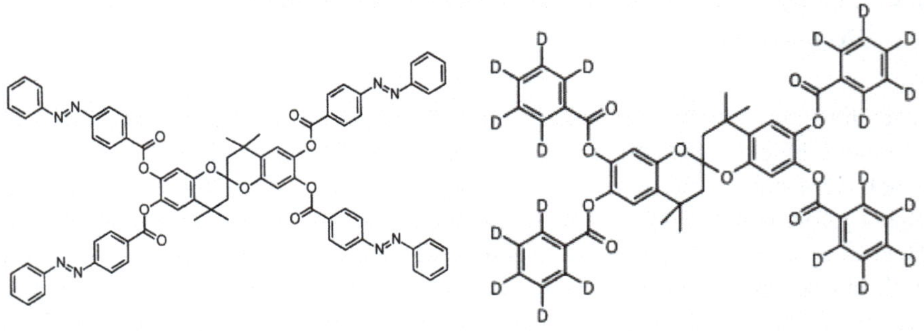

Fig. 7 High-T_g spirobichroman component "DH" (T_g = 382 K) with molar mass of M = 1205.3 g/mol (left) [69] and (deuterated) high-T_g component "SBC" (T_g = 356 K) with M = 809.0 g/mol (right) [70]

m-TCP is included. A weak shoulder is recognized on the low-frequency flank of the main relaxation. The latter is assigned to the relaxation of the high-T_g component DH (α_1-relaxation) and the former to that of m-TCP (α_2-relaxation). Like in the case of TPP/PS the α_2-spectra are broader, in particular on the low-frequency flank, and become wider upon decreasing temperature; i.e., FTS does not apply. Figure 8b shows the spectra of a mixture with $c_{m\text{-}TCP} = 34\%$ after subtracting the dc conductivity contribution. Two well-resolved relaxations with similar amplitudes are recognized. The relaxation at higher frequencies is again attributed to the α_2-process of m-TCP, while the slower one is assigned to the relaxation of the high-T_g component DH (α_1-relaxation). Yet, the signal amplitude of the latter is somewhat larger than that of the neat DH although the sample contains only 66% DH. Thus, as in the case of TPP/PS the results suggest that the α_1-relaxation contains a contribution from the low-T_g component m-TCP; i.e., a fraction of m-TCP molecules participates in the α-process of DH. One also recognizes a trend that the α_1-relaxation decreases with increasing temperature while the α_2-relaxation grows, suggesting that the contribution of m-TCP to the α_1-relaxation may disappear above some temperature. The spectra can again be interpolated by fit with a weighted sum of a Kohlrausch (α_1-process) and HN spectrum (α_2-process). The corresponding time constants are included in Fig. 10a. Similar dielectric spectra are observed in the mixture TPP/SBC (see Fig. 8c).

As discussed before, the NMR stimulated echo technique provides direct access to the reorientational correlation function $C_2(t)$. The stretching of the correlation function for neat m-TCP is $\beta_K = 0.62$, which is typical of neat glass formers (cf. also Fig. 1). In the binary mixtures, with decreasing m-TCP concentration the correlation functions become more and more stretched—see Fig. 9; for the 37% mixture again a quasi-logarithmic decay is recognized. In accordance with the manifestation of the dielectric spectra an increasing extent of dynamic heterogeneities of the low-T_g component is reflected by a highly stretched reorientational correlation function. The corresponding time constants are shown in Fig. 14a.

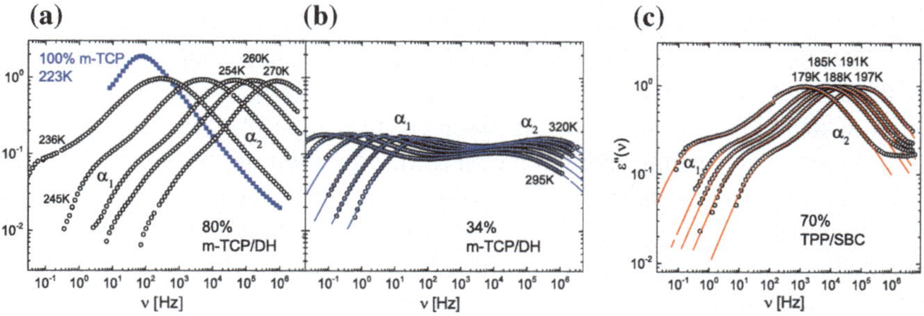

Fig. 8 **a** Dielectric spectra after subtracting dc conductivity contribution of 80% m-TCP/DH mixture (temperatures indicated); for comparison a spectrum of neat m-TCP at $T = 156$ K is shown [69]. **b** Spectra for a 34% mixture, lines: interpolations by superposition of a Kohlrausch and a HN susceptibility. **c** Spectra for a 70% TPP/SBC mixture [70]

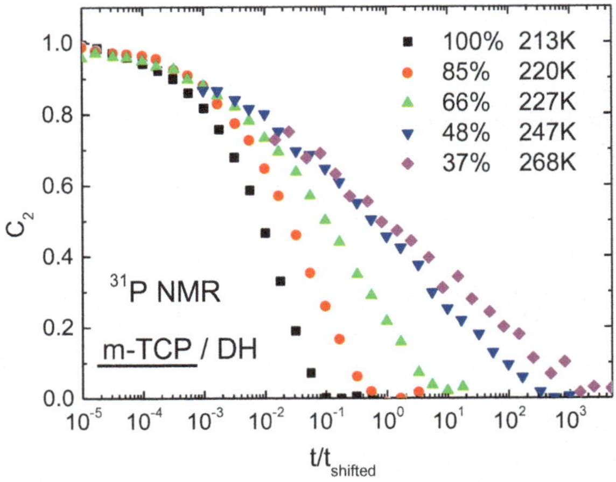

Fig. 9 Reorientational correlation function $C_2(t)$ for m-TCP in DH for different additive concentrations as indicated as a function of reduced time to allow direct comparison [69]

The dynamic heterogeneity reflected in broad dielectric spectra and highly stretched correlation functions C_2(t) is again seen in the ^{31}P NMR spectra of m-TCP in Fig. 10. At the lowest temperature typical asymmetric powder spectra determined by the CSA interaction are found, which signal that all m-TCP molecules are slow or immobile on the NMR timescale. At highest temperatures a central Lorentzian liquid line is observed, which indicates that all molecules perform fast isotropic reorientation. In between these limiting spectra, a concentration-dependent crossover from a solid-state spectrum to a liquid line is observed. While for neat m-TCP a gradual collapse from a powder spectrum to a Lorentzian line is discernible, one finds two-phase spectra in the mixtures. As discussed, such two-phase spectra are a direct indication of a broad distribution of correlation times $G(ln\tau_{\alpha 2})$ of an isotropic reorientation, i.e., a signature of pronounced dynamic heterogeneities. Interestingly, two-phase spectra are also observed at temperatures below T_{g1} (attributed to the dynamics of the high-T_g component DH, cf. Sect. 2.3) clearly indicating the presence of isotropic m-TCP reorientation in a matrix of virtually immobilized DH molecules.

The main relaxations of binary glass formers may also be studied by measuring the ^{31}P NMR spin-lattice relaxation rate $1/T_1$ which is plotted versus the temperature for the case of the m-TCP/PMMA mixtures in Fig. 11 [50]. While for neat m-TCP a narrow and high relaxation peak is observed it successively broadens and lowers its amplitude upon adding the polymer PMMA. In addition, the maximum shifts to higher temperatures indicating a slowing-down of the dynamics due to the anti-plasticizer effect by adding PMMA. The strong broadening and decreasing of the amplitude of the relaxation curve is a direct consequence of the emergence of the pronounced dynamic heterogeneities in the binary glass former described by the distribution $G(\ln\tau_{\alpha 2})$ becoming broader and broader at low m-TCP concentrations. In Fig. 11b we consider the case that m-TCP provides the high-T_g component. It is mixed with the (deuterated) low-T_g additive methyl tetrahydrofuran (MTHF; $T_g = 96$ K). Now, no broadening is observed. Actually, a slight opposite effect is

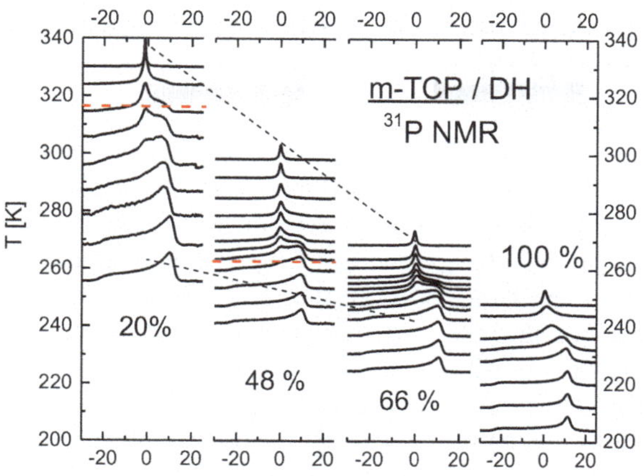

Fig. 10 1D ^{31}P NMR spectra of m-TCP mixed with DH for different concentrations as indicated. The baseline of the spectra corresponds to the temperature on vertical axis. The dashed lines mark the temperature interval in which two-phase spectra are observable in the mixtures, while this is not the case for 100% m-TCP. The red dashed lines indicate T_{g1} (associated with DH dynamics) [69]

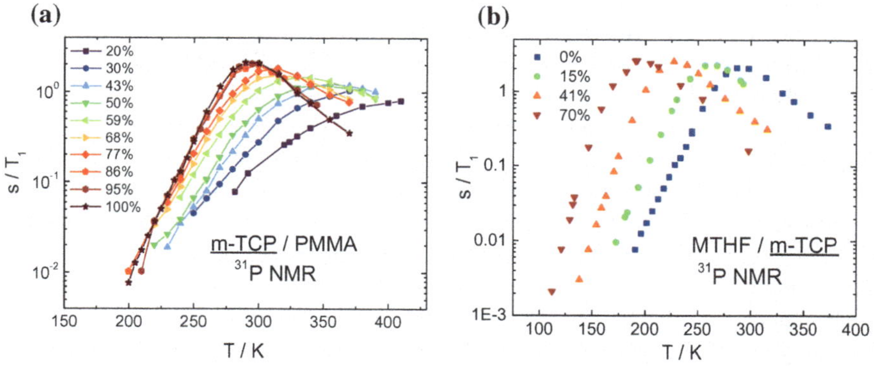

Fig. 11 ^{31}P NMR spin-lattice relaxation rate $1/T_1$ as a function of temperature for different additive concentrations: **a** m-TCP in the mixture m-TCP/PMMA, lines are guides-for-the-eye; **b** m-TCP in the mixture MTHF/m-TCP [50]

recognized: The relaxation maximum weakly increases with MTHF being added, and it shifts to lower temperatures due to the plasticizer effect of MTHF. These findings once again demonstrate that, depending on whether being the high- or the low-T_g component, the dynamics of the component is very different: The high-T_g component displays dynamics close to that of neat glass formers, while the low-T_g component exhibits strong dynamic heterogeneities.

2.3 Time Constants and Glass Transition Temperatures in Asymmetric Binary Glass Formers

As demonstrated in the previous subsections, the binary mixtures TPP/PS, m-TCP/PMMA, m-TCP/DH, TPP/SBC, and MTHF/m-TCP show very similar evolutions of their dynamic susceptibility or correlation function. In the following we compare the time constant $\tau_{\alpha 1}(T, c_{\mathrm{add}})$ and $\tau_{\alpha 2}(T, c_{\mathrm{add}})$ determined by dielectric (circles) and NMR spectroscopy (squares)—see Fig. 12a [68–70]. Clearly, well-separated time constants are identified for the two components at all concentrations, and applying the definition $100\ \mathrm{s} = \tau(T_g)$, two glass transition temperatures can be extracted, T_{g1} and T_{g2} (Fig. 13). The lower the additive concentration, the stronger is the dynamic decoupling among the components. Most important, the time constants of PS collected by ^{2}H NMR (open squares) are essentially identical with those of the α_1-process determined for TPP by DS (open circles). This confirms the conclusion that the dielectric α_1-relaxation of the additive is associated with the dynamics of the high-T_g component. At high TPP concentrations $\tau_{\alpha 2}(T)$ exhibits a non-Arrhenius behavior; however, at low concentrations (20 and 10%, dark yellow and blue symbols, respectively, in Fig. 12a) a crossover to an Arrhenius law is found at low temperatures. This fragile-to-strong crossover occurs close to T_{g1}, i.e., at T_g of the high-T_g component PS in the mixture (cf. arrow in Fig. 12a). Quite similar results are reported for the non-polymeric system m-TCP/DH—see Fig. 12b [70]. Thus, at low temperatures for which the high-T_g matrix has become essentially rigid, the α_2-process shows high similarity with a β-process. However, as demonstrated by the echo and the 2D spectra, even below T_{g1} the α_2-process comprises an isotropic, liquid-like reorientational process with an extremely broad distribution $G(ln\tau_{\alpha 2})$. It is important to realize that by extracting the time constants $\tau_{\alpha 2}(T, c_{\mathrm{add}})$ from the corresponding relaxation maximum of the DS spectra, values are obtained, which must not be identified with a typical correlation time defined by the integral over a correlation function. As the spectral shape of the α_2-relaxation is strongly broadened on the low-frequency side and the correlation function $C_2(t)$ probed by NMR is extremely stretched, one expects "true" correlation times longer than those displayed in Fig. 12 [29].

Figure 13a and b display the corresponding $T_{g1}(c_{add})$ and $T_{g2}(c_{add})$; we also included results from DSC measurements, which are able to detect two "glass steps" [67]. While $T_{g1}(c_{add})$ steadily decreases with increasing c_{add} from its value for the pure high-T_g component down to values approaching that of the low-T_g component, $T_{g2}(c_{TPP})$ is significantly lower and exhibits a maximum at intermediate concentrations. Such a maximum, up to our knowledge, was not reported in the literature so far. However, it is a direct consequence of a fragile-to-strong transition in $\tau_{\alpha 2}(T)$ observed a low c_{add}. We emphasize that even at very high concentrations of the low-T_g component two separate main processes (cf. Fig. 8c), yet with rather close time constants, can be identified by DS. Thus, a merging of the two processes as suggested in Ref. [33] is not confirmed.

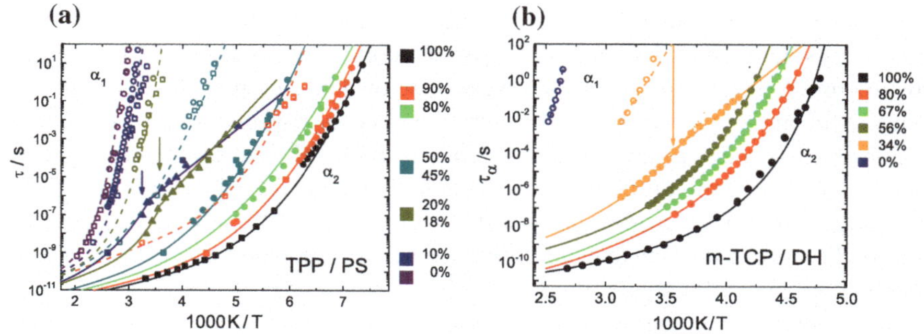

Fig. 12 Time constants of the mixtures TPP/PS (**a**) [67], and of the mixtures m-TCP/DH (**b**) [69]. High-temperature data of neat PS from Ref. [116]. Color code defines concentration (by mass percentage). Open symbols and dashed lines depict the α_1-relaxation, filled symbols and solid lines the α_2-relaxation (circles from DS, squares from NMR, triangles from combined NMR/DS). Arrows mark T_{g1}

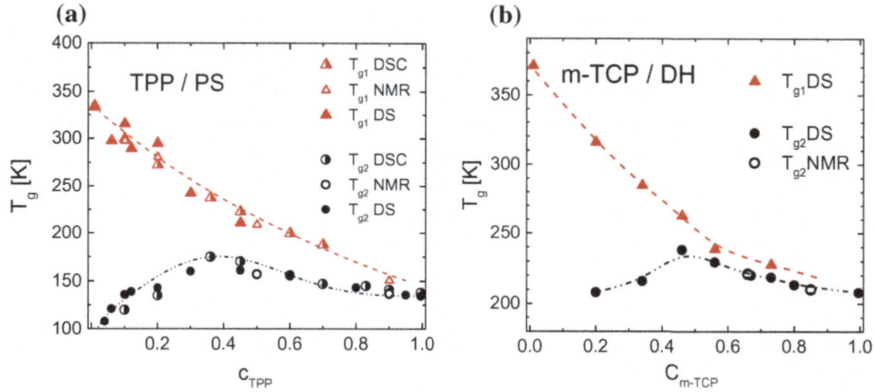

Fig. 13 Glass transition temperatures, T_{g1} and T_{g2} as function of the additive concentration revealed by DSC, DS, and NMR for the mixtures **a** TPP/PS [68] and **b** m-TCP/DH [69]. Lines serve as guides for the eye

2.4 Dynamics in Binary Glass Formers—General Discussion

Let us now summarize the results collected regarding the main relaxations in highly asymmetric binary glass formers. At all concentrations, two separated main relaxation processes are identified. Each involves liquid-like, isotropic reorientations of the component. The faster α_2-process is associated with the low-T_g component. The slower α_1-process is related to the reorientation of the high-T_g component. The corresponding time constants $\tau_{a1}(T)$ and $\tau_{a2}(T)$ and consequently their glass transition temperatures T_{g1} and T_{g2} differ the more the lower the TPP concentration is. There is a fundamental difference between the spectral evolution of the high-T_g component (α_1) and the low-T_g component (α_2): While the correlation function of the α_1-process is Kohlrausch-like, yet, somewhat more stretched with respect to the pure component, and FTS applies, the low-T_g component is governed by a broad distribution $G(\ln\tau_{\alpha 2})$ becoming extremely wide resulting in quasi-logarithmic correlation functions at low additive concentrations. As quite similar relaxation features are observed for glass formers in nano-confinement [46–50] and as theoretical papers point into this direction (cf. below), we conclude that the high-T_g component in an asymmetric mixture provides an intrinsic confinement for the low-T_g component. All in all, there is no fundamental difference between polymer-plasticizer systems and non-polymeric binary mixtures.

In the present context, it is worthwhile to discuss shortly the behavior of the important additive water which was studied in innumerous matrices (cf. Refs. [30, 31]). Considering polymer matrices [30, 31], the dynamics of water displays very similar behavior as observed in the present systems. For instance, a fragile-to-strong transition is found too, specifically, below T_{g1} the time constant $\tau_{\text{water}}(T)$ follow an Arrhenius dependence while a super-Arrhenius behavior is observed at $T > T_{g1}$ (cf. Fig. 14) [119]. As suggested by the presence of a well-defined peak in the structure factor [31] as well as by molecular dynamics (MD) simulations [117], the experimentally observed dynamic heterogeneities were discussed in terms of "water cluster" or "nano-phase separation" [31]. In contrast, a clear-cut extra peak was not observed in the case of THF as additive suggesting more homogeneously distributed THF molecules compared to water [120]. Moreover, the high similarity to water dynamics in nano-confinement as well as in protein matrices is stressed. Thus, similar phenomena are discussed as mentioned above for inorganic mixed glasses [108], and they may also apply to the present binary molecular glasses.

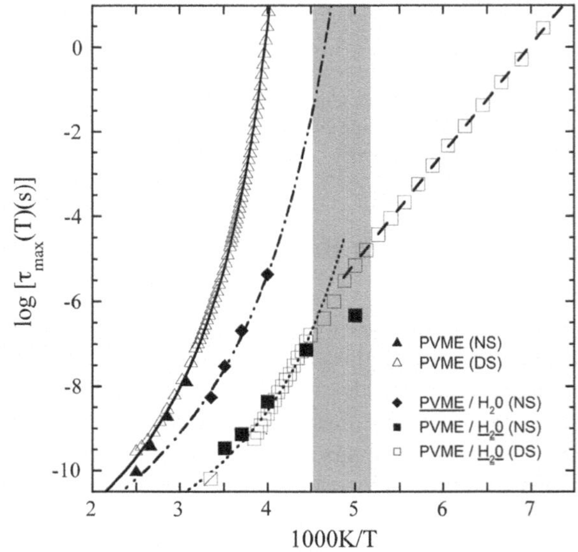

Fig. 14 Time constants by neutron scattering (NS, Q = 1Å^{-1}) and by dielectric spectroscopy of neat poly(vinyl methyl ether) (PVME, triangle) and of a 30 wt% water/PVME mixture: time constants of PVME (filled diamond) and of water (open and filled squares); shaded area indicates the width of the DSC step (Figure: courtesy J. Colmenero) [119]

We emphasize that in DS studies the kind of motion probed usually cannot be identified, and the notion "restricted motion" or "increase of water mobility," the letter derived from an increase of the relaxation strength with temperature, remained vague. Given the results of the component-selective NMR studies presented here, it is clearly demonstrated that the additive dynamics involves an isotropic liquid-like reorientation in a more or less arrested high-T_g matrix. A change of the relaxation strength is interpreted as a change of the corresponding population of (additive) molecules participating in the α_1- and α_2-process, respectively. The broad distribution $G(ln\tau_{\alpha 2})$ is directly reflected in two-phase NMR spectra (cf. Figs. 5 and 10), also observed in the case of (deuterated) water by ^{2}H NMR in a polymer matrix [119]. The extrapolation of the Arrhenius-like temperature dependence of $\tau_{\alpha 2}(T)$ at $T < T_{g1}$ allows to define a second glass transition temperature, T_{g2}, via the condition $\tau_{\alpha 2}(T_g) = 100$ s signaling the arrest of the low-T_g component. The fragile-to-strong transition of $\tau_{\alpha 2}(T)$ around T_{g1} leads directly to a maximum in $T_{g2}(c_{add})$ (cf. Fig. 13). As mentioned before, such a T_g-maximum was not reported before. Yet, most studies on polymer-plasticizer did not investigate the full concentration range.

Yet, for sake of completeness, we mention that very recently a somewhat different interpretation of the dynamics in binary glass formers was derived from essentially the same experimental basis [121]. The authors argue that the here discussed α_2-relaxation is actually a mixture of a β-relaxation and the "true" α_2-relaxation. By this interpretation a maximum in $T_{g2}(c_{add})$ does not appear. It is beyond the frame of the present contribution to discuss this interpretation. However, as emphasized before, the time constants extracted by DS from the relaxation maximum are no correlation times in a strict sense and the maximum in $T_{g2}(c_{add})$ may disappear if one can access them in a proper way which, of course, is very difficult facing an extremely stretched reorientational correlation function. Note that a (mean) correlation time given by the integral over the correlation function is always determined by its long-time decay (or equivalently by the low-frequency contribution of the dynamics). However, from a purely phenomenological point and as already discussed in Ref. [29], one can safely state that the liquid-like dynamics of the low-T_g component is governed by a very broad distribution of correlation times with essentially two cut-offs, $\tau_{\alpha 2}$ and $\tau_{\alpha 1}$, which allow to define two T_g values, T_{g1} and T_{g2}. In between these limits, the dynamics spans over a very large time window, in particular at low additive concentrations.

The binary glass formers discussed are dynamically highly "asymmetric"; i.e., their components exhibit a large T_g contrast. Reducing the T_g contrast, as done in the cases of MTHF/m-TCP or toluene/quinaldine (see Sect. 3), for example, still two relaxations (α_1 and α_2) are observable; however, it becomes difficult to spectrally disentangle them and to extract the corresponding relaxation times. Here, again component-selective NMR experiments may provide further information. The systems TPP/PS or m-TCP/DH are mixtures made from simple liquid components; i.e., no strong association effects are expected. In contrast, in the case of mixtures of associated liquids such as water/glycerol [122] or dimethyl sulfoxide/glycerol [123] the spectral shape of the dielectric response in the mixture virtually does not change at all. Alone the time constant changes as expected by the different T_g values of the components.

There are indications that some additive molecules are also involved in the α_1-process; their fraction grows with decreasing temperature. A possible scenario is sketched in Fig. 15 [117]. At high temperatures, all additive molecules (red circles) are decoupled from the dynamics of the high-T_g component (lines symbolizing a polymer). With decreasing temperature, more and more additive molecules are "condensed" at the high-T_g component and participate in the α_1-process. Finally, let us speculate what happens at $c_{add} \rightarrow 0$, i.e., at very low concentrations of the low-T_g component. It is well established that for this situation the additive serves as a probe for the dynamics of the matrix; similar time constants as those of the α_1-process are expected. Thus, one expects that the α_2-process disappears and only additive molecules (α_1-process) "attached" to the high-T_g component survive—as suggested by Fig. 15. However, covering quite low concentrations the disappearance of the α-relaxation was not observed so far (cf. Figure 13a. Possibly, at such low additive concentrations it may change its character and indeed become a secondary relaxation. Further experiments are needed to test this scenario.

Previously, the dynamic heterogeneities typical of binary glass formers were explained in terms of thermally driven concentration fluctuations [118–127] or so-called self-concentration effects [128, 129]. In the present case of dynamically very asymmetric mixtures, however, additionally confinement effects appear to come into play. The behavior of binary mixtures consisting of two Lennard-Jones particles with different size was also investigated by MD simulations [130, 131]. Here, a dynamical decoupling of both particle species was reported. The large particles exhibit a standard glass transition controlled by the cage effect, while the small particle still remain mobile within the arrested matrix of the large particles. As in our experiments, the dynamics of the small (fast) particles is qualitatively different compared to that of the large (slow) particles. The phenomenon was also investigated by the MCT where a dynamic decoupling of large and small particles is anticipated either, if the components differ sufficiently in size [110]. In that case the smaller particles retain mobility below T_g of the larger ones and undergo a localization transition in the confinement

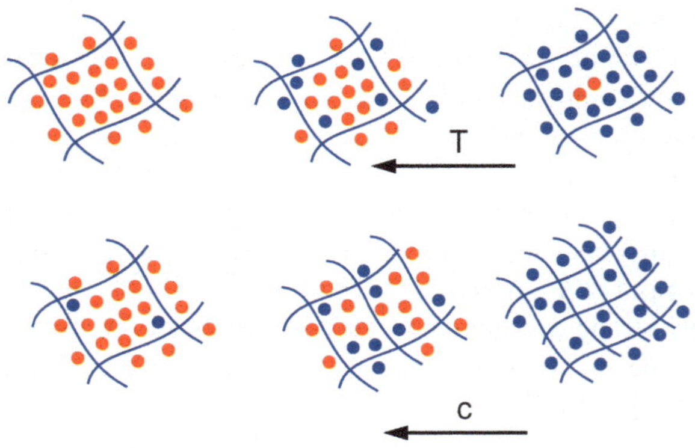

Fig. 15 Schematic scenario for the dynamics in binary glass formers as a function of temperature (top) or additive concentration (bottom); red circles: mobile additive molecules, blue circles: additive molecules "condensed" and thus more or less immobilized along the polymer segments (lines). Adapted from Ref. [117]

of a frozen matrix. In this respect, work on MCT in so-called quenched-annealed systems [132] and simulations for a Lorentz-gas model [134] may be relevant. In such systems, MCT predicts higher-order singularities [132, 133] in the vicinity of which the correlation functions of the smaller molecules become extremely broad and quasi-logarithmic decay curves are expected as observed in simulations [131]—and in our experiments.

3 β-Processes in Binary Glass Formers

Secondary processes are also observed in binary glass formers, and their phenomenology turns out to be even more puzzling [10, 32–39]. We investigated the above-discussed mixed glasses also below T_g where they exhibit a pronounced β-process. For example, neat PS does not show indications of a β-relaxation, but TPP displays one with an activation energy $E_\beta/R = 23.5\ T_g$ similar to that of the β-process in toluene [75]. As revealed by DS as well as by NMR, in the mixtures TPP/PS this β-relaxation is observed down to low additive concentrations. The time constants down to about $c_{TPP} = 45\%$ are essentially the same, only below 45% a weak change is observed—see Fig. 16a. Very similar results are found for TPP/SBC [70] and also for toluene/quinaldine mixtures—see Fig. 16b [134].

Regarding the NMR investigation of the β-relaxation in the non-polymeric system TPP/SBC, we present ^{31}P Hahn-echo (TPP) and ^{2}H solid-echo spectra (SBC) in Fig. 17 [70]. As in the case of pure glass formers (cf. Fig. 2), the inter-pulse delay t_p of the echo sequence is systematically enlarged. Therefore, the sensitivity on spatially highly restricted angular displacement is enhanced [89–92, 96]. In Fig. 17a the ^{31}P echo spectra are shown for neat TPP, and the ^{2}H echo spectra for neat SBC at a similar

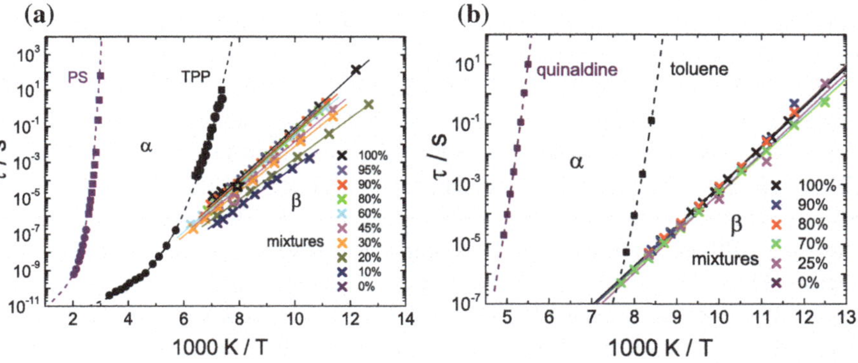

Fig. 16 **a** Time constants τ_α of the main relaxation in pure PS and TPP, and τ_β of the β-process in the TPP/PS mixtures as a function of reciprocal temperature (by NMR and DS) [68]. **b** Corresponding data for the toluene/quinaldine mixtures measured by DS [135]. Color code gives the mass concentration of the low-T_g component

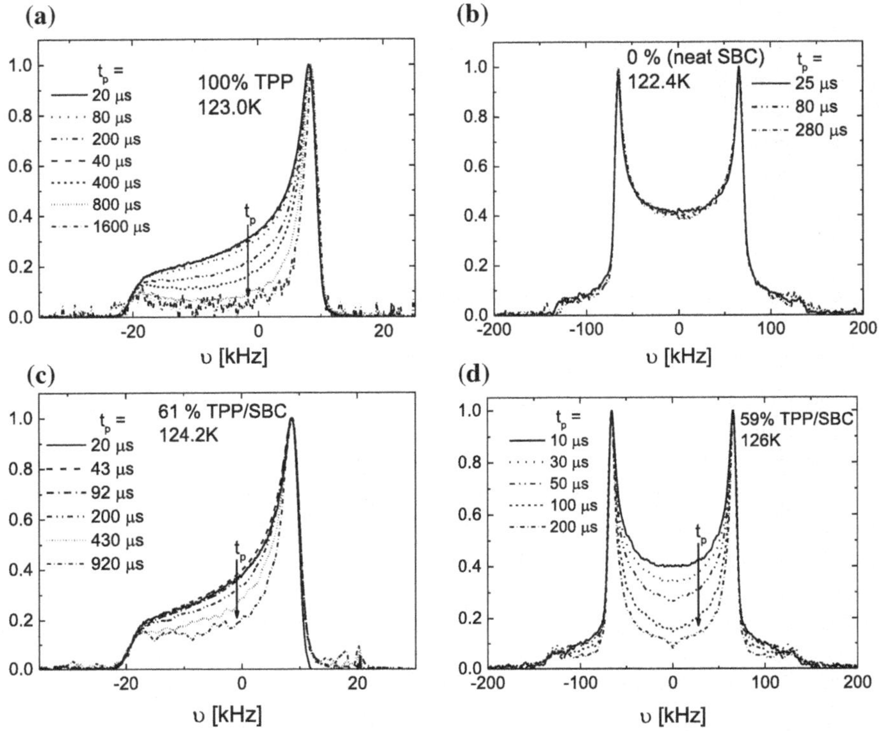

Fig. 17 Echo spectra of the mixture TPP/SBC at different inter-pulse delays t_p: **a** neat TPP: ^{31}P NMR spectra at $T = 123.0$ K; **b** neat SBC: ^{2}H NMR spectra at $T = 122.4$ K; **c** $c_{TPP} = 61\%$: ^{31}P NMR spectra at $T = 124.2$ K; **d** $c_{TPP} = 59\%$: ^{2}H NMR spectra $T = 126$ K [70]

temperature are displayed in Fig. 17b. In the case of TPP, the spectral effects of the β-process are well recognized: With increasing the inter-pulse delay t_p, the spectral intensity decreases mainly in the center of the spectra, while the singularities, i.e., the intensity at the edges of the spectrum, are essentially unaffected. In the case of SBC, no slow β-process is dielectrically observed [70] and no changes of the ^{2}H spectra with t_p are expected—as observed. In the mixtures, the β-process (introduced by TPP) is identified by the Hahn-echo spectra for all concentrations down to $c_{TPP} = 10\%$. An example is shown in Fig. 17c for $c_{TPP} = 61\%$. Clearly, the same features as in neat TPP are reproduced at very similar temperatures. Figure 17d shows the corresponding ^{2}H spectra of SBC. Now spectral changes typical of a β-process are observed in contrast to neat SBC (Fig. 17b), and they look very similar to those observed for neat toluene (cf. Fig. 2). It appears that the β-process in the mixtures involves the SBC molecules, too.

The same phenomenon is observed in the NMR spectra of the TPP/PS-d$_3$ mixtures—see Fig. 18 [93]. Typical echo spectra reflecting highly hindered reorientations are found. In addition, the spectra, shown for different temperatures, demonstrate that

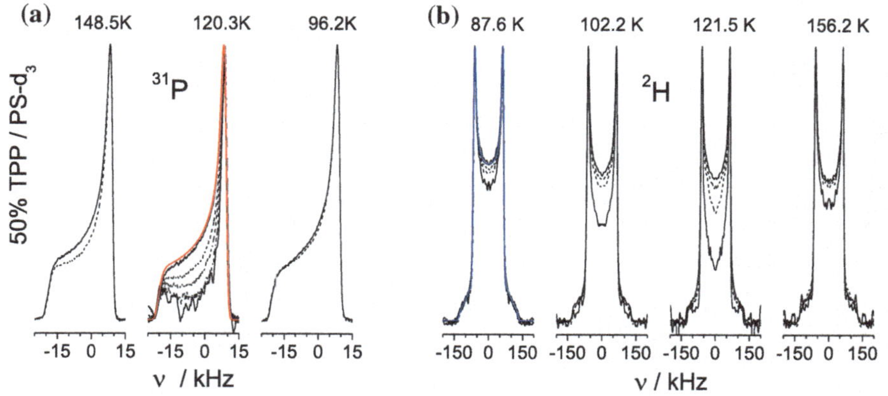

Fig. 18 **a** ^{31}P Hahn-echo spectra of TPP/PS-d$_3$ glass at indicated temperatures for $c_{TPP} = 50\%$, 20 μs (solid line) and 200 μs (dashed line) each. At $T = 120.3$ K long inter-pulse delays were applied additionally ($t_p = 400$ μs, 800 μs (dashed lines), 1600 μs (solid line)). A fit by a CSA powder spectrum is included for the shortest t_p value (solid red line). **b** Corresponding ^{2}H NMR spectra with $t_p = 20$ μs, 200 μs (solid lines), 40 and 80 μs (dashed lines). For the lowest temperature and shortest t_p a fit with a Pake spectrum is included (solid blue line) [93]

the spectral effects upon changing the echo pulse delay t_p disappear at high as well as low temperatures. This is a characteristic feature of a motional process being fast or slow, respectively, compared to the NMR time window. Remarkably, no change in the spectral width in the fast and slow exchange, respectively, is recognized clearly indicating that the β-process involves spatially highly restricted reorientation. For example, with a uncertainty of 2.5% of the spectral coupling constant δ, one can approximate within the wobbling-within-a-cone model a small-angle motion with a half-open angle smaller than 10° [70, 102].

For the characterization of the subtle spectral changes caused by the β-process, we introduced the ratio $R(t_p)$ of the spectral intensity in the center with respect to that of the singularities [70, 93, 94]. In Fig. 19a $R(t_p)$ normalized by $R(t_p = 0)$ measured at a temperature for which the line shape changes are largest are displayed, for both, ^{31}P NMR (a) and ^{2}H NMR (b) for TPP/SBC on a reduced parameter $t_p\delta$ where δ denotes the corresponding spectral NMR coupling constant (being different for ^{2}H or ^{31}P NMR). Regarding TPP, for concentrations $c_{TPP} \geq 34\%$, the ratio $R(t_P)$ decays to zero on a similar time scale. For $c_{TPP} = 10$ and 21% the R value does not decay to zero any longer; a finite long-time value is revealed. A finite R_∞ value signals that only a fraction of TPP molecules takes part in the secondary relaxation [88, 89]. For $c_{TPP} \to 0\%$ the β_2-process, of course, has to disappear. Apparently, its disintegration sets in at $c_{TPP} < 25\%$. Similarly, for ^{2}H NMR on SBC the decay takes place on a longer time scale at lowest concentration; i.e., the disintegration of the β_2-process is also recognized in the SBC signal. Such results were also obtained in the case of TPP/PS [93] and for toluene mixed with picolin and Acroclor 1254 [94]. Hence, the "islands of mobility" concept [107], refuted in the case of neat systems [88–90],

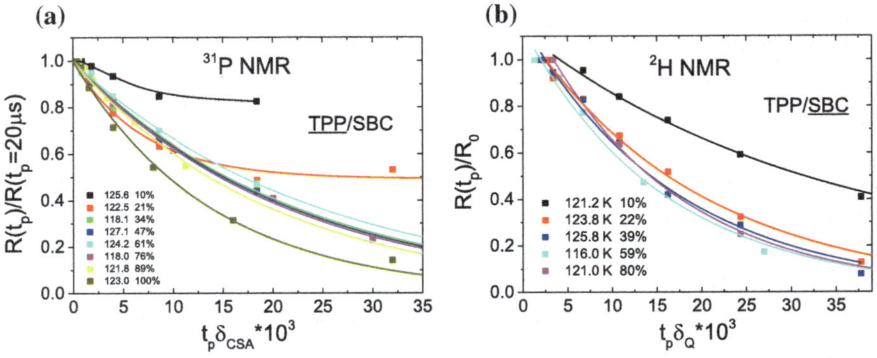

Fig. 19 **a** Normalized spectral shape parameter $R(t_p)$ for all concentrations of the mixture TPP/SBC measured by ^{31}P NMR on the TPP component ($\delta_{CSA} = 20$ kHz). **b** Normalized $R(t_P)$ measured by ^{2}H NMR on the SBC component ($\delta_Q = 135$ kHz). Solid lines are guides for the eye [70]

appears to be a valid description in the case of binary mixtures (and possibly in polymers): Below a certain concentration "islands of rigidity" appear [93].

Measurements of the spin-lattice relaxation rate $1/T_1(T)$ probing the spectral density in the MHz regime also provide information on the β-process [93]. Figure 20 shows $1/T_1(T)$ of both components PS-d_3 (^{2}H) and TPP (^{31}P) in the pure system (full symbols) and in the mixture (open symbols). Pure PS-d_3 displays a rather weak temperature dependence below T_g as no β-process is active. Above T_g the relaxation becomes faster due to the onset of the α-process. In the case of TPP, a well-resolved α-maximum is found which is located at lower temperatures because T_g is much lower than that of PS. Below T_g the temperature dependence of $1/T_1$ becomes weaker, yet, it is much stronger than in the case of PS-d_3 due to the β-process controlling the relaxation. As the spin-lattice relaxation rate probes the spectral density at the Larmor frequency ω_L $1/T_1 \propto \varepsilon''(\omega_L, T)$ is expected to hold. One can check this by extrapolating the dielectric loss to the Larmor frequency ($\nu_L = 161.98$ MHz) of the ^{31}P NMR experiment [93]. As indicated by the dashed line in Fig. 20, indeed, the temperature dependence of $1/T_1$ is well reproduced by that of ε''. In the mixture the temperature dependences $1/T_1(T)$ of TPP and PS-d_3 now proceed parallel below T_g, whereas that of TPP agrees with that of pure TPP. In accordance with the DS results, we conclude, first, that the relaxation behavior of TPP does not change in the mixture below T_g. Second, the PS relaxation in the mixture exhibits the same temperature dependence at low temperatures in contrast to neat PS. This is a strong indication that the polymer segments participate in the β-process, thus confirming the results from the NMR echo experiments. Above T_{g2} a maximum is found for TPP which reflects the isotropic reorientation of the TPP molecules in the mixture (α_2-process). A similar maximum occurring yet at higher temperatures is found for PS-d_3. This difference directly reflects the decoupling of the primary dynamics of the components: In the mixture PS reorients much slower than TPP.

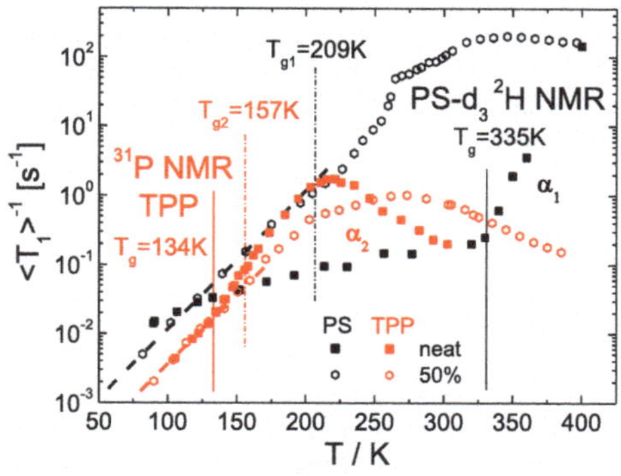

Fig. 20 Spin-lattice relaxation rates of TPP (^{31}P NMR; red symbols) and PS-d$_3$ (^{2}H NMR; black symbols) as neat components (full symbols) and in the mixture 50% TPP/PS-d$_3$ (open symbols). Glass transition temperatures are indicated by vertical lines, and dashed lines indicate temperature dependence of the dielectric loss extrapolated to the NMR frequency [93]

In summary, NMR echo experiments with their high resolution of small-angle processes demonstrate that the β-process in binary glass formers displays the same characteristics as in neat glasses. The NMR line shape changes are very similar to those observed in neat systems and are observed for both components. They can be explained by angular displacements below $\pm$ 10°; i.e., a spatially highly hindered motion is revealed. In addition, the ^{2}H spin-lattice relaxation of the high-T_g component (PS or SBC) in the mixture assumes the same temperature dependence as that of the low-T_g component. Actually, the fact that the time constant $\tau_\beta(T)$ does not change with concentration might indicate a "local" process controlled by some structural particularity of the TPP molecule. Yet, two observations point to another interpretation: (i) the "enslavement" of the high-T_g component SBC and (ii) the disintegration of the process at low TPP concentration. We take these facts as an indication that the β-process is a cooperative process which involves all molecules of the mixture at least above some critical concentration allowing cooperativity to be established. This agrees with recent MD simulations mentioned before which associate the β-process with string-like motions surviving in the glass [109]. Still, these far-reaching conclusions are to be regarded as preliminary, further systems have to be studied. In particular, one has to definitely rule out that the (weak) intermolecular dipole–dipole interaction between the deuterons of SBC or PS and the protons of TPP can be the cause of the ^{2}H NMR findings. Notwithstanding, generalizing our results, the NMR findings imply that in the case of a mixture of polar type A (displaying no β-process) glass former and a non-polar type B glass former (displaying a β-process) the relaxation strength of the β-process is expected to grow with adding more and more type A component provided that the dielectric relaxation strength can be directly interpreted in molecular terms. Therefore, the β-process is still puzzling and a challenge for further research.

Acknowledgements Financial support by the Deutsche Forschungsgemeinschaft (DFG) through the projects RO 907/10,11,15,19 is acknowledged. We also appreciate the cooperation with H.-W. Schmidt and K. Kreger (Macromolecular Chemistry II, Bayreuth) regarding the synthesis of the non-polymeric high-T_g glass formers.

Literature

1. Götze W (1999) J Phys: Condens Matter 11:A1
2. Ediger MD (2000) Annu Rev Phys Chem 51:99
3. Lunkenheimer P, Schneider U, Brand R, Loidl A (2000) Contemp Phys 41:15
4. Dyre J (2006) Rev Mod Phys 78:953
5. Blochowicz T, Brodin A, Rössler EA (2006) Adv Chem Phys 133:127
6. Cavagna A (2009) Phys Rep 476:51
7. Binder K, Kob W (2011) Glassy materials and disoredered solids. World Scientific, New Jersey
8. Berthier L, Biroli G (2011) Rev Mod Phys 83:587
9. Richert R (2011) Annu Rev Phys Chem 62:65
10. Floudas G, Paluch M, Grzybowski A, Ngai KL (2011) Molecular dynamics of Glass-forming systems. Springer, Berlin
11. Wolynes PG, Lubchenko V (eds) (2012) Structural glasses and supercooled liquids, Wiley, Hoboken
12. Petzold N, Schmidtke B, Kahlau R, Bock D, Meier R, Micko B, Kruk D, Rössler EA (2013) J Chem Phys 138:12A510
13. Schmidtke B, Petzold N, Kahlau R, Rössler EA (2013) J Chem Phys 139:084504
14. Sillescu H (1999) J Non-Cryst Solids 243:81
15. Berthier L, Biroli G, Bouchaud J-P van Saarloos W (2011) Dynamical heterogeneities in glasses, colloids and granular matter. Oxford Press, Oxford
16. Ediger MD, Harrowell P (2012) J Chem Phys 137:080901
17. Stillinger FH, Debenedetti PG (2013) Annu Rev Condens Matter Phys 4:263
18. Donati C, Glotzer SC, Poole PH, Kob W, Plimpton SJ (1999) Phys Rev E 60:3107
19. Doliwa B, Heuer A (2000) Phys Rev E 61:6898
20. Karmakar S, Dasgupta C, Sastry S (2009) Proc Natl Acad Sci 106:3675
21. Puosi F, Leporini D (2012) J Chem Phys 136:164901
22. Brun C, Ladieu F, l´Hote D, Tarzia M, Biroli G, Bouchaud J-P (2011) Phys Rev B 84:104204
23. Lunkenheimer P, Michl M, Bauer Th, Loidl A (2017) Eur Phys J Special Topics 226:3157
24. Utracki LA (1989) Polymer alloys and blends: thermodynamics and rheology. Hanser Verlag, Munich
25. Hains PJ, Williams G (1975) Polymer 16:725
26. Desando MA, Walker S, Baarschers WH (1980) J Chem Phys 73:3460
27. Scandola M, Ceccorulli G, Pizzoli M (1987) Polymer 28:2081
28. Nakazawa M, Urakawa O, Adachi K (2000) Macromolecules 33:7898
29. Bingemann D, Wirth N, Gmeiner J, Rössler EA (2007) Macromolecules 40:5379
30. Cerveny S, Alegria A, Colmenero J (2008) Phys Rev E 77:031803
31. Goracci G, Arbe A, Alegria A, Su Y, Gasser U, Colmenero J (2016) J Chem Phys 45:1676
32. Blochowicz T, Karle C, Kudlik A, Medick P, Roggatz I, Vogel M, Tschirwitz C, Wolber J, Senker J, Rössler E (1999) J Phys Chem B 103:4032
33. Blochowicz T, Schramm S, Lusceac S, Vogel M, Stühn B, Gutfreund P, Frick B (2012) Phys Rev Lett 109:035702
34. Blochowicz T, Rössler EA (2004) Phys Rev Lett 92:225701
35. Ediger MD, Lutz TR, He Yiyong (2006) J Non-Crystal Solids 352:4718
36. Kessairi K, Capaccioli S, Prevosto D, Lucchesi M, Rolla P (2007) J Chem Phys 127:174502

37. Cangialosi D, Alegria A, Colmenero J (2008) J Chem Phys 128:224508
38. Blochowicz T, Lusceac SA, Gutfreund P, Schramm S, Stühn B (2011) J Phys Chem B 115:1623
39. Capaccioli S, Kessairi K, Shahin M, Prevosto D, Lucchesi M (2011) J Non-Cryst Solids 357:251
40. Shirata Y (2000) J Mater Chem 10:1
41. Kremer F, Huwe A, Arndt M, Behrens P, Schwieger W (1999) J Phys Cond Matter 11:A175
42. Scheidler P, Kob W, Binder K (2000) Europhys Lett 52:277
43. Lusceac SA, Koplin C, Medick P, Vogel M, Brodie-Linder N, LeQuellec C, Alba-Simionesco C, Rössler EA (2004) J Phys Chem B 108:16601
44. Alba-Simionesco C, Coasne B, Dosseh G, Dudziak G, Gubbins KE, Radhakristhan R, Sliwinska-Bartkowiak M (2006) J Phys: Condens Matter 18:R15
45. Lodge TP, Wood ER, Haley JC (2006) J Polymer Sci Part B Polymer Phys 44:756
46. McKenna GB (2007) Confit III Summary and perspectives on dynamics in confinement. Eur Phys J Spec Top 141:291
47. Schönhals A, Goering H, Schick C, Frick B, Mayorova M, Zorn R (2007) Eur Phys J Special Topics 141:255
48. Buntkowsky G, Breitzke H, Adamczyk A, Roelofs E, Emmler T, Gedat E, Grunberg B, Xu YP, Limbach HH, Shenderovich I, Vyalikh A, Findenegg AG (2007) Phys Chem Chem Phys 9:484
49. Gradmann S, Medick P, Rössler EA (2009) J Phys Chem B 113:8443
50. Bock D, Petzold N, Kahlau R, Gradmann S, Schmidtke B, Benoit N, Rössler EA (2015) J Non-Cryst Solids 407:88
51. Schmidt-Rohr K, Spiess HW (1991) Phys Rev Lett 66:3020
52. Schiener B, Chamberlin RV, Diezemann G, Böhmer R (1997) J Chem Phys 107:7746
53. Tracht U, Wilhelm M, Heuer A, Feng H, Schmidt-Rohr K, Spiess HW (1998) Phys Rev Lett 81:2727
54. Vidal Russel E, Israeloff NE (2000) Nature 408:695
55. Blochowicz T, Rössler EA (2005) J Chem Phys 122:224511
56. Papon A, Montes H, Hanafi M, Lequeux F, Guy L, Saalwächter K (2012) Phys Rev Lett 108:065702
57. Pizzoli M, Scandola M, Ceccorulli G (1987) Eur Polym J 23:843
58. Savin DA, Larson AM, Lodge TP (2004) J Polym Sci B Polym Phys 42:1137
59. Miwa Y, Usami K, Yamamoto K, Sakaguchi M, Sakai M, Shimada S (2005) Macromolecules 38:2355
60. Lipson JEG, Milner ST (2006) J Pol Sci Part B Pol Phys 24:3528
61. Gaikwad AN, Wood ER, Ngai T, Lodge TP (2008) Macromolecules 41:2502
62. Leroy, E, Alegria, A, Colmenero, J (2002)
63. Mpoukouvalas K, Floudas G (2008) Macromolecules 41:1552
64. Schmidt-Rohr K, Spiess HW (1994) multidimensional solid-state NMR and polymers. Academic Press, New York
65. Vogel M, Rössler E (1998) J Phys Chem A 102:2102
66. Medick P, Vogel M, Rössler E (2002) J Magn Reson 159:126
67. Kahlau R, Bock D, Schmidtke B, Rössler EA (2014) J Chem Phys 140:044509
68. Bock D, Kahlau R, Pötzschner B, Körber T, Wagner E, Rössler EA (2014) J Chem Phys 140:094505
69. Pötzschner B, Mohamed F, Lichtinger A, Bock D, Rössler EA (2015) J Chem Phys 143:154506
70. Pötzschner B, Mohamed F, Bächer C, Wagner E, Lichtinger A, Minikejew R, Kreger K, Schmidt H-W, Rössler EA (2017) J Chem Phys 146:164503
71. Johari G, Goldstein M (1970) J Chem Phys 53:2372
72. McCrum NG, Read BE, Williams G (1991) Anelastic and dielectric effects in polymer solids. Wiley, New York
73. Wu L (1991) Phys Rev B 43:9906
74. Garwe F, Schönhals A, Lockwenz H, Beiner M, Schröter K (1996) Donth E Macromol 29:247

75. Kudlik A, Benkhof S, Blochowicz T, Tschirwitz C, Rössler E (1999) J Mol Struct 479:201
76. Kremer F, Schönhals A (2002) Broadband dielectric spectroscopy. Springer, Berlin
77. Gainaru C, Böhmer R, Kahlau R, Rössler EA (2009) J Chem Phys 131:184510
78. Floudas G, Paluch M, Grzybowski A, Ngai KL (2011) Molecular dynamics of glass-forming systems. Springer, Berlin
79. Kahlau R, Dörfler T, Rössler EA (2013) J Chem Phys 139:134504
80. Jakobsen B, Niss K, Maggi C, Olsen NB, Christensen T, Dyre JC (2011) J of Non-Cryst Solids 357:267
81. Gainaru C, Rivera A, Putselyk S, Eska G, Rössler EA (2005) Phys Rev B 72:174203
82. Pohl RO, Liu X, Thompson E (2002) Rev Mod Phys 74:991
83. Ngai KL, Paluch M (2004) J Chem Phys 120:857
84. Mierzwa M, Pawlus S, Paluch M, Kaminska E, Ngai KL (2008) J Chem Phys 128:044512
85. Capaccioli S, Paluch M, Prevosto D, Wang Li-Min, Ngai K L (2012) J Phys Chem Lett 3:6
86. Rössler R; Taupitz M, Richert R, Blumen A (eds) (1994) Disorder effects on relaxational processes, glasses, polmers, proteins. Springer, Berlin
87. Arbe A, Colmenero J, Frick B, Monkenbusch M, Richter D (1998) Macromolecules 31:4926
88. Vogel M, Rössler EA (2000) J Phys Chem B 104:4285
89. Vogel M, Rössler EA (2001) J Chem Phys 114:5802
90. Vogel M, Rössler EA (2001) J Chem Phys 115:10883
91. Vogel M, Tschirwitz C, Schneider G, Koplin C, Medick P, Rössler EA (2002) J Non-Cryst Solids 307:326
92. Vogel M, Medick P, Rössler EA (2005) Annu Rep NMR Spectrosc 56:231
93. Bock D, Kahlau R, Micko B, Pötzschner B, Schneider GJ, Rössler EA (2013) J Chem. Phys 139:064508
94. Micko B, Tschirwitz C, Rössler EA (2013) J Chem Phys 138:154501
95. Micko B, Lusceac SA, Zimmermann H, Rössler EA (2013) J Chem Phys 138:074503
96. Vogel M, Rössler E (2000) J Magn Reson 147:43
97. Wagner H, Richert R (1998) J Non-Crystalline Solids 242:19
98. Micko B, Kruk D, Rössler EA (2013) J Chem Phys 138:074504
99. Fragiadakis D, Roland CM (2014) Phys Rev E 89:052304
100. Williams G, Watts DC (1971) NMR basic principles and progress. In: Diehl P, Flick E, Kosfeld E (eds) Springer, Berlin, vol 4, p 271
101. Hinze G, Fujara F, Sillescu H (1995) Chem Phys Lett 232:154
102. Körber Th, Mohamed F, Hofmann M, Lichtinger A, Willner L, Rössler EA (2017) Macromolecules 50:1554
103. Körber Th, Rössler E A unpublished results
104. Tu W, Valenti S, Ngai KL, Capaccioli S, Liu YD, Wang LM (2017) J Phys Chem Lett 8:4341
105. Schmidt-Rohr K, Kulik AS, Beckham HW, Ohlemacher A, Pawelzik U, Boeffel C, Spiess HW (1994) Macromolecules 27:4733
106. Kulik AS, Beckham HW, Schmidt-Rohr K, Radloff D, Pawelzik U, Boeffel C, Spiess HW (1994) Macromolecules 27:4746
107. Johari GP (2002) J Non-Cryst Solids 307:317
108. Le Losq C, Neuville DR, Chen W, Florian P, Massiot D, Zhou Z, Greaves GN (2017) Sci Rep 7:16490
109. Yu H-B, Richert R, Samwer K (2017) Sci Adv e1701577
110. Bosse J, Kaneko Y (1995) Phys Rev Lett 74:4023
111. Böhmer R, Diezemann G, Hinze G, Rössler EA (2001) Prog Nucl Magn Reson Spectr 39:191
112. Fujara F, Wefing S, Spiess HW (1986) J Chem Phys 84:4579
113. Resing H (1965) J Chem Phys 43:669
114. Kambour RP, Kelly JM, McKinley BJ, Cauley BJ, Inglefield PT, Jones AA (1998) Macromolecules 21:2937
115. Rössler E, Taupitz M, Börner K, Schulz M, Vieth HM (1990) J Chem Phys 92:5847
116. He Y, Lutz TR, Ediger MD, Ayyagari C, Bedrov D, Smith GD (2004) Macromolecules 37:5032
117. Schramm S (2011) PhD thesis, University Darmstadt

118. Xiang TX, Anderson BD (2005) Pharm Res 22:1205
119. Capponi S, Arbe A, Cerveny S, Busselez R, Frick B, Embs JP, Colmenero J (2011) J Chem Phys 134:204906
120. Busselez R, Arbe A, Cerveny S, Capponi S, Colmenero J, Frick B (2012) J Chem Phys 137:084902
121. Valenti S, Cappaccioli S, Ngai KL (2018) J Chem Phys 148:2018
122. Jensen MH, Gainaru C, Alba-Siomionesco C, Hecksher T, Niss K (2017) Phys Chem Chem Phys 20:1716
123. Flämig M, Gabrielyan L, Rössler E A, to be published
124. Floudas G, Steffen W, Fischer EW (1992) J Chem Phys 99:695
125. Zetsche A, Fischer EW (1994) Acta Polym 45:168
126. Kumar SK, Colby RH, Anastasiadis SH, Fytas G (1996) J Chem Phys 105:3777
127. Kant R, Kumar SK (2003) Macromolecules 36:10087
128. Chung GC, Kornfield JA, Smith SD (1994) Macromolecules 27:964
129. Lodge TB, McLeish TCB (2000) Macromolecules 33:5278
130. Voigtmann T, Horbach J (2009) Phys Rev Lett 103:205901
131. Moreno AJ, Colmenero J (2006) J Chem Phys 125:164507
132. Krakoviack V (2005) Phys Rev Lett 94:065703
133. Kurzidim J, Coslovich D, Kahl G (2011) J Phys: Condens Matter 23:234122
134. Höfling F, Franosch T, Frey E (2006) Phys Rev Lett 96:165901
135. Minikejew R, Rössler E A, unpublished results

Depolarized Dynamic Light Scattering and Dielectric Spectroscopy: Two Perspectives on Molecular Reorientation in Supercooled Liquids

J. Gabriel, F. Pabst, A. Helbling, T. Böhmer and T. Blochowicz

Abstract Broadband dielectric spectroscopy (BDS) can be considered the standard and most widespread method to experimentally access molecular reorientation in supercooled liquids, as it covers a range of time constants from sub picoseconds corresponding to the highly fluid liquid to several thousand seconds below the glass transition temperature. In a similar fashion, depolarized dynamic light scattering (DLS) is able to probe molecular reorientation. A comparable range of time scales is covered by combining Tandem Fabry Perot Interferometry (TFPI) and Photon Correlation Spectroscopy (PCS) with recent multispeckle techniques allowing to access even the non-ergodic regime below T_g. Thus, DLS represents an alternative route to cover the full range of glassy dynamics. Moreover, due to the fact that both methods couple to different molecular properties, extra information in particular on the motional mechanism behind a certain dynamic process can be obtained by comparing experimental data from both techniques. In the present work we explore this approach for several examples, including ionic liquids and monohydroxy alcohols, and discuss the implications for different relaxation processes. For instance in the case of supercooled ionic liquids, i.e., molten salts, which are liquid at room temperature, the combination of both techniques allows to unambiguously disentangle the contribution of molecular reorientation from other polarization features that often mask reorientation in the dielectric spectra, and a detailed analysis reveals indications for a crossover in the motional mechanism involved in the α-relaxation. In monohydroxy alcohols we discuss the appearance of the Johari-Goldstein β-process in both techniques and what the observations imply for the underlying motional mechansim. Furthermore, we consider the Debye relaxation, which is frequently observed in the dielectric spectra of monoalcohols and is usually ascribed to transient supramolecular structures. Here, such a comparison of data reveals molecular details about the conditions under which the supramolecular structures are formed.

J. Gabriel · F. Pabst · A. Helbling · T. Böhmer · T. Blochowicz (✉)
Institute of Condensed Matter Physics, Hochschulstr. 6, 64289 Darmstadt, Germany
e-mail: thomas.blochowicz@physik.tu-darmstadt.de

F. Kremer and A. Loidl (eds.), *The Scaling of Relaxation Processes*,
Advances in Dielectrics, https://doi.org/10.1007/978-3-319-72706-6_7

1 Introduction

A glass is formed when a liquid is supercooled sufficiently fast so that crystallization is bypassed. On further cooling, eventually characteristic relaxation times exceed typical laboratory time scales and finally the system is trapped in an out-of equilibrium non-ergodic glassy state. As the liquid is supercooled, its structure essentially remains the same, apart from a general continuous increase in density. At the same time, however, viscosity and typical time constants, e.g., of molecular reorientation increase dramatically, often by more than ten orders of magnitude in a narrow temperature interval. As microscopic dynamics and vibrational contributions in the terahertz regime essentially do not depend on temperature throughout this process, a two step correlation decay develops, which is considered as the signature of glassy dynamics [21], although it actually is observed already above the melting temperature in many substances [81] and thus rather seems to be a feature of the liquid state.

As structural relaxation continuously slows down, a growing time window opens up in which various secondary relaxation modes are observed, as is sketched in Fig. 1. Finally, at temperatures around the glass transition temperature T_g, dynamics is observed on all accessible timescales, i.e., while the various dynamic modes superimpose and therefore in many cases cannot be distinguished at high temperatures, they separate at low temperatures and give experimental access to their particular distribution of relaxation times. Thus, the experimental challenge is two-fold: First, the fluctuating dynamic variable has to be experimentally accessed over a large dynamic range, ideally between, say 10^{-6} and 10^{13} Hz, and second, the microscopic origin of each observed process has to be understood. Moreover, there are even additional contributions, which are usually slower than the main structural relaxation, like conductivity and polarization contributions in dielectric spectroscopy, or contributions arising from supramolecular structures in case of particular interactions in the liquid, like hydrogen bonding in water or alcohols. Although even when dealing with seemingly simple molecular liquids the situation is quite complex as all of these

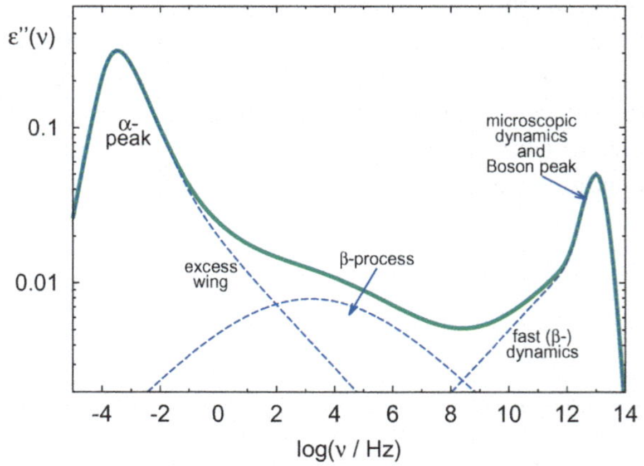

Fig. 1 Schematic representation of the dielectric loss, or more general, the imaginary part of a molecular dynamic susceptibility of a supercooled liquid around the glass transition temperature

contributions have to be disentangled and the molecular origin understood in order to fully elucidate the dynamic behavior.

Dielectric spectroscopy has long been routinely used as a broadband method to access molecular reorientation and data on many substances are available to date that even cover the full dynamic range of interest [56, 59, 64]. By contrast, depolarized dynamic light scattering, by which molecular reorientation is accessible in a similar fashion by combining several different techniques [9], so far is less frequently applied. In particular, only comparatively few attempts were made to produce data sets that cover the full dynamic range [67, 68, 81]. One of the reasons may be the fact, that many of the substances of interest show relatively weak anisotropic scattering and thus experiments become more difficult. However, a systematic comparison of broadband spectra from both dielectric spectroscopy and depolarized light scattering, not only provides complementary information on the reorientational dynamics and its scaling properties, but also reveals aspects that are not accessible otherwise. It is the aim of the present chapter, to explore the possibilities of such a comparison of broadband data from both methods.

The chapter is structured as follows: After a brief introduction of both methods in Sect. 2 with a few technical aspects of depolarized light scattering in Sect. 2.1 that arise when dealing with weakly depolarizing molecules, and some background discussion on comparing both methods in Sect. 2.3, we present two examples, where such a comparison proves particularly useful. One concerns the relaxation behaviour of monohydroxy alcohols (Sect. 3). Here, we discuss the Johari-Goldstein β-relaxation, the molecular nature of which is still matter of debate: While originally Johari and Goldstein suggested the picture of "islands of mobility" as origin for the β-relaxation [47, 48, 51], later on NMR investigations indicated a small-angle motion of almost all molecules as a reason [95]. It turns out that a comparison of dielectric and light scattering results can shed new light on this old problem. A second feature in the same substance class is the famous Debye relaxation frequently observed in the dielectric relaxation of monohydroxy alcohols, which is already discussed in the chapter "Scaling of Suprastructure and Dynamics in Pure and Mixed Debye Liquids" in this book. The Debye relaxation in monoalcohols is usually attributed to transient supramolecular structures, in particular transient linear chains, that reorient in a manner similar to the end-to-end vector in a short chain polymer [33]. Based on this idea, however, the signature of such a process is also expected in depolarized light scattering and a detailed comparison of its manifestation in both methods, reveals details of its origin. The second example concerns molecular reorientation in room temperature ionic liquids (Sect. 4), i.e., molten salts that are liquid at room temperature. Here, dc-conductivity and electrode polarization effects often mask the contributions due to molecular reorientation in the dielectric spectra. Accordingly, there is no agreement on their interpretation in the literature [74, 75, 83]. A comparison with light scattering results helps to unambiguously identify the processes and allows to discriminate temperature dependent changes in the reorientational mechanism involved.

2 Methods: Light Scattering and Dielectric Spectroscopy

Usually, molecular dynamics is captured in terms of time correlation functions. In the simplest case of, say, an atomic liquid or a hard-sphere model system, where only translational degrees of freedom are relevant, a description of the system in terms of number densities $\varrho(\mathbf{r}, t) = \sum_{i=1}^{N} \delta(\mathbf{r} - \mathbf{r}_i)$ is appropriate. After spatial Fourier transform this turns into $\varrho(\mathbf{q}, t) = \sum_{i=1}^{N} \exp i\, \mathbf{q}\mathbf{r}_i$ and the corresponding correlation function is known as the intermediate scattering function $F(\mathbf{q}, t) = \langle \varrho^*(\mathbf{q}, t)\varrho(\mathbf{q}, 0)\rangle$, and is routinely accessed, for example, by quasielastic neutron scattering on microscopic time- and lengthscales [5].

For molecular liquids of course the orientational degrees of freedom are important and it turns out that orientational correlation functions are even easier accessible in laboratory experiments, for example by depolarized dynamic light scattering or dielectric spectroscopy. For a vectorial quantity like the permanent molecular dipole moment, as it is probed in dielectric spectroscopy, the corresponding single-molecule correlation function is $C_\mu(t) = \mu^2\langle \mathbf{u}_i(0)\mathbf{u}_i(t)\rangle$ with $\mathbf{u}_i = \boldsymbol{\mu}/\mu$ being the unit vector along the molecular dipole moment. In depolarized light scattering, on the other hand, fluctuations of the anisotropic molecular polarizability are probed, i.e., a tensorial quantity. In the general case of arbitrarily shaped molecules and polarizability the corresponding correlation function is expressed in terms of Wigner rotation matrices [6]. However, in many cases, like, e.g., for linear or for symmetric top molecules, the treatment can be simplified considerably and with a unit vector $\mathbf{u}_i$ pointing along the principal axis of highest symmetry a correlation function $C_2 = \langle P_2(\mathbf{u}_i(0)\mathbf{u}_i(t))\rangle$ is obtained, with P_2 signifying the second order Legendre Polynomial. Therefore the single molecule reorientational correlation functions related to both methods can be represented in a common fashion as:

$$C_\ell(t) = \langle P_\ell(\mathbf{u}_i(0)\mathbf{u}_i(t))\rangle\,, \tag{1}$$

with $\ell = 1$ for dielectric spectroscopy and $\ell = 2$ for depolarized light scattering and with $P_\ell(\cos\theta)$ being the corresponding Legendre Polynomial of order ℓ.

One complication that arises in the treatment of both methods is that experimentally only collective correlation functions are accessible, i.e., the measured correlations not only contain the above-mentioned self part, but also cross correlations, so that the collective correlation functions have to be written as:

$$\Phi_\ell(t) = \frac{\sum_{i,j} \left\langle P_\ell(\mathbf{u}_i(0)\mathbf{u}_j(t))\right\rangle}{\sum_{i,j} \left\langle P_\ell(\mathbf{u}_i(0)\mathbf{u}_j(0))\right\rangle} \tag{2}$$

In general it is not an easy task to relate the collective and single molecule quantities, and usually it is argued that the effect of cross terms in the correlation functions is negligible in supercooled liquids. However, in cases with particular interactions like hydrogen bonding liquids, cross correlations may be all important, as for example in the discussion of the origin of the Debye contribution in the dielectric spectra of monohydroxy alcohols, which is thought to arise from supramolecular structures formed by hydrogen bonding and which will be discussed further below.

Some of the experiments are carried out in the time and some in the frequency domain. So, for example, photon correlation spectroscopy probes fluctuations of the polarizability in the time domain by the time autocorrelation function of the scattered electric field, which in many cases is well represented by $\Phi_2(t)$, while with a Tandem-Fabry-Perot interferometer one would measure the spectral density of the same fluctuations $\hat{S}(\omega)$, just on different timescales. Both quantities are related by a Fourier transform, see also the Wiener-Khinchin theorem [43]:

$$\hat{S}(\omega) = \int_0^\infty \Phi(t)\, e^{-i\omega t}\, \mathrm{d}t, \tag{3}$$

and thus can be combined in a straight forward manner. On the other hand, in a linear response experiment a susceptibility $\hat{\chi}(\omega)$ is obtained, like in dielectric spectroscopy, where the polarization is connected to the applied electric field by the dielectric susceptibility $\hat{P}(\omega) = \hat{\chi}(\omega)\hat{E}(\omega)$. Such a dynamic susceptibility is related to the corresponding spectral density of equilibrium fluctuations by the fluctuation dissipation theorem [5, 17, 43], which reads in the classical limit (i.e., $\hbar\omega \ll k_B T$):

$$\hat{\chi}(\omega) = \frac{C}{k_B\,T}\left(1 - i\,\omega\,\hat{S}(\omega)\right), \tag{4}$$

where C depends on the details of the experimental method that is used. Basically Eq. (4) states, that in the linear regime the response to an external perturbation is governed by the same molecular mechanisms as fluctuations about the equilibrium value. In particular for the imaginary part of the susceptibility $\hat{\chi}(\omega) = \chi'(\omega) - i\chi''(\omega)$ this relation reads:

$$\chi''(\omega) = \frac{C\,\omega}{k_B T}\int_0^\infty \Phi(t)\,\cos(\omega t)\,\mathrm{d}t \tag{5}$$

Due to the general validity of this relation, light scattering data, which actually represent direct measurements of fluctuations and their spectral densities, are often represented as generalized susceptibility spectra and will also be presented in that form further below.

2.1 Depolarized Dynamic Light Scattering

Dynamic light scattering is sensitive to the fluctuations in the polarizability, a second order tensorial quantity. Polarizability may vary due to long range fluctuations in density, which contribute to the diagonal elements of the tensor, which result in isotropic scattering, or due to fluctuations in molecular orientations, which modulate the off-diagonal elements and contribute to anisotropically scattered light. Experimentally, both contributions can be distinguished by selecting the polarization of the scattered with respect to the incident radiation: in vertical-vertical (VV) polarized geometry one is sensitive to both isotropic and anisotropic components of the scattered field, while in vertical-horizontal (VH) depolarized geometry only contributions due to anisotropic scattering play a role [6] and molecular reorientation is probed. We note that in the expression for the anisotropic scattering still the intermediate scattering function, which reflects translational motion, enters as a prefactor [6]. However, in molecular liquids $F(\mathbf{q}, t)$ is probed in the limit $\mathbf{q} \approx 0$, as the wavelength of light λ is much larger than intermolecular distances, and so translational motion does not play a role and $F(\mathbf{q}, t)$ does not decay on the timescale of the local reorientational motion and is usually not considered any further unless the scattering medium is expected to contain structures on the scale of the wavelength. Moreover, as also shear modes couple to the anisotropic scattering especially in the high frequency region, one usually selects the scattered light in backscattering geometry, where these modes do not contribute. Thus, molecular reorientation is best accessible in depolarized backscattering geometry.

When discussing the scattering mechanism, in principle, also next order interaction induced effects have to be considered, by which a molecular polarizability fluctuates due to interactions (e.g., collisions or dipolar interactions) with neighboring molecules. Concerning the relevance of contributions to the scattered light due to the latter mechanism there has been considerable controversy in the past [20, 65]. However, currently it is widely accepted that the mechanism behind depolarized scattering is the reorientation of the polarizability tensor in the case of optically anisotropic molecules [20].

Photon correlation spectroscopy When randomly distributed scattering centers are irradiated with coherent light from a laser beam, the scattered waves superimpose to form a speckle pattern, which reflects the time dependent spatial variations of the fluctuating polarizability of the medium under study. In photon correlation spectroscopy (PCS) usually the autocorrelation function of the fluctuating intensity of one such speckle, or coherence area is recorded:

$$G(t) = \langle I(0)I(t)\rangle / \langle I\rangle^2 = \frac{1}{\langle I\rangle^2} \lim_{T\to\infty} \frac{1}{T} \int_0^T I(t')I(t'+t)\,\mathrm{d}t'. \tag{6}$$

In the usual experiment the fluctuating intensity of roughly one speckle is coupled into a single-mode optical fiber, and the correlation function is calculated with a suitable hardware correlator (see Fig. 2) in a time range between 10^{-9} and about 10^3 s with currently available devices. By doing so, the ensemble average $\langle \dots \rangle$ indicated in Eq. (6) is technically replaced by a time average (right hand side of Eq. 6) and thus, strictly speaking, the whole technique can only be applied when the sample is ergodic and both averages are equivalent.

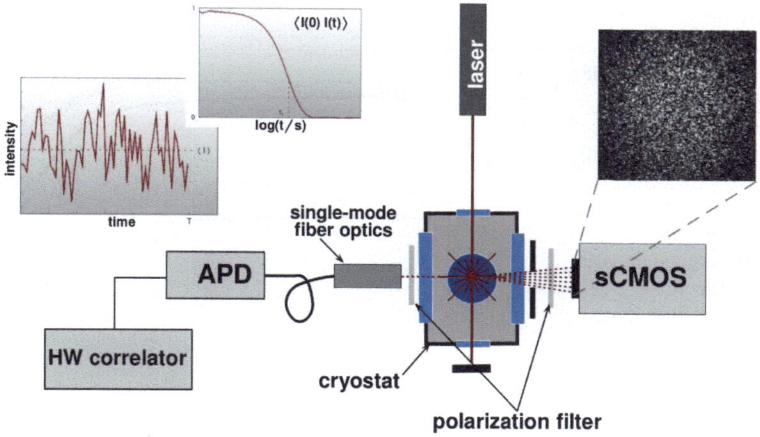

Fig. 2 Schematic representation of the photon correlation setup. The setup is equipped with a single mode optical fiber and a hardware correlator to cover timescales between 2 ns and some 10^3 s in ergodic samples and can also employ an sCMOS camera to record full speckle patterns and thus access particularly long times and also non-ergodic samples

The important physical quantity, which contains the information about molecular motions in the sample, is contained in the autocorrelation function of the scattered electric field $g(t) = \langle E(0)E^*(t)\rangle / \langle E^2\rangle$. In case the scattered field can be regarded as a Gaussian random variable the Siegert relation [85] connects the autocorrelation functions of the scattered intensity and the electric field via

$$G(t) = 1 + \Lambda\, |g(t)|^2 \tag{7}$$

Here, Λ is a spatial coherence factor, which ideally equals 1 if the detector selects an area that is smaller or equal to one coherence area $A_{\mathrm{coh}} \approx \lambda^2 R^2/(\pi\, a^2)$ for a scattering volume of radius a at a distance R [66].

In a PCS experiment with substances that show only weak anisotropic scattering, one faces the problem, that not only scattered light by the sample is recorded, but additionally, due to slight reflections of the incident beam on the light path by sample cell windows or the vacuum shroud, some of the primary laser light will superimpose with the weak scattering signal. This situation is called partial heterodyning, in contrast to homodyne scattering, where only scattered light by the sample is recorded. The effect of this is two-fold: First of all, the experiment becomes much more sensitive to all kinds of vibrations and other artifacts: any shift in the optical path difference between the so-called local oscillator field and the scattered light on the order of the laser wavelength becomes visible in the data. Second, in case of partial heterodyning it becomes more difficult to obtain the field autocorrelation function $g(t)$ from the intensity correlation data.

In order to treat the first problem, we took a series of measures to improve the setup, like reducing vibrations by replacing the liquid nitrogen cooling system with

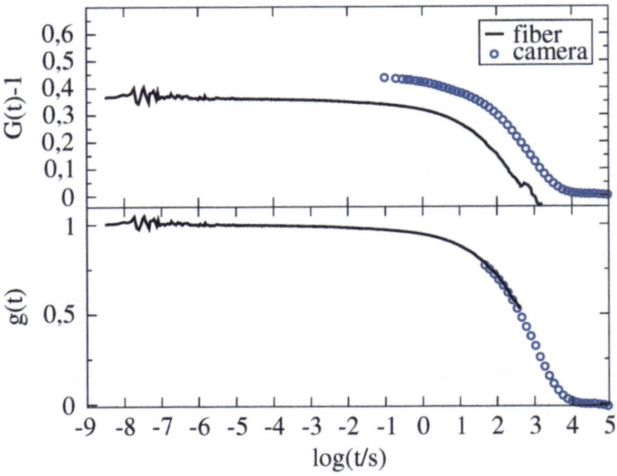

Fig. 3 Combination of data sets from photon correlation spectroscopy using a single mode optical fiber and a hardware correlator (black line) with one obtained with the multispeckle technique (points) obtained for 1-phenyl-1-propanol at 191 K. The original intensity correlation functions $G(t) - 1$ are shown along with the electric field correlation function $g(t)$

a gas cooling system, replacing the turbomolecular pumping station with an ion getter pump, and by stabilizing the temperature in the laboratory to better than 0.2 K, which significantly enhances the long time stability of the laser system. In addition, a custom made sample cell was used that was optimized to reduce reflections of the laser beam as it passes through. Altogether these improvements lead to a data quality, that allows us to Fourier transform the time domain light scattering data to represent them in terms of generalized susceptibility spectra and thus obtain a direct comparison with the dielectric permittivity.

In order to treat partial heterodyning we follow the derivation of Bremer et al. [12]. Starting from a very general ansatz for the intensity correlation function of the scattered light:

$$G(t) = \Lambda C^2 |g(t)|^2 + 2\Lambda C(1 - C)|g(t)| + 1 \qquad (8)$$

the degree of heterodyning is characterized by a constant $C = \langle I_s \rangle / (\langle I_{LO} \rangle + \langle I_s \rangle)$, which quantifies the influence of the local oscillator I_{LO} with respect to the intensity I_s scattered by the sample. In the limit of fully homodyne scattering $C \approx 1$ and the conventional Siegert relation is recovered. It turns out that C can be determined independently from the intercept $G(0)$ and from an estimate of the decorrelation due to microscopic dynamics, which can be obtained, e.g., from Tandem Fabry Perot interferometry data [31]. This allows to extract the field autocorrelation function even in the case of partial heterodyning and thus to obtain in good approximation the reorientational correlation function $g(t) \approx \Phi_2(t)$ even for weak anisotropic scattering molecules.

At temperatures close to the glass transition measurements become very time consuming in the standard PCS experiment, as the time average in Eq. (6) requires about two orders of magnitude more time than the longest lag time accessed by the correlation function. In that case and in cases where the sample becomes non-ergodic so that time and ensemble average are no longer equivalent, *multispeckle light scattering* offers the possibility to directly access ensemble averaged correlation functions. By simultaneous measurements of several speckles with a camera and subsequent correlation of each speckle, an average over different Fourier components of the scattering volume can be performed and thus an effective ensemble averaged correlation function can be obtained [3]. In the ergodic regime speckle (sp) and time (t') average can even be combined to enhance the signal-to-noise ratio at long times:

$$G(t) = \frac{\langle\langle I_{\text{sp}}(t')I_{\text{sp}}(t+t')\rangle_{\text{sp}}\rangle_{t'}}{\langle\langle I_{\text{sp}}(t')\rangle_{\text{sp}}\rangle_{t'}^2} \tag{9}$$

Such kind of averaging not only gives tremendous advantages in terms of photon count statistics but also provides correct amplitudes of the correlation function even if the full correlation decay cannot be observed in the experimental time window. Fig. 3 shows an example of a combination of a conventional and multispeckle data set in the case of 1-phenyl-1-propanol, whereby lag times are accessible up to 10^5 s.

Tandem Fabry Perot Interferometry While the time-dependent fluctuations of the optical anisotropy are accessible with the photon correlation technique, Tandem Fabry Perot Interferometry (TFPI) probes the spectral density $S(\omega) = \text{Re}(\hat{S}(\omega))$ of the same fluctuations, if depolarized backscattering geometry is chosen. From this spectral density the generalized susceptibility is readily calculated via

$$\chi''_{\text{DLS}}(\omega) = \frac{1}{\hbar}\frac{S_{\text{DLS}}(\omega)}{n(\omega,T)+1} \approx \frac{\omega}{k_B T} S_{\text{DLS}}(\omega) \tag{10}$$

with $n(\omega, T) = (\exp(\hbar\omega/k_B T) - 1)^{-1}$ being the Bose factor. The indicated approximation holds in the classical limit of $k_B T \gg \hbar\omega$, where Eqs. (4) and (5) are recovered.

Experimentally, the challenge is to probe the spectral density in the GHz to THz frequency range, i.e., to obtain a spectral resolution of a few GHz for incident radiation, which is on the order of 500 THz. The main component of such a setup is a pair of parallel mirror plates with separation L known as Fabry Perot interferometer (FPI). Such an FPI transmits incident radiation of wavelength λ if the condition $\lambda = 2L/n$ is met, with n being the transmission order, i.e., an integer number. The corresponding transmission function reads:

$$T(\lambda) = \frac{T_0}{1 + (4F^2/\pi^2)\sin^2(2\pi\nu L/c)} \tag{11}$$

Here, ν is the frequency of the radiation and F is the *finesse*, a quantity that characterizes the mirror quality, i.e., their reflectivity and flatness. Eq. (11) shows equidistant transmission peaks, as seen in the upper part of Fig. 4, and the distance $\Delta\nu = c/(2L)$ between adjacent transmission peaks is known as free spectral range (FSR) and the spectral resolution is given by the width of the transmissions peaks $\delta\nu = \Delta\nu/F$. In order to operate the FPI as a spectrometer, the mirror distance L is scanned to vary the frequency of the transmission, typically between 0.5 and 2 μm.

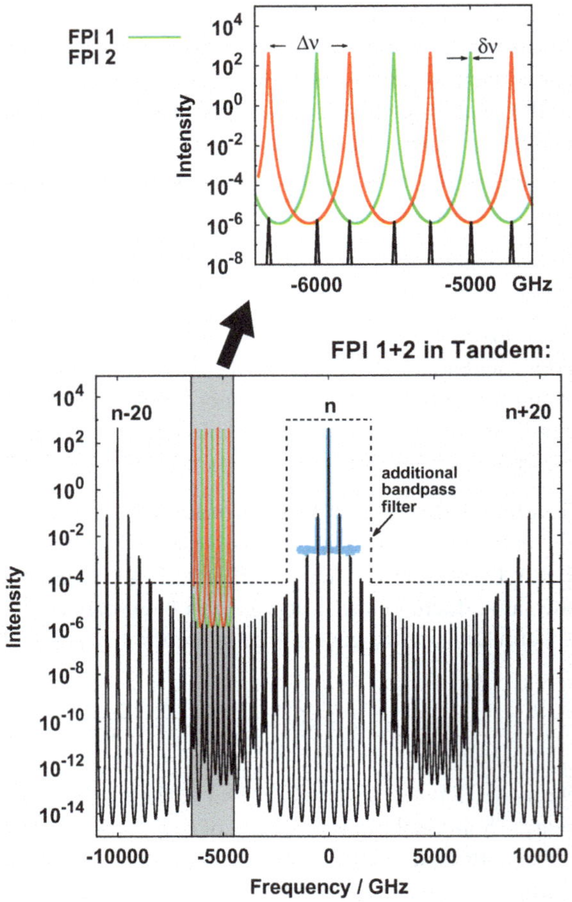

Fig. 4 Transmission function of a tandem Fabry Perot interferometer. Upper part shows the transmission lines of each single FPI, lower part the combined transmission function. Blue line experimental data for mirror spacing $L_1 = 0.3$ mm and $L_2 = 0.95\,L_1$. Full transmission in tandem mode would again occur at orders $n \pm 20$, which is suppressed by an additional bandpass filter

Practically, the use of such a configuration to scan a continuous light source is highly limited by the fact that the FSR is rather small and adjacent interference orders contribute significantly to the transmitted light intensity. The solution is, to simultaneously operate two FPIs tuned to obtain a common transmission frequency, however, with slightly different mirror spacings L_1 and L_2 and thus with slightly different FSRs, so that higher order transmission only occurs after several transmission orders of one single FPI, i.e., when higher orders of both FPIs coincide, as shown in the lower part of Fig. 4. In such a configuration, which is known as Tandem FPI, both FPIs are scanned simultaneously and it turns out that a common shift of the transmission frequency can be achieved, when $\delta L_1/\delta L_2 = L_1/L_2$ is maintained while changing the mirror spacing. In that way, the resulting effective FSR, i.e., the interval between two lines of full transmission, at n and $n \pm 20$ in our example in Fig. 4, is greatly enhanced compared to a single FPI while roughly the spectral resolution $\delta\nu$ stays the same. In particular, the combined effective FSR is large enough

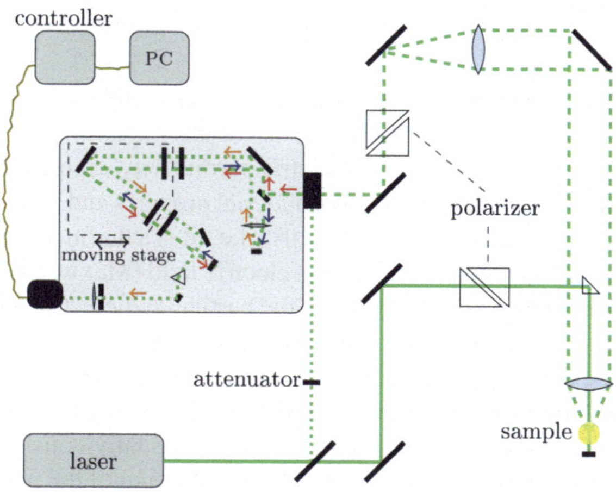

Fig. 5 Schematic representation of the Tandem Fabry Perot setup. Part of the laser beam is fed directly into the TFPI as reference and is used for stabilization. The main beam passes a polarizer and a small prism before it is focused into the sample placed in a temperature-controlled environment. The backscattered light is defocused, passes the prism and is focused again before it goes through the analyzer and enters the TFPI. Inside the TFPI the optical path is simplified, each FPI is passed three times by the scattered light, as indicated by the arrows, and the wavelength is selected according to the respective mirror plate separation. The respective intensities are recorded with a photo diode

so that higher order transmissions at $n \pm 20$ may be suppressed by applying conventional dielectric multilayer bandpass filters, which are known to be necessary to avoid artificial features in the resulting spectra [2, 36, 90].

In practice of course it is a non-trivial issue to maintain parallel mirrors during the scan and at the same time to be able to substantially change the FSR to scan various spectral ranges. Moreover, the scanning stage requires active vibration isolation and sufficient temperature stability to be reliable in operation. The most widely applied and commercially available device of this kind is the scanning multipass tandem Fabry Perot interferometer constructed by J. Sandercock, which is also used in the present work (Fig. 5). Typically several FSRs are combined to obtain the DLS spectral density in a frequency range from about 200 MHz to about 3 THz, as is seen in the lower part of Fig. 10. More details of this technique can be found [9, 58].

2.2 Dielectric Spectroscopy

In dielectric spectroscopy the reorientation of permanent molecular dipoles is probed by applying an electric field $\mathbf{E}(t)$ and monitoring the time-dependent polarization response $\mathbf{P}(t)$, which due to relaxations occurring in the material depends on the past evolution of the electric field. In case of a linear, homogeneous and isotropic dielectric medium one may write if the applied field is a periodic function of time:

$$\mathbf{P}(\omega) = \varepsilon_0\,(\hat{\varepsilon}(\omega) - 1)\,\mathbf{E}(\omega) = \varepsilon_0\hat{\chi}(\omega)\,\mathbf{E}(\omega) \tag{12}$$

this defines the frequency dependent dielectric susceptibility $\hat{\chi}(\omega)$ or the dielectric permittivity $\hat{\varepsilon}(\omega)$ with ε_0 being the permittivity of the vacuum.

It is important to note here, that the dielectric permittivity, or its time domain counterpart the dielectric response function, is a material property and as such independent of the particular size and shape of the dielectric, i.e., it is a thermodynamically intensive quantity. The polarization as well as the electric field (Maxwell field) of Eq. (12) on the other hand do depend on the shape of the particular dielectric under consideration due to the long-range dipole-dipole interactions. Therefore complications arise, when $\hat{\varepsilon}(\omega)$ is to be related to a microscopic equilibrium property of the system, like the autocorrelation function of molecular dipole moments $C_\mu(t) = \langle \boldsymbol{\mu}(t) \cdot \boldsymbol{\mu}(0) \rangle$.

To establish a link to equilibrium fluctuations one usually considers fluctuations $\delta\boldsymbol{M}(t)$ of the macroscopic dipole moment, which is the sum of the permanent molecular moments $\boldsymbol{M} = \sum_{i=1}^{N} \boldsymbol{\mu}_i(t)$. Those fluctuations can be expressed in terms of a correlation function $\langle \delta\boldsymbol{M}(t) \cdot \delta\boldsymbol{M}(0) \rangle$, which is directly related via the fluctuation dissipation theorem to the response function of the system to an external vacuum field E_0, which is entirely due to an external charge distribution in contrast to the Maxwell field used in Eq. (12). Therefore, any relation between $\langle \delta\boldsymbol{M}(t) \cdot \delta\boldsymbol{M}(0) \rangle$ and intensive quantities like $\hat{\varepsilon}(\omega)$ will depend on the shape of the dielectric, as $\langle \delta\boldsymbol{M}(t) \cdot \delta\boldsymbol{M}(0) \rangle$ itself is not intensive. But as soon as such a relation is established for a particular sample shape (usually a sphere or an infinitely extended sample is used [60]) again a shape independent relation may be obtained by reducing $\langle \delta\boldsymbol{M}(t) \cdot \delta\boldsymbol{M}(0) \rangle$ to a microscopic correlation function.

For such a microscopic correlation function one may consider n dipole moments in a spherical volume around one reference dipole $\boldsymbol{\mu}_i$ and then treat the remaining $N - n$ molecules as a dielectric continuum. The volume is chosen small compared to the overall sample size so that its average moment only depends on short-range interactions. Such a microscopic correlation function can be written as:

$$\Phi_{\mathrm{mic}}(t) = \frac{\left\langle \boldsymbol{\mu}_i(0) \cdot \sum_j^n \boldsymbol{\mu}_j(t) \right\rangle}{\left\langle \boldsymbol{\mu}_i(0) \cdot \sum_j^n \boldsymbol{\mu}_j(0) \right\rangle} \tag{13}$$

and is identical with Eq. (2) for $\ell = 1$. Now, the fluctuation dissipation theorem can take a form which is again independent of the sample shape:

$$\left(\hat{\varepsilon}(\omega) - \varepsilon_\infty\right) \frac{g(\omega)}{\Delta\varepsilon} = 1 - i\omega \int_0^\infty \Phi_{\mathrm{mic}}(t)\, e^{-i\omega t}\, \mathrm{d}t \tag{14}$$

Here, $g(\omega)$ represents an internal field factor, which is calculated in slightly different ways depending on the particular approach used (see, e.g., [19, 26, 37] or [11, 101] for reviews). In several cases, however, $g(\omega) \approx 1$ holds in reasonable approximation.

Although the above relation is independent of the macroscopic sample shape, i.e., $\Phi_{\mathrm{mic}}(t)$ contains only short range correlations, Eq. (13) still reflects the fact that what is measured in dielectric spectroscopy is a *collective* quantity. In fact, the microscopic correlation function may be split into an autocorrelation and a cross correlation part:

$$\Phi_{\mathrm{mic}} = \frac{1}{g_K\,\mu^2}\left(\left\langle \boldsymbol{\mu}_i(0)\cdot\boldsymbol{\mu}_i(t)\right\rangle + \left\langle \boldsymbol{\mu}_i(0)\cdot\sum_{i\neq j}\boldsymbol{\mu}_j(t)\right\rangle\right) \tag{15}$$

with the Kirkwood-Fröhlich correlation factor

$$g_K = 1 + \langle \boldsymbol{\mu}_i(0)\cdot\sum_{i\neq j}\boldsymbol{\mu}_j(0)\rangle/\mu^2 \tag{16}$$

quantifying the amount of static cross-correlations [28]. Typically g_K takes on values larger than 1 in case of preferentially parallel orientation of dipole moments, whereas for preferentially anti-parallel orientation values of $g_K < 1$ are obtained.

Usually, the time-dependent cross correlation term in Eq. (15) is either neglected or it is argued that the time dependence of cross correlations can be approximated by that of the dipolar auto correlation function itself [103]. However, in liquids with particular interactions, like hydrogen bonding, where supramolecular structures are expected, such cross correlations may not be negligible at all and may require careful consideration.

2.3 Combination of Methods

Thus, to first approximation, an interpretation of data when comparing dielectric spectroscopy and depolarized ligth scattering, may be based on considering the single molecule correlation functions of different order ℓ:

$$C_\ell(t) = \langle P_\ell(\mathbf{u}_i(t)\mathbf{u}_i(0))\rangle,$$

There is no general relation between correlation functions of different order ℓ, as such a relation will depend on the geometry of the underlying stochastic process. However, in that way, a comparison of data from different methods becomes sensitive to the motional mechanism underlying the observed dynamics: For example, in the case of isotropic rotational diffusion the correlation times are different by a factor of three $\tau_1/\tau_2 = 3$, while in the case of reorientation by large angle random jumps the correlation function becomes independent of ℓ [6]. Another example is, if molecular reorientation is restricted to small angles, what could be expected for secondary processes like the JG β relaxation [94], then an expansion of Eq. (1) to leading order in θ reveals that the relative amplitude of that small angle process should be larger by a factor of three for an $\ell = 2$ correlation function compared to $\ell = 1$, so that for such a small angle process for the respective susceptibilities

$$\chi''_{\mathrm{BDS}}(\omega) = 3\chi''_{\mathrm{DLS}}(\omega) \tag{17}$$

is expected to hold [7, 14, 57].

However, such considerations all rely on certain assumptions. The first one is, that the molecular polarizability tensor is well enough approximated by a "symmetric top" geometry, i.e., one assumes that one principal axis is largely different from the other two, that are similar. Furthermore, the dipole moment and the symmetry axis of the anisotropic polarizability are either aligned, which is a rather special situation that will hold only in very rare cases, or are in a fixed orientation with respect to each other. This requires that the molecule does not posses significant internal degrees of freedom and additionally it is necessary that the reorientational process considered is isotropic, i.e., without any preferred axis of reorientation, so that, on average, the correlation functions $\langle P_\ell(\cos\theta(t))\rangle$ refer to the same angle θ in both methods. In that latter case the symmetry requirements for the polarizability tensor are less strict, as even a weighted average over the reorientation of different principal axes will show the same dynamics.

Another important point is, in how far interaction induced effects influence the results of the light scattering experiment. Although it seems to be consensus that reorientation of the optical anisotropy is the dominant mechanism for depolarized scattering in optically anisotropic molecules [20, 57] interaction induced effects may play a more pronounced role on shorter time scales than on the scale of the α-relaxation [65]. For long time scales interaction induced effects seem to be negligible even in cases where anisotropy is smaller: For example in glycerol with a comparatively small anisotropy reorientation is probed in DLS in the regime of the α-relaxation [13]. However, in the time range of the β-process even a small contribution of an interaction induced mechanism may have a non-negligible effect and cannot be entirely excluded, even if on longer timescale reorientation is clearly dominant.

A last point concerns the fact, that when a comparison of data is based on Eq. (1) one refers to a single molecule correlation function and cross correlations are neglected. However, both methods actually provide collective correlation functions, and how collective and single molecule correlations are related is not easy to determine. Static crosscorrelations can be quantified via the Kirkwood correlation factor, that was introduced in Eq. (16) for dielectric spectroscopy, and is accessible via the Kirkwood-Fröhlich relation [53] from the dielectric relaxation strength. For light scattering a similar quantity can be defined:

$$g_K^{DLS} = 1 + \sum_{i \neq j} \langle P_2(\mathbf{u}_i \cdot \mathbf{u}_j)\rangle, \tag{18}$$

which is more difficult to access, e.g., through relating the depolarized Rayleigh ratio with the polarizability anisotropy [4].

But intermolecular cross-correlations are also known to influence the dynamics. A theoretical treatment of the problem by Madden and Kivelson [60] suggests that this can also be quantified by using the Kirkwood factor of static cross-correlations g_K and a dynamic correlation factor J_K in order to relate a single-molecule correlation time τ_s with the collective time τ_c:

$$\tau_c \approx \tau_s \cdot g_K / J_K. \tag{19}$$

As $J_K \approx 1$ is expected in simple liquids [60], cross correlations effectively lead to a slowing down of the collective relaxation if $g_K > 1$. We note that the above relation is supported by experimental findings in hydrogen bonding liquids as well as by molecular dynamics simulations, see [61, 100] and references therein. In the present context this implies that when data from different methods are compared and cross correlations play an important role, then the observed relaxation times may be affected in a different manner by crosscorrelations quantified by g_K^{BDS} and g_K^{DLS}, respectively.

3 Example I: Monohydroxy Alcohols

Many monohydroxy alcohols show interesting dynamic features, in particular the well-known Debye-process, which has since long been observed in dielectric spectra [16]. Most monoalcohols are easily supercooled and have the ability to form hydrogen-bonds. But due to the fact, that there is only one OH-group per molecule, monoalcohols do not form OH-bonding networks, like other alcohols, but simpler supra-molecular structures, which are thought to be the reason for the observed peculiar dynamic features. In particular, it is assumed that monohydroxy alcohols form transient chains by H-bonding [33], and there is support for this picture by NMR, shear-viscosity measurements, neutron scattering, computer simulations [16, 33, 34, 87] and recently also by DLS experiments [32]. The idea is that each molecule within a chain is transiently connected to a neighboring molecule in a manner schematically depicted in in Fig. 6. These chains fluctuate on the one hand due to motions of the chain as a whole and on the other hand because of the finite lifetime of the H-bonds.

To gain further knowledge about the nature of the relaxations occurring in these alcohols, we apply a simple strategy, which directly follows from the considerations of the experimental methods in the previous section: As one can expect that the structural α- and secondary-relaxation and the relaxation of supra-molecular structures are probed in a different manner by the two methods DLS and BDS, a combination of both techniques allows to clearly separate both types of processes and to gain insight into the underlying motional mechanisms.

In many monoalcohols the most prominent process in the dielectric spectra is the Debye peak, which can be envisaged to first approximation as the reorientation of the end-to-end vector dipole moment created by the molecular dipole moments of the individual molecules in a transient chain. As indicated for the case of 1-propanol

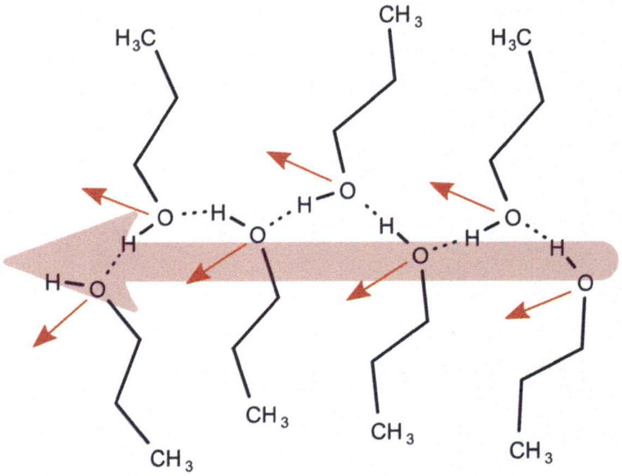

Fig. 6 Two dimensional schematic representation of an H-bonded chain structure formed by 1-propanol molecules. Red arrows show the direction of the single-molecule dipole moment and its projection onto the backbone of the chain [46]

in Fig. 6, basically the molecular moments are localized at the OH-groups and give a strong component parallel to the chain backbone, which adds up to form the end-to-end vector dipole moment. The fluctuation of this structure in many cases is the leading relaxation process in BDS and is due to fluctuations of the direction and/or the size of this end-to-end vector moment, which again is facilitated by formation and breaking of the H-bonds [33].

In order to separate the relaxation of this supramolecular structure from the α- and secondary-relaxation we benefit from the fact that DLS probes molecular reorientation by being sensitive to the reorientation of the molecular optical anisotropy tensor. In the following, we will give two examples of monohydroxy alcohols, namely 1-propanol and 1-phenyl-1-propanol as illustrated in Fig. 7. 1-propanol can simply be pictured as a hydrocarbon chain with an OH-group at one end, while 1-phenyl-1-propanol has an additional phenyl-ring located close to the OH-group and orthogonal to the chain. Both molecules are expected to be largely different in their ability to form supramolecular structures, and have separately been investigated with BDS by other groups [42, 52, 54].

In particular simulations indicate that besides linear chains [49] and closed-loop structures [49, 88] with various mean sizes also brushes are possible structures in monoalcohols [86]. The actual behavior of course depends both on the amount of sterical hindrance with respect to the formation of chains and on the temperature. It turns out that at high temperatures ring-like structures, where the component of the dipole moment that is parallel to the backbone is canceled, are more favoured. At low temperatures linear chain structures with an effective end-to-end dipole, which may even enhance the dipolar strength by assembling into brush-like structures [86], are more likely.

The idea of ring and chain like structures was discussed in detail by Dannhauser [49], who investigated a series of octanols and considered the temperature dependent Kirkwood correlation factors (Eq. 16 in Sect. 2.2). As discussed in the context of Eq. (16), if linear chain structures dominate $g_K > 1$ is observed, while in the

Fig. 7 Structure of the investigated monohydroxy alcohols [46]

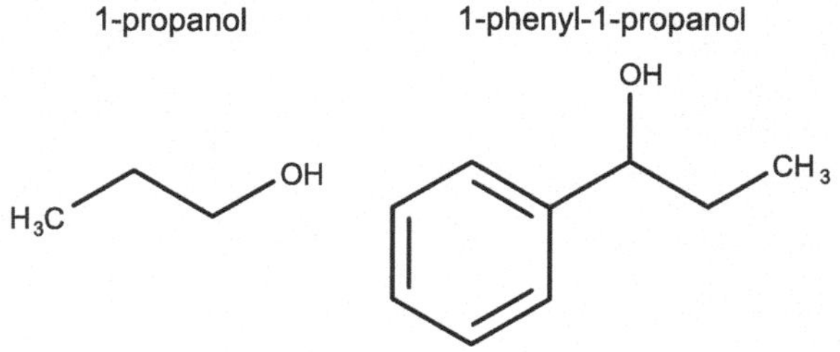

case of preferentially anti-parallel orientation of dipole moments as in the case of closed-loop structures, $g_K < 1$ is obtained. At temperatures, where $g_K(T)$ changes from below to above one an equilibrium of coexisting ring and chain structures can be envisaged. Such a scenario was also reported in 5-methyl-3-heptanol [49] and was recently investigated by non-linear dielectric spectroscopy [88, 106], where the results indicated that a high electric field strength forces the ring structure to break open. In that context the authors argue that the fluctuations in the end-to-end vector dipole moment originate in the opening and closing of ring structures.

In 1-propanol chains and brush-like structures where found by computer-simulations and a modeling of chain distributions [86]. The idea of chains and brushes, where some OH-groups are not connected to one but to two other OH-groups, was further confirmed by neutron diffraction [87]. In 1-phenyl-1-propanol Johari et al. argued that steric hindrance due to the phenyl-ring and the enhanced disorder created by the mixture of dextro- and levo-forms of the molecule will entirely prevent larger H-bonded structures to form [52], while in an earlier publication Johari and Dannhauser suggested that only short chains are present in 1-phenyl-1-propanol [50]. By comparing DLS and BDS data, we will also shed some light on this question further below.

The general situation of relaxation in monohydroxy alcohols is demonstrated in Fig. 8, with a typical dielectric loss spectrum of 1-propanol at 101 K, which is shown to contain three relaxation components. In many cases the strongest contribution is the Debye relaxation. For a long time there has been an ongoing debate on the molecular origin of this process [16], but presently many share the view that the origin are fluctuations of the end-to-end vector dipole moment of transient H-bonded structures. The general view is that the Debye process is not to be identified with the structural α-relaxation, which Debye originally must have had in mind as he developed his Debye relaxation model [24]. It turns out that in most cases the Debye process at least partially covers the structural α-relaxation governed by single molecular motion related to the viscosity. Both processes obey the Vogel Fulcher Tammann (VFT) temperature dependence.

Usually, the α-relaxation is followed by a secondary β-relaxation typical for glass forming liquids with a much weaker and Arrhenius-like temperature dependence. After supercooling the liquid below T_g leading to freezing of the structural relaxation, this process remains to be seen in the dielectric time window [55]. In cases when the considered molecules are rigid with respect to the dielectric probe, such a process

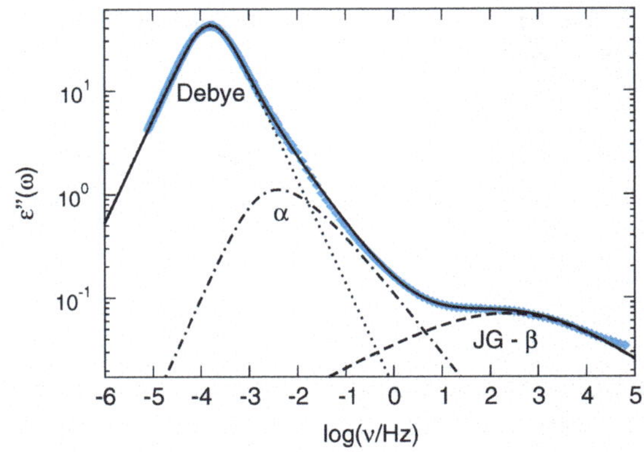

Fig. 8 Example of a typical dielectric loss spectrum of a monohydroxy alcohol, here 1-propanol at 101 K, showing the dominant Debye relaxation, the structural α- and Johari-Goldstein β-relaxation

is clearly identified and named after Johari and Goldstein as JG β-relaxation. In their seminal contribution [51] they pointed out, that even in rigid molecules so-called islands of mobility [38] in an otherwise slow or frozen matrix may cause such secondary relaxations [47, 48] and that these processes are due to localized motions in the glass [70, 71]. However, later on NMR investigations suggested a different interpretation by showing for several cases, that in the glassy state small angle reorientation of nearly all molecules underlies the JG β-relaxation [96].

To obtain a new perspective on all three processes it is useful to quickly recap what may be expected in a DLS experiment in relation to the BDS data on the background of the discussion in Sect. 2.3. For the Debye-process, for which one may assume rotational diffusion as molecular mechanism when picturing the reorientation of an end-to-end vector, the discussion in Sect. 2.3 implies that because of the angle sensitivity of both methods the DLS time constants are expected to be a factor of three faster than the ones from BDS. However, apart from few exceptions [42], reports on DLS of monoalcohols are rare in the literature. Although Debye-like contributions were previously reported in light scattering data of water [44] and of 2-ethyl-4-methylimidazole [99], only recently a Debye process was unambiguously identified in DLS of a mono alcohol, namely 5-methyl-2-hexanol [32]. In all cases, the mentioned factor of three in the relaxation times was approximately consistent with the reported data.

In the case of secondary processes there are only rare DLS reports in the literature [29]. Previous light scattering work dedicated to comparing secondary relaxations in BDS and DLS comes to the conclusion, that while the so-called high-frequency wing (HFW), which is frequently observed in dielectric spectra, is also recovered in light scattering, a Johari Goldstein β peak may not be visible at all in DLS [14]. This is surprising, because according to the discussion in Sect. 2.3 an $\ell = 2$ correlation function should be even more sensitive to a small angle process than an $\ell = 1$ function, so that the relative amplitude of such a process should be even larger in DLS compared with its dielectric counterpart (see Eq. 17) [7, 14].

3.1 Results: BDS and DLS in Monoalcohols

Broadband dielectric spectroscopy. Fig. 9 shows the temperature dependent spectra of our example liquids 1-propanol and 1-phenyl-1-propanol measured by BDS. The 1-propanol data shown in the upper plot for frequencies $\nu \geq 10^{-2}$ Hz were previously published by Kudlik et al. [54, 55] and are joined with Fourier transformed time-domain data obtained in a modified Sawyer-Tower setup already described in detail elsewhere [62, 72]. All three processes, which were mentioned above, can be observed in the data of 1-propanol in a frequency range covering 15 orders of magnitude and a temperature range from 87 K up to 273 K. The fit model is a sum of a KWW function for the Debye-Process and a convolution of two correlation time distributions for the α- and β-relaxation according to the Williams-Watts approach [102]. In the time domain this simplifies significantly and reads:

$$\Phi(t) = \Delta\varepsilon_D \, e^{-(t/\tau_D)^{\beta_D}} + \Delta\varepsilon_{\alpha\beta} \, \Phi_\alpha(t) \cdot \left[(1-k_\beta) + k_\beta \Phi_\beta(t)\right]. \qquad (20)$$

We note, that β_D is included to take account of a possible broadening of the Debye-like process (in case of 1-propanol $\beta_D = 1$) and k_β indicates the strength of the JG-relaxation in the Williams-Watts apporach. For more details on the applied relaxation time distributions and the shape parameters see Refs. [31] and [8]. The respective fit results are seen as solid lines.

In the lower plot in Fig. 9 seemingly a different situation occurs in the case of a non-typical monohydroxy alcohol 1-phenyl-1-propanol. The dielectric loss is remarkably lower and no clear Debye-Process is discernible. At first glance the data only show a quite narrow α-relaxation followed by a weak and strongly separated secondary γ-relaxation peak. Here, the situation becomes more clear when BDS and DLS measurements are compared further below. So far we only point out that earlier BDS measurement by Johari et. al [52] showed an additional, well resolved β-relaxation between our α- and γ-peak. In our data this β-relaxation is much weaker, in fact only a high frequency wing is visible in the data. The reason for this discrepancy so far is unclear and needs further investigation.

Depolarized light scattering. Due to the small optical anisotropy of 1-propanol the intensity of the depolarized scattered light is rather low. Thus, obtaining well resolved spectral densities especially in the long time regime, where the photon correlation technique is applied, in general is quite a difficult task, and some of the problems and experimental details were discussed in Sect. 2.1.

DLS experiments on 1-propanol and 1-phenyl-1-propanol were performed with setups described in Sect. 2.1 and in [10, 30] and data are presented in Fig. 10. Concerning literature data only 1-propanol was investigated earlier by Hansen et al. [42] in an interesting experiment, where BDS and PCS data were obtained simultaneously and correlation times could directly be compared. However, no detailed analysis of the lineshape of the correlation functions was given.

The upper panel in Fig. 10 shows the α- and β-relaxation of 1-propanol. Despite of weak anisotropic scattering correlation functions and corresponding spectral den-

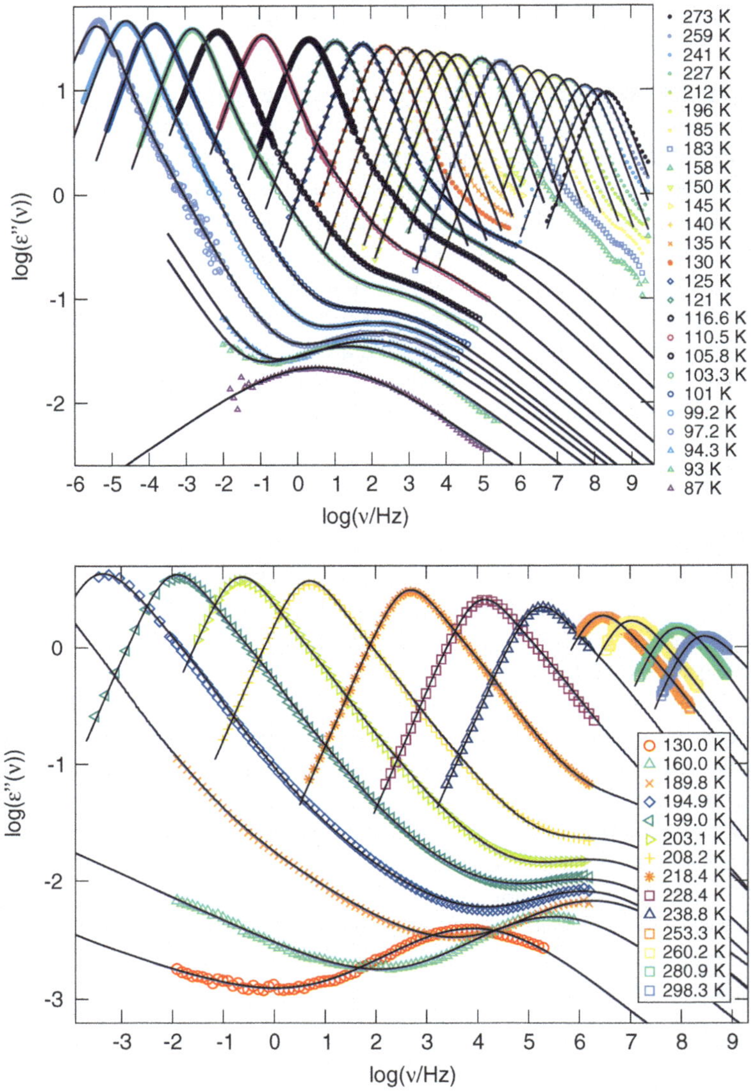

Fig. 9 (Upper panel) Frequency dependent dielectric spectra from Kudlik et al. ($\nu \geq 10^{-2}$ Hz) [54, 55] joined with Fourier transformed time-domain data ($\nu < 10^{-2}$ Hz) of 1-propanol with fits for Debye, α- and β-relaxation. Figure reprinted with permission from [31], © 2017 American Chemical Society. (Lower panel) Frequency dependent dielectric spectra of 1-phenyl-1-propanol

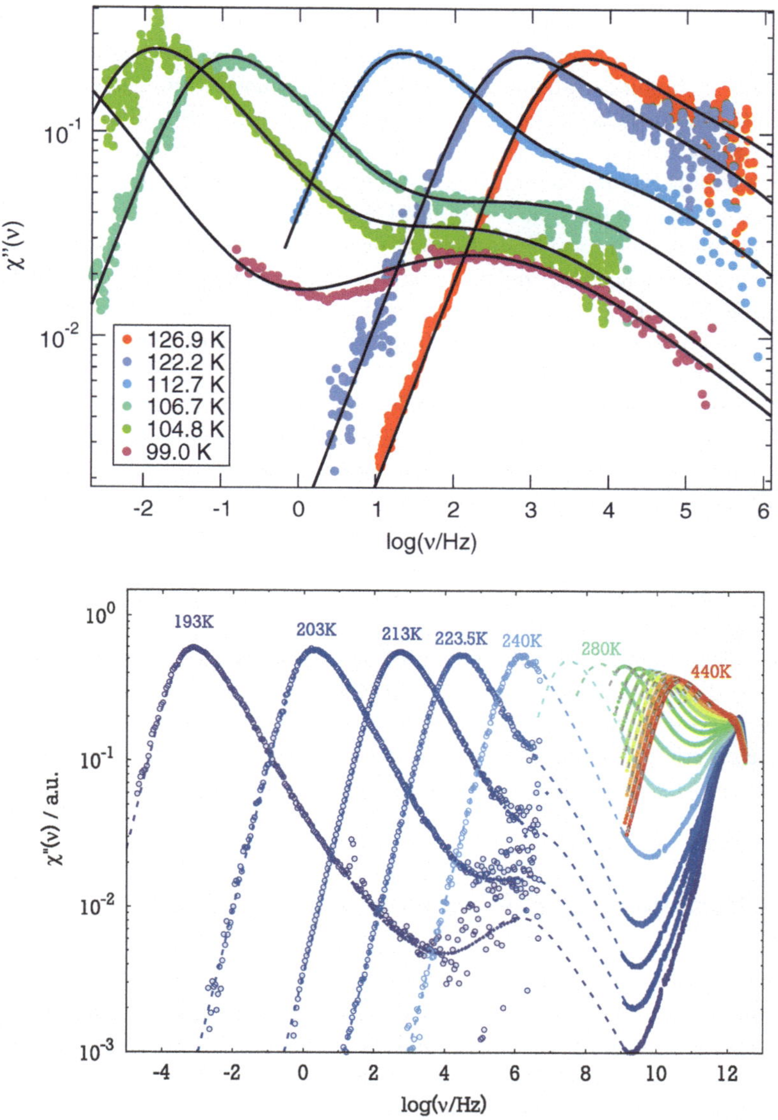

Fig. 10 Upper panel: Fourier transform of the electric field correlation functions $g_1(t)$ with fits for α and β relaxation (solid lines) of 1-propanol. Reprinted with permission from [31], © 2017, American Chemical Society. Lower panel: Combined data of 1-phenyl-1-propanol from PCS (open diamonds) and TFPI (filled circles) experiments. For comparison some BDS data of the secondary relaxation in 1-phenyl-1-propanol from Fig. 9 taken at the same temperatures are shown (solid diamonds) as guide to the eye

sities were obtained in a temperature range between 99 and 127 K. In the lower panel of Fig. 10 the DLS susceptibility of 1-phenyl-1-propanol is presented. The Fourier transformed photon correlation data, which were obtained from merged data sets of fiber- and camera-PCS (see Sect. 2.1), are shown together with TFPI measurements, which results in a time window that covers 18 orders of magnitude with sufficiently high resolution. At highest temperatures only a single, broadened relaxation peak is seen besides the microscopic dynamics. On cooling, as the dynamics slows down, this peak splits up into several contributions: Around T_g on can distinguish from high to low frequencies the microscopic dynamics and the Boson-peak, the fast β-relaxation, the γ-process and finally β- and α-relaxation. The β-relaxation simply appears as a power-law deviation on the high frequency slope of the α-process also known as high-frequency wing.

Interestingly, the low frequency flank of the α-relaxation shows no indication of any Debye contribution in DLS, despite the data having quite high resolution due to the comparatively strong depolarization of the phenyl ring. For the fits in Fig. 10 again Eq. (20) is used, however in a slightly modified form: As the β-process only expresses as a high frequency wing in the data, α-peak and wing are described together by a extended generalized gamma (GGE) distribution [8], and the γ-relaxation by a separate peak, in the same way as the β-relaxation was described in 1-propanol. As the DLS data are too noisy in the region of the γ-process to allow for an independent fit, the information of the γ-peak is used as obtained from the dielectric data. To emphasize that indeed BDS and DLS data are identical within experimental uncertainty in the frequency range between β- and γ-process, BDS data are superimposed and depicted as solid diamonds. And it turns out that the data of both methods almost match perfectly in that region.

3.2 Discussion: α-, β-Relaxation and Debye Peak

We start by comparing BDS and DLS data of 1-propanol based on the correlation times of the different processes that are observed and are depicted in Fig. 11. The relaxation times of the Debye- and the α-process follow a VFT temperature behaviour. The time constants of the α-relaxation in DLS and BDS are the same within experimental uncertainties indicated by error bars, which in many cases are smaller than the symbol sizes. The relaxation times of the JG β-process are also equal in BDS an DLS and follow an Arrhenius temperature dependence with an activation energy of $\Delta E_a/k_B \approx 2500\,\mathrm{K}$. Above T_g the temperature dependence becomes somehow stronger, at even higher temperatures, however, the error bars increase significantly and a quantitative statemant becomes difficult. The symmetric β-peak can well be described by the same temperature independent distribution of activation energies for both methods, which leads to the same spectral shape in BDS and DLS within the limit of experimental uncertainty. As α- and β-process turn out to be identical in both methods, Fig. 12 presents a direct comparison of the data. Both data sets are described with the same fit model, except that there is no discernible

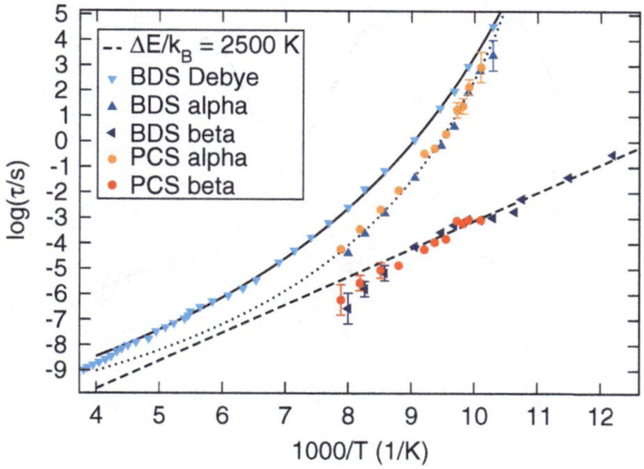

Fig. 11 Arrhenius plot of 1-propanol with dielectric relaxation data (triangles) and PCS data (circles). τ_D and τ_α data are fitted with a Vogel-Fulcher-Tammann law and the JG β-process is well described by an Arrhenius law with an activation energy of 2500 K$\cdot k_B$. Unless explicitly indicated uncertainties are smaller than the size of the symbols. Figure reproduced with permission from [31], © 2017 American Chemical Society

Debye contribution in the DLS data. The dashed black lines show the BDS fits for α- and β-process without the Debye contribution, and it becomes obvious, that α- and β-peak have the same shape, relative strength and position. Therefore, the correlation functions of α- and β-relaxation are identical in both methods and, if one follows the reasoning outlined in Sect. 2.3 about comparing $\ell = 1, 2$ correlation functions, the mechanism behind the JG process in 1-propanol cannot be small angle reorientation but large angles have to be involved like in the model of isotropic random jumps. This in turn favors the picture of islands of mobility over the idea that all molecules more or less equally take part in the β-process.

We note here, that recently similar observations were made in 5-methyl-2-hexanol [32], in particular the analysis reveals that only the Debye contributions shows a significantly different expression in both experimental methods, as a strong Debye peak in the dielectric loss is either absent (as in 1-propanol [31]) or much weaker (as in 5-methyl-2-hexanol [32]) in DLS. Especially in case of primary monoalcohols the mobility of the depolarizing unit turns out to be too large to leave sufficient correlation at times longer than τ_α to be observed in an $\ell = 2$ correlation function, which requires a significantly smaller average angle of motion for a full decay [31, 32]. Thus, the above findings may reflect a more general behavior in monoalcohols, which could be used to disentangle α/β-relaxation and Debye-process in the dielectric spectra, based on the notion of the α/β-relaxation in DLS, where the Debye contribution is either absent or much weaker than in the case of BDS. We will make use of this idea in the following to be able to discern a possible Debye contribution in 1-phenyl-1-propanol.

In 1-phenyl-1-propanol the additional phenyl ring represents a steric hindrance to the formation of larger H-bonded structures. However, the question is open, whether

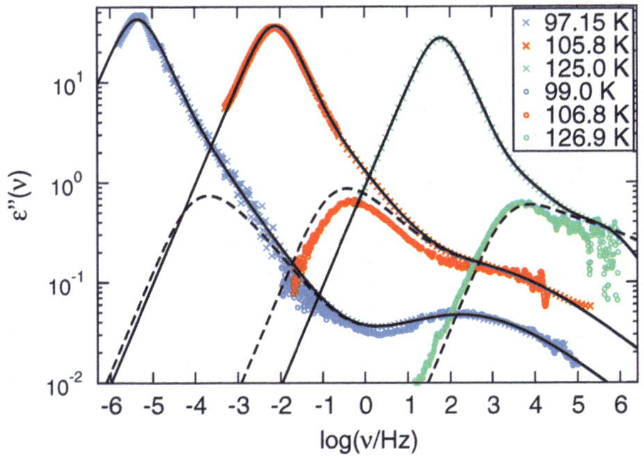

Fig. 12 Direct comparison of dielectric (circles) and depolarized light scattering data (diamonds) of 1-propanol at similar temperatures, which underlines the finding that α- and β-relaxation are found to be identical in both methods with respect to their line shape, time constants and strength. Dashed lines show the $\alpha\beta$-part of the fit to the dielectric data. Figure reproduced with permission from [31], © 2017 American Chemical Society

Fig. 13 Demonstration of the similarity of α-relaxation in 1-propanol (1P) and 1-phenyl-1-propanol (1P1P) by comparison of Fourier transformed PCS correlation functions. 1-phenyl-1-propanol data are vertically shifted to match the dielectric α-peak. In both cases the α-process lineshape is identical and there is no sign of Debye contribution in DLS

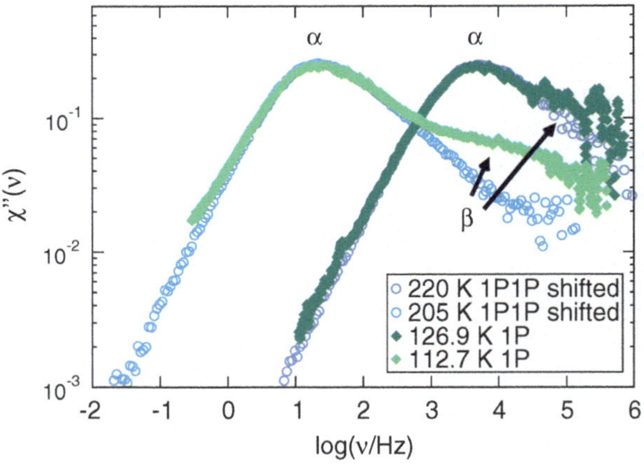

this just reduces the Debye contribution and makes it faster compared to propanol, because the supramolecular units are reduced in size, or whether a Debye contribution vanishes altogether and the phenyl groups prevent significant chain formation, as was argued in Ref. [52]. At first we compare the α-process, as it is seen in DLS of 1-propanol and 1-phenyl-1-propanol in Fig. 13. This is interesting, because in the former case the low depolarized scattering is mainly due to the optical anisotropy of the alkyl chain, while in the latter the anisotropic scattering by the phenyl ring dominates and provides a rather strong depolarized signal. In both cases, however, the α-process lineshape is identical and no indication of a Debye contribution is seen. Thus, although both molecules are not strictly rigid, the α-relaxation shape

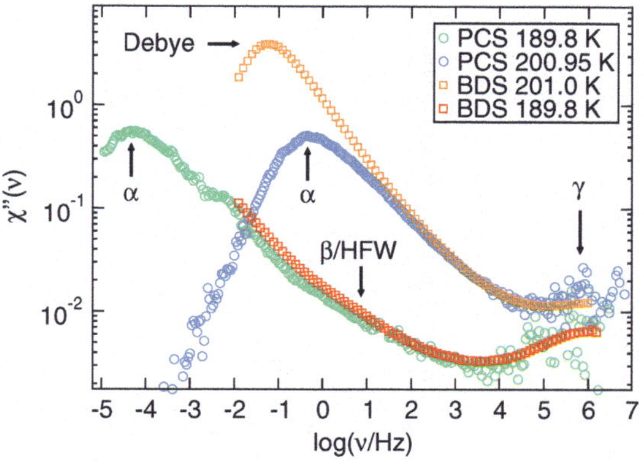

Fig. 14 Direct comparison of dielectric (square) and depolarized light scattering data (circles) of 1-phenyl-1-propanol at similar temperatures, which underlines the finding that α- and β-relaxation appear to be almost identical in both methods with respect to their line shape. Dashed black line shows a high frequency wing contribution in both methods

and thus the underlying correlation time distribution does not depend on which part of the molecule dominantly causes the depolarized scattering. This observation is reassuring when one argues in terms of $\ell = 1, 2$ correlation functions, as outlined above.

Next we attempt a direct comparison of DLS and BDS data for 1-phenyl-1-propanol in Fig. 14. At first glance the main relaxation peak in both methods is significantly different, while the spectra seem to coincide in the minimum region between the high frequency wing of the main relaxation and the γ-process. And a comparison with Fig. 12 reveals a striking resemblance in both cases, if one interprets the main peak seen in the dielectric spectra of 1-phenyl-1-propanol as a Debye-contribution which is slightly faster and significantly smaller than in the case of 1-propanol. Therefore, in the following, we will assume, based on the above mentioned observation made for 1-propanol and 5-methyl-2-hexanol that DLS and BDS data only differ in the Debye contribution, that the same holds true in case of 1-phenyl-1-propanol. Then, the fact, that the data perfectly match in the minimum region as seen in Fig. 14 confirms this assumption and implies that α- and γ-process are identical in both methods and that the Debye contribution can be extracted in the BDS data from the knowledge of the α-process in DLS.

Thus, reliable time constants, strength and shape parameters can be obtained for all processes by combining the two methods. The resulting time constants are shown in Fig. 15. The Debye time τ_D is only slightly slower than τ_α. Moreover it turns out that the resulting Debye process is much weaker than in 1-propanol and is broadened with a parameter $\beta_D = 0.85$. All of these findings imply that shorter chains are formed due to the steric hindrance brought about by the phenyl ring. Because of the smaller end-to-end vector dipole moment the strength decreases, and as the supramolecular structure is smaller, it is still sensitive to heterogeneous local environments, which causes the broadening of the Debye contribution. This agrees well with the previous findings that DLS probes the Debye relaxation on a more local scale and thus also observes a distribution of Debye relaxation times [32]. We also note here, that a

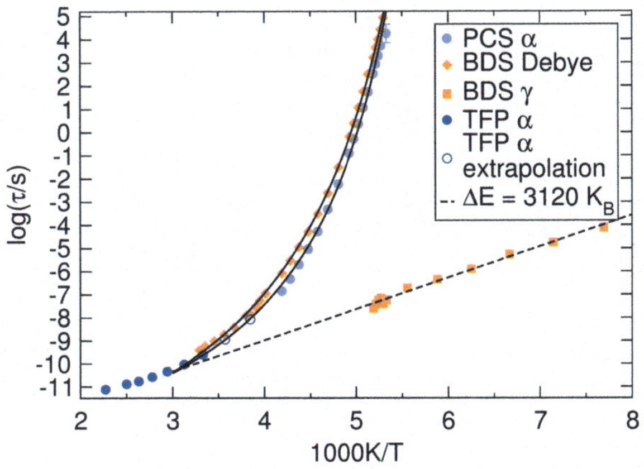

Fig. 15 Arrhenius plot for 1-phenyl-1-propanol: time constants of the α-Process (circles) were obtained from light scattering data and used to extract γ-relaxation (squares) and Debye-Process (diamonds) from BDS. τ_D and τ_α are interpolated with a Vogel-Fulcher-Tammann law and the γ-relaxation is well described by an Arrhenius law with an activation energy of 3120 K$\cdot k_B$. Unless explicitly indicated uncertainties are smaller than the size of the symbols

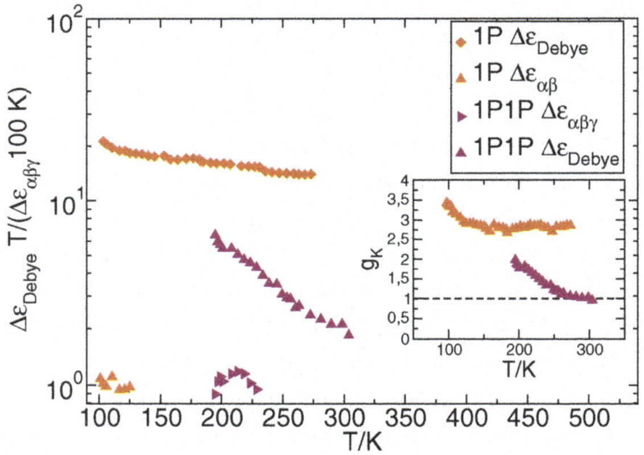

Fig. 16 Comparison of the relative relaxation strength and its temperature dependence of 1-propanol (1P) an 1-phenyl-1-propanol (1P1P). The relaxation strength is multiplied with temperature to remove the trivial Curie temperature dependence and is renormalized to the $\alpha\beta\gamma$-relaxation strength at 100 K. Inset shows the Kirkwood-correlation factor $g_K(T)$ for both substances

broadening of the Debye-peak was recently observed in several alcohols and was investigated by temperature stimulated depolarization current [1].

In order to compare the chains formed in 1-phenyl-1-propanol and 1-propanol in more detail, we examine the relaxation strength of both substances shown in Fig. 16. For easier comparison the trivial Curie temperature dependence of the dielectric relaxation was removed by multiplying the dielectric strength with T and renormalizing it to the $\alpha\beta\gamma$-relaxation strength at 100 K. It turns out that the Debye-Process in

1-propanol is much stronger than in 1-phenyl-1-propanol with $\Delta\varepsilon_D$ showing only a weak temperature dependence until it increases when approaching T_g. As the formation of supramolecular structures is reflected in the intermolecular crosscorrelations, it is worthwhile to consider the Kirkwood correlation factor, which is presented in the inset of Fig. 16. The temperature dependent density necessary to calculate g_K was extrapolated from the data in Ref. [93]. As already discussed in Sect. 2.2, the Kirkwood factor is larger than 1 if linear chains dominate in the formation of structures. This is the case for 1-propanol in the complete temperature range of observation. In 1-phenyl-1-propanol, on the other hand, $g_K \approx 1$ or is even slightly smaller at high temperatures and increases on cooling. This is similar to the behavior of g_K reported by Dannhauser for 5-methyl-3-heptanol [23], where an equilibrium of linear chains and closed-loop structures causes g_K to cross over from values smaller to values larger than one as temperature decreases. In the present case one can envisage a similar behavior, however, with rather small closed-loop structures, like, e.g., pairs at high and short linear chains at low temperatures. How this behavior depends on the exact relative position of OH group and phenyl ring in the molecule is currently under investigation.

4 Example II: Room Temperature Ionic Liquids

Room temperature ionic liquids (RTILs), i.e. molten salts which are liquid at room temperature, attracted wide interest lately because of their unique properties like negligible vapor pressure, high chemical stability, high conductivity and a large electrochemical window, which make them promising for various applications. However, RTILs are found to be also highly interesting from a fundamental perspective. For example, nanostructuring of neat RTILs was observed [18], originating from a complex interplay of polar entities and apolar alkyl-chains. Such nanostructures may influence the molecular dynamics in different ways: If ion-pairs, i.e., electrically neutral species, are formed [27], conductivity may be influenced beyond its simple viscosity dependence. If, on the other hand, alkyl-chains segregate into nanodomains [69], a dynamical signature slower than the processes originating from unsegregated ions is expected.

Whilst measurement of dc-conductivity in RTILs via dielectric spectroscopy is a straightforward task, information about dipolar reorientation is – in contrast to alcohols in the foregoing section – difficult to access. This is due to the fact that any peak in $\epsilon''(\omega)$, eventually originating from dipolar motions, is obscured by the large conductivity contribution and/or electrode polarization effects. Thus, there is no agreement in the literature on how the dielectric spectra of RTILs should be interpreted regarding their molecular origin: In general charge transport is assumed to be the predominant contribution [45, 73, 78], frequently described in the framework of the Dyre model of charge carrier hopping [25], sometimes with an additional process, the origin of which is not clear [39, 40], and is either assigned to electrode polarization [77, 79], or to a Maxwell-Wagner-Sillars polarization due to chain aggregation

[41]. Also, in some cases electrode polarization is thought to be the main process observed in dielectric spectra [74] or rotational dynamics, mostly of the cation, with an additional dc conductivity contribution is considered [22, 83, 97, 105]. The key advantage of studying RTILs with a combination of BDS and DLS lies in the fact that DLS is insensitive to conductivity or polarization effects and only probes reorientational dynamics. In combining the data obtained with both experimental methods it is thus possible to unambiguously identify the reorientational part of the dielectric spectrum. After that the usual comparative analysis of the spectra can be performed and information about the rotational mechanism of the ions and about secondary relaxations can be deduced.

In the first section we describe the procedure of identifying the reorientational part of the dielectric spectrum for a prototypical aprotic short chain RTIL. In the second part we show for the example of an protic ionic liquid, where the dipolar relaxation is less separated from the electrode polarization process, that a model of hopping charge carriers is still useful to evaluate the dielectric spectra. In the third and fourth sections we discuss the implications of the BDS-DLS comparison for the rotational mechanism of the ions and for the secondary relaxations, respectively. Most of this was recently published in Ref. [63].

4.1 Results: Aprotic, Ammonium-Based RTIL

As a first example we show for N,N-Diethyl-N-methyl-N-(2-methoxyethyl)ammonium bis(trifluoromethylsulfonyl)imide (DEME-NTf_2), which is an aprotic RTIL with an ammonium-based cation and the common NTf_2 anion (see Fig. 17), how the dielectric spectrum can be disentangled by combining it with light scattering data.

For two temperatures the dielectric data are displayed in the permittivity representation in Fig. 18. As already mentioned before, $\epsilon''(\omega)$ (full symbols) is rather featureless due to the large conductivity contribution. However, the conductivity free representation of $\epsilon''(\omega)$ can be calculated from $\epsilon'(\omega)$ (to which a dc-conductivity makes no contribution) in good approximation via [11, 104]

$$\epsilon''_{\mathrm{der}}(\omega) = -\frac{\pi}{2}\frac{\mathrm{d}\epsilon'(\omega)}{\mathrm{d}\ln\omega} \tag{21}$$

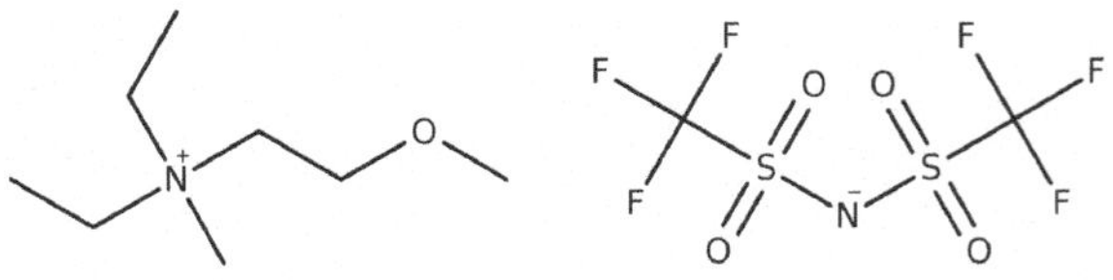

Fig. 17 Chemical structure of N,N-Diethyl-N-methyl-N-(2-methoxyethyl)ammonium bis(trifluoromethylsulfonyl) imide (DEME-NTf_2)

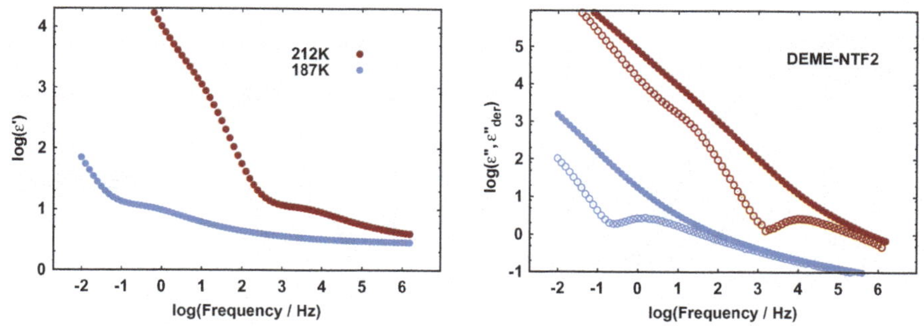

Fig. 18 Dielectric data of DEME-NTF2 in permittivity representation for two temperatures. Real part on the left and imaginary part on the right. Full symbols depict the raw data, open symbols is the conductivity free representation (see Eq. 21)

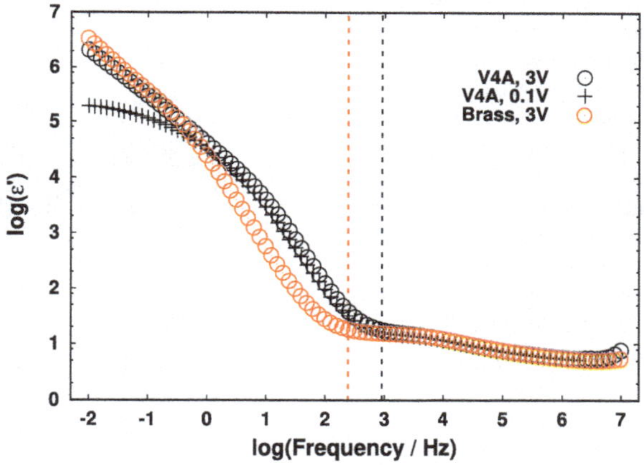

Fig. 19 DEME-NTF2 at 212 K measured with different electrode materials, brass and stainless steel (V4A), and different voltages. Dashed lines indicate the onset of electrode polarization for both materials

$\epsilon''_{\mathrm{der}}(\omega)$ (open symbols) on the other hand exhibits a clear peak on the high frequency side, corresponding to the step in $\epsilon'(\omega)$. We will show in the following how this process can be identified as dipolar reorientation.

But first, we turn to the processes at the low frequency side, which manifest as a tremendous increase in both representations, which is the typical form of electrode polarization. In contrast to interfacial polarization due to structures within the liquid, electrode polarization should be sensitive to the electrode material [84]. The kink in the polarization process is sometimes ascribed to a nonlinearity [98]. Varying both, electrode material and applied voltage, confirms these identifications as can be seen in Fig. 19. The dashed lines correspond to the onset of electrode polarization of the respective electrode material, indicating that the brass cell is more suitable for dielectric measurements of this RTIL because the electrode poalrization is shifted away from the process on the high frequency side in contrast to the stainless steel cell.

We compare the process underlying the step in $\epsilon'(\omega)$ and the peak in $\epsilon''_{\mathrm{der}}(\omega)$, respectively, directly to the dynamic light scattering data, which probes molecular

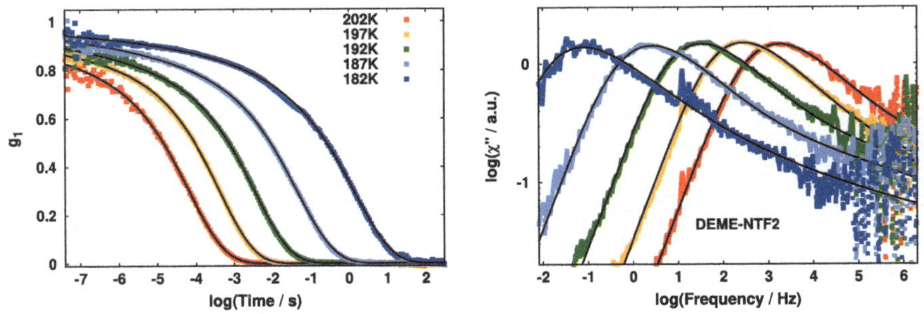

Fig. 20 Left: Correlation decays of DEME-NTF2 measured via PCS for selected temperatures in g_1-representation. Solid lines are fits with a GGE function [63]. Right: Fourier transform and corresponding fit of the time-domain data from the left hand side

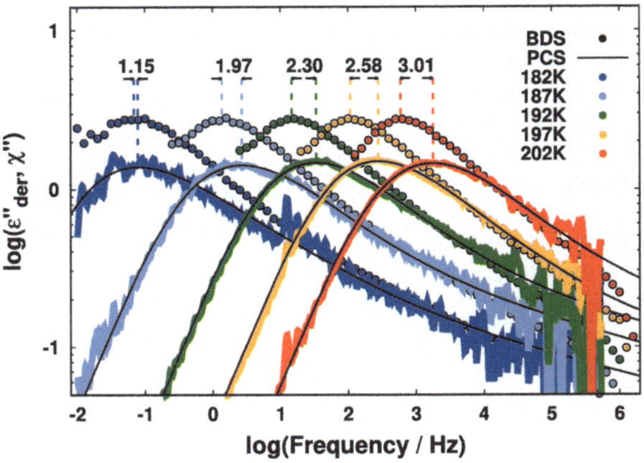

Fig. 21 Combined light scattering (lines) and dielectric spectra (full circles) for selected temperatures. Apart from similar peak shapes the position of the respective peaks is almost identical for low temperatures and differs by a factor of three at higher temperatures. This shift in the peak positions is discussed in Sect. 4.3

reorientations only. This is done for convenience in the frequency domain. Thus, the PCS data are Fourier transformed, as can be seen in Fig. 20 for selected temperatures.

The comparison of the light scattering data $\chi''(\omega)$ (lines) with the conductivity-free representation of the dielectric loss $\epsilon''_{\text{der}}(\omega)$ (full circles) is shown in Fig. 21 for the same temperatures measured in both experimental methods. It can be seen that the shape of the peaks is similar, but more importantly, the position of the peak maxima almost coincides for the lowest temperature and differs at maximum by a factor of 3 for the highest temperature. A thorough discussion of this shift is presented in Sect. 4.3. However, it becomes clear by this comparison that the origin of the dielectric process under consideration is due to reorientational motion, as this is the only process probed by light scattering in the present substance.

When considering molecular reorientation, one might ask which entity is probed, with possible candidates being the cation, the anion or – as proposed also in a recent model [35] – ion pairs. To answer this question, the dipole moment of the reorienting unit is calculated in the following using the Onsager formula and the density at room temperature taken from Ref. [80], which is extrapolated into the supercooled state similar to comparable RTILs [89, 92]. By assuming a separate reorientation of the ions, the same number density of cations and anions and the dipole moment of the anion $\mu_{\text{anion}} = 0.299$ D from the literature, we get the dipole moment of the cation $\mu_{\text{cation}} = (2.04 \pm 0.23)$ D. In comparison to the dipole moment of the most stable and the second most stable conformer of the DEME cation in the gas phase of 2.67 and 2.13 D, respectively, found in a recent simulation [91], this is a very reasonable value, especially when taking the Kirkwood factor g (see Eq. 16) into account, which is found for comparable ILs to lie between 0.78 and 0.88 [82]. This means that reorientation of individual ions is probed, predominantly of the cations due to their higher dipole moment.

Having now identified all processes in the dielectric spectrum, three corresponding regions are marked with roman numbers in Fig. 22: Zone I is the nonlinear part of the electrode polarization. The transition to zone II (linear part of electrode polarization) is marked by a subtle kink in all representations. Zone III is due do dipolar reorientation of the ions. The dashed line indicates that the peak maximum in the conductivity-free representation $\epsilon''_{\text{der}}(\omega)$ coincides with the inflection point in $\epsilon'(\omega)$ and the onset of the plateau, marking the dc-conductivity, in $\sigma'(\omega)$. Thus, it follows for the times assigned to these three processes:

$$\tau_{\epsilon''_{\max}} = \tau_\sigma = \tau_{\epsilon'_i} \tag{22}$$

τ_σ will be subject of the next section, where another RTIL is discussed, for which these three processes are not separated as clearly as in the case of DEME-NTf_2.

4.2 Results: Protic, Imidazolium-Based RTIL

As a second example we consider the protic imidazolium-based RTIL 1-(2-Hydroxyethyl)-3-methylimidazolium tetrafluoroborate (HEMIM-BF4), as depicted in Fig. 23.

As can be seen in Fig. 24, no obvious peak in the conductivity-free representation $\epsilon''_{\text{der}}(\omega)$ appears in zone III, which would possibly mark the dipolar relaxation. This is due to the fact that electrode polarization sets in already at such high frequencies that this process is overshadowed. Thus, a peak comparison of BDS and DLS data, like it was done for DEME-NTf_2 in the foregoing section, is impossible. Therefore a more indirect way has to be pursued, exploiting the fact that the onset of the dc-conductivity plateau is located where the peak position would be in $\epsilon''_{\text{der}}(\omega)$ (see Eq. 22).

A common model for evaluating the conductivity in ion conductors is the Dyre model [76], describing the hopping of ions through a disordered material [25]:

$$\sigma^*(\omega) = \sigma_0 \frac{i\omega\tau_\sigma}{\ln(1 + i\omega\tau_\sigma)} \tag{23}$$

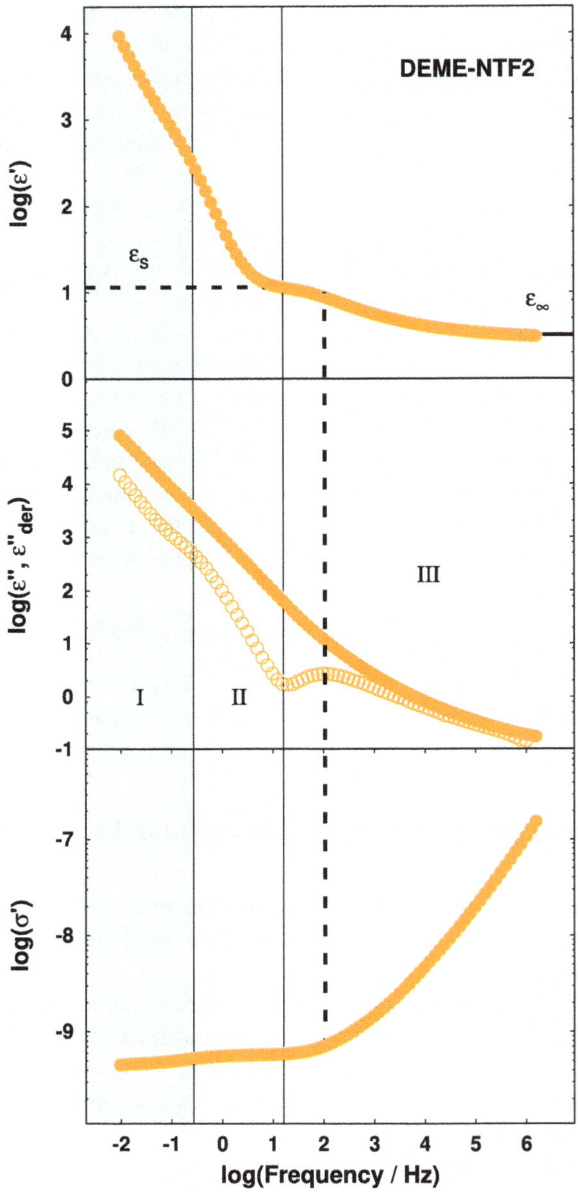

Fig. 22 Dielectric spectra of DEME-NTf$_2$ at 197 K in 3 different representations. Full symbols represent the raw data and open symbols the conductivity free representation $\epsilon''_{der}(\omega)$ calculated via Eq. (21). Zone I and II indicate the nonlinear and linear part of the electrode polarization, respectively. Zone III is due to the reorientational motion of the ions. The vertical dashed line indicates that the peak maximum in $\epsilon''_{der}(\omega)$ coincides with the onset of dc-conductivity, as well as with the inflection point in $\epsilon'(\omega)$, expressed in Eq. (22)

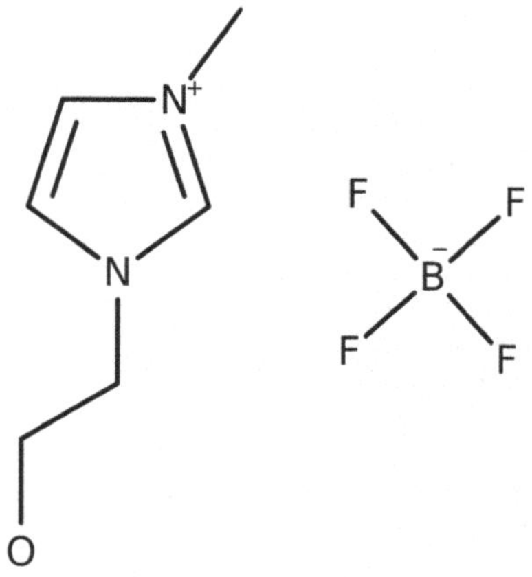

Fig. 23 Chemical structure of 1-(2-Hydroxyethyl)-3-methylimidazolium tetrafluoroborate (HEMIM-BF4)

σ_0 is the dc-conductivity, i.e. the plateau height in $\sigma'(\omega)$, and τ_σ is the characteristic hopping time corresponding to the crossover from the dc-plateau to the ac-conductivity. A fit of Eq. (23) to $\sigma'(\omega)$ is also shown in Fig. 24 as well as a dashed line indicating the characteristic hopping time τ_σ. To show that this time again corresponds to the rotational time, we compare it for all temperatures measured, to DLS times in Fig. 25. It can be seen that the times coincide again for low temperatures, whereas they differ by a factor of approx. 3 at high temperatures, just as it was the case in DEME-NTF2. This means that the dipolar reorientation can be identified in this way even though the peak in $\epsilon''_{\mathrm{der}}(\omega)$ is masked by electrode polarization.

However, these findings also have implications on the dynamical mechanism under consideration: The Dyre model identifies τ_σ with the hopping time of the ions, thus involving translational dynamics, whereas we demonstrated that DLS measurements identify the same time for rotational motion. This implies that translational and rotational motions are strongly coupled in those RTILs. In fact, a recent model proposed by Gainaru et al. [35] states that the conductivity process in ionic liquids takes place through a reorientational step of ions escaping their cage formed by surrounding counterions. This is conceptional in line with our findings here.

4.3 Discussion: The Mechanism of α- and β-Relaxation in RTILs

We now turn to the shift in the relative position of the dielectric and light scattering peak to each other depending on the temperature, already mentioned before. We start with a look at the literature, where similar observations were made for different RTILs [39, 40]. Therein the authors calculated the diffusion constant from τ_σ, so that $D \propto 1/\tau_\sigma$ and τ_α was taken from DLS measurements. They found a fractional Debye-Stokes-Einstein (fDSE) interrelation between diffusion constant and τ_α

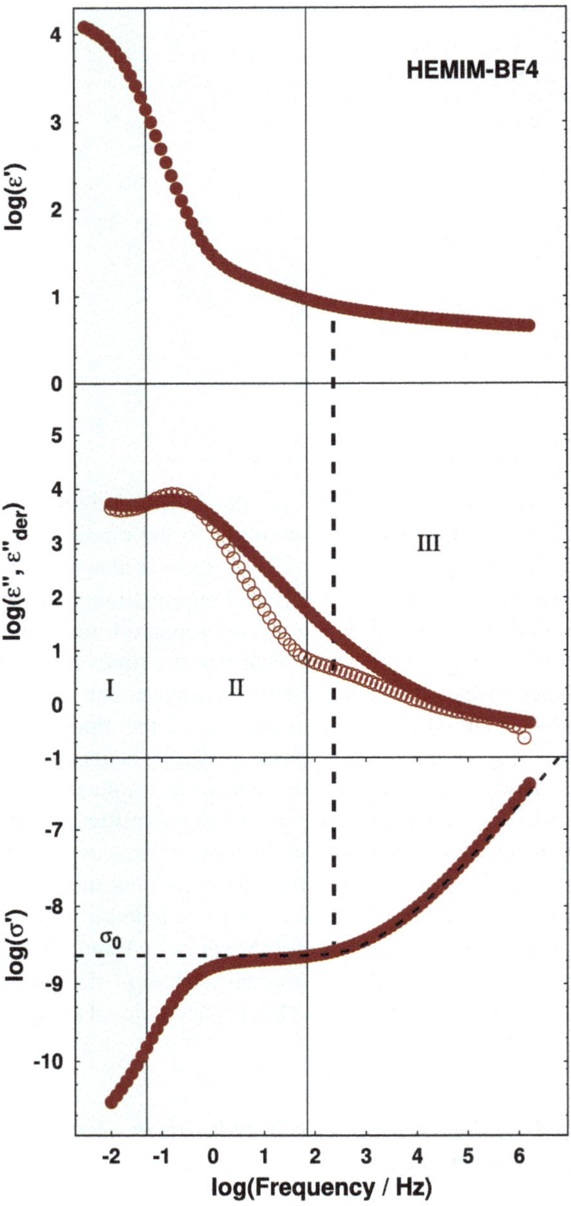

Fig. 24 Dielectric spectra of HEMIM-BF_4 at 210 K in 3 different representations. Full symbols represent the raw data and open symbols the conductivity free representation $\epsilon''_{\mathrm{der}}(\omega)$ calculated via Eq. (21). Zone I and II indicate the nonlinear and linear part of the electrode polarization, respectively. Zone III is due to the reorientational motion of the ions. The fit of Eq. (23) (model of ions hopping through a disordered medium) to the conductivity representation is indicated by a dashed line. Black dashed vertical line indicates the position of τ_σ obtained by this fit. No peak is resolved in $\epsilon''_{\mathrm{der}}(\omega)$ at this position as it was the case in Fig. 22, because here electrode polarization sets in at higher frequencies

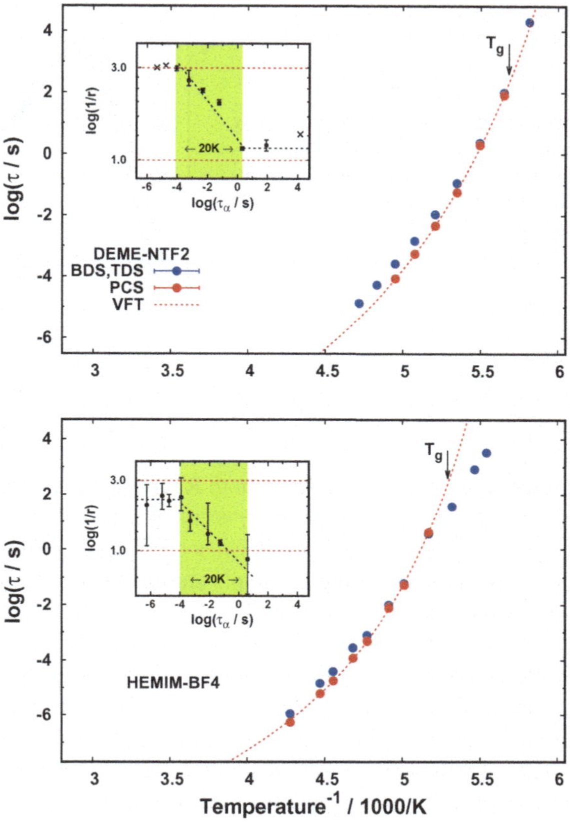

Fig. 25 Time constants $\tau = 1/(2\pi f_{max})$ for DEME-NTf$_2$ (upper panel) and HEMIM-BF$_4$ (lower panel). Data from light scattering is indicated in red and from dielectric spectroscopy in blue. Dashed lines are VFT fits to the LS data. Insets: Ratio τ_1/τ_2 in dependency of τ_2. Crosses mark the ratio of τ_1 to the VFT fit extrapolating the LS data. Reprinted from [63] with permission, © 2017 Elsevier B.V

$$R = D\tau_\alpha \propto \tau_\alpha^\delta \tag{24}$$

whereas they determined the decoupling exponent to be $\delta = 0.10 \pm 0.01$.

Since $D\tau_\alpha \propto \tau_\alpha/\tau_\sigma \propto \tau_\alpha^\delta$, we directly used the characteristic time of the dielectric process to calculate δ. In this way we found for DEME-NTf$_2$ $\delta = 0.10 \pm 0.02$ and for HEMIM-NTf$_2$ $\delta = 0.10 \pm 0.03$, respectively. Thus it seems that the value of 0.10 is universal to a wide variety of RTILs. However, the direct comparison of the dielectric and DLS data allows a more fundamental understanding of the decoupling process than the detour via the diffusion coefficient D: As outlined in Sect. 2.3 dielectric and light scattering correlation functions are proportional to the first and second rank Legendre polynomial, respectively. Thus, the time constants obtained from both methods are interrelated via $1 \leq \frac{\tau_1}{\tau_2} \leq 3$, where the factor 3 is the limit of rotational diffusion, i.e. small angle rotation and the factor 1 is the limit of random large-angle jumps. Based on these considerations the transition from approx. 3 at high temperatures to approx. 1 at low temperatures could be interpreted as a transition from small to large angle rotational steps as the temperature is lowered towards the glass transition temperature. It is noteworthy that this transition takes places within a

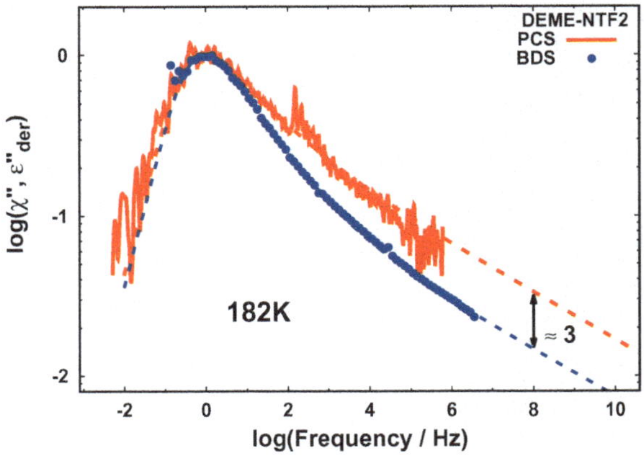

Fig. 26 Master-plot of DEME-NTF$_2$ at 182 K. Dashed lines are GGE fits. The factor three indicates the origin of the high frequency wing may be small angle rotations. Reprinted from [63] with permission, © 2017 Elsevier B.V

temperature range of 20 K for both RTILs and everything points in the direction that outside this range the ratio τ_1/τ_2 is constant near 3 or 1 for high and low temperatures, respectively, as can be seen in the insets of Fig. 25. We note that in the case of HEMIM-NTf$_2$ the dielectric time constants start to deviate from VFT-behavior at T$_g$. This is a known property of protic RTILs, but is beyond the scope of the present discussion.

In the light of the model of Gainaru et al. [35], already introduced in Sect. 4.2, one could evaluate the dielectric and light scattering spectra to find a hint of the small angle movement of the ions in the cage of surrounding counterions prior to their escape. Therefore the high frequency tail of the spectrum is considered, which is often assumed to have its origin in small angle rotations of the molecules [15, 68]. Then, for frequencies $\omega \gg 1/\tau_\alpha$ the spectrum is expected to be three times more intense in $l = 2$ than in $l = 1$ techniques, as already mentioned in Sect. 2.3. It is shown in Fig. 26 that the high frequency wing is indeed approx. three times stronger in the light scattering spectrum ($l = 2$) than in the dielectric spectrum ($l = 1$), pointing towards small angle motion as its origin.

By contrast, for another RTIL not mentioned yet, namely Octylmethylimidazolium bis(trifluoromethylsulfonyl)imide (OMIM-NTf$_2$), we found no deviation of light scattering and dielectric spectra at the high frequency side, as can be seen in Fig. 27. This implies that the high frequency behavior is not universal in RTILs and possibly, the deviations of the dielectric and light scattering spectra in the high frequency regime are not due to the particular motional mechanism involved but due to the fact that one has a two-component system and that a different ratio of the contributions from cations and anions to the polarizability and the dipole moment may be the reason. We note that the derivative performed in calculating the conductivity-free representation $\epsilon''_{\rm der}(\omega)$ (Eq. 21) is expected to be a rather accurate approximation on the high frequency side of a broad peak [104], thus making a misinterpretation due to this aspect of the analysis unlikely. Of course, to clarify the reason for individual deviations in the high frequency behavior of dielectric and light scattering data, further investigations are needed.

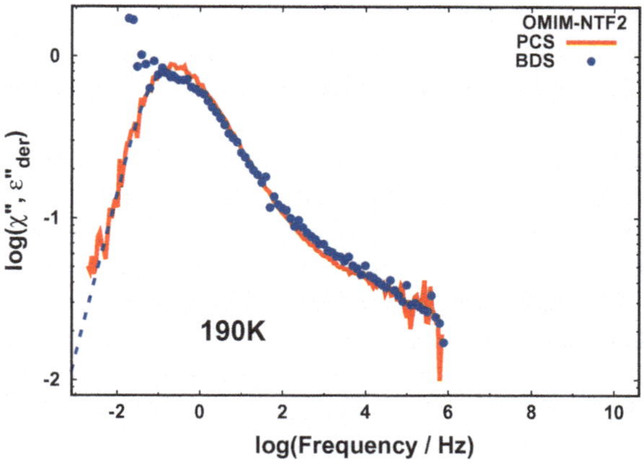

Fig. 27 Master-plot of OMIM-NTf$_2$ at 190 K. The dielectric (BDS) and light scattering data (PCS) coincide also at the high frequency wing, giving no indication of small angle motions prior to the α-process

5 Conclusions

Besides broadband dielectric spectroscopy, which for a long time has been the standard technique to follow glassy dynamics from the sub-picosecond timescale up to several thousand seconds, depolarized dynamic light scattering proves to be a suitable tool to provide broadband information on molecular reorientation in a comparable time window by combining different light scattering techniques, in particular photon correlation spectroscopy and Tandem Fabry-Perot interferometry. This holds true even for molecules with comparatively weak anisotropic scattering, if the appropriate measures are taken to stabilize the photon correlation setup with respect to vibrations and if partially heterodyne scattering, which is inevitable in such cases, is properly taken into account. Concerning the Tandem Fabry Perot interferometer an additional bandpass filter has to be applied to suppress higher transmission orders and prevent the spectral folding effect. When in addition multispeckle spectroscopy is applied for very long times and at temperatures even below the glass transition, then the dynamic susceptibility and corresponding correlation functions can reliably be obtained from the sub picoseconds timescale up to roughly 10^5 s.

However, as the optical anisotropy and the permanent dipole moment are different molecular properties, which are probed in each respective experiment, DLS does not simply represent an alternative route to obtain the same information about molecular reorientation, but substantial additional information can be gained by a detailed comparison of data sets from both techniques. The reasons include that optical anisotropy and permanent dipole moment may be situated at different locations in the molecule, that a different order Legendre polynomial correlation function is probed by each method, that intermolecular cross-correlations play a different role in both techniques or that light scattering is insensitive to free charges in the sample, which proves particularly useful in ionic systems.

In the present chapter we have given examples for exploring these possibilities. One example focuses on the sensitivity of such a comparison of methods for the underlying motional mechanism of an observed relaxation process and addresses the nature of the Johari Goldstein β-relaxation in several monohydroxy alcohols. From the shape of the susceptibility of α and β-relaxation and a comparison of relative amplitudes one can conclude that in the discussed alcohols in the supercooled liquid state the motional mechanism for the JG relaxation must be closer to large angle random jumps than to small angle reorientation of all molecules. Thus, in these systems the picture of islands of mobility as originally suggested by Johari and Goldstein is favored over that of small angle reorientation.

A second aspect of the relaxation behavior in monoalcohols is the effect of cross correlations in BDS and DLS. Here we consider the Debye relaxation, which is observed in the dielectric spectra of many monoalcohols and is usually attributed to the formation of transient supramolecular chains. Clearly, for supramolecular structures cross correlations become important. It turns out that, in contrast to α- and β-process, the Debye peak appears to be very different in DLS and BDS measurements. For example, in the case of the primary alcohol 1-propanol a Debye contribution does not appear at all in the DLS spectra, due to the comparatively large mobility of the alkyl tail, while in a secondary alcohol the Debye contribution is evident also in the DLS data. While in 1-propanol a strong Debye contribution is seen in BDS, the combination of DLS and BDS results is used in the case of 1-phenyl-1-propanol to identify a small Debye process in a primary alcohol, where an additional phenyl ring acts as a steric hindrance for the chain formation. Here, only rather short chains are formed and the reorientation of the end-to-end vector is close to the α-relaxation time and a spectral broadening of the Debye contribution indicates that the reorienting supramolecular structures locally probe a heterogeneous environment. Thus, the picture of transient chains as source for the Debye relaxation is confirmed and the combination of DLS and BDS provides details on the chain formation on a molecular level.

Finally, in the presence of electric charges where conductivity and charge polarization effects dominate the dielectric spectra, as is the case for room temperature ionic liquids, the contribution of molecular reorientation can unambiguously be identified and the relation between charge transport and molecular reorientation can be elucidated by comparing DLS and BDS data. It turns out that in agreement with previous findings in case of simple protic and aprotic ionic liquids considered in our example, rotation and charge transport are strongly coupled and that a change in the motional mechanism can be identified on cooling. At the same time a clear statement concerning the secondary processes remains difficult, because contributions of each ionic species may play a different role in both experiments, which requires further investigations.

Overall, a combination of BDS and DLS proves to be a powerful tool to provide detailed insight into the dynamics in complex glass forming systems by using the fact that both methods provide a truly different perspective on molecular reorientation. When certain experimental requirements are fulfilled, this approach not only yields a more complete picture of the scaling properties of the dynamic processes involved, but also sheds light onto the underlying molecular mechanisms. Thus it is worthwhile to apply the present approach to an even wider variety of complex materials, which will be the subject of future work.

Acknowledgements We cordially thank Ernst Rößler, Bayreuth, for providing the data on 1-propanol from Refs. [54, 55] and for making the dielectric time domain setup and the Tandem Fabry-Perot interferometer available to us. We are grateful to Catalin Gainaru, Dortmund, for stimulating discussions about alcohols and ionic liquids. Financial support by the Deutsche Forschungsgemeinschaft under Grant No. BL 923/1 and within FOR 1583 under Grant No. BL 1192/1 is gratefully acknowledged.

References

1. Arrese-Igor S, Alegria A, Colmenero J (2017) On the non-exponentiality of the dielectric Debye-like relaxation of monoalcohols. J Chem Phys 146(11):114,502
2. Barshilia HC, Li G, Shen GQ, Cummins HZ (1999) Depolarized light scattering spectroscopy of $Ca_{0.4}K_{0.6}(NO_3)_{1.4}$: A reexamination of the "knee". Phys Rev E 59:5625
3. Bartsch E, Frenz V, Kirsch S, Schärtl W, Sillescu H (1997) Multi-speckle autocorrelation spectroscopy - a new strategy to monitor ultraslow dynamics in dense and nonergodic media. Prog Coll Polym Sci 104:40–48
4. Battaglia MR, Cox TI, Madden PA (1979) The orientational correlation parameter for liquid CS_2, C_6H_6 and C_6F_6. Mol Phys 37:1413–1427
5. Bee M (1988) Quasielastic neutron scattering. Adam Hilger, Bristol
6. Berne BJ, Pecora R (1976) Dynamic light scattering. Wiley, New York
7. Blochowicz T, Kudlik A, Benkhof S, Senker J, Rössler E, Hinze G (1999) The spectral density in simple organic glassformers: Comparison of dielectric and spin-lattice relaxation. J Chem Phys 110(24):12,011–12,022
8. Blochowicz T, Tschirwitz C, Benkhof S, Rössler EA (2003) Susceptibility functions for slow relaxation processes in supercooled liquids and the search for universal relaxation patterns. J Chem Phys 118(16):7544–7555
9. Blochowicz T, Brodin A, Rössler EA (2006) Evolution of the dynamic susceptibility in supercooled liquids and glasses. Adv Chem Phys 133:127–256
10. Blochowicz T, Gouirand E, Schramm S, Stühn B (2013) Density and confinement effects of glass forming m-toluidine in nanoporous Vycor investigated by depolarized dynamic light scattering. J Chem Phys 138(114):501
11. Böttcher CJF, Bordewijk P (1978) Theory of electric polarization II: dielectrics in time-dependent fields. Elsevier, Amsterdam, London, New York
12. Bremer LGB, Deriemaeker L, Finsy R, Gelade E, Joosten JGH (1993) Fiber optic dynamic light scattering, neither homodyne nor heterodyne. Langmuir 9(8):2008–2014
13. Brodin A, Rössler EA (2005) Depolarized light scattering study of glycerol. Eur Phys J B 44(1):3–14
14. Brodin A, Bergman R, Mattsson J, Rössler EA (2003a) Light scattering and dielectric manifestations of secondary relaxations in molecular glassformers. Eur Phys J B 36:349–357
15. Brodin A, Bergman R, Mattsson J, Rössler EA (2003b) Light scattering and dielectric manifestations of secondary relaxations in molecular glassformers. Eur Phys J B-Condensed Matter Complex Syst 36(3):349–357
16. Böhmer R, Gainaru C, Richert R (2014) Structure and dynamics of monohydroxy alcohols - milestones towards their microscopic understanding, 100 years after Debye. Phys Rep 545:125–195
17. Callen HB, Welton TA (1951) Irreversibility and generalized noise. Phys Rev 83(1):34–40
18. Caminiti R, Gontrani L (2014) The structure of ionic liquids, vol 193. Springer

19. Cole RH (1965) Correlation function theory of dielectric relaxation. J Chem Phys 42:637
20. Cummins HZ, Li G, Du W, Pick RM, Dreyfus C (1996) Origin of depolarized light scattering in supercooled liquids: orientational fluctuation versus induced scattering mechanisms. Phys Rev E 53:896–904
21. Cummins HZ, Li G, Hwang YH, Shen GQ, Du WM, Hernandez J, Tao NJ (1997) Dynamics of supercooled liquids and glasses: comparison of experiments with theoretical predictions. Z Phys B Condens Mat 103:501–519
22. Daguenet C, Dyson PJ, Krossing I, Oleinikova A, Slattery J, Wakai C, Weingärtner H (2006) Dielectric response of imidazolium-based room-temperature ionic liquids. J Phys Chem B 110(25):12,682–12,688
23. Dannhauser W (1968) Dielectric study of intermolecular association in isomeric octyl alcohols. J Chem Phys 48:1911
24. Debye P (1929) Polar Mol. Chemical Catalog Co., New York
25. Dyre JC (1988) The random free-energy barrier model for ac conduction in disordered solids. J Appl Phys 64(5):2456–2468
26. Fatuzzo E, Mason PR (1967) A theory of dielectric relaxation in polar liquids. Proc Phys Soc London 90:741
27. Fraser KJ, Izgorodina EI, Forsyth M, Scott JL, MacFarlane DR (2007) Liquids intermediate between "molecular" and "ionic" liquids: Liquid ion pairs? Chem Commun 37:3817–3819
28. Fröhlich H (1958) Theory of dielectrics. Clarendon Press, Oxford
29. Fytas G (1989) Relaxation processes in amorphous poly(cyclohexyl methacrylate) in the rubbery and glassy state studied by photon correlation spectroscopy. Macromolecules 22:211–215
30. Gabriel J, Blochowicz T, Stühn B (2015) Compressed exponential decays in correlation experiments: the influence of temperature gradients and convection. J Chem Phys 142(10):104,902
31. Gabriel J, Pabst F, Blochowicz T (2017) Debye-process and β-relaxation in 1-propanol probed by dielectric spectroscopy and dynamic light scattering. J Phys Chem B 121:8847–8853
32. Gabriel J, Pabst F, Helbling A, Böhmer T, Blochowicz T (2018) On the nature of the Debye-process in monohydroxy alcohols: 5-methyl-2-hexanol investigated by depolarized light scattering and dielectric spectroscopy. arXiv:1805.04664
33. Gainaru C, Meier R, Schildmann S, Lederle C, Hiller W, Rössler EA, Böhmer R (2010) Nuclear-magnetic-resonance measurements reveal the origin of the Debye process in monohydroxy alcohols. Phys Rev Lett 105(258):303
34. Gainaru C, Figuli R, Hecksher T, Jakobsen B, Dyre JC, Wilhelm M, Böhmer R (2014) Shear-modulus investigations of monohydroxy alcohols: evidence for a short-chain-polymer rheological response. Phys Rev Lett 112(098):301
35. Gainaru C, Stacy EW, Bocharova V, Gobet M, Holt AP, Saito T, Greenbaum S, Sokolov AP (2016) Mechanism of conductivity relaxation in liquid and polymeric electrolytes: direct link between conductivity and diffusivity. J Phys Chem B 120(42):11,074–11,083
36. Gapinski J, Steffen W, Patkowski A, Sokolov AP, Kisliuk A, Buchenau U, Russina M, Mezei F, Schober H (1999) Spectrum of fast dynamics in glass forming liquids: Does the "knee" exist? J Chem Phys 110:2312
37. Glarum SH (1960) Dielectric relaxation of polar liquids. J Chem Phys 33:1371
38. Goldstein M (1969) Viscous liquids and the glass transition: a potential energy barrier picture. J Chem Phys 51(9):3728–3739
39. Griffin P, Agapov AL, Kisliuk A, Sun XG, Dai S, Novikov VN, Sokolov AP (2011) Decoupling charge transport from the structural dynamics in room temperature ionic liquids. J Chem Phys 135(11):114,509
40. Griffin PJ, Agapov AL, Sokolov AP (2012) Translation-rotation decoupling and nonexponentiality in room temperature ionic liquids. Phys Rev E 86(2):21,508
41. Griffin PJ, Holt AP, Wang Y, Novikov VN, Sangoro JR, Kremer F, Sokolov AP (2014) Interplay between hydrophobic aggregation and charge transport in the ionic liquid methyltrioctylammonium bis (trifluoromethylsulfonyl) imide. J Phys Chem B 118(3):783–790
42. Hansen C, Stickel F, Berger T, Richert R, Fischer EW (1997) Dynamics of glass-forming liquids. iii. comparing the dielectric alpha- and beta-relaxation of 1-propanol and o-terphenyl. J Chem Phys 107(4):1086–1093

43. Hansen JP, McDonald IR (1976) Theory of simple liquids. Academic Press, London, New York
44. Hansen JS, Kisliuk A, Sokolov AP, Gainaru C (2016) Identification of structural relaxation in the dielectric response of water. Phys Rev Lett 116(23):237601
45. Hensel-Bielowka S, Wojnarowska Z, Dzida M, Zorebski E, Zorebski M, Geppert-Rybczyska M, Peppel T, Grzybowska K, Wang Y, Sokolov AP, Paluch M (2015) Heterogeneous nature of relaxation dynamics of room-temperature ionic liquids $(EMIm)_2[Co(NCS)_4]$ and $(BMIm)_2[Co(NCS)_4]$. J Phys Chem C 119(35):20,363–20,368
46. www.chemaxon.com, copyright (2017) Marvinsketch 17.8.0
47. Johari G (2002) Localized molecular motions of β-relaxation and its energy landscape. J Non-Cryst Solids 307–310:317–325
48. Johari GP (1976) Glass transition and secondary relaxations in molecular liquids and crystals. Annals of the N Y Acad Sci 279:117–140
49. Johari GP, Dannhauser W (1968) Dielectric study of intermolecular association in sterically hindered octanol isomers. J Phys Chem 72(9):3273–3276
50. Johari GP, Dannhauser W (1972) Effect of pressure on dielectric polarization of 1-phenyl-1-propanol. High Temp-High Press 4(2):199–206
51. Johari GP, Goldstein M (1970) Viscous liquids and the glass transition II: secondary relaxations in glasses of rigid molecules. J Chem Phys 53(6):2372–2388
52. Johari GP, Kalinovskaya OE, Vij JK (2001) Effects of induced steric hindrance on the dielectric behavior and h bonding in the supercooled liquid and vitreous alcohol. J Chem Phys 114(10):4634–4642
53. Kirkwood JG (1939) The dielectric polarization of polar liquids. J Chem Phys 7:911
54. Kudlik A, Tschirwitz C, Benkhof S, Blochowicz T, Rössler E (1997) Slow secondary relaxation processes in supercooled liquids. Europhys Lett 40(6):649–654
55. Kudlik A, Benkhof S, Blochowicz T, Tschirwitz C, Rössler E (1999) The dielectric response of simple organic glass formers. J Mol Struct 479:201–218
56. Köhler M, Lunkenheimer P, Goncharov Y, Wehn R, Loidl A (2010) Glassy dynamics in mono-, di- and tri-propylene glycol: from the α- to the fast β-relaxation. J Non-Cryst Solids 356:529–534
57. Lebon MJ, Dreyfus C, Guissani Y, Pick RM, Cummins HZ (1997) Light scattering and dielectric susceptibility spectra of glassforming liquids. Z Phys B 103(3–4):433–439
58. Lindsay SM, Anderson MW, Sandercock JR (1981) Construction and alignment of a high performance multipass vernier tandem fabry-perot interferometer. Rev Sci Instruments 52:1478
59. Lunkenheimer P, Schneider U, Brand R, Loidl A (2000) Glassy dynamics. Contemporary Phys 41(1):15–36
60. Madden P, Kivelson D (1984) A consistent molecular treatment of dielectric phenomena. Adv Chem Phys 56:467
61. Malo de Molina P, Alvarez F, Frick B, Wildes A, Arbe A, Colmenero J (2017) Investigation of the dynamics of aqueous proline solutions using neutron scattering and molecular dynamics simulations. Phys Chem Chem Phys 19(27):739
62. Mopsik FI (1984) Precision time-domain dielectric spectrometer. Rev Sci Instrum 55:79
63. Pabst F, Gabriel J, Weigl P, Blochowicz T (2017) Molecular dynamics of supercooled ionic liquids studied by light scattering and dielectric spectroscopy. Chem Phys 494:103–110
64. Pardo LC, Lunkenheimer P, Loidl A (2007) Dielectric spectroscopy in benzophenone: the β-relaxation and its relation to the mode-coupling Cole-Cole peak. Phys Rev E 76(030):502(R)
65. Patkowski A, Steffen W, Nilgens H, Fischer EW, Pecora R (1997) Depolarized dynamic light scattering from three low molecular weight glass forming liquids: a test of the scattering mechanism. J Chem Phys 106:8401
66. Pecora R (ed) (1985) Dynamic light scattering - applications of photon correlation spectroscopy. Plenum, New York
67. Petzold N, Rössler EA (2010) Light scattering study on the glass former o-terphenyl. J Chem Phys 133(124):512
68. Petzold N, Schmidtke B, Kahlau R, Bock D, Meier R, Micko B, Kruk D, Rössler EA (2013) Evolution of the dynamic susceptibility in molecular glass formers: results from light scattering, dielectric spectroscopy, and NMR. J Chem Phys 138:12A510

69. Pott T, Méléard P (2009) New insight into the nanostructure of ionic liquids: a small angle X-ray scattering (SAXS) study on liquid tri-alkyl-methyl-ammonium bis (trifluoromethanesulfonyl) amides and their mixtures. Phys Chem Chem Phys 11(26):5469–5475
70. Power G, Johari GP, Vij JK (2003) Relaxation strength of localized motions in d-sorbitol and mimicry of glass-softening thermodynamics. J Chem Phys 119(1):435–442
71. Power G, Vij JK, Johari GP (2006) Orientation polarization from faster motions in the ultraviscous and glassy diethyl phthalate and its entropy. J Chem Phys 124(4):044,513
72. Rivera A, Blochowicz T, Gainaru C, Rössler EA (2004) Spectral response from modulus time domain data of disordered materials. J Appl Phys 96(10):5607–5612
73. Rivera A, Brodin A, Pugachev A, Rössler EA (2007) Orientational and translational dynamics in room temperature ionic liquids. J Chem Phys 126(11):114,503
74. Samet M, Levchenko V, Boiteux G, Seytre G, Kallel A, Serghei A (2015) Electrode polarization versus Maxwell-Wagner-Sillars interfacial polarization in dielectric spectra of materials: characteristic frequencies and scaling laws. J Chem Phys 142(194):703
75. Sangoro J, Iacob C, Serghei A, Friedrich C, Kremer F (2009) Universal scaling of charge transport in glass-forming ionic liquids. Phys Chem Chem Phys 11(6):913–916
76. Sangoro JR, Kremer F (2011) Charge transport and glassy dynamics in ionic liquids. Acc Chem Res 45(4):525–532
77. Sangoro JR, Serghei A, Naumov S, Galvosas P, Kärger J, Wespe C, Bordusa F, Kremer F (2008) Charge transport and mass transport in imidazolium-based ionic liquids. Phys Rev E 77(5):51,202
78. Sangoro JR, Iacob C, Serghei A, Friedrich C, Kremer F (2009) Universal scaling of charge transport in glass-forming ionic liquids. Phys Chem Chem Phys 11(6):913–916
79. Sangoro JR, Iacob C, Naumov S, Valiullin R, Rexhausen H, Hunger J, Buchner R, Strehmel V, Kärger J, Kremer F (2011) Diffusion in ionic liquids: the interplay between molecular structure and dynamics. Soft Matter 7(5):1678–1681
80. Sato T, Masuda G, Takagi K (2004) Electrochemical properties of novel ionic liquids for electric double layer capacitor applications. Electrochim Acta 49(21):3603–3611
81. Schmidtke B, Petzold N, Kahlau R, Rössler EA (2013) Reorientational dynamics in molecular liquids as revealed by dynamic light scattering: from boiling point to glass transition temperature. J Chem Phys 139(084):504
82. Schröder C, Rudas T, Steinhauser O (2006) Simulation studies of ionic liquids: orientational correlations and static dielectric properties. J Chem Phys 125(24):244,506
83. Schröder C, Hunger J, Stoppa A, Buchner R, Steinhauser O (2008) On the collective network of ionic liquid/water mixtures. ii. decomposition and interpretation of dielectric spectra. J Chem Phys 129(184):501
84. Serghei A, Tress M, Sangoro JR, Kremer F (2009) Electrode polarization and charge transport at solid interfaces. Phys Rev B 80(184):301
85. Siegert AJF (1943) On the fluctuations in signals returned by many independently moving scatterers. MIT Rad Lab Rep 465:1–14
86. Sillren P, Bielecki J, Mattsson J, Borjesson L, Matic A (2012) A statistical model of hydrogen bond networks in liquid alcohols. J Chem Phys 136(9):094514
87. Sillren P, Swenson J, Mattsson J, Bowron D, Matic A (2013) The temperature dependent structure of liquid 1-propanol as studied by neutron diffraction and EPSR simulations. J Chem Phys 138(21):214501
88. Singh LP, Richert R (2012) Watching hydrogen-bonded structures in an alcohol convert from rings to chains. Phys Rev Lett 109(16):167802
89. Song D, Chen J (2014) Density and viscosity data for mixtures of ionic liquids with a common anion. J Chem Eng Data 59(2):257–262
90. Surovtsev NV, Wiedersich JAH, Novikov VN, Rössler E, Sokolov AP (1998) Light-scattering spectra of fast relaxation in glasses. Phys Rev B 58(14):888
91. Takekiyo T, Imai Y, Abe H, Yoshimura Y (2012) Conformational analysis of quaternary ammonium-type ionic liquid cation, N, N-diethyl-N-methyl-N-(2-methoxyethyl) ammonium cation. Adv Phys Chem 2012:829523

92. Tariq M, Forte PAS, Gomes MFC, Lopes JNC, Rebelo LPN (2009) Densities and refractive indices of imidazolium- and phosphonium-based ionic liquids: effect of temperature, alkyl chain length, and anion. J Chem Thermodyn 41(6):790–798
93. Vercher E, Orchillés AV, Miguel PJ, Martínez-Andreu A (2007) Volumetric and ultrasonic studies of 1-ethyl-3-methylimidazolium trifluoromethanesulfonate ionic liquid with methanol, ethanol, 1-propanol, and water at several temperatures. J Chem Eng Data 52(4):1468–1482
94. Vogel M, Rössler E (2000) Effects of various types of molecular dynamics on 1D and 2D ^{2}H NMR studied by random walk simulations. J Mag Res 147(1):43–58
95. Vogel M, Rössler E (2000) On the nature of slow β-process in simple glass formers: A ^{2}H NMR study. J Phys Chem B 104:4285–4287
96. Vogel M, Tschirwitz C, Schneider G, Koplin C, Medick P, Rössler E (2002) A 2h nmr and dielelectric spectroscopy study of the slow beta-process in organic glass formers. J Non-Cryst Solids 307–310:326–335
97. Wakai C, Oleinikova A, Ott M, Weingärtner H (2005) How polar are ionic liquids? Determination of the static dielectric constant of an imidazolium-based ionic liquid by microwave dielectric spectroscopy. J Phys Chem B 109(36):17,028–17,030
98. Wang Y, Sun CN, Fan F, Sangoro JR, Berman MB, Greenbaum SG, Zawodzinski TA, Sokolov AP (2013) Examination of methods to determine free-ion diffusivity and number density from analysis of electrode polarization. Phys Rev E 87(4):42,308
99. Wang Y, Griffin PJ, Holt A, Fan F, Sokolov AP (2014) Observation of the slow, Debye-like relaxation in hydrogen-bonded liquids by dynamic light scattering. J Chem Phys 140(10):104,510
100. Weingärtner H, Nadolny H, Oleinikova A, Ludwig R (2004) Collective contributions to the dielectric relaxation of hydrogen-bonded liquids. J Chem Phys 120(11):692
101. Williams G (1972) Use of the dipole correlation function in dielectric relaxation. Chem Rev 72:55–69
102. Williams G, Watts DC (1971) Analysis of molecular motion in the glassy state. Trans Farad Soc 67:1971–1989
103. Williams G, Cook M, Hains PJ (1972) Molecular motion in amorphous polymers. J Chem Soc: Farad Trans II 68:1045–1050
104. Wübbenhorst M, van Turnhout J (2002) Analysis of complex dielectric spectra. I. One-dimensional derivative techniques and three-dimensional modelling. J Non Cryst Solids 305(1–3):40–49
105. Wulf A, Ludwig R, Sasisanker P, Weingärtner H (2007) Molecular reorientation in ionic liquids: a comparative dielectric and magnetic relaxation study. Chem Phys Lett 439(4):323–326
106. Young-Gonzales AR, Richert R (2016) Field induced changes in the ring/chain equilibrium of hydrogen bonded structures: 5-methyl-3-heptanol. J Chem Phys 145(7):

Relaxation Processes in Liquids and Glass-Forming Systems: What Can We Learn by Comparing Neutron Scattering and Dielectric Spectroscopy Results?

Arantxa Arbe and Juan Colmenero

Abstract In this chapter, we first introduce the main concepts related to quasielastic neutron scattering (QENS) techniques and the way they can be connected to dielectric spectroscopy (DS). This is not obvious, because they access different correlation functions. The dielectric permittivity measured by DS reflects the orientational dynamics of the molecular dipoles in a very broad temperature/frequency range, while, thanks to the transfer of energy ($\hbar\omega$) and momentum ($\hbar Q$) dependence of the measured intensities, QENS provides information about nuclear positions with space/time resolution. In particular, QENS on protonated samples follows the self-correlation function of the hydrogens. Next, we describe the general findings from both techniques relative to the α-relaxation in glass-forming systems. From the comparison of the results, we define a Q-value ($Q^\star$) at which the timescale of the α-process measured by QENS and DS become similar and compile its values from the literature for diverse systems ranging from polymers and low-molecular weight glass-forming systems to water and water solutions. The results are discussed in a phenomenological way in terms of structural and dynamic parameters. Thereafter, we show that in the case of a simple diffusive process, a simple approach based on molecular hydrodynamics and a molecular treatment of DS allows expressing $Q^\star$ in terms of a many-body magnitude—a generalized Kirkwood parameter—and a single-molecule magnitude—the hydrodynamic radius. The application of these ideas to liquid water and water solutions is presented. Finally, we explore the possibility of extending this kind of treatment to the more complex subdiffusive case.

A. Arbe · J. Colmenero
Centro de Física de Materiales (CFM) (CSIC–UPV/EHU) – Materials Physics Center (MPC), Paseo Manuel de Lardizabal 5, 20018 San Sebastián, Spain
e-mail: a.arbe@ehu.eus

J. Colmenero (✉)
Departamento de Física de Materiales, UPV/EHU, and Donostia International Physics Center, Apartado 1072, 20080 San Sebastián, Spain
e-mail: juan.colmenero@ehu.eus

F. Kremer and A. Loidl (eds.), *The Scaling of Relaxation Processes*,
Advances in Dielectrics, https://doi.org/10.1007/978-3-319-72706-6_8

1 Introduction

The correlation functions observed by quasielastic neutron scattering (QENS) and dielectric spectroscopy (DS) are rather different. QENS techniques allow one access to either the density–density correlation function, when coherent scattering is measured, or the self-correlation function of the atoms in the incoherent case. On the other hand, DS observes the frequency-dependent complex dielectric permittivity $\varepsilon^{\star}(\omega)$, which reflects the orientational dynamics of the molecular dipoles in the sample. Moreover, in contrast to DS, QENS provides space/time resolution and a direct microscopic interpretation of the results through the well-known van Hove formalism. Therefore, when a given relaxation process is observed by QENS or by DS, different information is obtained. However, both methods should reflect—perhaps in a different way—the same underlying physics.

In the case of the glass-forming dynamics, incoherent QENS results of many different systems well above their glass-transition temperatures show a clear dependence of the relaxation time on the momentum transfer Q, implying that the main dynamical process—the so-called α-relaxation—as it is measured by QENS, is a diffusive-like process. On the other hand, taking into account the cases where the α-relaxation has been investigated by QENS and DS, it seems that the time-domain dielectric relaxation function and the incoherent scattering function show a similar non-exponential (non-Debye) shape, at least in the high-frequency/temperature range where both techniques overlap. Moreover, the temperature dependence of the relaxation time of these two functions is also almost the same in such a dynamic range. Then, in principle, it is always possible to find a Q-value (we will call it $Q^{\star}$) at which the timescale of the α-process measured by QENS and DS become similar. In the Gaussian approximation, $Q^{\star}$ also allows easily calculating the mean squared atomic displacement at the timescale of the dielectric α-relaxation $[\langle r^2(\tau_{DS})\rangle]$. In this chapter, we first introduce the main concepts related to QENS techniques and the way they can be connected to DS. Thereafter, we describe the general findings from both techniques relative to the α-relaxation in glass-forming systems and give a compilation of the values of $Q^{\star}$ for many different systems, ranging from water and water solutions to low molecular weight glass-forming systems and polymers. The results are discussed in a phenomenological way in terms of structural and dynamic parameters of the different systems. Second, we show that in the case of a simple diffusive process, a simple approach based on molecular hydrodynamics and a molecular treatment of DS allows expressing $Q^{\star}$ in terms of a many-body magnitude—a generalized Kirkwood parameter—and a single-molecule magnitude—the hydrodynamic radius. We present the recent application of these ideas to liquid water and water solutions. The possibility of extending this kind of treatment to the more complex subdiffusive case is finally explored.

2 Quasielastic Neutron Scattering

2.1 Neutron Cross Sections and Correlation Functions

Neutrons interact with the atomic nuclei [78, 105]. Since the involved interactions are important only at very short distances ($|\mathbf{r}| \approx 10^{-14}$m), they can be approximately described by the so-called Fermi pseudopotential

$$V(\mathbf{r}) = -\frac{2\pi\hbar^2}{m} b_\alpha \delta(\mathbf{r}). \tag{1}$$

Here m is the mass of the neutron and b_α is the scattering length, that can be positive, negative or complex. It depends on the isotope α considered (α: H, D, C, O, ...) and the relative orientation of the neutron–nuclear spin pairs. Table 1 shows the mean values of b_α for isotopes commonly found in polymers and other glass-forming materials.

In neutron scattering experiments, we measure cross sections. The double-differential scattering cross section $\partial^2\sigma/\partial\Omega\,\partial\hbar\omega$ is the number of neutrons scattered into a solid angle comprised between Ω and $\Omega + d\Omega$ and which have experienced a change in energy $\hbar\omega$, with respect to the total number of incident neutrons [78]. If the energy exchanged is zero or smaller than the energy resolution of the spectrometer, the scattering event is considered to be elastic. If it is not zero and the neutron has led to an excitation in the sample, the scattering is inelastic; otherwise, it is known as quasielastic neutron scattering (QENS). QENS signal manifests as a broadening around the elastic line and can reflect a variety of motions. On the other hand, the difference between the wavevectors of the scattered ($\mathbf{k}$) and incident ($\mathbf{k_o}$) neutron determines the scattering vector $\mathbf{Q}$ (equivalently, the momentum transfer $\hbar\mathbf{Q}$). In elastic scattering events or if the energy transfer is very small, the modulus of the scattering vector is given by $Q = 2k_o \sin(\theta/2)$ (θ: scattering angle). The double-differential scattering cross section is determined by the (time dependent) distribution of scattering centers in the sample. It can be expressed as (see, e.g., [78]):

$$\frac{\partial^2\sigma}{\partial\Omega\,\partial\hbar\omega} \propto \sum_{\alpha,\beta} \overline{b}_\alpha \overline{b}_\beta S_{coh}^{\alpha\beta}(\mathbf{Q},\omega) + \sum_\alpha \overline{\Delta b_\alpha^2} S_{inc}^{\alpha}(\mathbf{Q},\omega). \tag{2}$$

Table 1 Values of the average scattering lengths $\overline{b_\alpha}$, their squares $\overline{b_\alpha}^2$, and their deviations $\overline{\Delta b_\alpha^2}$ for different isotopes α

Isotope α	$\overline{b_\alpha}$/fm	$\overline{b_\alpha}^2$/fm^2	$\overline{\Delta b_\alpha^2}$/fm^2
^{1}H	−3.7406	13.992	638.78
^{2}H (D)	6.6710	44.502	16.322
^{12}C	6.6511	44.237	0
^{16}O	5.8030	33.675	0

There, the indexes α and β run over all the possible kinds of isotopes in the sample (α, β: H, D, C, O, ...). In $\partial^2\sigma/\partial\Omega\partial\hbar\omega$ we can identify a coherent and an incoherent contribution. The latter arises from the random distribution of the deviations of the scattering lengths from their mean value, $\overline{\Delta b_\alpha^2} = \overline{b_\alpha^2} - \overline{b_\alpha}^2$. From Table 1 we can see that for hydrogen, the value of $\overline{\Delta b_H^2}$ weighing the incoherent contribution to the scattered intensity is huge compared to that of any other common isotope, as well as to the product $\overline{b}_\alpha\overline{b}_\beta$ of any pairs of isotopes weighing the coherent term. Therefore, in the case of samples containing hydrogens, the measured cross sections are dominated by the incoherent contribution from the protons. In this chapter, we will focus on this kind of samples, for which we can assume that

$$\frac{\partial^2\sigma}{\partial\Omega\partial\hbar\omega} \propto S_{inc}^H(\mathbf{Q}, \omega). \qquad (3)$$

The interpretation of the incoherent scattering function $S_{inc}(\mathbf{Q}, \omega)$—in the following we will drop the superscript 'H'—is straightforward within the formalism of van Hove. $S_{inc}(\mathbf{Q}, \omega)$ is related, via Fourier transformation, with the incoherent intermediate scattering function $F_s(\mathbf{Q}, t)$ and the self-part of the van Hove correlation function $G_s(\mathbf{r}, t)$. In the classical limit, $G_s(\mathbf{r}, t)$ can be written as:

$$G_s(\mathbf{r}, t) = \langle\frac{1}{N}\sum_i^N \delta\{\mathbf{r} - [\mathbf{r}_i(t) - \mathbf{r}_i(0)]\}\rangle. \qquad (4)$$

Here $\mathbf{r}_i(t)$ is the position vector of the ith hydrogen at time = t (N: total number of hydrogens). Thus, $G_s(\mathbf{r}, t)d\mathbf{r}$ is the probability that, given a hydrogen at the origin at time $t = 0$, the same hydrogen is found within the volume $d\mathbf{r}$ at position $\mathbf{r}$ at time t. Incoherent scattering relates to single-particle motions.

From Table 1, it can also be deduced that substituting H by D this incoherent contribution is drastically reduced. This allows 'hiding' hydrogens for neutrons by selective deuteration.

2.2 Sources and Instrumentation

Neutron scattering experiments are realized in large facilities including a neutron source (either a nuclear reactor or a spallation source) and dedicated instrumentation. Table 2 lists the most widely used neutron facilities (currently operating or close to operation).

Based on the different principles and configurations employed to analyze the energy transfer during the experiment, several kinds of QENS spectrometers can be distinguished.

In a time-of-flight (ToF) spectrometer, the double-differential cross section is directly determined by the number of neutrons arriving in each detector as a function

Table 2 High-level neutron facilities in the world (neutron landscape around 2020) [1]

Facility	Location	Info Web-page
ESS	Lund, Sweden	https://ess-scandinavia.eu/
ISIS	Oxford, UK	www.isis.stfc.ac.uk/
ILL	Grenoble, France	www.ill.eu/
FRM II	Garching, Germany	www.frm2.tum.de/
JCNS	FRM II, SNS & ILL	www.jcns.info/jcns/
SINQ	Villigen, Switzerland	www.psi.ch/sinq/
RID	Delft, Netherlands	https://tnw.tudelft.nl/
BNC	Budapest, Hungary	www.bnc.hu/
FLNP	Dubna, Russia	https://flnp.jinr.ru/25/
PNPI	Gatchina, Russia	www.pnpi.spb.ru/
SNS	Oak Ridge, USA	https://neutrons.ornl.gov/facilities/SNS/
LANSCE	Los Alamos, USA	https://lansce.lanl.gov/
NCNR	Gaithersburg, USA	https://rrdjazz.nist.gov/
HFIR	Oak Ridge, USA	https://neutrons.ornl.gov/facilities/HFIR/
NRC	Chalk River, Canada	https://cins.ca//
ANSTO	Sydney, Australia	www.ansto.gov.au/
CSNS	Beijing, China	https://csns.ihep.ac.cn/
ISSP	Tokai, Japan	www.issp.u-tokyo.ac.jp/labs/neutron/index-e.html/
KURRI	Osaka, Japan	www.rri.kyoto-u.ac.jp/en/
J-PARC	Tokai, Japan	https://j-parc.jp/en/
BARC	Mumbai, India	www.barc.ernet.in/

of time. For this, the beam has to be pulsed. From continuous sources like reactors, pulses are usually produced either by combination of choppers or choppers and crystal monochromator. The incident energy is then also selected by such methods. The final energy is determined from the time of arrival in the detectors and the flight path. For pulsed sources, like spallation sources, the geometry is usually ‘inverted.’ A ‘white’ pulse hits the sample, and only neutrons of a fixed energy are accepted by the analyzer system. We note that using backscattering (BS) geometry (Bragg angles close to 180°) optimizes the energy resolution; therefore, monochromatization is usually performed by perfect crystals in such a geometry.

According to these principles, QENS spectrometers can be categorized into three types: direct-geometry time of flight (ToF) [representative examples: IN5, IN6 (ILL); DCS (NCNR); TOFTOF (FRM II); FOCUS (PSI); PELICAN (ANSTO); LET (ISIS)], indirect-geometry time of flight (BS-ToF) [e.g., IRIS, OSIRIS (ISIS); BASIS (SNS); DNA (J-PARC)], and indirect-geometry without time of flight [like IN16B (ILL); HFBS (NCNR) or SPHERES (FRM II)]. The latter are also commonly referred to as 'backscattering' (BS) spectrometers. In general, the Q-range covered is rather similar, spanning between 0.2 and 2 Å^{-1}. This means that the explored length scales are in the range ≈1–30 Å, i.e., local scales around typical inter- and intramolecular distances. BS spectrometers usually achieve the highest energy resolution (in the order of 1 μeV) but offer a limited time window (less than 2 decades). The dynamic range is extended in BS-ToF instruments but with poorer resolution, in the order of 10–100 μeV. ToF spectrometers are finally optimized for probing ps-dynamical processes. To access the whole QENS dynamic range ($\approx 10^{-13} - 10^{-9}$ s), it is usually necessary to combine results from more than one spectrometer. We note that even in that case the dynamic range covered by QENS techniques is rather poor in comparison with, for instance, broadband dielectric spectroscopy. However, QENS has other advantages, mainly: (i) It is an actual microscopic technique; (ii) it provides spatial/time resolution; and (iii) there is a classical formalism—the van Hove formalism—allowing direct interpretation of the scattering results in terms of atomic dynamics.

Though also sensitive to the QENS signal, the Neutron Spin Echo (NSE) technique is based on a very different principle. In NSE, the spin of the neutron is used as an 'internal clock' to encode the energy transfer in the scattering process [84]. Applying precession magnetic fields before and after the scattering event, the polarization of the neutron depends only on the velocity difference of each neutron individually, irrespective of its initial velocity. This technique is unique since it delivers the information directly in the time domain, allowing for a deconvolution of the experimental results from resolution effects by simple division. Another advantage of NSE is that the accessed Q- and dynamic ranges are usually more extended than those offered by 'conventional' QENS methods. However, in some cases, NSE might not be the best choice for the investigation of incoherent scattering. This is because the measuring procedure involves polarization analysis and strongly suppresses incoherent contributions (by a factor 1/3). Therefore, the use of NSE is usually not advisable to study self-motions of hydrogens in H/D labeled samples where the aim is to 'hide' by deuteration an undesired component (e.g., methyl groups) [5]. In such cases, the coherent contribution revealing the corresponding partial dynamic structure factor might reach values comparable to one-third of the incoherent cross section. Since in the measuring procedure intensities of coherent and incoherent origin contribute with opposite sign, the final NSE signal might even completely vanish in such cases. For this kind of experiments, 'regular' QENS (e.g., ToF, BS) is the right choice to follow H self-motions. Due to the above-mentioned peculiarities, NSE is thus usually considered aside QENS techniques. NSE spectrometers can be found in different facilities, like the ILL (IN15 and WASP), FRM II (J-NSE and RESEDA), NCNR, or SNS.

3 Dielectric Experiments: What's Measured?

The theoretical basis and experimental details of the broadband dielectric spectroscopy techniques are widely described in other chapters of this book. Our aim here is to briefly summarize the dynamic variables obtained by means of these methods and to discuss their actual physical meaning. Dielectric relaxation measurements of the orientational depolarization processes in glass-forming systems are usually carried out in the frequency domain, though in some cases time-domain techniques are also employed [2, 20–23]. The measured magnitude is the frequency-dependent complex dielectric permittivity $\varepsilon^\star(\omega) = \varepsilon'(\omega) - i\varepsilon''(\omega)$, which can also be expressed as Laplace transformation of a time-domain relaxation function $\varphi(t)$ as:

$$\frac{\varepsilon^\star(\omega) - \varepsilon_\infty}{\Delta\varepsilon} = \Phi^\star(\omega) = \int_0^\infty \left[-\frac{d\varphi(t)}{dt}\exp(-i\omega t)\right] dt. \tag{5}$$

Here, ε_∞ is the high-frequency limit value of the permittivity, $\Delta\varepsilon$ is the relaxation strength ($\Delta\varepsilon = \varepsilon_s - \varepsilon_\infty$, where ε_s is the static value of the permittivity). Thereby, $\Phi(\omega)$ is a normalized relaxation function. In this expression, $\varphi(t)$ represents the decay of the polarization of a fully polarized sample after removing the electric poling field, i.e., the response to a step-like excitation, which would correspond to the autocorrelation function of the total dielectric polarization.

Taking into account the well-known property of the Laplace transformation: $[i\omega\mathscr{L}_{i\omega}\{f(t)\} = \mathscr{L}_{i\omega}\{\dot{f}(t)\} + f(0)]$, expression 5 can also be written in terms of $\varphi(t)$ itself as

$$\Phi^\star(\omega) = 1 - i\omega \int_0^\infty \left[-\varphi(t)\exp(-i\omega t)\right] dt. \tag{6}$$

By analyzing either $\Phi(\omega)$ or $\varphi(t)$ by means of different phenomenological models, we can obtain two types of parameters: (i) a relaxation time (we will call it τ_{DS}) characterizing the timescale of the dielectric relaxation process under consideration and (ii) one or more 'shape parameters,' which measure the deviations of $\varphi(t)$ from a single exponential decay or of $\Phi(\omega)$ from the corresponding Debye-like process. In some particular cases, the shape parameters can be understood as some kind of measure of a distribution of relaxation times with actual physical meaning. Obviously, the interest uses to be focused on the dependence of the relaxation time and shape parameters as functions of the control parameters as, for example, temperature or pressure.

One important question that not always is taken into consideration is that the dipolar relaxation in a condensed matter system (either solid or liquid) is by nature a collective many-molecule (many-dipole) relaxation. As it has already been mentioned, $\varphi(t)$ is the autocorrelation function but of the total dielectric polarization of the sample. This implies that $\varphi(t)$ can be written as

$$\varphi(t) = \frac{\langle \mathbf{M}(0)\mathbf{M}(t)\rangle}{\langle|\mathbf{M}(0)|^2\rangle} \tag{7}$$

where

$$\mathbf{M}(t) = \sum_{i=1}^{N} \boldsymbol{\mu}_i(t). \tag{8}$$

Here μ_i is the dipole moment of the ith molecule and N the number of molecules of the system. The average is over an equilibrium ensemble. Obviously, expressions 7 and 8 indicate that not only self-dipolar relaxations $\langle \boldsymbol{\mu}_i(0)\boldsymbol{\mu}_i(t)\rangle$ but also cross-terms $\langle \boldsymbol{\mu}_i(0)\boldsymbol{\mu}_j(t)\rangle$ can play an important role in $\varphi(t)$. Because of the long-range nature of the dipolar interactions, the orientational motions of different dipoles (molecules) become correlated over large distances, i.e., the dielectric relaxation is collective.

The collective orientational relaxation in the long-wavelength limit, i. e., including all molecules of the system, was first studied taking concepts and ideas from the continuum description of electrostatics. In these approaches, the dipolar molecule is replaced by a cavity of some simple shape which includes a point dipole in the center. On the other hand, the condensed matter system is replaced by a frequency-dependent dielectric continuum. The interaction between the dipole and the dense liquid is considered through a time-dependent reaction field, which arises from the electric polarization of the liquid by the rotating dipole. A classical example of this approach is, for instance, the work of Nee and Zwanzig [88]. This type of approaches provide more or less simple expressions for the orientational correlation functions that can be directly compared with experimental results. However, the main drawback is that they do not directly take into account the inter-molecular interactions that are present in a dense dipolar system. If these interactions are important for a given relaxation process, the continuum model results to be not adequate.

The development of molecular theories for orientational correlations in general is a difficult task. At the end, this is mainly due to the fact that the orientational part of the interaction potential is not trivial. In spite of these difficulties, several theoretical approaches have been developed over last 30 years (see the book [24] as a recent review). Of particular relevance is the detailed molecular theory developed by Madden and Kievelson [81]. We would like also to mention the—in some way complementary approach to that of Madden and Kievelson—developed by Bagchi and Chandra [25] that is focused on the collective orientational motions at intermediate length scales.

One relevant theoretical question concerning collective dipolar relaxation is the relationship between the dielectric relaxation time deduced from the measurements (collective) and that corresponding to the self-part of the dipolar relaxation function (single-particle rotational relaxation time). This has often been called the 'macro–micro' relation.

For the rotational diffusion limit, the theory of Madden and Kievelson [81] predicts a 'macro–micro' relation given by:

$$\tau_{DS} = \tau_{\ell=1}\frac{3\varepsilon_s}{2\varepsilon_s+1}\tilde{g} \tag{9}$$

where $\tau_{\ell=1}$ is the single-molecule reorientation time τ_ℓ related to dielectric spectroscopy (angular moment, ℓ: $\ell = 1$), ε_s is the above-introduced low-frequency limit of the dielectric permittivity and $\tilde{g}$ is a generalized Kirkwood parameter $\tilde{g} = g_K/j_K$, being g_K the static Kirkwood parameter and j_K a complex dynamic coupling parameter (see, for instance, Ref. [81]).

It is worthy of remark that an expression formally equal to Eq. 9 also works in the framework of the above mentioned theory for non-diffusional motions and even for the case of polarizable polar molecules [81]. The only difference in such cases with respect to Eq. 9 is the meaning and expression of the $\tilde{g}$ factor. We note that Eq. 9 reduces to the well-known Glarum–Powles relation [66, 94] for the simple case $\tilde{g} = 1$.

4 Phenomenological Results Corresponding to $T > T_g$ in Glass-Forming Systems

4.1 Incoherent Quasielastic Neutron Scattering

Incoherent quasielastic neutron scattering data corresponding to many different glass-forming systems—including polymers—show that the incoherent intermediate scattering function $F_s(Q, t)$ displays a two-steps time decay, the slower one ($t \gtrsim 1$ ps being associated to the α-relaxation. A representative example can be found in Fig. 1. This slow decay is usually described by means of the well-known Kohlrausch–Williams–Watts (KWW) [75, 113] function:

$$F_s(Q, t) = A(Q) \exp\left[-\left(\frac{t}{\tau}\right)^{\beta}\right] \quad (10)$$

Here $0 < \beta \leq 1$ is the stretching parameter measuring the deviations from a single exponential decay and τ the incoherent relaxation time. The phenomenological factor $A(Q)$ is a generalized Lamb–Mössbauer factor (LMF) accounting for the fast decay of $F_s(Q, t)$:

$$A(Q) = \exp\left(-\frac{\langle u^2 \rangle}{3} Q^2\right) \quad (11)$$

where $\langle u^2 \rangle$ is the mean squared displacement associated to the fast decay step. The origin of this decay, in addition to vibrational contributions—the so-called fast dynamics—is still under debate in the scientific community. During the past decades, the 'fast dynamics' has been ascribed to conformational motions in asymmetric double-well [65], or in soft [68] potentials. In the framework of the mode-coupling theory (MCT) [67] it is attributed to a rattling in a cage formed by neighbor atoms. It has also been connected to free volume in Ref. [89] and for polymers, to the short-time regime of the segmental relaxation [42].

The relaxation time τ uses to show dispersion in Q, i. e., it can be described by a power law $\tau \propto Q^{-n}$, suggesting some kind of diffusive behavior for the self-atomic motions involved in the α-relaxation. We note that $n = 2$ and $\beta = 1$ correspond to a simple diffusion process with the mean-squared atomic displacement $\langle r^2(t) \rangle = 6Dt$

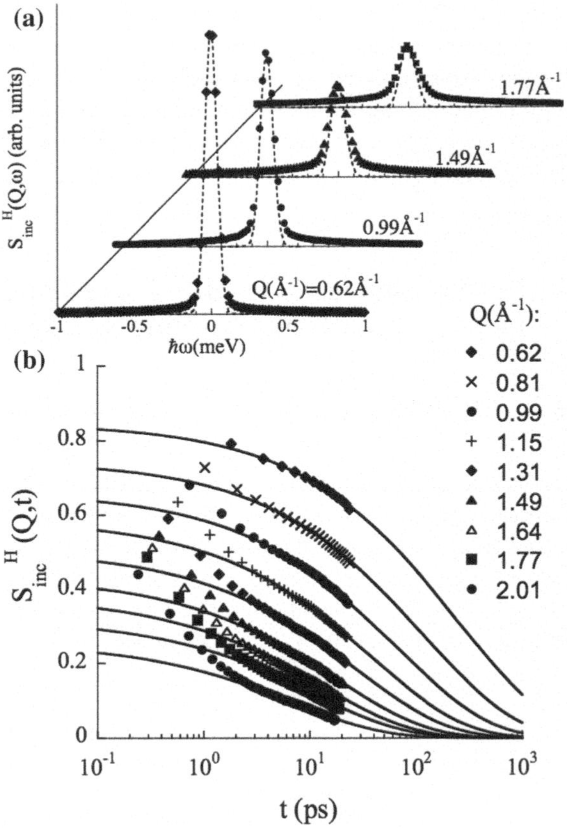

Fig. 1 **a** Spectra obtained by a ToF spectrometer for poly(vinyl methyl ether) (PVME) at about 150 K above its T_g at the indicated Q-values. The dashed lines show the resolution function. **b** Fourier transformed data deconvoluted with the instrumental resolution. The solid lines are KWW fits of the second step. (Reprinted from [39], with the permission of AIP Publishing.)

(D being the diffusion coefficient) and a Gaussian functional form of the intermediate scattering function

$$F_s(Q,t) = A(Q)\exp\left[-Q^2\frac{\langle r^2(t)\rangle}{6}\right]. \tag{12}$$

In the case of glass-forming polymers, an approximate value of $n \approx 2/\beta$ is usually observed for $Q \lesssim 1\,\text{Å}^{-1}$, in both, experimental and molecular dynamics (MD) simulations results [12, 14, 44, 47, 56, 60, 91, 97]. We note that this Q-dependence of τ—which was reported for the first time in Ref. [44]—can be understood in the framework of an anomalous diffusion model in the Gaussian approximation. In that case $F_s(Q,t) = A(Q)\exp[-Q^2\langle r^2(t)\rangle/6]$ (Eq. 12) with $\langle r^2(t)\rangle = 6\tilde{D}t^\beta$ results in a KWW function for $F_s(Q,t)$ (Eq. 10) with $\tau = \tilde{D}^{-1/\beta}Q^{-2/\beta}$. Here $\tilde{D}$ is a generalized diffusion coefficient with dimensions of $[\tilde{D}] \equiv L^2t^{-\beta}$. This type of subdiffusive model—which was proposed by us in Ref. [44]—in fact can be formally understood in the continuous-time random walk (CTRW) theoretical framework, first introduced by Montroll and Weiss [86]. A simple case of the CTRW arises when one assumes that the jump length and jump time random variables are decoupled and that the jump length variances are always finite (so that the central limit theorem applies in the limit of a large sequence of jump lengths); however, the jump times may be arbitrarily long, so that they obey a Lévy distribution with its characteristic long tail. Thus the jump length distribution ultimately becomes Gaussian with a finite jump length variance while the mean waiting time between jumps diverges on account of the underlying Lévy waiting time distribution [90].

At values of $Q \gtrsim 1\,\text{Å}^{-1}$—which for many polymers marks the range of the first maximum Q_{max} of the static structure factor $S(Q)$—deviations from the law $\tau(Q) \propto Q^{-2/\beta}$ have been reported for different polymers [14, 16, 30, 40, 48–51, 56, 60, 62, 64, 91, 109]. MD simulation studies [48] indicate that these deviations take place when the non-Gaussian parameter $\alpha_2(t) = 3\langle r^4(t)\rangle/(5\langle r^2(t)\rangle^2) - 1$ starts to show significant values ($\alpha_2 \gtrsim 0.2$). In this Q-region, the Q-dependence of the KWW relaxation time crosses over toward another power law, Q^{-n}, with $n \approx 2$ at high Q. This is illustrated in Fig. 2. However, in this high Q-range this law cannot be due to a single diffusion process. This crossover has been interpreted in terms of the anomalous jump-diffusion model [14, 16, 30, 36, 40, 47, 49, 50, 60] and in terms of the well-known mode coupling theory (MCT) [48]. Also, assuming it to reflect an underlying crossover from homogeneous to heterogeneous dynamics, the apparent contradiction between neutron scattering and relaxation techniques concerning dynamical heterogeneity of the α-relaxation was rationalized in Ref. [48].

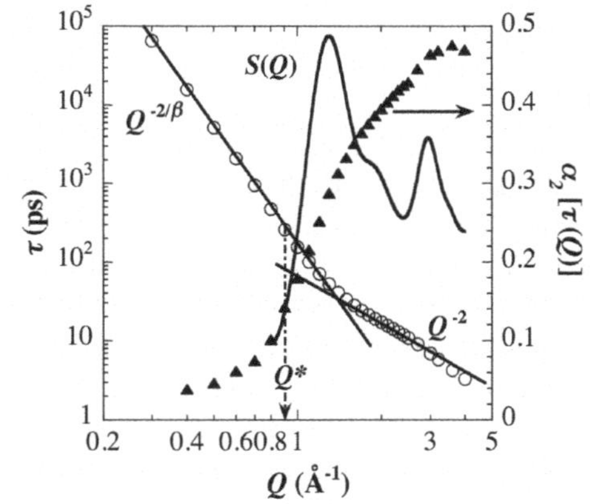

Fig. 2 Q-dependence of the characteristic time $\tau(Q)$ of the α-relaxation for polyisoprene (PI) obtained from the slow decay of the incoherent intermediate scattering function of the main chain protons (circles). The solid lines through the points show the Q-dependences indicated. The Q-dependence of the value of the non-Gaussian parameter at $\tau(Q)$ is also included (triangles) as well as the static structure factor $S(Q)$ on the linear scale in arbitrary units. The data correspond to MD-simulations published in Ref. [48]. The value of $Q^{\star}$ reported in Ref. [13] for PI from QENS and DS experiments is indicated by the dashed-dotted arrow

In the case of low-molecular weight (LMW) glass-forming systems, the Q-dependence of the incoherent scattering function has been considered in the literature only for a few systems [27, 35, 37, 99, 103, 111, 114, 115]. Moreover, the available data usually do not cover an extended Q-range. An exception is the case of glycerol, which has been studied by QENS and MD-simulations by different groups. As it is discussed in Ref. [99], for $Q \gtrsim 0.3\,\text{Å}^{-1}$, the Q-dependence of the incoherent relaxation time in glycerol follows a similar behavior than that found in polymers and above described. We note that glycerol is considered as a 'strong' glass and polymers 'fragile' glasses in the Angell's classification [8]. A similar behavior has been reported for a LMW counterpart (the monomer 1-PI) of the glass-forming polymer polyisoprene (PI) [99] and for the trimer (3VAc) of the glass-forming polymer poly(vinyl acetate) (PVAc) [108].

As can be seen in Fig. 3 the differences with respect to polymers emerge in the low-Q range where in the case of glycerol and 1-PI another crossover of $\tau(Q)$ toward $\tau(Q) \propto Q^{-2}$ is observed. At the same time, the β-parameter also increases toward 1. This implies that $\langle r^2(t)\rangle \approx 6Dt$, thereby meaning that there is a crossover from subdiffusive behavior in the Gaussian approximation toward normal diffusion at very low Q-values. This is in agreement with the very well-known long-time diffusion limit of the α-relaxation in LMW glass-forming systems. However, this crossover cannot be observed in glass-forming polymers because in those systems, the long-time limit of the α-relaxation is not the molecular (macromolecular) diffusion—as in the case of the LMW systems—but the well-known Rouse regime dictated by the connectivity of the macromolecular chains. We note that in the case of Rouse dynamics $\langle r^2(t)\rangle \propto t^{0.5}$ and $\tau(Q) \propto Q^{-4}$ [55, 98]. Since in the α-relaxation regime of polymers, the β-values found are usually in the range 0.4–0.5, it results basically impossible to detect any crossover from $\tau(Q) \propto Q^{-2/\beta}$ with $\beta \approx 0.4$–0.5 toward the Rouse range [96].

On the other hand, for some LMW glass-forming systems like ortho-terphenyl (OTP) the reported neutron scattering results correspond to a KWW $F_s(Q,t)$ with $\beta < 1$ (typically in the range 0.6–0.7) but at the same time the reported Q-dependence of the relaxation time in the relatively low-Q range seems to correspond to a normal diffusion $\tau(Q) = D^{-1}Q^{-2}$ [93]. This behavior could be understood in terms of an additional distribution of diffusion coefficients beyond the distribution of local jumps involved in the jump diffusion models [14]. This distribution of D should correspond to some kind of heterogeneous dynamic behavior and obviously would imply a non-Gaussian behavior of $F_s(Q,t)$ [12]. Heterogeneities induced by mixing with trehalose were also assumed at the origin of the decrease in the value of the β-parameter and in the power exponent of the characteristic time observed for glycerol in Ref. [35]. Ethylene glycol on the other hand showed a continuous power-law dependence of the characteristic time with exponent close to Gaussian behavior in the range $0.3\,\text{Å}^{-1} \leq Q \leq 1.5\,\text{Å}^{-1}$; deviations toward a weaker dispersion could be envisaged at the highest Q-values investigated [103].

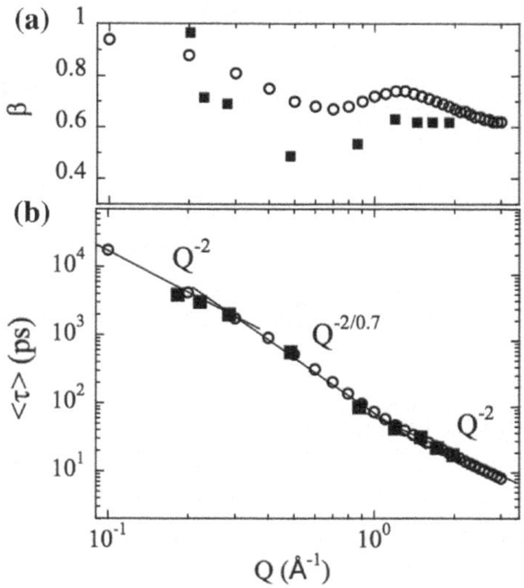

Fig. 3 Q-dependence of the stretching parameter β **a** and the average relaxation time **b** (circles: from the MD-simulations on a 1-PI LMW (PI-monomer) system reported in Ref. [99]; squares: experimental data of glycerol from Ref. [114] multiplied by 50 in (**b**)). (Reproduced from [99] with permission)

4.2 *Dielectric Spectroscopy*

The α-relaxation of glass-forming polymers has been investigated by dielectric spectroscopy from the beginning of the development of these techniques (see, for instance, Ref. [82] and references therein). However, the extension of the measurements to the high-frequency range needed to overlap with neutron scattering results has only been possible more recently. We note that in this high-temperature/frequency range the measured relaxation process is usually some kind of merging of the low-temperature α-relaxation and β-relaxation—being this latter the so-called secondary relaxation which is observed in the low-temperature range of most polymers. As it has been mentioned in Sect. 3, the dielectric measurements of the α-relaxation are usually carried out in the frequency domain. The magnitude usually considered is the imaginary part of the complex permittivity $\varepsilon''(\omega)$. The observed $\varepsilon''(\omega)$ peak uses to be rather broad, clearly deviating from a Debye peak. The $\varepsilon''(\omega)$ peak is habitually described in terms of empirical expressions such as Havriliak-Negami [70] or Cole-Davidson [53, 54] functions. It can be demonstrated [4] that under certain restrictions, the most general Havriliak–Negami relaxation function

$$\Phi^{\star}_{HN}(\omega) = \frac{1}{(1+(i\omega/\omega_c)^{\alpha})^{\gamma}} \tag{13}$$

can be considered as corresponding to a KWW function for $\varphi(t)$ in the time domain. Under these conditions, the permittivity can be written as

$$\varepsilon^{*}_{HN}(\omega) \approx \varepsilon_{\infty} + \frac{\Delta\varepsilon}{[1+(i\omega/\omega_c)^{\alpha}]^{\gamma(\alpha)}} - i\frac{\sigma}{\omega} \tag{14}$$

where ω_c is a characteristic frequency of the order of τ^{-1} and $-i\sigma/\omega$ is a term giving account for the DC conductivity. In this expression $\gamma(\alpha) = 1 - 0.812(1 - \alpha)^{0.387}$. The relationship among these parameters and those of the KWW equation are given [4] by

$$\beta^{1.23} = \alpha\gamma(\alpha) \tag{15}$$

$$\log_{10}(\tau) = -\log_{10}(\omega_c) - 2.6(1-\beta)^{0.5}\exp(-3\beta). \tag{16}$$

A careful investigation of dielectric results of different polymers by using the above described Eqs. [3] suggests that in general the β-values change from T_g until about $1.2T_g$, from where the shape remains more or less constant (see Fig. 4). In most polymers, the high-temperature value of β is close to 0.5.

On the other hand, the temperature dependence of the dielectric α-relaxation of polymers follows a non-Arrhenius behavior that usually is parametrized in terms of the well-known Vogel–Fulcher (VF) equation [58, 112]

$$\tau = \tau_o \exp\left(\frac{B}{T - T_o}\right). \tag{17}$$

At the VF temperature T_o the relaxation times would become infinite. We note that modern broadband dielectric spectroscopy allows covering more up to 18 orders of magnitude in frequency [80] thereby allowing to explore the relaxation time in a very wide temperature range.

The results obtained in the case of LMW glass-forming systems are qualitatively speaking rather similar to those corresponding to polymers. The dielectric

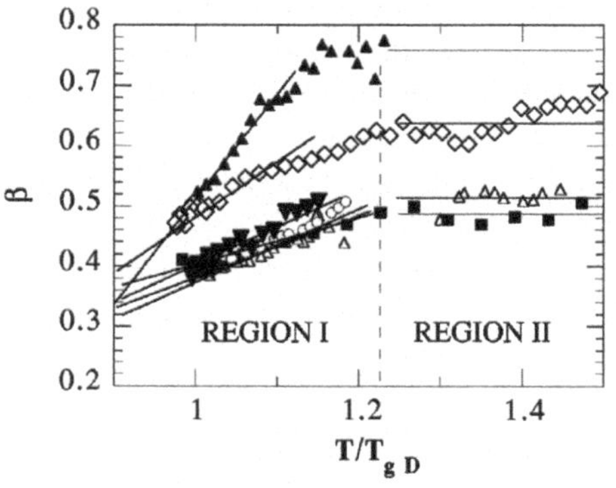

Fig. 4 Temperature dependence of the non-exponentiality of the segmental dynamics of the polymers investigated: polybutadiene (PB) (circles), poly(vinyl methyl ether) (PVME) (filled squares), poly(vinyl ethylene) (PVE) (empty triangles), poly(vinyl acetate) (PVAc) (empty diamonds), phenoxy (PH) (filled inverted triangles), and polysulfone (PSF) (filled triangles). The lines in region I are linear fitting of the data in the interval (1.0 , 1.1) whereas in region II they are to guide the eye. (Reproduced from [3] with permission)

α-relaxation time also follows the non-Arrhenius temperature dependence that can be described by the Vogel–Fulcher equation—at least approximately. On the other hand, the dielectric data of $\varepsilon''(\omega)$ are usually described in terms of the Cole–Davidson phenomenological function

$$\Phi^{\star}_{CD}(\omega) = \frac{1}{(1 + (i\omega/\omega_c)^{\alpha})} \tag{18}$$

instead of the Havriliak–Negami function. We note that the LMW systems usually display the so-called excess wing process [79] which restricts the fitting range of the α-relaxation $\varepsilon''(\omega)$ peaks. In any case, the temperature dependence of the shape parameter is similar to that corresponding to polymers. The values of the shape parameter increase with temperature until a temperature about $1.2T_g$ and then remain more or less constant [80]. However, in the case of the LMW systems, the values of β at high temperature use to be larger than those characteristic for most of the polymers ($\beta \approx 0.5$). Typical examples are reported in Ref. [101] for glycerol (a rather strong glass-former in the Angell's classification [8]) and propylene carbonate which is a fragile glass-former. We note that most polymers are fragile glass-formers in the framework of this classification. In the case of glycerol a β_{CD} high-temperature value of $\approx$0.75 has been reported [80] and for propylene carbonate a value of $\approx$0.9 [80]. Similar values are found for other LMW glass-forming systems. It is worthy of remark that the Cole–Davidson β_{CD}- values are not exactly the same as as the β_{KWW} ones. There are different methods relating these two magnitudes [106]. By using these methods, the values of $\beta_{CD} \approx 0.75$ and $\beta_{CD} \approx 0.9$ transform in the KWW-β values $\approx$0.83 and $\approx$0.93 respectively.

The existence of two dynamical ranges in supercooled liquids can be understood invoking the idea that the free diffusion at high temperature is slowed down dramatically when the temperature is lowered and therefore the activated processes become relevant for the dynamics in the range approaching T_g—a point of view compatible with the energy landscape picture [7, 34]. The high-temperature limit of the β-value in case of simple diffusion would be 1, but in general should be related to the details of the dynamics of each material. A limiting value smaller than unity would indicate that, even at high temperatures, there exists some kind of correlation in the reorientation of the dipolar units. This would be expected in polymers due to the unavoidable connectivity of the main chain, explaining the different ranges of β-values reported at high temperature for LMW and polymeric glass-forming systems. This kind of argument would also allow rationalizing the high value of β for polysulfone (PSF) (see Fig. 4) with respect to those shown by other polymers: Since in PSF the main dipolar units (SO_2) in the backbone chain are well separated—four phenyl rings between each couple of dipoles—the correlations in the dipole moment orientations seem to be negligible at high temperatures. Nevertheless, other factors as the coupling between the dipolar motions and the segmental dynamics would also determine the particular high-temperature value of the non-exponential parameter. We also note that the data reported in Ref. [3] and shown in Fig. 4 seem to indicate a common β-value close to 1/3 at the VF temperature, and these results are similar to those found for a series of different model systems mimicking the glass-forming system behavior, like Ising spin glasses (see [3] and references therein).

5 Relating Neutron Scattering and Dielectric Results in Glass-Forming Systems at $T > 1.2T_g$ and Related Systems

Taking into account what it is described in the previous section, it is obvious that incoherent neutron scattering and BDS techniques follow different correlators. Moreover, while a microscopic interpretation of incoherent scattering results can be straightforwardly achieved through the van Hove formalism, this is more difficult in the case of BDS results. Therefore, in principle, it is not an easy task to correlate the results obtained either by QENS or by BDS corresponding to a given relaxation.

In spite of these difficulties, there are several investigations on the α-relaxation in different glass-forming polymer systems which show compatible results [10, 13, 15, 17, 43, 45, 52, 64, 109]. The magnitudes that are usually compared are the dielectric relaxation function $\varphi(t)$ and the incoherent intermediate scattering function $F_s(Q,t)$ at different Q-values. It is generally found that $\varphi(t)$ and $F_s(Q,t)$ show a similar non-exponential (non-Debye) shape, at least in the high-frequency/temperature range where both techniques overlap. We remark that the careful investigation of dielectric results carried out in Ref. [3] shows that, despite the change found for the shape of the relaxation function from T_g until about $1.2T_g$, above $1.2T_g$ this shape remains more or less constant (see Fig. 4) and similar to what it is estimated by QENS. Moreover, the temperature dependence of the characteristic times of either $\varphi(t)$ or $F_s(Q,t)$ is also compatible in that high-temperature range [10, 13, 15, 17, 43, 52, 64, 109]. Under these conditions and taking into account that usually the characteristic time of $F_s(Q,t)$ depends on Q (see previous sections), then it is always possible to find a Q-value (we have called it $Q^\star$) at which $\tau_{DS}(T) \approx \tau_{QENS}(Q^\star, T)$. In this way, $Q^\star$ means a kind of link between dielectric and QENS measurements of glass-forming dynamics well above T_g.

Table 3 compiles the $Q^\star$-values reported in the literature (or calculated in this work starting from published data) for different systems including polymers and LMW systems. For polymers, most of the values are in the range $0.65\,\text{Å}^{-1} \leq Q^\star \leq 1\,\text{Å}^{-1}$. Exceptions are poly(ethyl methacrylate) (PEMA), polystyrene (PS) and high-order members of the family of poly(alkylene oxides) (PAOs) (namely the hexyl, octyl, and decyl members PHO, POO, and PDO, respectively), for which the $Q^\star$ value is lower. We note that all these polymers present voluminous side groups (like phenyl rings in the case of PS and relatively long alkylene chains in PAOs); interestingly enough, for PAOs the value of $Q^\star$ decreases with increasing length of the side group. There seems thus to be a correlation between the $Q^\star$-value and the length/volume of the lateral groups in polymers.

Table 3 Values of the scattering vector at which the incoherent QENS characteristic times match the DS times ($Q^{\star}$), the one corresponding to the first structure factor peak determined by neutron diffraction on deuterated samples (Q_{max}) and that where a crossover from Gaussian to non-Gaussian behavior is observed in the characteristic times for incoherent scattering (Q_c)

System	$Q^{\star}$(Å^{-1})	Q_{max}(Å^{-1})	Q_c(Å^{-1})
Homopolymers			
Polybutadiene (PB)	0.80[a][11, 12]	1.4 [57]	>0.8 [12]
Poly(vinyl chloride) (PVC)	0.83 [15]	1.2 [46]	<0.4 [15]
Polyisoprene (PI)	0.90 [13]	1.3 [6, 48]	1.0 [14, 16, 48]
Poly(ethylene propylene) (PEP)	1.0[a][91]	1.3 [92]	0.5 [91]
head-to-head Polypropylene (hhPP)	0.65 [76, 91]	1.2[b]	0.6 [91]
Polyisobutylene (PIB)	1.0 [50]	1.0 [50, 74, 95]	≈1.2 [12, 50, 52, 56]
Poly(vinyl ethylene) (PVE)	0.80[a][12, 71]	0.96 [96]	>0.8 [12, 96]
Poly(vinyl methyl ether) (PVME)	1.0 [40]	1.0[c] [40]	1.5 [47]
Poly(vinyl acetate) PVAc	0.65 [109]	0.65 [110]	<0.3 [109]
Poly(methyl methacrylate) (PMMA)	0.85[a][60]	0.87 [61]	0.65 [60]
Poly(ethyl methacrylate) (PEMA)	0.60 [62]	0.60 [18, 62]	<0.6 [62]
Poly(butylene oxide) (PBO)	0.65 [64]	0.75 [63]	0.9 [51]
Poly(hexylene oxide) (PHO)	0.57 [64]	0.50 [63]	<0.9 [64]
Poly(octylene oxide) (POO)	0.53 [64]	0.40 [63]	<0.9 [64]
Poly(decylene oxide) (PDO)	0.42 [64]		<0.9 [64]
Polystyrene (PS)	0.55 [59]	0.6[72]	<0.5 [59]
Low-molecular weight systems			
H_2O	0.67 [19]	1.95 [19]	0.9 [19]
Glycerol	1.0[a][37, 101, 114]	1.4 [37]	1.2 [37]
Ethylene glycol	0.75[a][102, 103]	1.55 [26]	1.3 [103]
Ortho-terphenyl (OTP)	0.6[a][69, 93]	1.4 [28]	

[a]This work, based on the following cited references
[b]Unpublished; from neutron diffraction and X-rays on hydrogenated polymer
[c]From X-rays

The presence of side groups leads to separation of the backbones in polymers. In amorphous materials like glass-forming polymers, the short-range order is reflected by broad peaks in the structure factor $S(Q)$. $S(Q)$ is directly accessible by neutron diffraction on fully deuterated samples, where all atomic correlation pairs are equally weighted (D and C have nearly identical scattering lengths for neutrons, as can be seen in Table 1). In the Q-region below approx. 2 Å^{-1}, the structure factor of main-chain polymers like polybutadiene (PB) [57], poly(vinyl chloride) (PVC) [46], molten polyethylene (PE) [9], or poly(ethylenene oxide) [33], shows a single peak. According to its temperature dependence, the origin of this first maximum of $S(Q)$ in main-chain polymers is predominantly due to inter-main-chain correlations [57].

From its position Q_{max}, an approximate value of the average inter-chain distance can be calculated in the Bragg approximation as $d \approx 2\pi/Q_{max}$. The value of Q_{max} in these polymers is in the range 1.2–1.5 Å^{-1} ($d \approx$ 4–5 Å). In polymers with side groups the situation is more complicated. Usually more than one peak appear in this region below 2 Å^{-1} and unraveling their microscopic origin is usually impossible without the help of MD simulations [6, 61]. Paradoxically, the situation becomes again simpler if the side groups occupy a large fraction of the volume, since a nano-phase separation of entropic origin is then induced where side groups tend to form nano-domains separated by the backbones. A 'pre-peak' appearing in the structure factor at low Q-values is the signature of such a nano-segregation. Its position reveals the inter-main-chain typical correlation distances (equivalently, the inter-side-groups nano-domains correlations) [18, 29, 87]. Thus, if no MD simulation help is at hand, the position Q_{max} of the first structure factor peak or 'pre-peak' can be used as a first approach to the characterization of the inter-main-chain correlations in polymers.

It has been indeed suggested a possible correlation between the values of $Q^{\star}$ and Q_{max} [48, 64]. To check it, Table 3 shows also the values of Q_{max} and Fig. 5 explores such a correlation. In this figure, we have represented with filled symbols the values corresponding to either main-chain polymers or polymers with only methyl groups as side groups, and with open symbols the values corresponding to polymers with bulkier side groups. For the latter, within the uncertainties involved, we can deduce a correlation between $Q^{\star}$ and Q_{max} which is close to their equivalence $Q^{\star} = Q_{max}$;

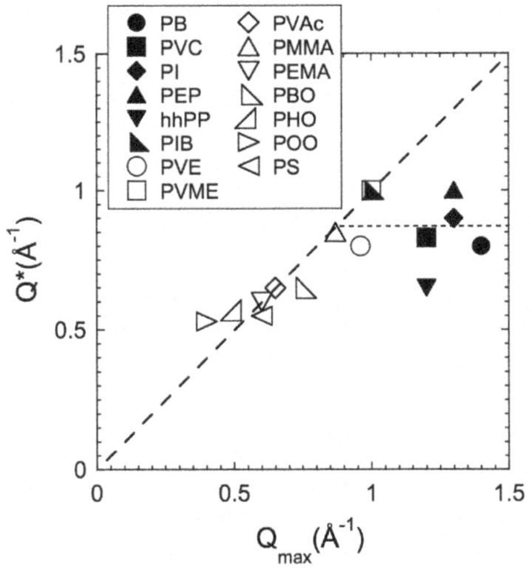

Fig. 5 Value of the scattering vector at which the incoherent QENS characteristic times match the DS times ($Q^{\star}$) versus that corresponding to the first structure factor peak determined by neutron diffraction on deuterated samples (Q_{max}) for the polymers indicated. For PVME and hhPP, the value of Q_{max} has been determined from X-rays and neutron diffraction on protonated samples, respectively. The dashed line shows the $Q^{\star} = Q_{max}$ correlation and the dotted line, an average constant value

however, polymers without side groups or only methyl groups as side groups present rather similar values of $Q^{\star} \approx 0.9\ \mathring{A}^{-1}$ independently of the position of their first maximum in $S(Q)$.

Regarding LMW systems, the $Q^{\star}$-values found are also scattered within the range $0.60\,\mathring{A}^{-1} \leq Q^{\star} \leq 1\,\mathring{A}^{-1}$. No correlation can be identified between $Q^{\star}$ and Q_{max} for these materials, at least with the information at hand.

Another interesting question is whether the dynamics probed by a given technique—in this case, dielectric spectroscopy—would be dominated by Gaussian microscopic events or rather be more sensitive to the heterogeneous underlying processes [48]. For polymers, where $\beta \approx 0.5$, this requires discerning whether τ_{DS} lays in the Q-region where Gaussian ($\tau(Q) \propto Q^{-2/\beta}$) or non-Gaussian behavior ($\tau(Q) \propto Q^{-2}$ approx.) is observed, or, equivalently, to discriminate if $Q^{\star}$ is lower or higher than the above-mentioned Q-value (let us call it Q_c) at which the crossover from Gaussian to non-Gaussian behavior takes place. With these ideas in mind, we have also compiled information about the values of Q_c in Table 3. The uncertainties involving the determination of these values are sometimes rather large, and in many cases only an upper or lower limit can be deduced from the reported QENS results. In view of the collected information about this parameter, it is not possible to find any clear correlation between Q_c and Q_{max}. However, Fig. 6 at least allows discriminating for which system dielectric techniques probe Gaussian ($Q^{\star} < Q_c$) or non-Gaussian ($Q^{\star} > Q_c$) dynamics. It follows that for PEP, PMMA, PVC, and PVAc, and probably PEMA, the dielectrically probed dynamics is non-Gaussian ('heterogeneous'), while for PVME, PIB, PB, PVE, hhPP, PI, and PBO is Gaussian ('homogeneous'). For high-order PAOs, the situation cannot be discerned with the data at hand.

As for the LMW systems, water at room temperature shows simple diffusion (exponential time dependence, $\beta = 1$ and Gaussian Q-behavior, $\tau \propto Q^{-2}$) which

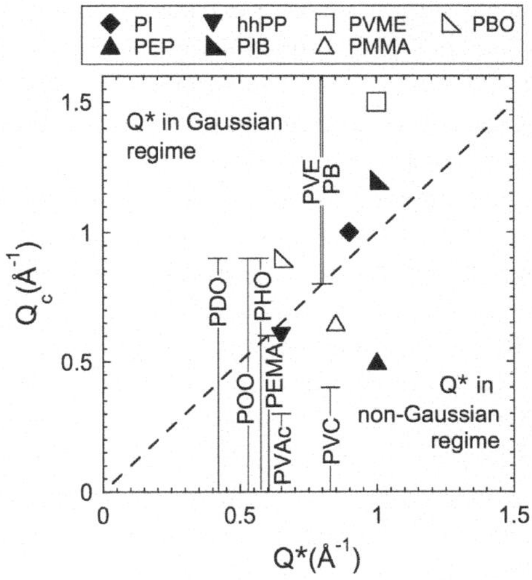

Fig. 6 Value of the scattering vector at the crossover between the Gaussian and non-Gaussian regimes (Q_c) as function of the scattering vector at which the incoherent QENS characteristic times match the DS times ($Q^{\star}$) for the glass-forming polymers indicated. The dashed line separates the cases where $Q^{\star}$ lays in the Gaussian regime (upper right zone) and in the non-Gaussian regime (right down zone)

extends down to $Q \approx 0.9\,\text{Å}^{-1}$; thus, τ_{DS} is contained in the simple diffusion regime. The dielectric time for glycerol is located in the regime where, despite non-exponentiality is found ($\beta \approx 0.75$), the Q-dependence of the characteristic time reveals Gaussian behavior (in this case, the deviations occur at $Q_c \approx 1.2\,\text{Å}^{-1}$). This is also the situation of ethylene glycol, as can be seen in Table 3. These results in glycerol and ethylene glycol would suggest the observation of anomalous diffusion in the range around $Q^\star$. In the case of OTP, the value of $Q^\star$ is in the region where the incoherent characteristic time follows a Q^{-2}-dependence (which persists, as reported in Ref. [93], beyond the value of $Q_{max} = 1.4\,\text{Å}^{-1}$); in this region, the β-parameter is smaller than 1 (around 0.6), and therefore the atomic displacements are not distributed according to a Gaussian function. In this case, as above commented, the reason might be sought in distribution effects due to a variety of mobilities associated to the different hydrogens.

In the framework of the Gaussian approximation for the atomic displacements (Eq. 12), we have shown above that the intermediate scattering function in glass-forming systems can be written as $F_s(Q,t) = \exp[-(t/\tau)^\beta]$ with $\tau = \tilde{D}^{-1/\beta} Q^{-2/\beta}$. In such conditions, the definition of $Q^\star$ implies that $\tau_{DS} \equiv \tilde{D}^{-1/\beta} Q^{\star -2/\beta}$ and thereby $Q^\star = [\tau_{DS}^\beta \tilde{D}]^{-1/2}$. Obviously, this expression reduces to that corresponding to the case of normal diffusion when $\beta \to 1$. It is straightforward to demonstrate that the mean squared displacement of the atoms at the time corresponding to dielectric relaxation $\langle r^2(\tau_{DS})\rangle = 6\tilde{D}\tau_{DS}^\beta$ becomes

$$\langle r^2(\tau_{DS})\rangle = \frac{6}{Q^{\star 2}} \tag{19}$$

independent of β. Obviously this equation also applies in the case of normal diffusion ($\beta = 1$). The square root of Eq. 19 gives thus the mean atomic displacement at the characteristic time for dipolar relaxation $\xi = \sqrt{6}/Q^\star = \sqrt{\langle r^2(\tau_{DS})\rangle}$. For those systems where the Gaussian approximation holds for the dynamics probed by dielectric spectroscopy, we have calculated this magnitude. The resulting values are represented in Fig. 7 as function of the estimated average inter-molecular distance defined from the first static structure factor peak $d = 2\pi/Q_{max}$. The average displacements at such time are scattered within the approximate interval $2.5\,\text{Å} \leq \xi \leq 4\,\text{Å}$. The atomic displacements involved in the structural relaxation which are needed to lead to the dipolar relaxation seem thus to be of a few Ångstroms independently of the presence or absence of chain connectivity. These values are in the order or smaller than the inter-molecular average distance phenomenologically estimated d, which ranges between the minimum value of 3.2 Å for water and the maximum of 8.4 Å for PBO. In fact, only for water the reported values of ξ and d nearly coincide.

Considering also results on water in concentrated solutions, we can see in Table 4 that again the values deduced for $Q^\star$ are in a similar range as those found for glass-forming systems. In the case of water in different environments, a general reported feature is the stretching of the intermediate scattering function together with a Q^{-2}-dependence of the characteristic time for the diffusional component. These findings—analogous to those reported for OTP—have been interpreted in

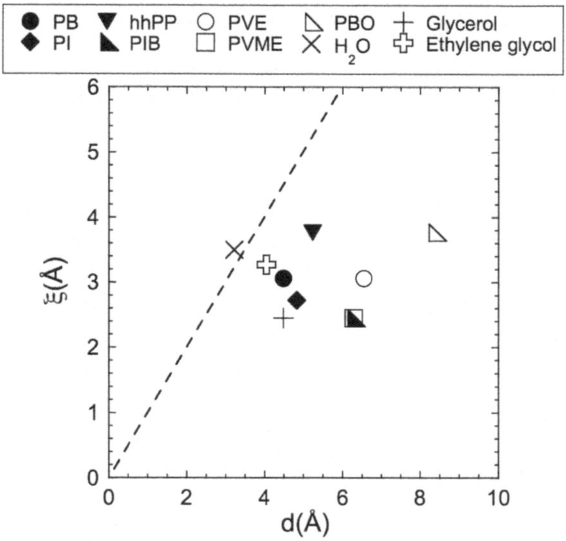

Fig. 7 Value of the magnitude $\xi = \sqrt{6}/Q^{\star}$ as function of $d = 2\pi/Q_{max}$ calculated from the position of the first maximum of the structure factor for the different systems indicated. As for the systems represented the Gaussian approximation is fulfilled at times close to τ_{DS}, ξ represents the mean displacement of hydrogen atoms at the characteristic time for dipolar relaxation; d can be considered as a phenomenological approach to the average inter-molecular distance. The dashed line shows the function $\xi = d$

Table 4 Values of the scattering vector at which the incoherent QENS characteristic times match the DS times ($Q^{\star}$) and of the ratio between the static and dynamic Kirkwood parameter $\tilde{g}$ for water in bulk and concentrated aqueous solutions

System	$Q^{\star}$(Å^{-1})	$\tilde{g}$
H_2O	0.67 [19]	1.3 [19]
H_2O in proline	0.78 [85]	0.97 [85]
H_2O in poly(vinyl pyrrolidone) (PVP)	1.0 [38]	0.6
H_2O in PVME	1.0 [40]	0.6
H_2O in amorphous polyamide (PA)	0.60 [77]	1.64
H_2O in semicrystalline PA	1.2 [77]	0.4
H_2O in myoglobin	1.0 [73]	0.6

terms of broad underlying distributions of mobilities [38, 40, 77, 85]. The resulting non-Gaussian behavior of the intermediate scattering function prevents, in principle, applying Eq. 19 to deduce a reliable value of the average displacements during dipolar relaxation; however, we may infer that the general increase of the value of $Q^{\star}$ in the presence of solute reflects a reduction of the mean atomic displacement of water hydrogens at the dielectric relaxation time. The only exception is the case of water in amorphous polyamide [77].

6 Theoretical Interpretation

6.1 The Simplest Case: Translational and Rotational Diffusion

In the case of simple translational diffusion, we have shown that $Q^\star = [\tau_{DS} D]^{-1/2}$. In the framework of a molecular hydrodynamic approximation [19] and taking into account the Stokes–Einstein relationship, the translational self-diffusion coefficient can be written as $D = K_B T/(6\pi a_T \eta)$, where a_T is the molecular hydrodynamic radius for translation, η is the 'solvent' viscosity, K_B the Boltzmann constant, and T the temperature. On the other hand, the single-particle (molecule) reorientation time τ_ℓ is given by $\tau_\ell = 8\pi\eta a_R^3/[\ell(\ell+1)K_B T]$ with a_R meaning a hydrodynamic radius but now for rotation. For the reorientation of dipoles—i. e., the magnitude related to dielectric spectroscopy—the angular moment, ℓ, is $\ell = 1$. Then, we obtain the following relationship: $D\tau_{\ell=1} = \frac{2}{3}\frac{a_R^3}{a_T}$ which for $a_R \approx a_T = a$ becomes $D\tau_{\ell=1} \approx \frac{2}{3}a^2$. As it has been above described (see Sect. 3) in the Madden and Kievelson theory [81] the relationship between the relaxation time measured using dielectric spectroscopy, τ_{DS}, and the microscopic single-particle time, $\tau_{\ell=1}$, can be expressed as $\tau_{DS} = \tau_{\ell=1} 3\varepsilon_s(2\varepsilon_s+1)^{-1}\tilde{g}$ (Eq. 9) with $\tilde{g}$ a generalized Kirkwood parameter (see Sect. 3). For $\varepsilon_s \gg 1$ this expression can be approximated by $\tau_{DS} = \frac{3}{2}\tau_{\ell=1}\tilde{g}$. Taking into account these expressions, it is straightforward to show that $Q^\star$ can be expressed as $Q^\star = [a^2\tilde{g}]^{-1/2}$, i. e., in terms of a 'single-molecule' magnitude—the effective hydrodynamic radius a—and a factor, $\tilde{g}$, measuring the strength of many-body effects on dipolar relaxation. On the other hand, starting from experimental values of $Q^\star$ and a, we can try to get information about $\tilde{g}$—a magnitude not very easy to evaluate only from dielectric measurements.

In reference [19], this formalism was applied to water dynamics at 300 K, where the dominant dynamic process in the time scale of the dielectric Debye peak ($\tau_{DS} \approx 8$ ps) is a diffusion process. The comparison between the incoherent neutron scattering and dielectric spectroscopy indicates that $Q^\star \approx 0.67\,\text{Å}^{-1}$ [19, 85]. This value seems to be independent of temperature in the range 270–300 K [19]. On the other hand, $a_T \approx a_R \approx 1.3\,\text{Å}$ [19, 32, 104]. With these values, the above expression delivers $\tilde{g} \approx [Q^\star a]^{-2} \approx 1.3 \pm 0.2$ [85]. We note in passing that this value differs from that previously reported in Ref. [19]. The reason is that the value previously given corresponds to $\tilde{G} \equiv G_K/J_K$, where G_K is the Kirkwood parameter for a finite system $[G_K \approx \frac{3}{2}g_K]$ [100]. It is worthy of remark that MD simulations carried out by means of the SPC/E force field [32] reported a value of $\tilde{G} \approx 2.25$, which should correspond to $\tilde{g} \approx 1.5$, i. e., in the range of that deduced from $Q^\star$.

In Ref. [85] this investigation was extended to water in water solutions of an aminoacid, proline, at two different concentrations corresponding to 31 and 6.4 molecules of water per proline molecule. In both cases, the neutron scattering results showed that the slow dynamic process of water in the solutions was diffusive ($\tau \propto Q^{-2}$). In the most concentrated solution (6.4 water molecules per proline molecule) indications of a possible distribution of diffusive processes were also reported. As it has been described in Sect. 5, the reported value of $Q^\star$ was the same for both solutions, $Q^\star \approx 0.78\,\text{Å}^{-1}$, and larger than that corresponding to pure water. If we consider the hydrodynamic radius of water in the solutions similar to that in pure water at the same temperature ($a \approx 1.3\,\text{Å}$), this value of $Q^\star$ corresponds to $\tilde{g} \approx 0.97$ (see Table 4), smaller than that from pure water. We note that $\tilde{g} \approx 1$ basically means no coupling among water dipoles in the presence of proline. This suggests that proline disrupts the water H-bond network, in agreement with Ref. [116].

As we have previously mentioned, for concentrated water solutions and also for OTP stretched functional forms are reported together with a Q^{-2}-dependence of the characteristic time that corresponds to a simple diffusive process. Let us consider in detail this case, when the time dependence of the correlation function can be approximated by a KWW function (eq. 10) with a characteristic (let us call it 'phenomenological' time) τ_w. We will follow the same arguments as those invoked in Ref. [12] to address the question of the homogeneous versus heterogeneous nature of the α-relaxation in glass-forming polymers. In a heterogeneous scenario, the KWW function can be interpreted as arising from a superposition of different simple exponential relaxations weighted by a distribution of relaxation times $g(\ln \tau)$,

$$\varphi(t) = \exp\left[-\left(\frac{t}{\tau_w}\right)^\beta\right] = \int_{-\infty}^{+\infty} g(\ln \tau) \exp\left(-\frac{t}{\tau}\right) d(\ln \tau). \tag{20}$$

The function $g(\ln \tau)$ can be properly chosen such that the observed stretching is reproduced by this superposition of elementary events. For QENS, in the case of a distribution of local diffusivities associated to the different regions in the sample, the resulting incoherent scattering function would be expressed as

$$F_s(Q,t) = \exp\left[-\left(\frac{t}{\tau_w(Q)}\right)^\beta\right] = \int_{-\infty}^{+\infty} g(\ln D^{-1}) \exp\left(-\frac{Q^2 t}{D^{-1}}\right) d(\ln D^{-1}). \tag{21}$$

where $\exp(-Q^2 D t)$ is the intermediate scattering function corresponding to simple diffusion in the Gaussian approximation and D is the corresponding diffusion coefficient. With the distribution function $g(\ln D^{-1})$ a KWW time dependence of the experimentally observed $F_s(Q,t)$ function would now be obtained. We note that since now the 'stretched' variable is the product $Q^2 t$, the resulting incoherent scattering function becomes

$$F_s(Q,t) = \exp\left[-\left(\frac{Q^2 t}{D_w^{-1}}\right)^\beta\right]. \tag{22}$$

This means, we recover the experimentally observed dispersion of the characteristic time as $\tau_w = D_w^{-1} Q^{-2}$: The Q-dependence of the 'phenomenological' characteristic time is the same as that of the elementary diffusive process. We may assume that the distribution of diffusivities originates from different local viscosities and thus the distributed variable is the viscosity η. In analogous way, the distribution of viscosities will have a 'KWW' or 'macroscopic' viscosity η_w. The deduced magnitude from QENS experiments, D_w^{-1}, would be related with η_w via

$$D_w^{-1} = \frac{6\pi \eta_w a_T}{K_B T}. \tag{23}$$

Similarly, an underlying variety of viscosities would generate a distribution of elementary single-molecule reorientational times related to dielectric spectroscopy $\tau_{\ell=1}$, which would be represented by the 'characteristic microscopic time' $\tau_{\ell=1}^w = 4\pi \eta_w a_R^3 / K_B T$. Applying the 'macro–micro' relation (Eq. 9) with the usually assumed condition $\varepsilon_s \gg 1$, we deduce that the phenomenologically observed dielectric relaxation time in the KWW relaxation function would be given by

$$\tau_{DS}^w \approx \frac{3}{2}\tau_{\ell=1}^w \tilde{g} = \frac{6\pi \eta_w a_R^3}{K_B T}\tilde{g}. \tag{24}$$

On the other hand, the definition of $Q^\star$ is the same: The Q-value at which the 'phenomenologically' characteristic times observed by DS (τ_{DS}^w) and QENS ($\tau_w = D_w^{-1} Q^{-2}$) coincide: $Q^\star \equiv (D_w^{-1}/\tau_{DS}^w)^{1/2}$. Introducing in this definition the expressions above given for D_w^{-1} and τ_{DS}^w (Eqs. 23 and 24 respectively) we obtain

$$Q^\star = \left(\frac{a_T}{a_R^3 \tilde{g}}\right)^{1/2} \approx \frac{1}{\sqrt{a\tilde{g}}}. \tag{25}$$

This means, under the assumptions made, in the case of simple diffusion the same expression relates $Q^\star$ and $\tilde{g}$ independently of the presence of distributions associated to a variety of local frictions (effective local viscosities).

We can now apply this formalism to the values of $Q^\star$ reported in Table 4 corresponding to water in concentrated aqueous solutions. We note that in the case of PVP, the number of water molecules per monomer is about 4 and in the case of PVME about 1.4. In both cases, the reported value of $Q^\star$ is $Q^\star \approx 1\,\text{Å}^{-1}$, which would correspond to $\tilde{g} \approx 0.6$ for the same value of the hydrodynamic radius of water. We remind that $\tilde{g} = g_K/j_K$, with g_K being the static Kirkwood correlation function which is a measure of the degree of orientational correlation of the neighboring molecular dipoles in dense polar liquids. j_K is the so-called dynamic Kirkwood coupling, which is expected to be close to 1. Then, values of $\tilde{g} < 1$ as is obtained for water in PVME and PVP water concentrated solutions suggest some kind of anti-parallel dipolar correlation. We note that the value of $Q^\star$ for water in water solutions of myoglobin reported in Ref. [73] and included in Table 3 is also of the order of 1 as in the case of PVME and PVP solutions. This suggests similar orientational correlation of water molecules in these different solutions. We note in passing the universal behavior found for water dynamics in mixtures with very different hydrophilic substances reported from DS in Ref. [41]. The temperature dependence of the relaxation times namely exhibits a crossover from non-Arrhenius to Arrhenius behavior at the T_g range of the mixtures and universal behavior both above and below the crossover temperature. The observed crossover was attributed to the emergence of confinement effects.

As it was commented in Sect. 5, in the case of water absorbed in polyamides the value of $Q^\star$ strongly depends on the structural state of the polyamide: $Q^\star \approx 0.6\,\text{Å}^{-1}$ for amorphous polyamide (14wt% of water) and $Q^\star \approx 1.2\,\text{Å}^{-1}$ for semicrystalline polyamide (7wt% of water). These values also translate into very different values of $\tilde{g}$ for water in such systems: $\tilde{g} \approx 1.6$ for the amorphous case and $\tilde{g} \approx 0.4$ for the semicrystalline case. These values mean qualitatively different orientational correlation of absorbed water molecules in these host polymer matrices.

The case of OTP could be also considered in this framework. In the review [107], it is mentioned that for this molecule the hydrodynamics seems to work and the hydrodynamic radii for translation and rotation are, respectively, $a_T \approx 0.21$ nm and $a_R \approx 0.23$ nm and independent of temperature. If we apply the model with $a \approx 0.22$ nm, we find a $\tilde{g}$-value of $\approx$0.6. Thus, in OTP we would also find an anti-parallel dipolar alignment, according to these results.

For simple diffusion, the proposed model may be applied even without incoherent QENS measurements at hand. Namely, to determine the value of $Q^\star$ all we need to know is the value of the diffusion coefficient D –besides, obviously, that of τ_{DS}– since $Q^\star = [D\tau_{DS}]^{-1/2}$. The diffusion coefficient can be determined for instance from NMR measurements. We may consider the case of 1-butanol as an example. The temperature-dependent diffusion coefficient obtained from NMR was reported in Ref. [31]. From the comparison with dielectric results shown in the same review at 300 K, a $Q^\star$-value of $Q^\star \approx 0.17\,\text{Å}^{-1}$ is deduced. We note the extremely low value found in this case. Taking into account the hydrodynamic radius $a \approx 2.75$ Å reported for this molecule [31], we deduce $\tilde{g} \approx 4.6$. This remarkably high $\tilde{g}$-value suggests a high degree of orientational correlation of the neighboring molecular dipoles in this H-bonding system.

6.2 Generalization to Fractional (Subdiffusive) Diffusion

In the previous section, we have described that in the case of a simple translational and rotational diffusion process $Q^\star$ can be expressed in terms of a 'single-molecule' magnitude—the effective hydrodynamic radius a—and a factor $\tilde{g}$, measuring the strength of many-body effects on dipolar relaxation [Eq. 25]. We have also seen how this simple interpretation can be extended to the case when neutron scattering data suggest the existence of a distribution of diffusive processes. However, the phenomenology described in Sect. 4.1 indicates that in some cases the results obtained by QENS suggest the existence of an actual—or at least effective—anomalous subdiffusive process. Anomalous diffusions in general have attracted much interest over last 25 years (see, as an example, the review [83]). In this section, we will try to extend our results to the case of a subdiffusive behavior.

In the framework of different theoretical approaches to subdiffusion (see Ref. [83]), the generalized diffusion coefficient introduced in Sect. 4, $\tilde{D}$, can be expressed as $\tilde{D} = K_\beta/\Gamma(1+\beta)$. Here Γ is the Gamma-function and K_β is defined by $K_\beta \equiv \sigma^2/\tau_o^\beta$ in terms of the scales σ (length) and τ_o (time). In the case of $\beta = 1$, $\Gamma(2) = 1$, and $K_1 = D = K_B T/\xi_T$, where K_B is the Boltzmann constant, T the temperature and ξ_T the friction coefficient. As it was mentioned in the previous section, for a spherical particle moving in a fluid of viscosity η, $\xi_T = 6\pi\eta a$, with a the radius of the particle (Stokes' law). K_1 can also be expressed as $K_1 = \sigma^2 W_o$, with the elementary rate $W_o \equiv \tau_o^{-1}$ given by $W_o = K_B T/(\sigma^2\xi_T)$. In the case of subdiffusive behavior $K_\beta = \sigma^2 W_o^\beta$ and then

$$\tilde{D} = \frac{\sigma^2}{\Gamma(1+\beta)}\left[\frac{K_B T}{\sigma^2\xi_T}\right]^\beta . \tag{26}$$

Since for anomalous diffusion in Gaussian approximation $Q^\star = [\tau_{DS}^\beta \tilde{D}]^{-1/2}$ (see Sect. 5), it becomes

$$Q^\star = \left[\frac{\sigma^2}{\Gamma(1+\beta)}\left(\frac{K_B T}{\sigma^2\xi_T}\tau_{DS}\right)^\beta\right]^{-1/2} . \tag{27}$$

Now we can proceed as in the case of the simple diffusion and express τ_{DS} in terms of the $\tau_{\ell=1}$ and $\tilde{g}$. For $\varepsilon_s \gg 1$, $\tau_{DS} \simeq \frac{3}{2}\tau_{\ell=1}\tilde{g}$. On the other hand, in the simplest Debye theory $\tau_{\ell=1} = \xi_R/(2K_B T)$, where ξ_R is the friction coefficient for rotation, which for a spherical particle of radius a is $\xi_R = 8\pi\eta a^3$. Then, for such a kind of particle $\xi_R = \frac{4}{3}\xi_T a^2$. By assuming $\sigma = a$, expression 27 becomes after some algebraic manipulation

$$Q^\star = \frac{[\Gamma(1+\beta)]^{1/2}}{a\tilde{g}^{\beta/2}}. \tag{28}$$

Again for $\beta = 1(\Gamma(2) = 1)$, we recover the expression found for simple diffusion.

In spite of the simplicity of this expression, its practical application to glass-forming systems is not straightforward. In the case of LMW systems, the Q-dependence of the available incoherent neutron scattering results only seem to show the signature of an actual anomalous diffusion ($\tau \propto Q^{-2/\beta}$) in a very narrow intermediate Q-range (see Fig. 3). Then, it is not clear whether or not this behavior reflects an actual subdiffusive behavior associated to the α-process or if, in contrast, is only a signature of a crossover regime toward pure molecular diffusion at low Q. Conversely, in the case of polymer systems, the $Q^{2/\beta}$-law is well documented in a rather wide Q-range for many polymers as it has been already discussed. On the other hand, segmental subdiffusive behavior with $\beta = 0.5$ is observed in the case of the Rouse regime above discussed. However, the question arises: What is the realistic size of the segment (a) to be included in Eq. 28? Moreover, the approximation of 'spherical molecules' involved in the deduction of Eq. 28 is certainly rather strong and likely unphysical for the case of polymers. To explore the limits of applicability of Eq. 28 is beyond the scope of this chapter and will be the subject of future work.

7 Conclusions

Being techniques sensitive to different correlation functions, the comparison of the results from quasielastic neutron scattering and dielectric spectroscopy is not trivial. However, the phenomenology observed from both methods when applied to glass-forming systems—in particular to polymers—reveals compatible results regarding the stretching of the relaxation function and the temperature dependence of the characteristic times, at least in the high-temperature/frequency regime where the two techniques overlap.

We have compiled the values of the momentum transfer $Q^\star$ at which the QENS characteristic times match the one deduced from DS experiments. Glass-forming systems including polymers and smaller molecules, as well as water in bulk and in different concentrated solutions, have been considered. Most of the values, independently of the nature of the system, range between 0.6 and $1.0\,\text{Å}^{-1}$. For polymers, we have explored the possible relationship between this Q-value and the position of their first maximum Q_{max} in the structure factor $S(Q)$—which is determined by the inter-chain average distance. Main-chain polymers (without side groups) or polymers with only methyl groups as side groups present rather similar values of $Q^\star \approx 0.9\ Å^{-1}$ independently of the value of Q_{max}. However, for polymers with bulkier side groups we can deduce, within the uncertainties, a correlation between $Q^\star$ and Q_{max} which is close to their equivalence $Q^\star = Q_{max}$.

For those systems where the Gaussian approximation holds for the dynamics probed by dielectric spectroscopy, we have calculated the average atomic displacement at the characteristic time for dipolar relaxation. The values are of a few Ångstroms (about 3 Å) in systems of both, small molecules and polymers. These values are in the order or smaller than the average inter-molecular distance.

Trying to give a step beyond the pure phenomenology, we have presented a simple model based on molecular hydrodynamics and a molecular treatment of dielectric spectroscopy. In the framework of this model, $Q^\star$ can be expressed as a function of two parameters—the effective hydrodynamic radius and a parameter giving account for the many-body aspects of the dipolar relaxation. This model works for water and we have also applied it to the cases of water in concentrated solutions and to the low-molecular weight glass-former OTP, which shows diffusive-like behavior. From this application, we have deduced the degree of orientational correlation of the neighboring molecular dipoles in the different systems. In the case of water, the presence of solute seems to lead to disruption of the water H-bond network.

A possible generalization of the theoretical approach to the case of anomalous subdiffusive processes has also been worked out. The problems of the applicability of this generalized approach have also been discussed. The exploration of the practical application of the expression deduced for $Q^{\star}$ in that case will be the subject of future work.

Acknowledgements We are indebted to our collaborators in the Polymers and Soft Matter Group in San Sebastian and at the different neutron facilities (Jülich Center for Neutron Science, Institute Laue Langevin, Paul Scherrer Institute, etc.), who helped to shape many of the experiments we report on in this chapter. In particular, we we would like to thank Prof. Angel Alegría and Prof. Dieter Richter for helpful and lively discussions. We acknowledge financial support by the Spanish Ministry 'Ministerio de Economia y Competitividad,' code: MAT2015-63704-P (MINECO/FEDER, UE) and by the Basque Government IT-654-13 (GV).

References

1. https://neutronsources.org/
2. Alegría A, Goitiandía L, Tellería I, Colmenero J (1998) Recent Res Dev Macromol Res 3:49
3. Alegría A, Colmenero J, Mari PO, Campbell IA (1999) Phys Rev E 59:6888
4. Alvarez F, Alegría A, Colmenero J (1991) Phys Rev B 44:7306
5. Alvarez F, Arbe A, Colmenero J, Zorn R, Richter D (2002) Comput Mat Sci 25:596
6. Alvarez F, Colmenero J, Zorn R, Willner L, Richter D (2003) Macromolecules 36:238
7. Angell C (1997) Complex Behavior of Glassy Systems, vol 402. Springer Verlag, Heidelberg, Lecture Notes in Physics
8. Angell CA (1995) Science 267:1924
9. Arbe A, Colmenero J (2009) Phys Rev E 80:041805
10. Arbe A, Alegría A, Alvarez F, Colmenero J, Frick B (1993) Dynamics of the α-relaxation in glass-forming polymeric systems. Steinkopff, Darmstadt, Study by neutron scattering and relaxation techniques, p 24
11. Arbe A, Richter D, Colmenero J, Farago B (1996) Phys Rev E 54:3853
12. Arbe A, Colmenero J, Monkenbusch M, Richter D (1998) Phys Rev Lett 81:590
13. Arbe A, Alegría A, Colmenero J, Hoffmann S, Willner L, Richter D (1999) Macromolecules 32:7572
14. Arbe A, Colmenero J, Alvarez F, Monkenbusch M, Richter D, Farago B, Frick B (2002a) Phys Rev Lett 89:245701
15. Arbe A, Moral A, Alegría A, Colmenero J, Pyckhout-Hintzen W, Richter D, Farago B, Frick B (2002b) J Chem Phys 117:1336
16. Arbe A, Colmenero J, Alvarez F, Monkenbusch M, Richter D, Farago B, Frick B (2003a) Phys Rev E 67:051802
17. Arbe A, Colmenero J, Richter D (2003b) Broadband Dielectric Spectroscopy. Springer, Berlin
18. Arbe A, Genix AC, Colmenero J, Richter D, Fouquet P (2008) Soft Matter 4:1792
19. Arbe A, Malo de Molina P, Alvarez F, Frick B, Colmenero J (2016) Phys Rev Lett 117:185501
20. Arrese-Igor S, Alegría A, Colmenero J (2014a) Phys Rev Lett 113:078302
21. Arrese-Igor S, Alegría A, Colmenero J (2014b) ACS Macro Lett 3:1215
22. Arrese-Igor S, Alegría A, Colmenero J (2015) J Chem Phys 142:214504
23. Arrese-Igor S, Alegría A, Colmenero J (2017) J Chem Phys 146:114502
24. Bagchi B (2012) Molecular Relaxations in Liquids. Oxford University Press
25. Bagchi B, Chandra A (1991) Adv Chem Phys 80:1
26. Bako I, Grosz T, Palinkas G, Bellissent-Funel MC (2003) J Chem Phys 118:3215

27. Bartsch E, Debus O, Fujara F, Kiebel M, Sillescu H, Petry W (1991) Ber Bunsen Phys Chem 95:1146
28. Bartsch E, Bertagnolli H, Chieuxp P, David A, Sillescu H (1993) Chem Phys 169:373
29. Beiner M (2001) Macromol Rap Commun 22:869
30. Bhowmik D, Pomposo JA, Juranyi F, García Sakai V, Zamponi M, Arbe A, Colmenero J (2014) Macromolecules 47:93005
31. Böhmer R, Gainaru C, Richert R (2014) Phys Rep 545:125
32. Braun D, Boresch S, Steinhauser O (2014a) J Chem Phys 140:064107
33. Brodeck M, Alvarez F, Arbe A, Juranyi F, Unruh T, Holderer O, Colmenero J, Richter D (2009) J Chem Phys 130:094908
34. Buchanan M (1996) Nature 382:302
35. Busselez R, Lefort R, Guendouz M, Frick B, Merdrignac-Conanec O, Morineau D (2009) J Chem Phys 130:214502
36. Busselez R, Arbe A, Alvarez F, Colmenero J, Frick B (2011a) J Chem Phys 134:054904
37. Busselez R, Lefort R, Ghoufi A, Beuneu B, Frick B, Affouard F, Morineau D (2011b) J Phys Condens Matter 23:505102
38. Busselez R, Arbe A, Cerveny S, Capponi S, Colmenero J, Frick B (2012) J Chem Phys 137:084902
39. Capponi S, Arbe A, Alvarez F, Colmenero J, Frick B, Embs JP (2009) J Chem Phys 131:204901
40. Capponi S, Arbe A, Cerveny S, Busselez R, Frick B, Embs JP, Colmenero J (2011) J Chem Phys 134:204906
41. Cerveny S, Alegría A, Colmenero J (2008) Phys Rev E 77:031803
42. Colmenero J, Arbe A (1998) Phys Rev B 57:13508
43. Colmenero J, Alegría A, Alberdi JM, Alvarez F, Frick B (1991) Phys Rev B 44:7321
44. Colmenero J, Alegría A, Arbe A, Frick B (1992) Phys Rev Lett 69:478
45. Colmenero J, Arbe A, Alegría A (1994) J Non-Cryst Solids 172–174:126
46. Colmenero J, Arbe A, Coddens G, Frick B, Mijangos C, Reinecke H (1997) Phys Rev Lett 78:1928
47. Colmenero J, Arbe A, Alegría A, Monkenbusch M, Richter D (1999) J Phys Condens Matter 11:A363
48. Colmenero J, Alvarez F, Arbe A (2002) Phys Rev E 65:041804
49. Colmenero J, Arbe A, Alvarez F, Narros A, Richter D (2004) Pramana-J Phys 63:25
50. Colmenero J, Alvarez F, Khairy Y, Arbe A (2013a) J Chem Phys 139:044906
51. Colmenero J, Brodeck M, Arbe A, Richter D (2013b) Macromolecules 46:1678
52. Colmenero J, Alvarez F, Arbe A (2015) EPJ Web of Conferences 83:01001
53. Davidson DW, Cole RH (1950) J Chem Phys 18:1417
54. Davidson DW, Cole RH (1951) J Chem Phys 19:1484
55. Doi M, Edwards SF (1986) The Theory of Polymer Dynamics. Clarendon Press, Oxford
56. Farago B, Arbe A, Colmenero J, Faust R, Buchenau U, Richter D (2002) Phys Rev E 65:051803
57. Frick B, Richter D, Ritter C (1989) Europhys Lett 9:557
58. Fulcher GS (1925) J Am Ceram Soc 8:339
59. Gambino T, Alegría A, Arbe A, Colmenero J, Malicki N, Dronet S, Schnell B, Lohstroh W, Nemkovski K Macromolecules submitted
60. Genix AC, Arbe A, Alvarez F, Colmenero J, Farago B, Wischnewski A, Richter D (2006a) Macromolecules 39:6260
61. Genix AC, Arbe A, Alvarez F, Colmenero J, Schweika W, Richter D (2006b) Macromolecules 39:3947
62. Genix AC, Arbe A, Colmenero J, Wuttke J, Richter D (2012) Macromolecules 45:2522
63. Gerstl C, Brodeck M, Schneider GJ, Su Y, Allgaier J, Arbe A, Colmenero J, Richter D (2012a) Macromolecules 45:7293
64. Gerstl C, Schneider GJ, Fuxman A, Zamponi M, Frick B, Seydel T, Koza M, Genix AC, Allgaier J, Richter D, Colmenero J, Arbe A (2012b) Macromolecules 45:4394
65. Gilroy KS, Phillips WA (1981) Philos Mag B 43:5735
66. Glarum S (1960) J Chem Phys 33:1371

67. Götze W, Sjögren L (1992) Rep Prog Phys 55:241
68. Gurevich VL, Parshin DA, Pelous J, Schober HR (1993) Phys Rev B 48:16318
69. Hansen C, Stickel F, Berger T, Richert R, Fischer EW (1997) J Chem Phys 107:1086
70. Havriliak S, Negami S (1967) Polymer 8:161
71. Hofmann A, Alegría A, Colmenero J, Willner L, Buscaglia E, Hadjichristidis N (1996) Macromolecules 29:129
72. Iradi I, Alvarez F, Colmenero J, Arbe A (2004) Physica B Condens Matter 350:E881
73. Jansson H, Kargl F, Fernandez-Alonso F, Swenson J (2009) J Chem Phys 130:205101
74. Khairy Y, Alvarez F, Arbe A, Colmenero J (2014) Macromolecules 47:447
75. Kohlrausch F (1863) Pogg Ann Phys Chem 119:337
76. Krygier E, Lin G, Mendes J, Mukandela G, Azar D, Jones AA, Pathak JA, Colby RH, Kumar SK, Floudas G, Krishnamoorti R, Faust R (2005) Macromolecules 38:7721
77. Laurati M, Sotta P, Long DR, Fillot LA, Arbe A, Alegría A, Embs JP, Unruh T, Schneider GJ, Colmenero J (2012) Macromolecules 45:1676
78. Lovesey SW (1984) Theory of neutron scattering from condensed matter. Clarendon Press, Oxford
79. Lunkenheimer P, Loidl A (2003) Broadband Dielectric Spectroscopy. Springer, Berlin
80. Lunkenheimer P, Schneider U, Brand R, Loidl A (2000) Contemp Phys 41:15
81. Madden P, Kievelson D (1984) Adv Chem Phys 56:467
82. McCrum N, Read B, Williams G (1967) Anelastic. Dielectric Effects in Polymeric Solids, Wiley, London
83. Metzler R, Klafter J (2000) Phys Rep 339:1
84. Mezei F (1980) Neutron Spin Echo, vol 28. Springer-Verlag, Heidelberg, Lecture Notes in Physics
85. Malo de Molina P, Alvarez F, Frick B, Wildes A, Arbe A, Colmenero J (2017) Phys Chem Chem Phys 19:27739
86. Montroll EW, Weiss GH (1965) J Math Phys 6:167
87. Moreno AJ, Arbe A, Colmenero J (2011) Macromolecules 44:1695
88. Nee T, Zwanzig R (1970) J Chem Phys 52:6353
89. Novikov VN, Sokolov AP (2004) Nature 431:961
90. Paul W, Baschnagel J (1999) Stochastic Processes. Springer Verlag, Berlin Heidelberg From Physics to Finance
91. Perez-Aparicio R, Arbe A, Colmenero J, Frick B, Willner L, Richter D, Fetters LJ (2006) Macromolecules 39:1060
92. Perez-Aparicio R, Arbe A, Alvarez F, Colmenero J, Willner L (2009) Macromolecules 42:8271
93. Petry W, Bartsch E, Fujara F, Kiebel M, Sillescu B Hand Farago (1991) Z Phys B Condens Matter 83:175
94. Powles J (1953) J Chem Phys 21:633
95. Richter D, Arbe A, Colmenero J, Monkenbusch M, Farago B, Faust R (1998) Macromolecules 31:1133
96. Richter D, Monkenbusch M, Willner L, Arbe A, Colmenero J, Farago B (2004) Europhys Lett 66:239
97. Richter D, Monkenbusch M, Arbe A, Colmenero J (2005) Neutron Spin Echo in Polymer Systems, Adv. Polym. Sci., vol 174. Springer Verlag, Berlin Heidelberg, New York
98. Rouse PEJ (1953) J Chem Phys 21:1272
99. Sacristán J, Alvarez F, Colmenero J (2007) Europhys Lett 80:38001
100. Saiz L, Guardia E, Padro JA (2000) J Chem Phys 113:2814
101. Schneider U, Lunkenheimer P, Brand R, Loidl A (1998) J Non-Cryst Solids 235–237:173
102. Sengwa RJ, Kaur K, Chaudhary R (2000) Polym Int 49:599
103. Sobolev O, Novikov A, Pieper J (2007) Chem Phys 334:36
104. Sposito G (1981) J Chem Phys 74:6943
105. Squires GL (1996) Introduction to the theory of thermal neutron scattering. Dover Publication Inc., New York
106. Svanberg C, Bergman R (2001) J Non-Cryst Solids 283:225

107. Tölle A (2001) Rep Prog Phys 64:1473
108. Tyagi M (2007) Private commun
109. Tyagi M, Alegría A, Colmenero J (2005) J Chem Phys 122:244909
110. Tyagi M, Arbe A, Alvarez F, Colmenero J, González MA (2008) J Chem Phys 129:224903
111. Vispa A, Rovira-Esteva M, Ruiz-Martín MD, Busch S, Unruh T, Pardo LC, Tamarit JL (2014) J Phys: Conf Series 549:012013
112. Vogel H (1921) Phys Z 22:645
113. Williams G, Watts DC (1970) Trans Faraday Soc 66:80
114. Wuttke J, Chang I, Randl OG, Fujara F, Petry W (1996) Phys Rev E 54:5364
115. Wuttke J, Chang I, Fujara F, Petry W (1997) Physica B Condens Matter 234–236:431
116. Yu D, Hennig M, Mole RA, Li JC, Wheeler C, Strassle T, Kearley GJ (2013) Phys Chem Chem Phys 15:20555

The Scaling of the Molecular Dynamics of Liquid Crystals as Revealed by Broadband Dielectric, Specific Heat, and Neutron Spectroscopy

Andreas Schönhals, Bernhard Frick and Reiner Zorn

Abstract A combination of different complementary methods is employed to investigate scaling of the molecular dynamics of two different liquid crystals. Each method is sensitive to different kind of fluctuations and provides therefore a different window to look at the molecular dynamics. In detail, broadband dielectric spectroscopy is combined with specific heat spectroscopy and neutron scattering. As systems, the nematic liquid crystal E7 and a discotic liquid crystalline pyrene are considered. First of all, it was proven that both systems show all peculiarities which are characteristic for glassy dynamics and the glassy state. Especially for the nematic liquid crystal E7, it could be unambiguously shown by a combination of dielectric and specific heat spectroscopy that the tumbling mode is the underlying motional process responsible for glassy dynamics. Dielectric investigations on the discotic liquid crystalline pyrene reveal that at the phase transition from the plastic crystalline to the hexagonal columnar liquid crystalline phase, the molecular dynamics changes from a more strong to fragile temperature dependence of the relaxation rates. Moreover, a combination of results obtained by specific heat spectroscopy with data from structural methods allows an estimation of the length scale relevant for the glass transition.

Keywords Broadband dielectric spectroscopy · Specific heat spectroscopy
X-ray Photoelectron spectroscopy · Rod-like liquid crystal · Discotic liquid crystal

A. Schönhals (✉)
Bundesanstalt für Materialforschung und-prüfung (BAM), Unter den Eichen 87,
12205 Berlin, Germany
e-mail: andreas.schoenhals@bam.de

B. Frick
Institut Laue-Langevin, 71, Avenue Des Martyrs, Grenoble 38000, France

R. Zorn
Jülich Centre for Neutron Science, JCNS and Institute for Complex Systems ICS,
Forschungszentrum Jülich, 52425 Jülich, Germany

F. Kremer and A. Loidl (eds.), *The Scaling of Relaxation Processes*,
Advances in Dielectrics, https://doi.org/10.1007/978-3-319-72706-6_9

Abbreviations

BP	Boson Peak
$c_p^* = c_p' - i\, c_p''$	Complex heat capacity, c_p'—real part, c_p''—loss part
D	Fragility parameter
δT	Width of the glass transition
E_A	Activation energy
$\varepsilon * = \varepsilon' - i\varepsilon''$	Complex dielectric function, ε'—real part, ε''—loss part
f	Frequency
f_p	Relaxation rate
f_∞	Relaxation rate at infinite temperatures
HN	Havriliak–Negami function
k_B	Boltzmann constant
q	Scattering vector
$S(q,\omega)$	Dynamic structure factor
$S(q,t)$	Incoherent intermediate scattering function
T_g	Glass transition temperature
T_0	Vogel or ideal glass transition temperature
τ	Relaxation time
VDOS, $g(\omega)$	Vibrational density of states
VFT	Vogel–Fulcher–Tammann
ξ	Correlation length for the glass transition
ω	Angular frequency ($\omega = 2\pi f$)

1 Introduction

The glass transition and the underlying molecular mobility are current important problems of condensed matter physics [1–6]. It is not completely understood till today whether a glass-forming system freezes into the glassy state instead of solidifying as a crystal. As an experimental fact, a dramatic increase of both the viscosity and the structural relaxation time, τ (relaxation rate $f_p = 1/(2\pi\tau)$), is observed when a glass-forming system is cooled down. The corresponding relaxation process is called structural relaxation (α-relaxation) or dynamic glass transition. Microscopic and macroscopic timescales are bridged when decreasing the temperature by a factor of two toward the glass transition temperature T_g as τ increases by more than 14 orders of magnitude. In a temperature region close to the glass temperature, the temperature dependence of τ or f_p can be described by the so-called Vogel–Fulcher–Tammann equation [7–9] which reads

$$\log f_p = \log f_\infty - \frac{A}{T - T_0} \tag{1}$$

where f_∞ is the relaxation time for $T \to \infty$, and A is a constant related to the fragility parameter $D = A/(\ln(10\, T_0))$ [23, 24]. The fragility approach is among other approaches a useful scheme to classify glass-forming systems. Materials are called "fragile" if the temperature dependence of τ deviates strongly from an Arrhenius-type behavior and "strong" if $\tau(T)$ is close to the latter. T_0 is the so-called Vogel or ideal glass transition temperature, empirically found 30–70 K below the T_g measured by differential scanning calorimetry.

In a widely discussed concept, called cooperativity approach to the glass transition [10–13] this slowing down of the relaxation rates is related to the assumption that in the environment of a given molecule, other molecules have to rearrange cooperatively to reach the minimum of a complex energy landscape.

Besides the α-relaxation, further dynamical processes are observed which are important for the glass transition phenomena. Goldstein and Johari argued that a so-called slow β-relaxation can be considered as a precursor of the α-relaxation in cases where the process (true genuine β-relaxation) cannot be assigned to internal degrees of freedom [14, 15]. The temperature dependence of the relaxation time is characterized by an Arrhenius law

$$f_p = f_\infty \exp\left(-\frac{E_A}{k_B T}\right), \tag{2}$$

where E_A is the activation energy, and k_B is the Boltzmann constant.

A further molecular process in relationship to the glass transition is observed in the frequency range of THz. This process is due to excess contributions to the vibrational density of states (VDOS) $g(\omega)$ in addition to what is predicted by the Debye theory of sound waves, which gives $g(\omega) \sim \omega^2$. This process is called Boson Peak (BP) [16]. The molecular origin of the BP is under controversial discussion in the literature. There are two main groups of theoretical approaches. In the first group of theories, the so-called soft potential approach, the Boson Peak is considered to be different from sound waves and originates from peculiarities of the interatomic forces in the amorphous structure (groups of atoms subjected to a soft potential) [17–19]. In the second group of theories, it was argued that the VDOS of an amorphous material is in principle related to sound waves (frequency-broadened version of the Van Hove-singularity of crystalline systems [20]). In this scenario, the VDOS of the corresponding crystalline system is modified by random fluctuations of the force constants [21]. In addition to these approaches, in a numerical simulation it was recently shown that the Boson Peak can be also explained as a consequence of a local breaking of center-inversion symmetry [22].

A glass transition is not only be observed for the widely studied low molecular weight liquids, silica glasses, or polymers but also for systems with partial order and restricted mobility [23, 24]. Seki and Suga reported a glass transition of plastic crystals for the first time [25]. For systems with restricted mobility, due to a partial order, it is discussed till today which molecular processes are responsible for glassy dynamics. Further examples of glassy dynamics in plastic crystals are discussed for instance in [26–30]. It is worth to be noted that also discotic liquid crystals show a kind of glassy dynamics (see for instance [31–33]). Investigations on such systems like liquid crystals might be suited to study the glass transition phenomena in general.

Thermotropic liquid crystals are a class of soft matter materials which have structures between the completely ordered crystalline and the disordered isotropic state [34, 35]. They are assembled for instance from molecules which either have a stiff, rod-like mesogenic unit in their structure or for discotic liquid crystals by disk-like molecules [36]. In the crystal state, the molecules have a maximal (positional and translational) order and a minimal mobility, while in the liquid state the opposite is valid. Liquid crystalline (or mesomorphic) phases combine different aspects of crystals and liquids and order and mobility compete. For rod-like mesogenic molecules, the most important liquid crystalline states are nematic and smectic phases. In the nematic state, the molecules have the lowest liquid crystalline order. The molecules have no long-range order but all translational degrees of freedom. The long axes of the mesogenes are preferentially aligned with respect to a common unit vector the so-called nematic director. An order parameter S [35] can be introduced to characterize the degree of orientation. In smectic phases, the molecules arrange in layer-like structures. Besides an orientation, they have also some positional order where the degree of positional order depends on the type of the smectic phase.

Discotic liquid crystals consist of a stiff disk-like core surrounded by flexible alkyl side chains [36]. Whereas the former leads to π-stacking, the latter increases the solubility. At low temperatures discotic liquid crystals form mostly plastic crystalline phases, this can be followed by a hexagonal columnar mesophase at higher temperatures. Here, the disk-like molecules assemble in columns where the alkyl chains fill the intercolumnar space. Besides that, also nematic phases can be observed [37]. At even higher temperatures, a phase transition to a more or less isotropic phase is observed.

The most mesogens are polar molecules. Therefore, in particular broadband dielectric spectroscopy is a powerful method to study the molecular dynamics in liquid crystalline systems. For instance, some reviews are available in the literature with regard to dielectric properties of liquid crystals [38–40]. Till now, a theory for the dielectric response is only available for rod-like shaped mesogenic molecules. For anisotropic systems like nematic and smectic liquid crystals for rod-like mesogens, the complex dielectric function $\varepsilon^*(\omega)$ (ω—angular frequency) has to be described by a tensor. For the simplest case of the uniaxial nematic phase, this tensor has only two components $\varepsilon^*_{\|}(\omega)$ and $\varepsilon^*_{\perp}(\omega)$ parallel and perpendicular to the nematic director, respectively. In reference [38], the theoretical approach to the theory of the dielectric relaxation of liquid crystals is briefly outlined. A more complete discussion can be found in [41–43]. One has to consider that a linear mesogenic molecule has two components of its dipole moment vector. One is oriented parallel, and the second one is directed perpendicular to the long axis of the mesogen. In general, the dielectric response is due to correlation functions of polarization fluctuations [44]. In the case of liquid crystals, the correlation function for both components has to be calculated. In a semi-microscopic approach [41], the measured dielectric function parallel $\varepsilon^*_{\|}(\omega)$ and perpendicular $\varepsilon^*_{\perp}(\omega)$ to the director comprises different weighted sums of four underlying relaxation modes. The dielectric intensity of each mode depends on the macroscopic orientation of the sample. One relaxation mode is related to rotational fluctuations of the molecules around its short axis. This process contributes mainly to $\varepsilon_{\parallel}^{*}(\omega)$ and is called δ-relaxation. The other three modes are due to different tumbling modes around the long axis of the molecule. These three modes have almost the same relaxation rate and form one broad relaxation process related to $\varepsilon_{\perp}^{*}(\omega)$ which is observed at higher frequencies than the δ-relaxation.

According to this theoretical approach, the temperature dependence of the relaxation times should follow the Arrhenius equation for both relaxation processes. Due to the nematic potential [35, 45], the (apparent) activation energy of the δ-relaxation should be higher than that of the tumbling process which was confirmed experimentally [38, 46–50]. However, a combination of measurements carried out in a larger temperature range and a more detailed data analysis have shown that the temperature dependence of the relaxation time can be better described by the VFT equation. This is especially true for the isotropic phase (see for instance [51, 52]). But also in the liquid crystalline state deviations from the predicted Arrhenius-like temperature dependence have been reported [53–63]. A free volume approach is employed by Diogo and Martins [56] to discuss the VFT-like temperature dependence of the relaxation rate of the δ-process. As for the glass transition in general, the glassy dynamics of liquid crystals is not completely understood. Especially for linearly shaped liquid crystals the question which mode, the δ- or the tumbling, is responsible for the glass transition is hardly addressed. The same is valid for the glassy dynamics of discotic liquid crystals. Here, a combination of broadband dielectric, specific heat, and neutron spectroscopy is employed to investigate the molecular dynamics of a linear nematic liquid crystal and a discotic liquid crystal.

2 Experimental

2.1 *Materials*

2.1.1 Rod-like Liquid Crystal

As an example for the glassy dynamics of a liquid crystal consisting of rod-like molecules, the commercially available nematic mixture E7 is considered. E7 was obtained from Merck (Darmstadt, Germany). To avoid crystallization, E7 is mixture of four components 4-cyano-4′-pentyl-1,1′-biphenyl (51 wt%), 4-n-heptyl-4′-cianobiphenyl (25 wt%), 4,4′-n-octyloxycyanobiphenyl (16 wt%), and 4′-n-pentyl-4-cyanoterphenyl (8 wt%) [58, 64] (see Fig. 1). Figure 2 gives a DSC measurement (heating rate 10 K/min, second heating run) for E7. A glass transition can be observed at 211 K while the transition from the nematic to the isotropic state takes place at $T_{I/N} = 333$ K [58, 64].

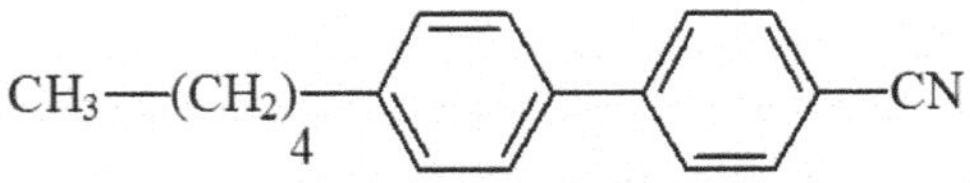

4-cyano-4’-pentyl-1,1’-biphenyl (51 wt-%)

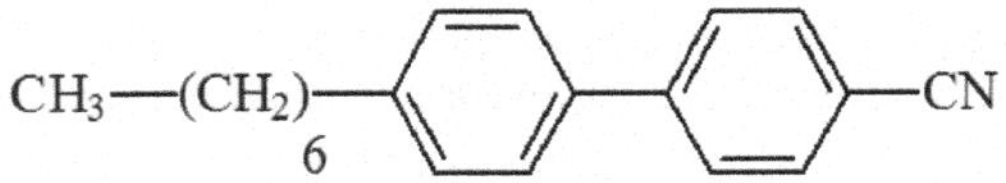

4-n-heptyl-4’-cianobiphenyl (25 wt-%)

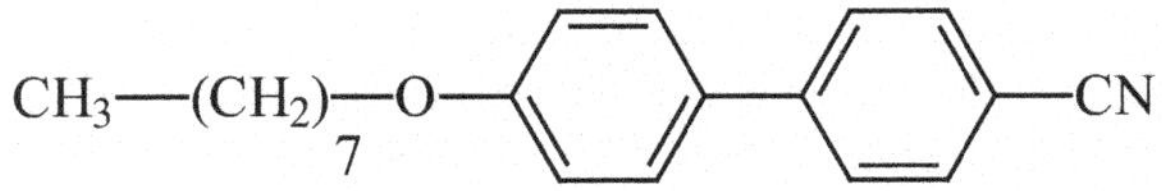

4,4’-n-octyloxycyanobiphenyl (16 wt-%)

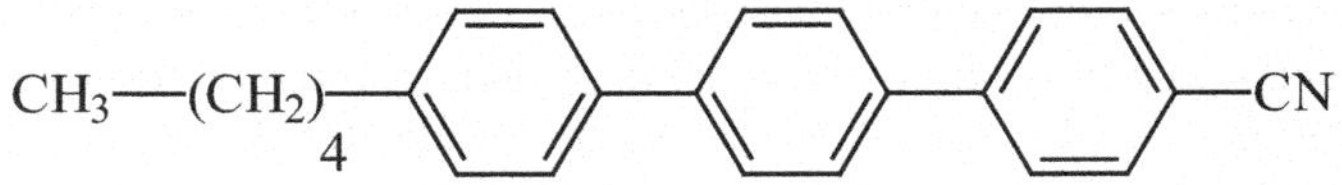

4’-n-pentyl-4-cyanoterphenyl (8 wt-%)

Fig. 1 Composition of the nematic liquid crystal E7

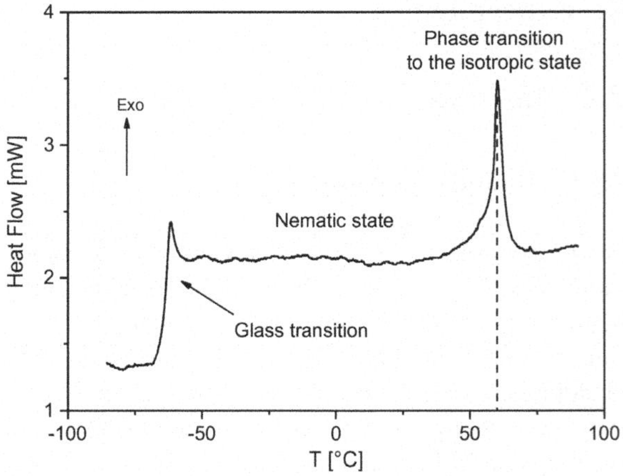

Fig. 2 DSC trace for E7 (heating rate 10 K/min, second heating). The figure was adapted from reference [64]

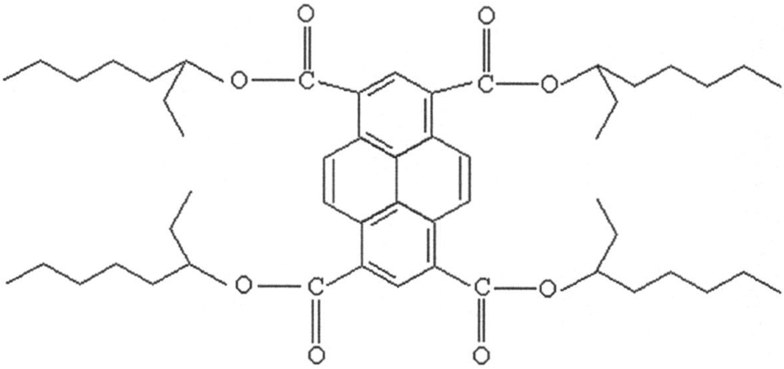

Fig. 3 Chemical structure of pyrene-1,3,6,8-tetracarboxylictetra(2-ethylhexyl)ester (Py4CEH)

2.1.2 Discotic Liquid Crystal

As an example for a discotic liquid crystal, the pyrene-based compound pyrene-1,3,6,8-tetracarboxylictetra(2-ethylhexyl)ester (Py_4CEH) is considered. The synthesis of Py_4CEH is discussed in reference [65], and the structure is given in Fig. 3. Py_4CEH has a plastic crystal phase at low temperatures. During heating, a phase transition from plastic crystal phase to a hexagonal columnar liquid crystalline phase is observed at $T_{cry,col} = 247$ K (see Fig. 4a) [66]. Upon further heating, the transition from the hexagonal columnar to the more or less isotropic state takes place at $T_{col,iso} = 369$ K. There is a small hysteresis for the transition from isotropic to liquid crystalline state during cooling $T_{col,iso} = 365$ K ($\Delta T = 4$ K, see Fig. 4). For the transition from the liquid crystalline to plastic crystal state, the hysteresis during cooling is larger $T_{cry,col} = 221$ K ($\Delta T = 26$ K). Figure 4b enlarges the heat flow in the temperature range between 120 and 200 K during cooling. A small step-like change is observed which might be considered as a sign for a glass transition.

2.2 *Broadband Dielectric Spectroscopy*

For a detailed discussion of broadband dielectric spectroscopy (BDS), see references [44, 67]. For the further discussion, one should keep in mind that BDS is sensitive to dipole fluctuations. The complex dielectric function $\varepsilon^*(\omega) = \varepsilon'(\omega) - i\varepsilon''(\omega)$ (ε'—real part, ε''—loss or imaginary part, $i = \sqrt{-1}$) was measured in the frequency range from 10^{-2} to 10^9 Hz by a combination of two different methods. For frequency from 10^{-2} to 10^6 Hz, Fourier correlation analysis was employed utilizing a high-resolution Alpha analyzer with an active sample head (Novocontrol®, Montabaur). A coaxial reflectometer based on the impedance analyser Agilent E4991 also provided by Novocontrol was used to measure $\varepsilon^*(\omega)$ in the frequency range from 10^6 to 10^9 Hz. In the latter case, the sample was treated as part of the inner conductor of the

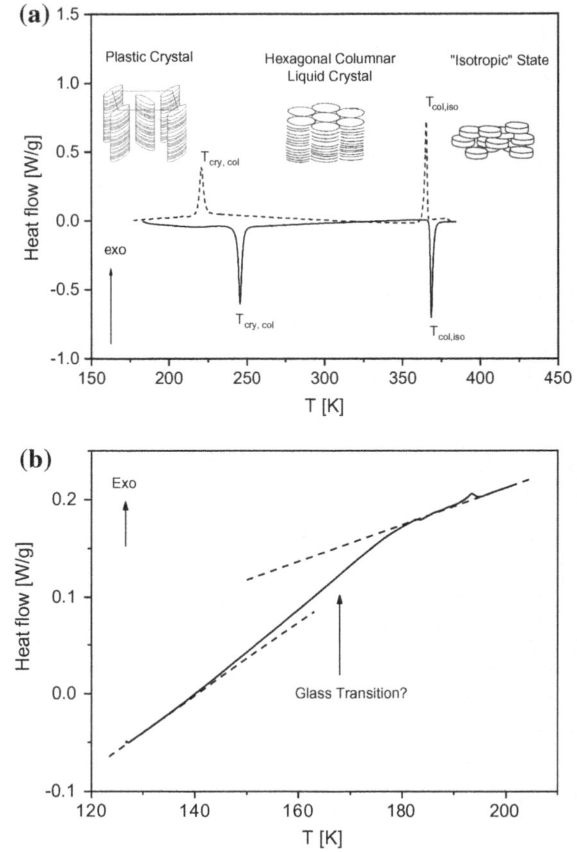

Fig. 4 **a** DSC thermogram of Py4CEH during cooling (dashed line) and heating (solid line) measured with a cooling/heating rate of 10 K/min. The figure was adopted from reference [66], **b** Heat flow versus temperature during cooling in the temperature range from 120 K to 200 K for Py4CEH. Cooling rate 10 K/min

coaxial line [67]. The samples were measured in parallel plate geometry (electrode diameters 20 mm in the low-frequency range and 6 mm for the high-frequency range). A thickness of the samples of 50 μm was maintained by fused silica spacers. For both experimental setups, the temperature was controlled by a Quatro Novocontrol® cryo-system with a temperature stability better than 0.1 K. Figure 5 gives as an example the dielectric loss of a non-oriented E7 sample in the frequency range from 10^6 to 10^9 Hz versus frequency and temperature in a 3D representation. The data were taken from reference [68]. In the isotropic state, only one process is observed as expected for conventional glass-forming systems. At the phase transition from the isotropic to the nematic state, the dielectric loss changes abruptly and two processes are observed which correspond to the δ-relaxation at lower frequencies and to the tumbling mode at higher frequencies.

The measured dielectric spectra were analyzed by fitting the model function of Havriliak–Negami (HN-function) [69] to the data. The HN-function reads

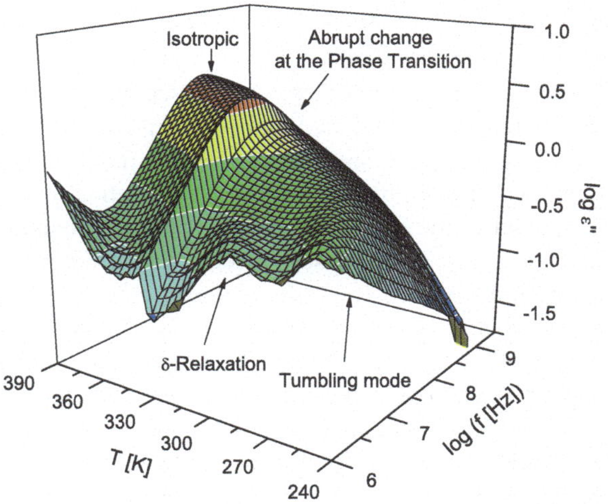

Fig. 5 Dielectric loss versus frequency and temperature in a 3D representation for E7. The figure was adopted from reference [68]

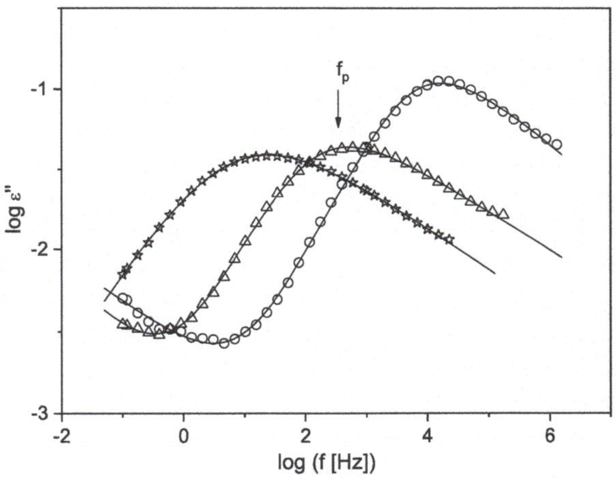

Fig. 6 Dielectric loss versus frequency for Py4CEH (asterisks—207 K, triangles—221 K, circles—233 K). The lines are fits of the HN-function including a conductivity contribution. Data were taken from reference [66]

$$\varepsilon^*(\omega) - \varepsilon_\infty = \frac{\Delta\varepsilon}{\left(1 + (i\omega/\omega_0)^\beta\right)^\gamma}. \tag{3}$$

$\Delta\varepsilon$ symbolizes the dielectric strength, and ω_0 is a characteristic frequency related to the frequency of maximal loss f_p (relaxation rate). ε_∞ describes the value of the real part ε' for $\omega \gg \omega_0$. β and γ are fractional parameters ($0<\beta\leq 1$ and $0<\beta\gamma\leq 1$) characterizing the symmetric and asymmetric broadening of the spectra. To treat conduction effects, the term $\sigma/(\omega^s\varepsilon_0)$ is added to the loss part of the HN-function, where σ is related to the DC conductivity. The parameter s ($0<s\leq 1$) describes ohmic, for $s=1$, and non-ohmic effects, for $s<1$, in the conductivity. Figure 6 provides an example for the fitting of the HN-function to data of Py_4CEH.

If several relaxation processes are observed in the accessed frequency window, a sum of HN-functions is fitted to the data.

2.3 Specific Heat Spectroscopy

Specific heat spectroscopy was carried out by a combination of temperature-modulated DSC (TMDSC) and AC chip calorimetry. The theoretical formulation of the specific heat spectroscopy in the frame of the linear response theory was done by Schawe [70]. In principle, a periodically modulated temperature program is applied to the sample. The modulated temperature program results in a modulated heat flow as response. The data can be described by a complex heat capacity $c_p^*(\omega) = c_p'(\omega) - ic_p''(\omega)$ (c_p'—real part, c_p''—loss or imaginary part, $\tan\delta = c_p''/c_p'$. If molecular fluctuations take place in the sample which are sensitive to entropy fluctuations like the glassy dynamics, a phase shift is observed between the frequency dependence of the temperature modulation and the heat flow.

Temperature-modulated measurements were carried out in the frequency range of 1–55 mHz by the so-called StepScan DSC (SSDSC) approach. SSDSC is a special variant of TMDSC [71]. In principle fast but small (ΔT ca. 1 K) heating steps are carried out followed by an isothermal period. The length of the isothermal period determines the frequency of the measurement in sense of a Fourier transformation. Possible underlying heating rates were adjusted to meet stationary conditions. As devices, a Perkin Elmer Pyris 1 and a Perkin Elmer Diamond DSC were employed. Gaussians were fitted to $\tan\delta$ to estimate the dynamic glass transition temperature. It should be noted that $\tan\delta$ has to be corrected for heat conductivity effects. Figure 7 gives an example for a TMDSC measurement for the rod-like liquid crystal E7 at the indicated frequencies in the nematic state. Figure 7a shows data for the real part of the specific heat capacity c_p' where Fig. 7b displays the corrected $\tan\delta_{cor}$.

AC chip calorimetry [72] was employed to estimate the complex heat capacity in a frequency range from 1 Hz to 10 kHz. In the differential setup, this method has a sensitivity of pJ/gK. As nanocalorimeter chips, the sensors XEN 3974 and XEN 39390 from Xensor Integration (Nl) were used (see reference [73]). The sensors consist of a silicon nitride (SiN) membrane with a protected small heated area. In that heated area, heaters as well as thermopiles are integrated. By the differential approach, the contribution of the chip itself to the measured heat capacity is minimized. The heat capacity of the sample C_S is then given by

$$C_s = \frac{i\omega C_{\mathrm{eff}}(\Delta U - \Delta U_0)}{P_0 S}, \tag{4}$$

where the approximation of thin films is used. The effective heat capacity of the empty sensor is described by $\bar{C} \equiv C_0 + G/i\omega$ where $G/i\omega$ is the heat loss through the surrounding atmosphere. S and P are the sensitivity of the thermopiles and the applied heating power, respectively. ΔU is the complex differential thermopile signal for an empty and a sensor with a sample, and ΔU_0 is the complex differential voltage measured for two empty sensors measured by a lock-in approach. The temperature scan mode was employed for Py_4CEH, which means that the frequency was kept constant while the temperature was scanned with a heating/cooling rate of 1–2.0 K/min, depending on the used frequency to ensure stationary conditions. For E7, isothermal

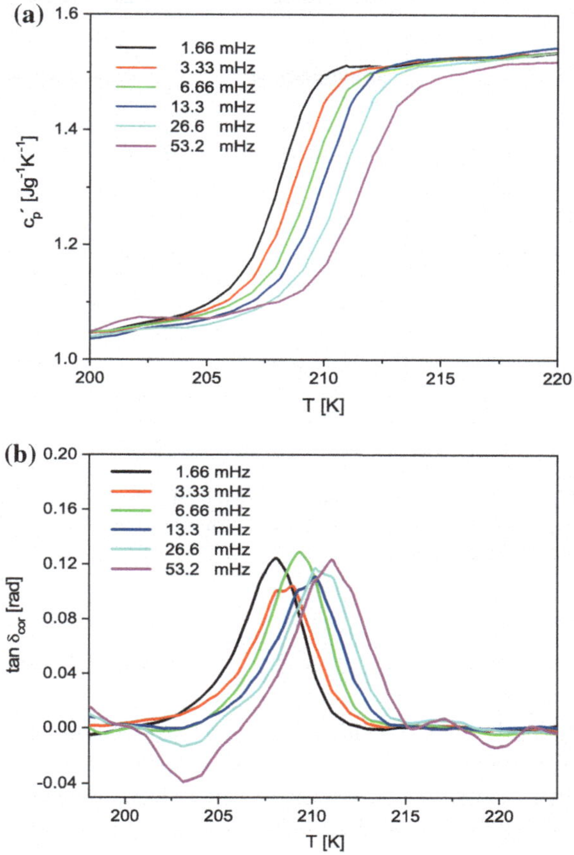

Fig. 7 **a** Real part of the complex specific heat spectroscopy c_p' for E7 at the indicated frequencies measured by TMDSC, **b**: Corrected loss factor of the complex heat spectroscopy tan δ_{cor} for E7 at the indicated frequencies measured by TMDSC

frequency scans were carried out where the temperature was changed in steps of 2 K. During the measurements, the heating power for the modulation was kept constant at about 25 μW, which ensures that the amplitude of the temperature modulation is less than 0.5 K. Figure 8 displays an AC chip calorimetry measurement for the liquid crystal E7.

2.4 *Neutron Spectroscopy*

Momentum and energy are exchanged between the neutrons and the nuclei of a material under investigation during a scattering process. Therefore, inelastic and quasielastic neutron scattering is sensitive to the space and time dependence of the molecular dynamics at microscopic length and time scales [74]. The experimental information is contained in the double differential cross section

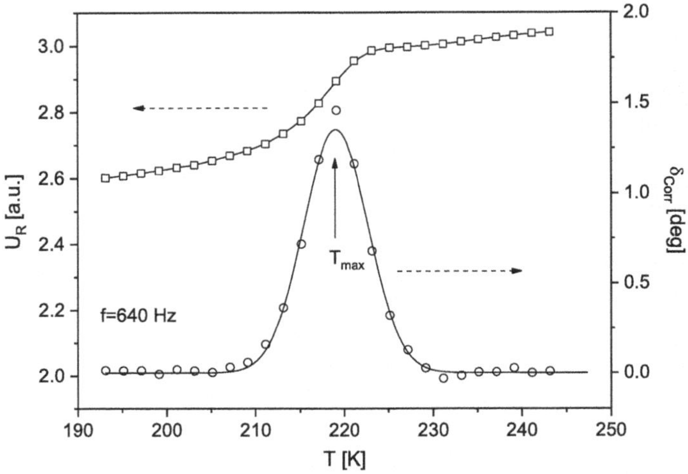

Fig. 8 Real part of the complex differential voltage U_R and corrected loss factor for E7 versus temperature measured by AC chip calorimetry at the indicated frequency. The figure was adopted from reference [64]

$$\frac{d^2\sigma}{d\Omega d\omega} = \frac{1}{4\pi}\frac{k_f}{k_i}(\sigma_{coh} S_{coh}(q,\omega) + \sigma_{inc} S_{inc}(q,\omega)). \quad (5)$$

k_i and k_f are the incident and final wave vectors of the neutron beam and q is the scattering vector. Ω denotes the solid angle of detection. The measured energy transfer as experimentally measured quantity is related to an angular frequency ω by $\omega = \Delta E/\hbar$. $S_i(q,\omega)$ denotes the incoherent and coherent dynamic structure factors (scattering functions, i = coh, inc) where σ_I are the corresponding scattering cross sections for coherent and incoherent scattering. Figures 1 and 3 show that structures of both mesogenic molecules contain a large number of hydrogen atoms. The incoherent scattering cross section for hydrogen is essentially higher than all coherent cross sections. Therefore, the scattering is mainly incoherent and dominated by the hydrogen nuclei.

Three different neutron scattering experiments are carried out. The low-frequency vibrational density of states was measured by inelastic neutron scattering using the time-of-flight spectrometer IN6 (wavelength λ = 5.12 Å). An overview about the molecular dynamics at a timescale of about 1 ns was obtained by elastic fixed window scans utilizing the neutron backscattering spectrometer IN10 (wavelength λ = 6.271 Å). Finally, quasielastic neutron scattering is carried out by combining time-of-fight (IN6) and backscattering (IN10) neutron spectroscopy. All used neutron spectrometers are operated by the Instiut Laue–Langevin (Grenoble/France). In all cases, the resolution of the spectrometers was obtained by measuring the sample at T = 2 K. The data were corrected for the scattering of the empty can. The program INX [75] was used for background and adsorption corrections as well as to obtain absolute values of the structure factors.

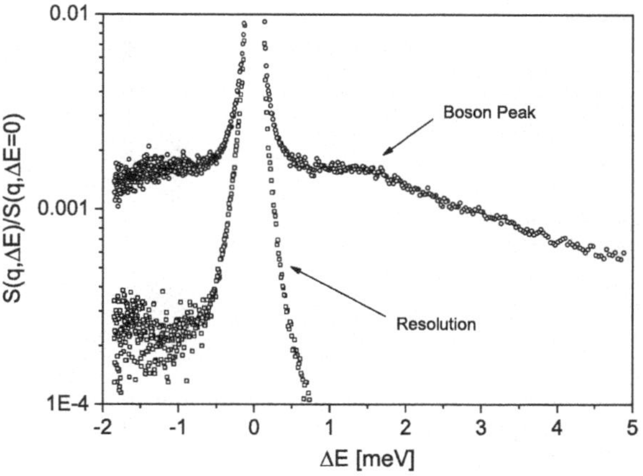

Fig. 9 IN6 spectra at a q vector of 1.66 Å^{-1}. Squares—resolution of IN6 measured for the sample at 2 K; circles—inelastic scattering measured at 100 K. The data are normalized to the height of the elastic line

Inelastic neutron scattering: To estimate the low-frequency density of states g(ω), the time focusing time-of-flight spectrometer IN6 is used. Figure 9 shows the incoherent dynamic structure factor for E7 measured at 100 K at a q value of 1.66 Å^{-1} in comparison to the resolution of IN6 measured for the sample at 2 K (S(q,ΔE) values were normalized to the height of the elastic line.). The data show a Boson Peak at an energy transfer of ca. 1.5 meV. The standard expression for the one-phonon scattering [76]

$$S(q,\omega) = e^{-2W(q)}\left(\delta(\omega) + \frac{\hbar q^2 g(\omega)}{2\bar{m}(-\omega)} \times \left(1 - \exp\left(\frac{\hbar\omega}{k_B T}\right)\right)^{-1}\right) \quad (6)$$

is used to extract g(ω) from the measured data. $e^{-2W(q)}$ is the Debye–Waller factor, and $\bar{m}$ denotes the average mass of an atom. In the experiment, the measured scattering is a convolution of the sample signal with the resolution function of the instrument. The convolution effect on the measured data can be neglected because the Boson Peak is a broad effect (see Fig. 9). Therefore, Eq. (6) can be approximated by [77]

$$S_{obs}(q,\omega) = R(q,\omega) \otimes S(q,\omega) \approx \\ e^{-2W(q)}\left(R(q,\omega) + \frac{\hbar q^2 g(\omega)}{2\overline{m}(-\omega)} \times \left(1 - \exp\left(\frac{\hbar\omega}{k_B T}\right)\right)^{-1}\right). \quad (7)$$

If the scattering is measured at two different temperatures, here 2 and 100 K, a system of linear equations is obtained. For that set of equations, g(ω) and R(ω) can be calculated.

Elastic fixed window scans: Elastic fixed window scans (ΔE ≈ 0) on a backscattering instrument are a powerful tool to obtain an overview about the molecular

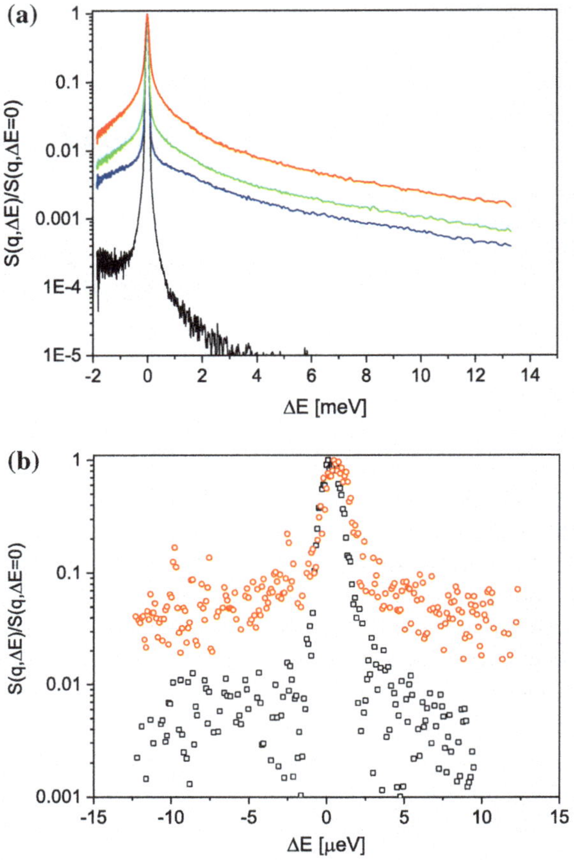

Fig. 10 **a** Neutron time-of-flight spectra from IN6 for PDMS at $q = 1.66$ Å^{-1}. The spectra were normalized by the peak value of the elastic line. Temperatures (bottom to top): 2 K (resolution), 200, 240, and 300 K, **b** Neutron backscattering spectra from IN10 for E7 at $q \approx 1.68$ Å^{-1}. The spectra were normalized by the peak value of the elastic line. Black—resolution measured at T = 5 K, red T = 240 K

dynamics at a timescale corresponding to the resolution of the used instrument. An effective mean-squared displacement $< \Delta u^2 >$ can be calculated from

$$\frac{I_{el}}{I_0} = \exp\left(-\frac{q^2\langle\Delta u^2\rangle}{3}\right), \tag{8}$$

assuming a Gaussian distribution for the displacements. I_{el} and I_0 are the elastically and the totally scattered intensities, respectively. The latter was approximated by the elastic scattering at $T \approx 0$ K.

Quasielastic neutron spectroscopy: Neutron time-of-flight and neutron backscattering spectroscopy were combined to cover, as for other methods, a broad time range by neutron scattering. Figure 10 depicts quasielastic spectra for E7 measured at IN6 (Fig. 10a) and IN10 (Fig. 10b). Both types of spectra show the characteristic quasielastic broadening at higher temperatures compared to the elastic line.

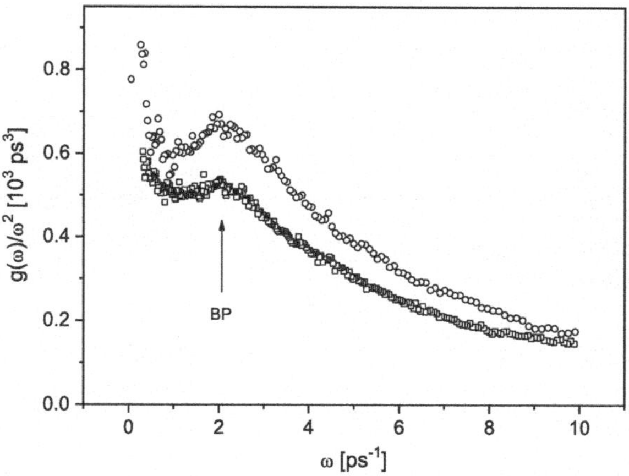

Fig. 11 Vibrational density of states normalized by ω^2 versus frequency: squares—E7; circles—poly(methyl phenyl siloxane). The data for the latter material were taken from reference [78]

Figure 10 shows that the data measured by IN6 (resolution 80–100 μeV) and IN10 (resolution ca. 1 μeV) cannot be compared directly in the energy domain because of the large difference in the energy resolution. As discussed above, in the energy domain the measured data are a convolution of the sample signal with the resolution of the instrument. Therefore, the data are Fourier transformed to the time domain which converts the convolution to a product. Absolute values of the incoherent intermediate time-dependent scattering function S(q,t) are obtained by dividing the measured $S_{obs}(q,t)$ by the Fourier transform of the resolution. By this procedure, both sets of data can be directly compared in the time domain.

3 Results

3.1 Rod-like Liquid Crystal E7

Figure 2 shows that the nematic liquid crystal undergoes a glass transition at a temperature of 211 K. In order to investigate whether also other characteristic features of glasses can be observed, inelastic neutron scattering is carried out to measure the low-frequency vibrational density of states. Figure 11 depicts the low-frequency density of states normalized by ω^2. A Boson Peak is observed in that representation at a frequency of ca. 2 ps^{-1}. The data for E7 are compared to the BP of the conventional glass-forming system poly(methyl phenyl siloxane) [78]. This comparison shows that the Boson Peak observed for the nematic glass E7 is similar to the BP of a conventional glass-forming material. This will give evidence to the hypothesis that the Boson Peak is due to sound waves because it could not be expected that a soft potential for an amorphous glass is similar to that for an nematic one.

In a limited frequency range, dielectric data for E7 have been already reported [57–64]. Figure 12 depicts the dielectric spectra of E7 in both the isotropic and liquid

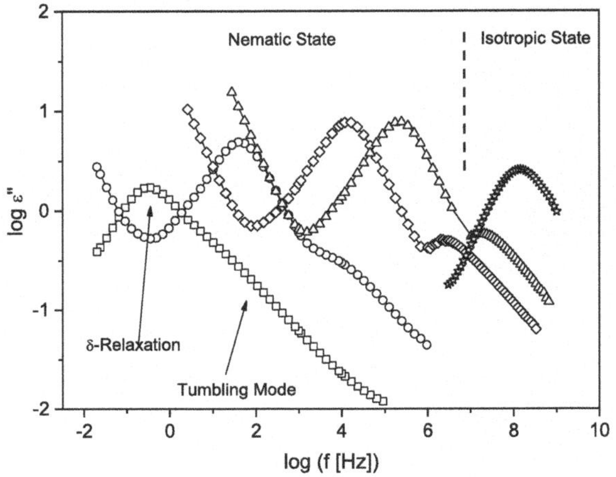

Fig. 12 Dielectric loss versus frequency for E7: squares—214.2 K, circles—220.1 K, rhombuses—252.2 K, triangles—274.2 K, asterisks—274.2. Lines are guides for the eyes. The figure was adapted from reference [68]

crystalline state. In the isotropic state, only one process is observed. This process splits into two relaxation modes in the nematic state. The process at lower frequencies is the δ-relaxation where the process is the tumbling mode.

In Fig. 13, the relaxation rates in the different states are plotted in the activation diagram. For the process observed in the isotropic state, the temperature dependence is curved when plotted versus 1/T, indicating glassy dynamics. Also, in the nematic state, the temperature dependencies are a nonlinear versus 1/T for both the δ-relaxation and the tumbling mode. Although pre-transition behavior is observed around the phase transition indicated by an unusual temperature dependence of f_p, at lower temperature $f_p(T)$ for both processes might follow the VFT dependence. At the glass transition temperature, the temperature dependencies of both processes seem to merge.

To analyze the temperature dependencies of the δ-relaxation and of the tumbling mode in more detail, a derivative method is utilized. This method is sensitive to the temperature dependence of $f_p(T)$ irrespectively of the prefactor f_∞. Taking the derivative of the VFT equation with respect to temperature, one obtains

$$\left(\frac{\mathrm{d}\,\log f_p}{\mathrm{d}T}\right)^{-1/2} = A^{-1/2}(T - T_0). \tag{9}$$

This means, in that representation, a VFT dependence will show up as a straight line when plotted versus temperature. The inset of Fig. 13 shows that the temperature dependence of the relaxation rates of the process in the isotropic state has to be described by a VFT equation. In Fig. 14, the derivative according to Eq. (9) is calculated for the δ-relaxation and the tumbling mode and plotted versus temperature. This figure shows that in the temperature range close to the glass transition temperature, the temperature dependence of the relaxation rates of both the δ-relaxation and the tumbling mode has to be described by a VFT equation. The temperature dependencies of the relaxation rates of both processes are essentially different, expressed

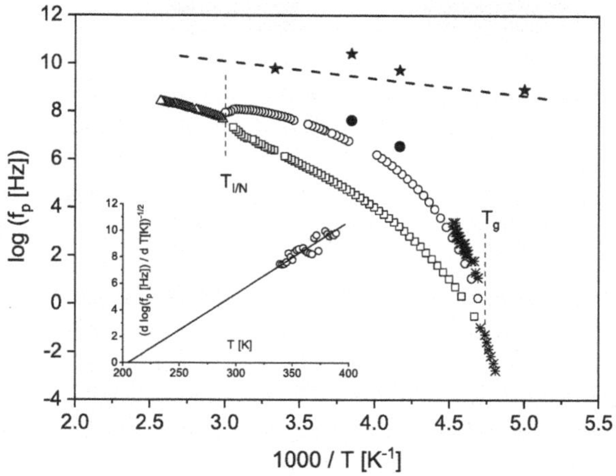

Fig. 13 Temperature dependence of the relaxation rates in the activation diagram for E7. Dielectric data: triangles—dielectric process in the isotropic state; squares—tumbling mode; circles—δ-relaxation. Thermal data—crosses. Neutron data: filled circles—slow process from quasielastic measurements; asterisks—fast process from quasielastic measurements (methyl group rotation). The dashed line is fit of the Arrhenius equation to the data of the methyl group rotation. The inset gives the derivative representation of the data for the process in the isotropic state. The line is a linear regression to the data

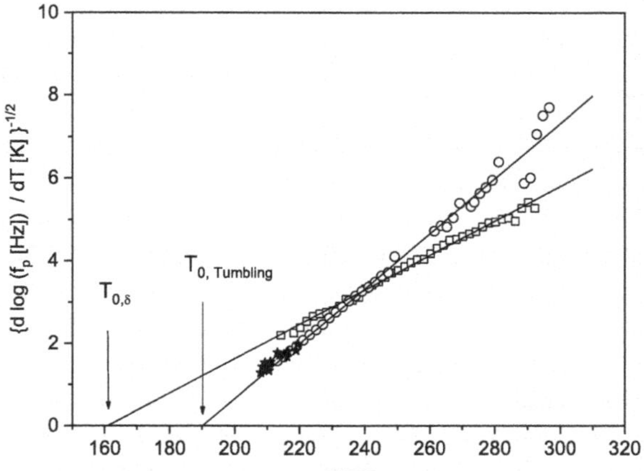

Fig. 14 Derivative $\left(\mathrm{dlog}\ f_p/\mathrm{dT}\right)^{-1/2}$ versus temperature for E7: circles—tumbling mode; squares—δ-relaxation; asterisks—thermal data. Lines are linear regression to the corresponding data. The figure was adapted from reference [68]

by different values of the Vogel temperature. The T_0 value found for the tumbling mode is ca. 30 K higher than that for the δ-relaxation. Furthermore, the temperature dependence of the relaxation rates of the tumbling is much more fragile than that of the δ-relaxation.

From Figs. 13 and 14, it cannot be concluded which process is responsible for glassy dynamics in the nematic crystal E7. It is generally accepted that a glass transition can be unambiguously identified by thermal measurements. Therefore, the relaxation rates measured by thermal heat spectroscopy are included in Fig. 13. This comparison shows that the thermal data are located close to the tumbling mode. The thermal data are further analyzed by applying the derivative technique to it (see Fig. 14). This analysis reveals that the relaxation rates estimated by specific heat spectroscopy have exactly the same temperature dependence as the tumbling mode. From these results, one has to conclude that the tumbling mode is responsible for glassy dynamics in the glass-forming liquid crystal E7.

In the light that the tumbling mode is assigned to glassy dynamics, the different temperature dependencies of the tumbling and the δ-relaxation can be discussed further. In principle, a similar decoupling of the temperature dependencies is observed for the α- and δ-relaxation in liquid-crystalline side-group polymers [79, 80] or for segmental and chain dynamics of polymers (see for instance [81]). Such a decoupling of glassy dynamics from other processes which should have the same temperature dependence of the relaxation rates from a theoretical point of view is also observed for low molecular glass-forming liquids (see for instance [82, 83]). It can be considered as a characteristic phenomenon of the glass transition which is due to different averaging processes for molecular motions having different length scales.

Figure 15 depicts the temperature dependence of the mean-squared displacement $\langle u^2 \rangle$ for E7 where in the inset of Fig. 15 the low temperature region is enlarged. At low temperatures, the increase of $\langle u^2 \rangle$ is due to vibrations. The observed linear temperature dependence is expected for a harmonic solid. The temperature dependence of $\langle u^2 \rangle$ changes in the temperature range around 110 K. This change in $\langle u^2 \rangle(T)$ indicates the onset of a molecular mobility at a timescale of ca. 1 ns. The chemical structure of E7 (see Fig. 1) contains a considerable amount of methyl groups. It is known from the literature that the methyl group rotation sets in at temperature around 100 K [84]. Therefore, this change in the temperature dependence of $\langle u^2 \rangle$ is assigned to the methyl group rotation. This will be discussed in more detail below. A further change in the temperature dependence of the mean-squared displacement is observed around 210 K which corresponds to the glass transition temperature measured by DSC. For that reason, this change in $\langle u^2 \rangle(T)$ is associated with the onset of the molecular mobility due to the glass transition.

Figure 16 depicts the incoherent intermediate time-dependent scattering function S(q,t) at 200 K at a q value of 1.7 $Å^{-1}$. The data in the time range from 0.1 ps correspond to measurements on IN6 while the data between 100 ps and 3 ns are obtained from measurements carried out at IN10. A single decay of S(q,t) is observed which is due to the methyl group rotation. A stretched exponential function

$$S(q,t) = DWF^* \left((1 - EISF) \exp\left(-\left(\frac{t}{\tau_M} \right)^{\beta_M} \right) + EISF \right) \tag{10}$$

is fitted to the data. Here, DWF symbolizes the Debye–Waller factor, and EISF is the Elastic Incoherent Structure Factor. τ_M is the relaxation time for methyl group rotation, and β_M the corresponding stretched exponential parameter. In a first fitting approach, all parameters were varied. From that analysis, a mean value is calculated

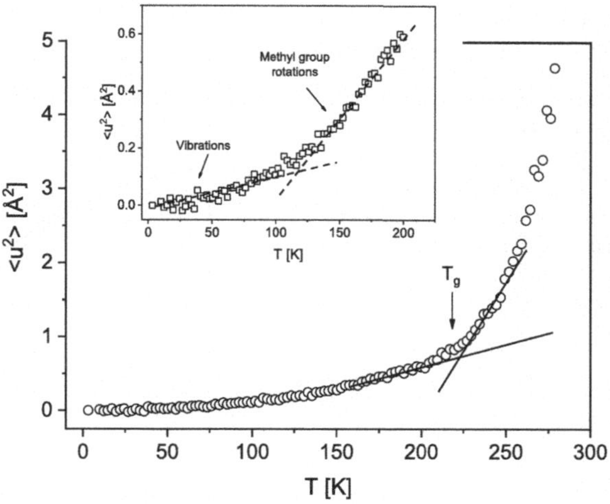

Fig. 15 Temperature dependence of the mean-squared displacement for E7. The inset enlarges the low-temperature range. Lines are guides to the eyes

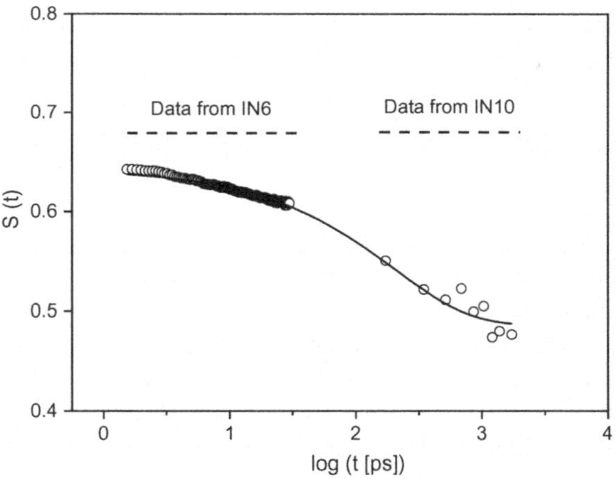

Fig. 16 Incoherent intermediate time dependent scattering function S(t) at 200 K at a q value of 1.7 Å^{-1} for E7. The line is a fit of Eq. (10) to the data

for β_M and fixed to 0.7 in the following second fit procedure. Figure 16 shows an example for the fitting procedure. From the relaxation time τ, the corresponding mean relaxation time $<\tau>$ is calculated by

$$\langle\tau\rangle = \frac{\tau}{\beta}\Gamma\left(\frac{1}{\beta}\right), \tag{11}$$

where Γ symbolizes the Gamma function. This procedure will enable a comparison of the neutron data with the relaxation rates measured by dielectric and specific heat spectroscopy.

Figure 17 shows S(t) at 260 K at a q value of 1.7 Å^{-1}. A two-step decay is observed for this temperature above the glass transition temperature. The process observed at shorter times corresponds to the methyl group rotation as discussed

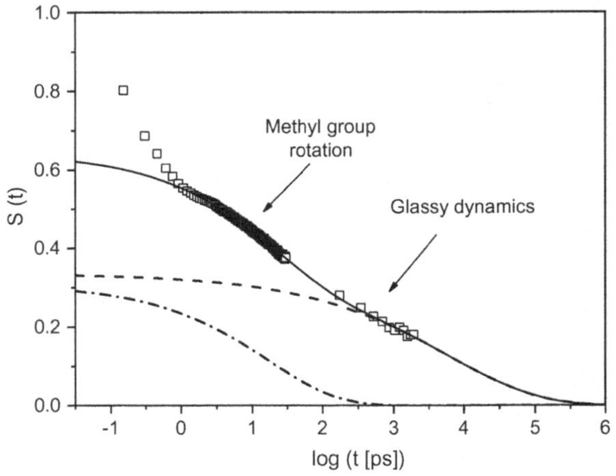

Fig. 17 Incoherent intermediate time-dependent scattering function S(t) at 260 K at a q value of 1.7 Å^{-1} for E7. The solid line is a fit of Eq. (12) to the data. The dashed and dashed–dotted lines are the contributions of glassy dynamics and the methyl group rotation, respectively

above. The process at longer times is assigned to the glassy dynamics of E7. These data are quantitatively analyzed by fitting two stretched exponentials to the data

$$S(q,t) = DWF^{*}\left((1-EISF)\exp\left(-\left(\frac{t}{\tau_M}\right)^{\beta_M}\right)+EISF\exp\left(-\left(\frac{t}{\tau_g}\right)^{\beta_g}\right)\right). \quad (12)$$

Here, τ_g and β_g are the relaxation time and the stretched exponential parameter for glassy dynamics. β_M was fixed to 0.7 during the fitting process. An example for the fit is given in Fig. 17 together with the contributions of the individual processes.

The relaxation rates obtained for the methyl group rotation are included in the relaxation map of E7 (see Fig. 13). The temperature dependence of the relaxation rates of the methyl group rotation follows the Arrhenius equation with an activation energy of 12.5 kJ/mol. This value is in agreement with values for the activation energy found for the methyl group rotation found in the literature [84]. This result additionally supports the molecular assignment of this relaxation process to the methyl group rotation.

The estimated relaxation rates for glassy dynamics are plotted versus inverse temperature in the relaxation map of E7 (see Fig. 13). These data points are located close to the relaxation rates of the dielectric tumbling mode. Because neutron scattering is sensitive to density fluctuation, this result furthermore supports the assignment of the tumbling to glassy dynamics.

3.2 Discotic Liquid Crystal Py_4CEH

The low-frequency vibrational density of states was not investigated for Py_4CEH. Nevertheless, it was recently shown by inelastic neutron scattering that a Boson Peak could be also observed for model systems based on triphenylenes [85]. Therefore, it

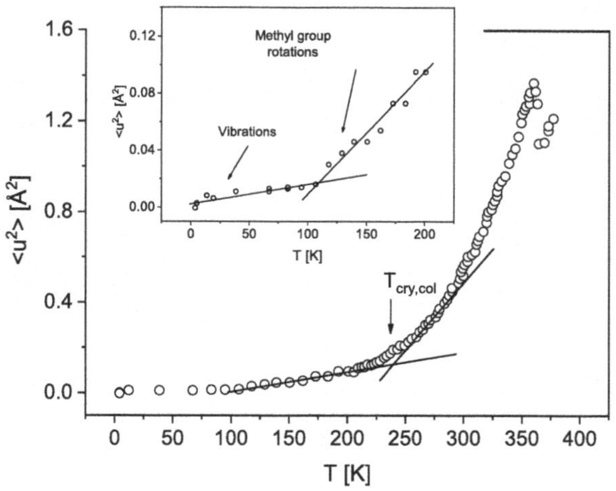

Fig. 18 Temperature dependence of the mean-squared displacement for Py_4CEH. The inset enlarges the low-temperature range. Lines are guides to the eyes. The data are taken from reference [86]

is expected that also Py_4CEH shows the excess contributions to the low-frequency vibrational density characteristic for glass-forming materials.

The temperature dependence of the mean-squared displacement measured for Py_4CEH is redrawn from reference [86] in Fig. 18. Comparing the absolute value of the mean-squared displacement obtained for Py_4CEH with that of E7, one has to conclude that $<u^2>$ is much smaller for Py_4CEH. This can be explained by the much denser structure of the plastic crystal phase of the pyrene in comparison with the nematic glass of E7.

The inset of Fig. 18 enlarges the temperature dependence of $<u^2>$ for low temperatures. Like for E7, the temperature dependence of $<u^2>$ is linear in the temperature range from 0 to 100 K. As discussed above, this temperature dependence originates from vibrations in a harmonic sold. Around 100 K, the temperature dependence of the mean-squared displacement changes. This change is assigned to the onset of the methyl group rotation because methyl groups are also present in the structure of Py_4CEH (see Fig. 3). Interestingly, no clear change in $<u^2>$ is observed in the temperature range around 170 K where DSC measurements show indications for a glass transition (see Fig. 4).

A further change in the temperature dependence of $<u^2>$ is observed around 247 K which is assigned to the phase transition from the plastic crystal to the hexagonal columnar liquid crystalline phase because the latter allows a higher molecular mobility compared to the plastic crystalline state. The change around 350 K is related to the phase transition from the hexagonal columnar liquid crystalline phase to the isotropic state.

Figure 19 depicts the dielectric loss for the discotic liquid crystal Py_4CEH. Different dielectrically active process can be observed. At low temperatures, a β-relaxation is observed in the plastic crystal state. At higher temperature than the β-relaxation but still in the plastic crystalline state, the α-relaxation becomes active which increases in its intensity at the phase transition from the plastic crystalline to hexagonal columnar

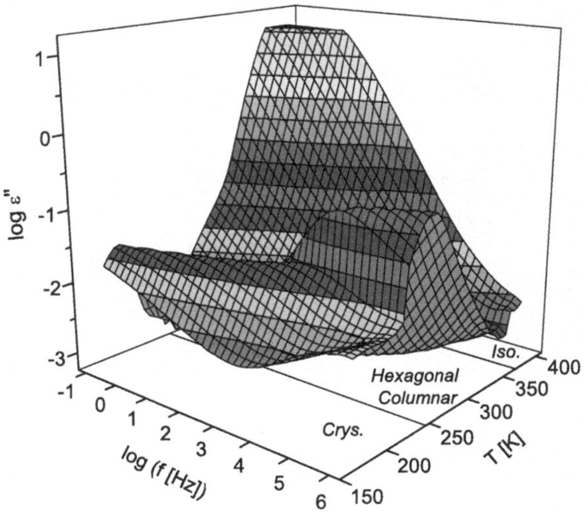

Fig. 19 Dielectric loss versus frequency and temperature in a 3D representation for Py_4CEH. The figure was adapted from reference [66]

liquid crystalline phase. At even higher temperatures, a conductivity contribution is observed as an increase of the dielectric loss.

Figure 20 presents the relaxation map for Py_4CEH. For the β-relaxation, an Arrhenius-like temperature dependence of the relaxation rates is found. For the activation energy, a value of 10.3 kJ/mol is found. This value points to localized fluctuations. Therefore, this process is assigned to localized fluctuations of methylene groups where also the ester units carrying a dipole should be involved. The observed low value for the activation energy, compared to a corresponding relaxation process found for polyethylene, might be due to a lower order resulting from the relatively short alkyl chains.

The temperature dependence of the relaxation rates of the α-relaxation is curved when plotted in the activation diagram. There seems to be a discontinuity of $f_p(T)$ at the phase transition temperature from the plastic crystalline phase to the liquid crystalline phase. The temperature dependence of the relaxation rates of the α-relaxation is further analyzed by the derivative technique introduced above (see Fig. 21). This analysis shows that the temperature dependence of the relaxation rates of the α-relaxation has two regimes, above and below $T_{cry,col}$ which have to be described by straight lines. This means that a kind of glassy dynamics takes place in both the plastic crystalline and the hexagonal columnar liquid crystalline phase. By fitting linear equations to the data in both phases, the corresponding Vogel temperatures and fragility parameters can be estimated. For the hexagonal columnar liquid crystalline phase, a Vogel temperature of $T_{0,col} = 140.1$ K is found, where $T_{0,cry} = 62.5$ K is estimated for the plastic crystalline state. This large difference in the Vogel temperatures evidences a considerable difference in the underlying molecular fluctuations. This is also expressed by different values of the fragility parameters $D_{cry} = 13.73$ and $D_{col} = 2.25$ for the plastic crystalline and the hexagonal columnar liquid crystalline phase, respectively. This means that the molecular mobility in the hexagonal

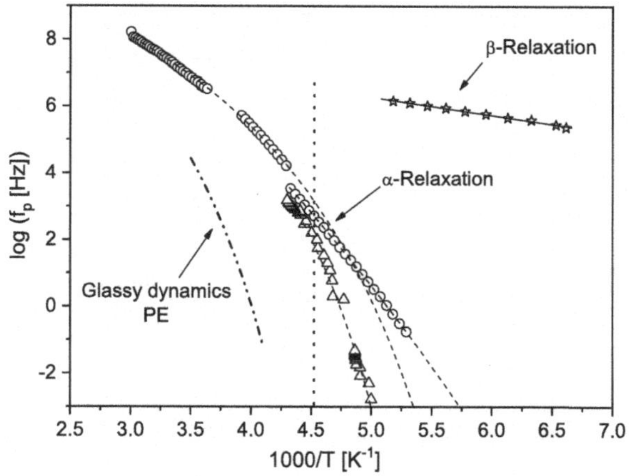

Fig. 20 Temperature dependence of the relaxation rates in the activation diagram for Py_4CEH: asterisks—dielectric β-relaxation; circles—dielectric α-relaxation; triangles—α-relaxation measured by specific heat spectroscopy. The solid line is a fit of the Arrhenius equation to the data. The dashed lines are fits of the VFT equation to the data in the plastic crystal and hexagonal columnar liquid crystalline phase. The dotted line indicates the phase transition temperature from the plastic crystalline to the hexagonal columnar liquid crystalline phase measured by DSC. The dot-dot-dashed line is the glassy dynamics of polyethylene (PE) where the data were taken from reference [88]. The figure was adapted from reference [66]

columnar liquid crystalline phase is much more fragile than the motional process taking place in the plastic crystalline phase. A similar behavior was also found for other discotic liquid crystal but with less extended data analysis [33, 87].

The relaxation rates measured by specific heat spectroscopy are included in Fig. 20. Firstly, as for E7 the data measured by TMDSC and AC chip calorimetry are collapsing into one curve. Secondly, in comparison to the dielectric data the relaxation rates measured by calorimetry are shifted to higher temperatures. This is especially true for the plastic crystalline state. Although the thermal data have a larger error than the dielectric ones, they are also subjected to the derivative procedure (see Fig. 21). Also, the thermal data obeys the VFT equation with a Vogel temperature $T_{0,\mathrm{ther}} = 192$ K which is much higher than the Vogel temperature obtained by the dielectric measurements. Discussing these results, one has to keep in mind that dielectric spectroscopy is sensitive to dipole fluctuations, while specific heat spectroscopy senses entropy fluctuations. The structure of Py_4CEH shows that the main dipole moment is due to the ester groups located close to the stiff core of the pyrene (see Fig. 3). It is further assumed that also in the plastic crystalline phase, the pyrene cores are arranged in columns, where the alkyl chains fill the intercolumnar space. The organization of the pyrene units into columns will further restrict the mobility of the ester groups. Therefore, the molecular mobility which can be detected by dielectric spectroscopy is much more localized than the fluctuations which can be detected by specific heat spectroscopy which is sensitive to the whole intramolecular space. A more localized process is characterized by a less fragile behavior with a smaller

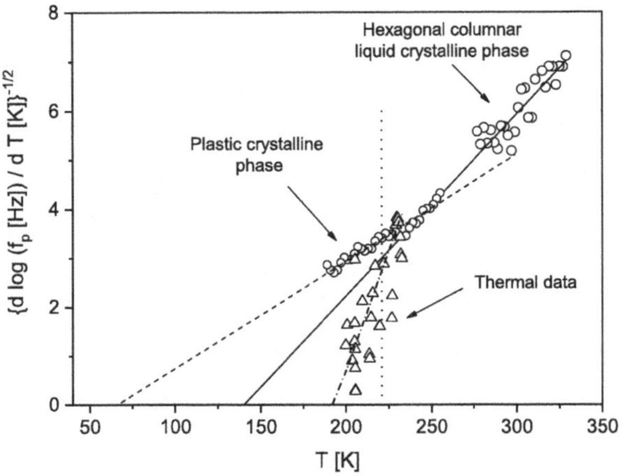

Fig. 21 Derivative $\left(\mathrm{dlog}\ f_p/dT\right)^{-1/2}$ versus temperature for Py_4CEH: circles—dielectric data; triangles—thermal data. The dotted line indicates the phase transition temperature from the plastic crystalline to the hexagonal columnar liquid crystalline phase measured by DSC. The solid and the dashed line are linear regressions to the data of the hexagonal columnar liquid crystalline and the plastic crystalline phase. The dash-dotted line is calculated from the relaxation rates of the glassy dynamics of polyethylene [88]. The figure was adapted from reference [66]

value of the Vogel temperature. This conclusion is further supported by the fact that for the hexagonal columnar liquid crystalline phase, which is less ordered than the plastic crystalline state, a higher value of the Vogel temperatures and a more fragile behavior is found.

The intercolumnar space which is sensed by specific heat spectroscopy is built by the arms of the pyrene which consist mainly of CH_2 and CH_3 groups. Therefore, data for the glassy dynamics of polyethylene taken from reference [88] were included in Fig. 20. From these data, the derivative was calculated according to Eq. (9) and included in Fig. 21. The calculated data for polyethylene matches completely data measured for Py_4CEH by specific heat spectroscopy. This coincidence means that both data sets have to be described by the same Vogel temperature. This means further that the α-process measured for Py4CEH by specific heat spectroscopy is due to similar molecular fluctuations also taking place in polyethylene.

As discussed in the introduction, the glass transition can be discussed in cooperativity approach where the Cooperatively Rearranging Regions (CRR) were introduced by Adam and Gibbs [10]. A CRR is the smallest spacial region which can change its configuration independently from the neighboring regions. Within the fluctuation approach to the glass transition [11], the size of a CRR characterized by the correlation length ξ can be calculated to

$$\xi = \sqrt[3]{\frac{k_B T_g^2 \Delta\left(\frac{1}{c_p}\right)}{\rho(\delta T)^2}}, \tag{13}$$

where $\Delta(1/c_p)$ is the step of the reciprocal specific heat capacity at T_g, where $c_V \approx c_p$ was assumed. δT is the width of the glass transition, and ρ is the density. δT and the heat capacity values can be estimated by temperature-modulated DSC. For Py_4CEH, δT was found to be 7.58 K while for $c_{p,Glass}$ and $c_{p,Liquid}$ 1.05 and 1.10 J/(g K) was estimated [66]. With these values, the correlation length is calculated to 0.78 nm. From small-angle X-ray measurements, a core–core distance of 1.7 nm is obtained [66]. Considering that a part of this distance belongs to the cores themselves (0.4–0.5 nm), one obtains ca. 0.8 nm for the intercolumnar space. This value obtained from structural data correlates well with the correlation length estimated from the thermal data.

4 Summary

A combination of broadband dielectric, specific heat, and neutron spectroscopy is employed to investigate the molecular dynamics in complex soft materials where order and mobility compete. Each of these complementary methods is sensitive to a different molecular fluctuation. Therefore, different aspects of the motional processes can be selectively addressed. As examples, two liquid crystalline systems are considered. Firstly, the rod-like system E7 forming a nematic liquid crystalline phase is considered while in the second case a disc-like mesogen is investigated which forms more complex structures like a plastic crystalline phase and a hexagonal columnar liquid crystalline state. As a first result, all peculiarities known for conventional glass-forming systems can be also observed for the complex systems studied here.

According to theoretical approaches, two relaxation processes are observed in the nematic state of the rod-like liquid crystal E7 where the relaxation rates of both relaxation processes have a VFT-like temperature dependence. Both processes merge at the glass transition temperature. Form this result, it could not be concluded which mode is responsible for glassy dynamics in E7. By applying specific heat spectroscopy, it could be unambiguously shown that the small-scale tumbling mode is responsible for glassy dynamics. This result is further confirmed by quasielastic neutron scattering.

For the discotic liquid crystal, an α-relaxation process is observed in the plastic crystalline and the hexagonal columnar phase. The fragility of the α-relaxation changes at the phase transition from the plastic crystalline to the hexagonal columnar liquid crystalline phase. In the plastic crystalline phase, the α-relaxation is more strong, while in the hexagonal columnar liquid crystalline phase, the α-relaxation is more fragile. From the fluctuation approach to the glass transition, the correlation length for glassy dynamics is calculated to 0.78 nm from the thermal data. This value corresponds well to the core–core distance estimated from small-angle X-ray measurements.

Acknowledgements The authors thank for fruitful collaborations with Prof. Dr. C. Schick and Dr. H. Huth (University Rostock). Further, Prof. Dr. M. Dionisio (Universidade Nova de Lisboa) and Dr. A. R. Bras (University Cologne) are thanked for cooperation. Dr. C. Krause is thankfully acknowledged for her work in the field of discotic liquid crystals. The Institut Laue–Langevin (Grenoble, France) is thanked for enabling the neutron scattering experiments. The financial support from the German Science Foundation (Deutsche Forschungsgemeinschaft, SCHO-470/21-1) is highly acknowledged.

References

1. Anderson PW (1995) Science 267:1615
2. Angel CA (1995) Science 267:1924
3. Debenedetti PG, Stillinger FH (2000) Nature 410:259
4. Dyre JC (2006) Rev Mod Phys 78:953
5. Kivelson S, Tarjus G (2008) Nat Mater 7:831
6. Ediger M, Harrowell P (2012) J Chem Phys 137:080901
7. Vogel H (1921) Physikalische Zeitschrift 22:645
8. Fulcher GS (1925) J Am Ceram Soc 8:339
9. Tammann G, Hesse WZ (1926) Anorg Allg Chem 156:245
10. Adam G, Gibbs JH (1965) J Chem Phys 43:139
11. Donth EJ (1992) Relaxation and thermodynamics in polymers: glass transition. Akademie Verlag, Berlin
12. Berthier L, Biroli G, Bouchaud J-P, Cipelletti L, El Masri D, Ladieu F, Pierno M (2005) Science 310:1797
13. Bauer Th, Lunkenheimer P, Loidl A (2013) Phys Rev Lett 111:225702
14. Johari GP, Goldstein MJ (1970) Chem Phys 53:2372
15. Johari GP (1976) Goldstein M, Simha R, The glass transition and the nature of the glassy state. Ann NY Acad Sci 279:117
16. Zallen R (1983) The physics of amorphous solids. Wiley, New York
17. Malinovsky VK, Novikov VN, Sokolov AP (1987) J Non-Cryst Solids 90:485
18. Buchenau U, Galperin YM, Gurevich VL, Parshin DA, Ramos MA, Schober HR (1992) Phys Rev B: Condens Matter Mater Phys 46:2798
19. Laird BB, Schober HR (1991) Phys Rev Lett 66:636
20. Taraskin SN, Loh YL, Natarajan G, Elliott SR (2001) Phys Rev Lett 86:1255
21. Schirmacher W, Diezemann G, Ganter C (1998) Phys Rev Lett 81:136
22. Milkus R, Zaccone A (2016) Phys Rev B 93:094204
23. Angell CA (1991) J Non-Cryst Solids 131–133:13
24. Angell CA (1997) J Res Natl Inst Stand Technol 102:171
25. Suga H, Seki S (1974) J Non-Cryst Solids 16:171
26. Wang LM, Richert R (2004) J Chem Phys 121:11170
27. Leslie-Pelecky DL, Birge NO (1994) Universal scaling of the relaxation near a model glass. Phys Rev Lett 72:1232
28. Kuchta B, Descamps M, Affouard F (1998) J Chem Phys 109:6753–6763
29. Benkhof S, Blochowicz T, Kudlik A, Tschirwitz C, Rössler E (2000) Ferroelectrics 236:193–207
30. Brand R, Lunkenheimer P, Loidl A (2002) J Chem Phys 116:10386–10401
31. Krause C, Yin H, Cerclier C, Morineau D, Wurm A, Schick C, Emmerling F, Schönhals A (2012) Soft Matter 8:11115
32. Groothues H, Kremer F, Schouten PG, Warman JM (1995) Adv Mater 7:283
33. Glüsen B, Kettner A, Kopitzke J, Wendorff JH (1998) J Non-Cryst Solids 241:113
34. de Gennes PG (1975) The physics of liquid crystals. Clarendon press, Oxford
35. Chandrasekhar S (1992) Liquid crystals. Cambridge University Press, Cambridge
36. Demus D, Goodby J, Gray GW, Spiess HW, Vill V (eds) (1998) Handbook of liquid crystals. Wiley-VCH, Weinheim

37. Laschat S, Baro A, Steinke N, Giesselmann F, Hägele G, Scalia G, Judele R, Kapatsina E, Sauer S, Schreivogel A, Tosoni T (2007) Angew Chem 119:4916
38. Kremer F, Schönhals A (2003) Molecular and collective dynamics of (polymeric) liquid crystals In: Kremer F, Schönhals A (eds) Broadband dielectric spectroscopy. Springer, Berlin
39. Moscicki K (1992) Dielectric relaxation of macromolecular liquid crystals. In: Collyer AA (ed) Liquid crystal polymers—from structure to applications. Elsevier, Amsterdam
40. Williams G (1989) In: Allen G, Bevington JC (eds) Comprehensive polymer science, vol 2. Pergamon, Oxford
41. Nordio PL, Rigatti G, Serge U (1973) Mol Phys 25:129
42. Williams G (1994) In: Luckhurst GR, Veracini CA (eds) The molecular dynamics of liquid crystals. Kluwer Academic Press, p 431
43. Tarroni R, Zannoni C (1991) J Chem Phys 95:4550
44. Schönhals A, Kremer F (2003) Theory of dielectric relaxation. In: Kremer F, Schönhals A (eds) Broadband dielectric spectroscopy. Springer, Berlin
45. Martin AJ, Meier G, Saupe A (1971) Chem Soc Faraday Symp 5:119
46. Kreul HG, Würflinger A (1992) Phys Rev A 45:8624
47. Urban S, Brückert T, Würflinger A (1994) Z Naturforsch 49a:552
48. Urban S, Gestblom B, Kresse H, Dabrowski A (1996) Z Naturforsch 51a:834
49. Rozanski SA, Stannarius R, Gorbatschow W, Kremer F (1996) Liq Cryst 20:59
50. Schönhals H, Zubowa H-L, Fricke R, Frunza S, Frunza L, Moldovan R (1999) Cryst Res Technol 34(1309):1999
51. Gottke SD, Cang H, Bagchi B, Fayer MD (2002) J Chem Phys 116:6339
52. Cang H, Novikov VN, Fayer MD (2003) Phys Rev Lett 90:197401
53. Zeller HR (1982) Phys Rev Lett 48:334
54. Benguigui L (1983) Phys Rev A28:1852
55. Benguigui L (1984) Phys Rev A29:2968
56. Diogo AC, Martins AF (1982) J Phys (Paris) 43:779
57. Zhong ZZ, Schuele DE, Gordon WL, Adamic KJ, Akins RB (1992) J Polym Sci B: Polym Phys 30:1443
58. Maschke U, Benmouna M, Coqueret X (2002) Macromol Rapid Commun 23:159
59. Roussel F, Buisine JM, Maschke U, Coqueret X (1997) Mol Cryst Liq Cryst 299:321
60. Viciosa MT, Nunes AM, Fernandes A, Almeida PL, Godinho MH, Dionisio MD (2002) Liq Cryst 29:429
61. Van Boxtel MCW, Wübbenhorst M, Van Turnhout J, Bastiaansen CWM, Broer DJ (2003) Liq Cryst 30:235
62. Van Boxtel MCW, Wübbenhorst M, Van Turnhout J, Bastiaansen CWM, Broer DJ (2004) Liq Cryst 31:1207
63. Bras AR, Viciosa MT, Rodrigues C, Dias CJ, Dionisio MD (2006) Phys Rev E 73:061709
64. Dionísio M, Brás AR, Henriques S, Casimiro T, Ricardo AA, Sotomayor J, Caldeira J, Santos C (2005) Electronic liquid crystals, Mar/21
65. Hassheider T, Benning SA, Kitzerow HS, Achard MF, Bock H (2001) Angew Chem Int Ed 40:2060
66. Krause C, Yin H, Cerclier C, Morineau D, Wurm A, Schick C, Emmerling F, Schönhals A (2012) Soft Matter 8:11115
67. Kremer F, Schönhals A (2003) Broadband dielectric measurement techniques. In: Kremer F, Schönhals A (eds) Broadband dielectric spectroscopy. Springer, Berlin
68. Brás AR, Dionísio M, Huth H, Schick C, Schönhals A (2007) Phys Rev E 75:061708
69. Havriliak S, Negami S (1967) Polymer 8:161; Havriliak S, Negami S (1966) J Polym Sci Part C: Polym Symp 14:99
70. Schawe JEK (1995) Thermochim Acta 260:1–16
71. Schick C (2002) In: Cheng SZD, Handbook of thermal analysis and calorimetry, vol 3. Elsevier, Amsterdam
72. Huth H, Minakov AA, Schick C (2006) J Polym Sci, Part B: Polym Phys 44:2996
73. http://www.xensor.nl/index.php/products/calorimeter-chips

74. Bée M (1988) Quasielastic neutron scattering. Principles and applications in solid state chemistry, biology and materials science. Adam Hilger, London
75. http://www.ill.eu/instruments-support/computing-forscience/cs-software/all-software/tofhr/inx/
76. Lovesey SW (1987) Theory of neutron scattering from condensed matter, vol 1. Oxford University, New York
77. Zorn R, Hartmann L, Frick B, Richter D, Kremer F (2002) J Non-Cryst Solids 307:547
78. Zorn R, Mayorova M, Richter D, Schönhals A, Hartmann L, Kremer F, Frick B (2008) AIP Conf Proc 982:79–84
79. Schick C, Sukhorukov D, Schönhals A (2001) Macromol Chem Phys 202:1398
80. Schönhals A, Wolff D, Springer J (1998) Macromolecules 31:9019
81. Schönhals A (2003) Molecular dynamics in polymer model systems. In: Kremer F, Schönhals A (eds) Broadband dielectric spectroscopy. Springer, Berlin
82. Sillescu H (1999) J Non-Cryst Solids 243:8
83. Ediger MD Ann Rev Phys Chem 51:99
84. Colmenero J, Moreno AJ, Alegria A (2005) Prog Polym Sci 30:1147
85. Krause C, Zorn R, Emmerling F, Falkenhagen J, Frick B, Huber P, Schönhals A (2014) Phys Chem Chem Phys 16:3724
86. Ndao M, Lefort R, Cerclier CV, Busselez R, Morineau D, Frick B, Ollivier J, Kityk AV, Huber P (2014) RCS Adv 4:59358
87. Elmahdy MM, Dou X, Mondeshki M, Floudas G, Butt HJ, Spiess HW, Müllen K (2008) J Am Chem Soc 130:5311
88. van den Berg O, Sengers WGF, Jager WF, Picken SJ, Wübbenhorst M (2004) Macromolecules 37:2460

The Calorimetric Glass Transition in a Wide Range of Cooling Rates and Frequencies

T. V. Tropin, J. W. P. Schmelzer, G. Schulz and C. Schick

Abstract The glass transition at common laboratory scan rates (K/min) has been a highly debated topic for the last decade. The continuous increase in the variety of available glass-forming materials and methods to characterize them maintains a research interest, as well as opens new perspective applications. In parallel, many different theoretical methods aimed at describing the glass transition have been proposed in the last 70 years. A general theory has yet to be developed and carefully tested. In the present chapter, we describe the results of theoretical and experimental investigations of the glass transition of a model polymer—polystyrene. State-of-the-art scanning calorimetry allows for measuring the temperature dependence of the isobaric heat capacity in an exceedingly wide range of cooling rates. Besides providing novel data on the glass transition of polymers at fast cooling rates, this allows for one to test the capabilities of convenient theoretical methods in modelling the kinetics of the glass transition under very different vitrifying conditions. The glass transition of atactic polystyrene was investigated at different cooling rates in the range of $q_c = 10^{-6}$–10^4 K/s. Dependencies of the glass transition temperature, T_g, and the shape of heat capacity, C_p, curves on q_c were obtained. Furthermore, we have applied a number of different theoretical methods to test their capability to model the glass transition kinetics for such a wide range of control parameter q_c. The list of investigated theoretical methods consists of the Tool–Narayanaswamy–Moynihan approach, Adam–Gibbs theory, an irreversible thermodynamics-based approach and some of their modern modifications. As a first step, we show that most of these methods are capable of fitting the cooling rate dependencies of the glass transition parameters (T_{g} and others). The model parameters in this case are close to literature data. Furthermore, we show that while fitting the $C_p(T)$ curves for a single cooling–heating experiment bears acceptable results, the parameters have to be changed

T. V. Tropin
Frank Laboratory of Neutron Physics, Joint Institute for Nuclear Research, Dubna, Russia

J. W. P. Schmelzer · G. Schulz · C. Schick (✉)
Institute of Physics and Competence Center CALOR, University of Rostock, Rostock, Germany
e-mail: christoph.schick@uni-rostock.de

C. Schick
Butlerov Institute of Chemistry, Kazan Federal University, Kazan, Russia

F. Kremer and A. Loidl (eds.), *The Scaling of Relaxation Processes*,
Advances in Dielectrics, https://doi.org/10.1007/978-3-319-72706-6_10

with respect to q_c, with their difference becoming significant for very slow or very fast cooling rates. Thus, none of the methods can be applied successfully to model and predict the kinetics of glass transition in a wide range of q. We compare the results of different methods and propose an expression for the relaxation time dependence on model parameters within an irreversible thermodynamics approach. Thus, we extend the experimental results for polystyrene and state that the presently applied theoretical methods are incapable of accurately describing the heat capacity temperature curves $C_p(T)$ for a wide range of cooling/heating rates, q. The present methods and expressions for relaxation time τ do not account for a certain additional effect spanning over different rates of temperature change, which has yet to be discovered.

Keywords Glass transition · Polymers · Polysterene (PS) · Calorimetry · Glass transition kinetics

Abbreviations

5PPE	5-phenyl-4-ether
ATHAS	Advanced thermal analysis system (data bank)
AC	Alternating current (calorimetry)
AG	Adam–Gibbs (theory)
CRR	Cooperatively rearranging regions
DFSC	Differential fast-scanning calorimetry
DSC	Differential scanning calorimetry
FSC	Fast-scanning (chip) calorimetry
GS	Gutzow–Schmelzer (method)
KAHR	Kovacs–Aklonis–Hutchinson–Ramos (method)
PS	Polystyrene
TNM	Tool–Narayanaswamy–Moynihan (method)
TMDSC	Temperature-modulated differential scanning calorimetry
VFT	Vogel–Fulcher–Tammann (law)

1 Introduction

The nature of a glassy state and the glass transition remains one of the open questions in science [1, 2]. Theoretical efforts for describing the structure of glass, as well as the transition from liquid to glass, are generally required. Many works with microscopic approaches have been already been published [3–5], yet even at a general macroscopic level, a number of questions is still open. The glass transition, not being a phase transition, has a number of general features that should be accounted for. Apart from nonlinearity and non-exponentiality of relaxation in the respective temperature range, one of the important properties is the dependence of glass transition temperature and overall course of the process on the dynamics of changing external control parameters. The pronounced dependencies of T_g on cooling rate or (for dynamic transition) applied frequency are well-known. Generally, this is the effect of dependence of systems properties on its thermal history.

Over the last few years, the experimental capabilities of calorimetry have been extremely widened [6]. Whereas for conventional instruments, only a short range of cooling rates and frequencies was available, at present, one can measure the glass transition of a material on several qualitatively different devices, covering the range from ultra-slow (e.g. ~10^{-6} K/s) to ultra-high (~10^6 K/s) cooling/heating rates [7]. This extension of experimental capabilities provides a possible pathway to verify existing theoretical methods that were developed for describing the glass transition kinetics by properly accounting for relaxation laws in the glass transition range [8]. In the present work, we report the results for this type of investigation for atactic polystyrene, a model polymer. The range of cooling rates was initially 5×10^{-6}–2 K/s with further heating at a constant rate of 0.5 K/s [8]. The range of the applied cooling rates was extended using fast-scanning calorimetry (FSC). The measurements of temperature dependence of C_p were made for q in the range of 0.1 to 10,000 K/s with constant heating rate of 5000 K/s. This allowed for the estimation of the glass transition temperature for high scanning rates, which we use in this work for evaluating the relaxation time dependence.

The modelling of the resulting kinetic curves was performed to clarify the applicability of common theoretical methods for describing the glass transition process. The most popular method for modelling the thermal glass transition is the Tool—Narayanaswamy–Moynihan (TNM) method [9]. In its present formulation, an arbitrary expression for the relaxation law may be used. We thus clarify the applicability of the Vogel–Fulcher–Tammann (VFT) law [10–12] and expressions obtained from the Adam–Gibbs theory [13]. Generally, a disagreement between modelling and experiment is observed. We demonstrate how the extent of disagreement can be decreased and show that the highest inconsistency occurs for a cooling rate q much smaller than the heating rate. Finally, basing on modelling the polystyrene glass transition at a fast cooling rate and on glycerol, we conclude that this problem has a general character.

2 Broad Band Dynamic Calorimetry

A brief description of fast-scanning chip calorimetry (FSC) and heat capacity spectroscopy by temperature-modulated differential scanning calorimetry (TMDSC) and AC calorimetry for a wide range of scanning rates and frequencies are presented. Finally, the scanning rate and frequency dependencies are summarized in a relaxation map.

2.1 Fast-Scanning Calorimetry (FSC)

Experiments on the cooling rate dependence of glass transition are often limited by the maximum possible cooling rate. Fast-scanning chip calorimetry (FSC) was developed to overcome this limitation of conventional differential scanning calorimeters (DSC) by a significant reduction of both the sample mass and heat capacity of the furnaces [14–17]. While the maximum cooling rate in special setups of conventional DSC is around few hundred K/min, state-of-the-art FSC permits glass transition studies at rates of up to 1,000,000 K/s, owing to the continuous improvement/miniaturization of the chip sensors [6, 14, 18].

The sensor of the commercial Mettler-Toledo Flash DSC 1 with a diameter of the circular heated area of about 500 μm (Fig. 1c left) allows for the controlled heating and cooling of samples with a typical mass of about 100 ng up to a few thousand K/s [19, 20]. For higher scanning rates, ultra-fast custom-made FSC devices with even smaller samples and sensor sizes can be used, allowing measurements up to 1,000,000 K/s. For illustration, Fig. 1c shows photographs of the rather large UFS 1 sensor of the Flash DSC 1 (left) and distinctly smaller sensors accommodating samples with a typical mass between 500 and 1 ng (left to right).

The fastest cooling of any calorimeter cell containing a sample is obtained if heat is not supplied, that is, the electrical power $P_0(T)$ in the heat-balance equation equals zero [see Eq. (1)]. The maximum possible cooling rate is therefore defined by the ratio between the heat flow rate away from the measuring cell ($P_{loss}(T)$) and the heat capacities of measuring cell (addenda) ($C_0(T)$) and sample ($C_s(T)$).

$$(C_0(T) + C_S(T))\frac{dT}{dt} = P_0(T) - P_{loss}(T). \tag{1}$$

At a given temperature, assuming conductive and convection losses only, $P_{loss}(T)$ is, in good approximation, proportional to the temperature difference between the cell and the heat sink. Linear cooling is realized by keeping the right-hand side of Eq. (1) at a constant negative value by controlling the heater power $P_0(T)$. This idea was successfully realized for non-adiabatic FSC [15] with extremely low addenda and sample heat capacities. It was shown that a gas is the optimum cooling agent to achieve the highest cooling rates [6, 17, 24, 25]. Furthermore, thermopiles are better suited than resistive thermometers because they do not need an electrical current, which always generates undesirable power. The technique described in [6, 15] is based on a single thin-film chip sensor and is capable of applying both controlled heating and controlled cooling at rates up to 10^6 K/s.

The ultra-fast-scanning calorimeters described in references [6, 15, 22, 26, 27] were successfully applied for the investigation of polymer crystallization, melting and reorganization [15, 24, 28–41]. The examples have shown that the effective range of controlled heating and cooling rates using different sensors [21] is 10^0–10^6 K/s. Unfortunately, at low rates, the signal-to-noise ratio and therefore the sensitivity are reduced. Typically, the use of the device is limited to rates higher than 10^2 K/s for polymers. Nevertheless, the scanning rate range between conventional DSC and this

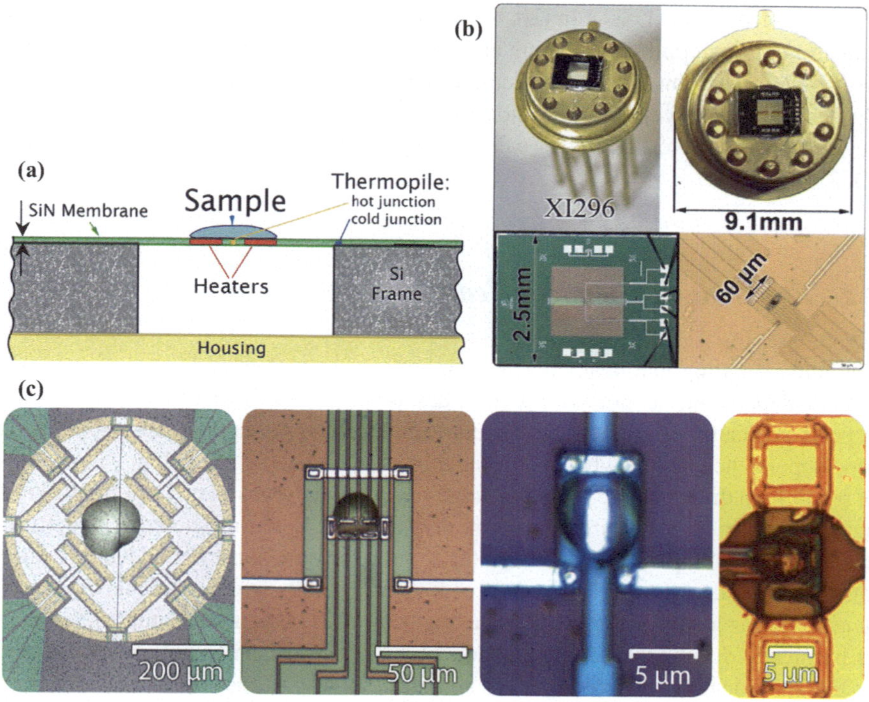

Fig. 1 **a** Schematic of a calorimetric chip sensor, **b** photographs of a sensor and **c** different variations of calorimetric sensors from Xensor Integration, NL [21]. Reproduced with permission of [22] (**a**), (**b**) and [23] (**c**)

chip-based technique is in the range of 10–100 K/s, and is of high industrial interest since it is often employed in polymer processing [42, 43]. This gap is bridged by the Mettler-Toledo Flash DSC 1 covering a heating rate range from 1 to 20,000 K/s and cooling rate range from 1 to 4000 K/s.

To improve the signal-to-noise ratio even at the lowest scanning rates, a differential scheme of two sensors with power compensation is introduced. The presence of an empty reference sensor reduces the influence of heat losses and the addenda heat capacity on the resulting data. It directly provides, under conditions of ideal symmetry of both sensors and perfect power compensation, the heat flow rate into the sample, which simplifies the ensuing heat capacity measurements/calculations [22, 44]. To improve the performance of chip calorimeters, particularly during phase transitions, and to simplify data treatment, the power-compensation scheme was first introduced by Rodriguez-Viejo et al. [45, 46] and Merzlyakov [47]. To improve the signal-to-noise ratio and the resolution of the device even under fast-scanning conditions, later on an analog power-compensation technique was implemented [22]. Details of the devices of interest and data treatment are reported for the differential fast-scanning calorimeter (DFSC) in [22, 27] and for the commercialized version, the

Mettler-Toledo Flash DSC 1 in [20, 48]. In summary, it is possible to study the glass transition during controlled heating and cooling at rates between 1 and 500,000 K/s with DFSC. The power-control circuit allows for isothermal experiments for times shorter than 1 ms, with over- or undershoots of less than 1 K. Specifications of the devices and the sensors as well as calibration procedures are available elsewhere [20, 22, 27, 48–51].

2.2 *Frequency-Dependent Heat Capacity*

The understanding that heat capacity can be time dependent and consequently frequency dependent has not always been accepted, e.g. [52]. The first hint for such dependency came from the measurements of ultrasound propagation in gases [53]. A time dependency occurs when heat is not simultaneously distributed into all degrees of freedom in the sample, e.g. translational first and rotational later. The sample's fast degrees of freedom are directly (thermally) coupled to the measured temperature. The change in temperature or heat in the sample becomes time dependent if heat is later distributed to slower degrees of freedom; as a result, time and frequency dependencies occur [54]. Consequently, heat capacity also becomes a complex function of frequency-dependent magnitude and phase [54, 55]. Heat can only be transferred to the slow degrees of freedom first and successively to the fast degrees of freedom in very particular cases [56–58]. The frequency-dependent heat capacity can be compared to admittance in dielectric spectroscopy and can be treated similarly [59].

Direct time-dependent enthalpy measurements in single component systems that are not undergoing a phase transition are not reported very often. One example is the isothermal temperature drift measurement in adiabatic calorimeters in the glass transition range, e.g. Suga et al. [60, 61] and references there in. A direct measurement of the heat flow rate during annealing a polymer sample below the glass transition was described in [62, 63].

Examples of such time-dependent heat flow rate measurements at very different time scales are shown in Fig. 2 for a Setaram Calvet calorimeter BT 2.15 and in Fig. 3 for a Mettler-Toledo Flash DSC1. The resulting sample temperature is illustrated in (a) after a programmed step in temperature, and the corresponding heating rate and the heat flow rate are shown in (b). An upward temperature step at 0.01 K/s is programmed in Fig. 2 and a downward step at 2000 K/s in Fig. 3. The heat flow rate curves for both instruments are shown for two temperatures in and outside the glass transition range of 5-phenyl-4-ether (5PPE), respectively.

For both instruments, the smearing of the temperature step and consequently the heating rate profile due to instrumental time constants can be seen. In the BT 2.15, the decay of the heat flow rate in the glassy state follows closely the heating rate profile since no relaxation process occurs on the timescale of the instrument. In the Flash DSC 1, at least one significant additional time constant exists between the thermometer and the sample, as seen from the significant longer relaxation of the heat flow rate in the glassy state compared to the cooling rate decay. For both

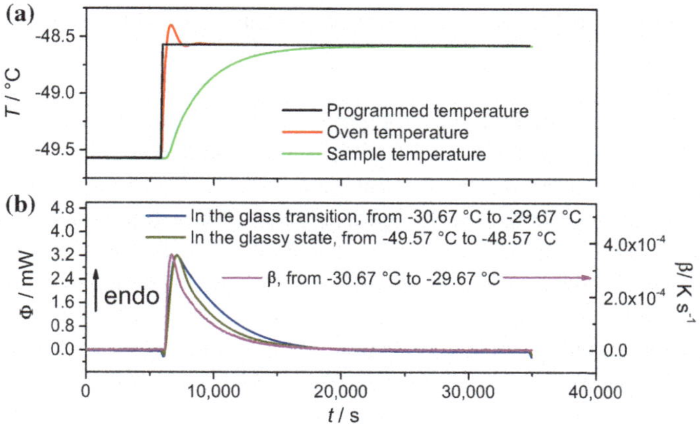

Fig. 2 Programmed upward temperature step of 1 K at heating rate 0.5 K min^{-1} (ca. 10^{-2} K/s) in a Setaram Calvet calorimeter BT 2.15. **a**—Programmed temperature (black), oven temperature (red) and measured sample temperature (green), **b**—heating rate (magenta, right axis) and heat flow rate in the glass transition region (blue) and in the glassy state (dark yellow, left axis)

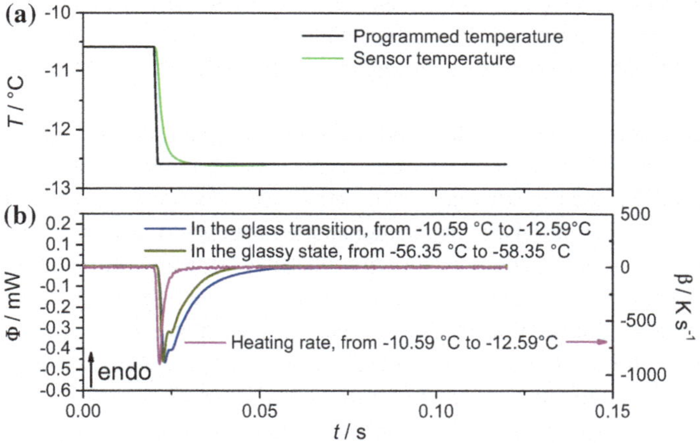

Fig. 3 Programmed downward temperature step of 2 K at cooling rate 2000 K/s in a Mettler-Toledo Flash DSC 1 calorimeter. **a**—Temperature program (black) and measured sensor temperature (green), **b**—heating rate (magenta, right axis) and heat flow rate in the glass transition region (blue) and in the glassy state (dark yellow, left axis)

instruments, the much longer decay of the heat flow rate due to the relaxation in the glass transition range is seen in the blue curves. Note the different timescales.

The time and frequency dependence of heat capacity is accessible from such measurements or measurements under periodic perturbations. Time domain and periodic calorimetric measurements are possible and described briefly next.

2.2.1 Temperature-Modulated Differential Scanning Calorimetry (TMDSC)

DSC is widely used in many fields, e.g. physics, chemistry and material science. It is commonly accepted as a standard tool for the investigation of phase transitions, chemical reactions and to determine heat capacity in a short time with reasonable accuracy. The theory of DSC is well established [64, 65]. Temperature modulation is an extension to DSC. The first frequency-dependent TMDSC measurements in the glass transition region were reported in [66]. Since we are interested in the frequency-dependent complex heat capacity $\left(c_p^*(T,\omega) = c_p^{'}(T,\omega) - ic_p^{''}(T,\omega)\right)$, we follow the TMDSC data treatment proposed by Schawe [67]. A frequency-dependent effective complex heat capacity $C_{p\,effective}(\omega)$ is available from Eq. (2) [68]:

$$C^*_{p\ effective}(\omega) = \frac{A_\Phi(\omega)}{A_\beta(\omega)} = \frac{A_\Phi(\omega)}{\omega A_T}, \tag{2}$$

with $\omega = 2\pi/t_p$ is the angular frequency, t_p is the period of oscillation, $A_\Phi(\omega)$ is the complex heat flow rate amplitude, A_β is the heating rate amplitude, and A_T is the temperature amplitude. One should keep in mind that the perturbation in modulation calorimetry is a scanning rate (heating/cooling) and not a temperature. After careful calibration [69, 70], the complex heat capacity is available from the effective complex heat capacity.

In TMDSC, the rate of heat flow and the temperature amplitudes are obtained by Fourier transformation of the measured signals. For periodic signals with a known frequency, this has the big advantage that only the response at the selected frequency is evaluated. Compared to DSC experiments, the influence of low frequency (instrumental drifts) and high-frequency (mainly electrical) noise is dramatically reduced by the frequency selective evaluation. Therefore, with the help of a Fourier transformation, a high accuracy for the C_p determination is achieved [71].

TMDSC with harmonic perturbation $T(t) = T_0 + \beta_0 t + A_T sin(\omega t)$ is available from different manufacturers. Since temperature scans at constant frequency are performed, for each single frequency, only one scan is needed. For very low frequencies, a single measurement can take a long time, and to repeat these time-consuming experiments for different frequencies is not convenient. To measure all frequencies of interest simultaneously is therefore very attractive. Not only will multi-frequency measurements save time, but it also guarantees a perfect match of sample properties and thermal history for all frequencies. As mentioned above, any temperature-time profile can be applied (periodic, harmonic, anharmonic, stochastic) [72] as long as the response is linear [68].

Multi-frequency measurements can be realized by a temperature-time profile consisting of an anharmonic perturbation containing more than one frequency. The rectangular or sawtooth perturbations are examples of anharmonic perturbations. If we assume a periodic heating $\beta(t)$ with n points per period and a heat flow rate $\Phi(t)$ with the same number of points per period, the effective heat capacity [see Eq. (2)]

can be calculated with the help of a Fourier transformation [73, 74] according to Eq. (3)

$$C_{p\_effective} = \frac{1}{m}\frac{\sum_{k=1}^{n}\Phi(t_k)\cos(\omega t_k) - i\sum_{k=1}^{n}\Phi(t_k)\sin(\omega t_k)}{\sum_{k=1}^{n}\beta(t_k)\cos(\omega t_k) - i\sum_{k=1}^{n}\beta(t_k)\cos(\omega t_k)}. \quad (3)$$

If the rate of heating is sinusoidal, then a single point in the heat capacity spectrum is obtained corresponding to the base frequency ω_0. If the periodic perturbation has more than one harmonic $\omega = k\omega_0$, k being an integer, heat capacity will be at the higher harmonics. With perturbations containing higher harmonics up to the limit defined by the transfer function of the instrument, the whole accessible spectrum of complex heat capacity can be obtained [69, 70]. Applying stochastic perturbations, as in TOPEM™ [72], is another elegant way to generate such heat capacity spectra. Nevertheless, some special software options are needed. Here we focus on an easy way to generate heat capacity spectra with basically any DSC on the market.

As already discussed in [73], the programmed temperature–time profile should be easy to program and it should encompass as many higher harmonics at comparable heating rate amplitudes as possible. The way to generate equal heating rate amplitudes for all harmonics is to have a step-wise increase (decrease) in temperature, which is repeated until the final temperature is reached. Realization of quasi-isothermal measurements is possible by programming a square wave or an asymmetric sawtooth [73]. In an ideal case, the resulting heating rate, $\beta = dT/dt$, is a delta function per step, which contains all higher harmonics at equal heating rate amplitudes.

2.2.2 Alternating Current Calorimetry (AC Calorimetry)

In AC calorimetry, an oscillating power is applied to the sensor and the corresponding temperature response is measured. Under quasi-adiabatic and quasi-static conditions [75, 76], the apparent heat capacity, c_p, is given by Eq. (2).

For AC calorimetry, the low frequency limit is defined by the violation of the condition of quasi-adiabaticity [75, 76] and the corresponding decrease of the signal-to-noise ratio. The heating rate amplitude, A_β, immediately becomes very large at high frequencies, and the temperature oscillations are no longer small. For A_T = 0.01 K and $\omega = 10^6$ rad s^{-1} (frequency, f = 160 kHz) the heating rate is A_β = 10,000 K/s, which means the calorimeter must be able to heat and, more importantly, cool the sample at a rate of 10,000 K/s. Such high cooling rates are accessible by fast-scanning chip calorimeters as discussed in 2.1 [6, 15, 22]. By minimizing sample and addenda heat capacity (nJ K^{-1}) and utilizing a gas as the cooling medium [17], fast cooling can be realized and the thin-film limit for heat capacity determination is satisfied by using thin (<1 μm) samples [77, 78]. Therefore, the inherent problems of methods of measuring effusivity [78, 79], such as the 3ω-method [80, 81], transient grating techniques [82], or other high-frequency thermal techniques [83] are not important here.

We developed a differential AC chip calorimeter with pJ K^{-1} sensitivity, which is based on a chip sensor [21]. Because of the small total heat capacity (addenda + sample) and the possibility to cool, high sensitivity is achieved, and AC measurements at relative high frequencies become possible [84]. The calorimeter allows for measurements in the frequency range of 1 Hz to 50 kHz [85–87] and with laser heating up to 1 MHz [88]. The basic equations for such a differential setup are discussed in detail in [85–87]. Measurements are commonly done at a single frequency while continuously changing the temperature (isochronal temperature scans). To get an almost uniform temperature distribution and small errors in temperature, the heating rate of the thermostat is limited to 5 K/min. The current calorimeter can be operated in air or in inert gas as nitrogen or helium at ambient or slightly reduced pressure in a gas with controlled humidity [89, 90] or even in high vacuum [87, 91]. The electrical scheme of the device is described in [85, 86].

In Fig. 4, data from a polystyrene film of 40 nm thickness are shown. Frequency sweeps from 80 to 480 Hz were performed, and the temperature was varied from 323 to 423 K in steps of 5 K. Here, we present the measured signal, which is proportional to sample heat capacity for thin films [85, 87]. Nevertheless, the glass transition temperature and changes of the signal are easily detected.

From the real part of the measured thermopile signal, the frequency-dependent glass transition temperature is determined as the half-step temperature, which equals the maximum of the imaginary part (peak in the phase). The phase angle is corrected for heat transfer contributions as described in [92].

For even higher frequencies (higher cooling rates), it is of particular interest to have low sample and addenda heat capacities as well as a small size of the heated area [17, 25]. Chip sensors with only one thermopile and focused laser heating allow for temperature oscillations at angular frequencies up to 10^6 rad s^{-1} at still detectable temperature amplitudes of about 1 mK as shown in [88]. Figure 5 shows the scheme of such a device.

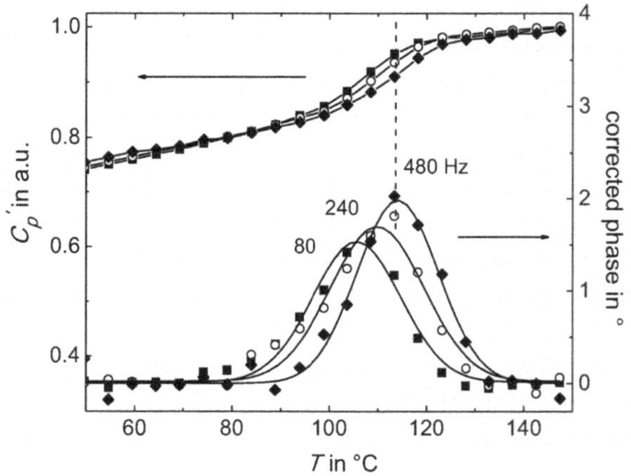

Fig. 4 Real part of complex heat capacity and corrected phase for a 40 nm thin polystyrene film for frequencies 80 Hz (filled square), 240 Hz (open circle) and 480 Hz (filled diamond). All data are taken from reference [85]

For the laser-heated AC chip calorimeter, a heating power on the order of 10 μW is adequate. A light beam is generated by a laser diode and is directed and shaped by a glass fibre onto the hot junction of the thermocouple. The fibre must be brought very close (≈10 μm) to the calorimeter chip to avoid enlargement of the spot size. The modulation depth is adjustable and can reach close to 100%. It is controlled by the oscillator output of the digital lock-in amplifier. The device can work as a single sensor AC calorimeter [26]. For details, see [88].

2.3 *Relaxation Map of Polystyrene from Calorimetry and Dielectric Relaxation Spectroscopy*

Advanced calorimetric techniques provide information about glass transition dynamics covering up to 11 orders of magnitude in frequency or cooling rate. From a combination of conventional DSC at low frequencies and slow heating rates with chip calorimeters allowing high cooling rates (see Sect. 2.1) and consequently high frequencies, two independent quantities are available. From TMDSC and AC calorimetry, the complex heat capacity as the entropy compliance is obtained within the linear response scheme. From scanning experiments at constant cooling rate, the heat capacity, in its classical definition as the temperature derivative of enthalpy, can be determined. The glass transition in the latter corresponds to the transition from the supercooled liquid to the non-equilibrium glass (vitrification), a typical nonlinear response experiment. The temperature of vitrification depends on cooling rate. In complex heat capacity, the glass transition is seen in the liquid state and the temperature of the observed relaxation process depends on frequency. The complex heat capacity $\left(c_p^*(T,\omega) = c_p'(T,\omega) - ic_p''(T,\omega)\right)$ and the so-called total heat capacity $(c_{p\,total}(T,\beta_0) = (1/m)dH(T,\beta_0)/dT$, with β_0 the underlying scanning rate and

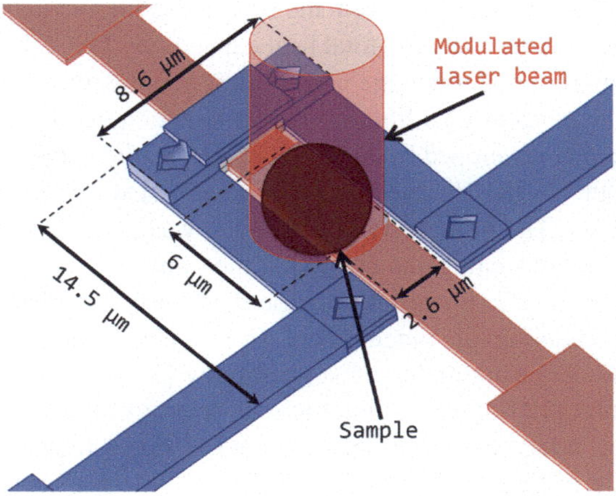

Fig. 5 Modulated laser beam shining on a sample on the hot junction of the thermocouple. The two heaters and the aluminium electric connections are shown in blue. The doped poly-Si stripes of the thermocouple are in red and the hot junction, made by an aluminium connection, in blue [88]

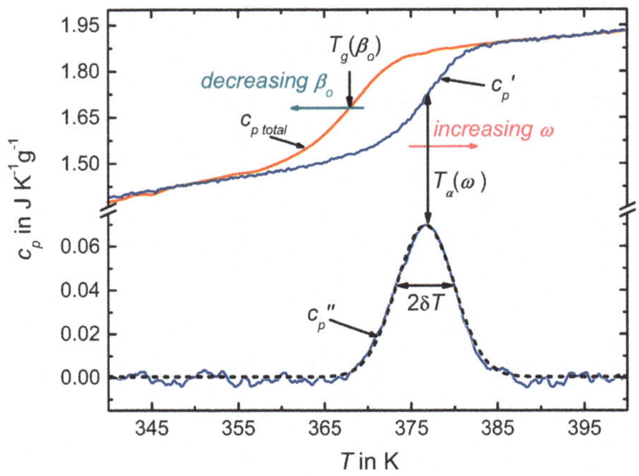

Fig. 6 Real and imaginary parts of the complex specific heat capacity, $c_p'(\omega)$, $c_p''(\omega)$, for polystyrene measured by TMDSC, and the total heat capacity, $c_{p\,total}(\beta_0)$, from the underlying linear scan at a cooling rate β_0. Measuring conditions: temperature modulation amplitude $A_T = 0.5$ K, angular frequency $\omega = 2\pi/P$, period $P = 60$ s, underlying scanning rate $\beta_0 = -0.25$ K min^{-1}. The dashed curve is a Gaussian fit to the imaginary part of specific heat capacity. The width of the Gaussian function equals $2\delta T$. All data are taken from reference [97]

H the enthalpy) are shown in Fig. 6. For a more detailed discussion of differences between these quantities, particularly in the case of thin films, see references [93–96].

For the complex heat capacity, as for other compliances, the relaxation process is described by the maximum, T_α, and the width of the peak in the imaginary part, $2\delta T$, and the step height of the step of the real part, Δc_p, which is related to the area under the peak in the imaginary part. The maximum of c_p'' coincides, as usual, with the half-step temperature of c_p'. As for other compliances, the position of both quantities depends on the measurement frequency and is commonly described by the Vogel–Fulcher–Tamman (VFT) function given by Eq. (17), and illustrated in Fig. 7.

For the transition from the supercooled liquid to the non-equilibrium glassy state, e.g. measured by the total heat capacity by DSC, the situation is more complex and different quantities are used to describe the glass transition; see, for example [98]. The concept of a limiting fictive temperature, T_f, as introduced by Tool [99] and further developed by Moynihan [100] is often applied and preferred because it has a sound basis. The calorimeter was calibrated according the GEFTA recommendation [101]. The influence of thermal lag was corrected by making use of the symmetry between heating and cooling at the same rate.

Figure 7 shows calorimetric data for polystyrene (PS) in a very wide frequency and cooling rate range. The $T_\alpha(\omega)$ from calorimetry is in reasonable agreement with the data from dielectric spectroscopy [102]. Only above the dielectric $\alpha\beta$ splitting point [104], the calorimetric data seem to follow the Arrhenius fit to the dielectric secondary β-relaxation, measured at lower temperatures [102]. For an analytical description, the calorimetric $T_\alpha(\omega)$ data were approximated by a separate VFT func-

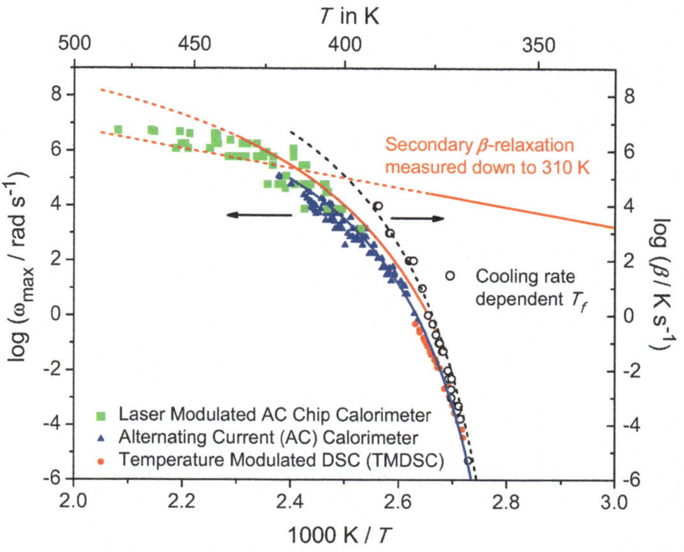

Fig. 7 Relaxation map for polystyrene PS168N. The coloured data points correspond to the segmental relaxation from calorimetric measurements (left axis). The black open circles are cooling rate dependent glass transition temperatures, T_f, from linear scan experiments (right axis). The curved red line represents the VFT fit to dielectric data of the segmental relaxation from Schönhals et al. [102]. The straight red line shows the Arrhenius fit to the data of the dielectric secondary β-relaxation process [102]. The curves are VFT fits Eq. (17) with the parameters given in Table 1. Reproduced with permission from reference [103]

Table 1 VFT parameters [Eq. (17)] for the curves shown in Fig. 7

Curve in Fig. 7	A	B/K	T_∞/K
T_α dielectric (red)	10.5	475.3	334.4
T_α calorimetric (blue)	10.2	388.0	341.5
T_f calculated (black) [103]	$\log\left[5 \cdot \left(-128.8 + 0.5 \cdot T - 4.8 \times 10^{-4} \cdot T^2\right)\right] + 10.2$	388.0	341.5

tion for 1000 K/ T between 2.4 and 2.75. This function was then recalculated to the cooling rate dependent $T_f(\beta)$ (black dashed line), see reference [103] for details. These VFT parameters were then used for detailed modelling of the vitrification in the next section.

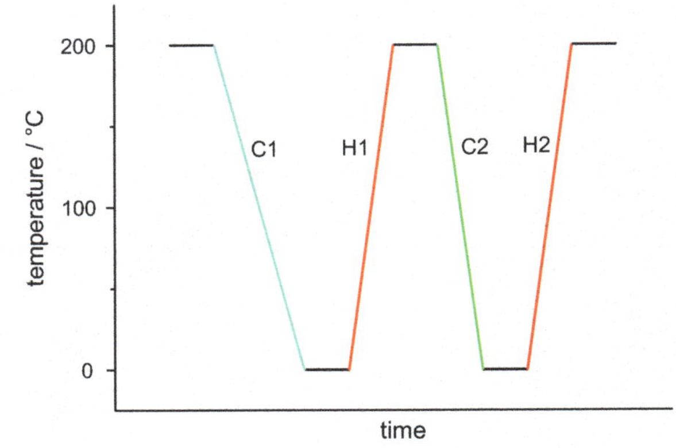

Fig. 8 Profile of temperature change during a single DSC measurement on polystyrene. C1, C2—first and second cooling runs, respectively. H1, H2—first and second heating runs, respectively

2.4 *Experiments Compared to Results from Modelling*

For the measurement, we have used atactic polystyrene (PS) due to its pronounced thermal stability. Even at extremely slow cooling rates, where the sample stays above the glass transition temperature for several days, no degradation of the polymer was observed. The sample is PS168N (M_w = 270,000 g/mol, M_n = 95,000 g/mol) from BASF, Germany. The bulk sample with a mass of 3.8 mg was put into a standard Perkin–Elmer DSC aluminium pan.

The experimental data were acquired with a Perkin–Elmer Pyris 1 DSC. The temperature profile used for each of the measurements is presented in Fig. 8. One measurement cycle went as follows: the system was cooled from liquid state at T = 200 °C with a certain cooling rate (C1, light blue curve in Fig. 8). This cooling rate, C1, was altered in every following run and covered a range between 5×10^{-6} K/s and 2 K/s. The next steps of the measurement did not change for each following run. The sample was always reheated with the same heating rate of 0.5 K/s back up to the initial temperature of 200 °C (H1). During heating, the system goes through the glass transition region a second time. The response of the system during this heating reflects the sample's thermal history, which in our case is cooling at C1. Thus, experimental heat capacity curves vary with the chosen cooling rate. The next two steps are cooling (C2) with 0.5 K/s and reheating with 0.5 K/s (H2). Performing these steps for each cooling rate C1, we obtain a reference curve for calculating the excess heat capacity. The reference curve must be measured and calculated after every run, because the measurement takes several days and even weeks for very low cooling rates, and it is unlikely that previously measured reference curves correspond to exactly the same experimental conditions.

The excess heat capacity curves were calculated using the Pyris software. For the reference curve (H2), the sample mass was set to zero as a baseline curve in the software. This experimental procedure has advantages regarding the stability of the measurement because of missing sample changes between measurements with sample and reference, respectively. Furthermore, determination of the excess heat

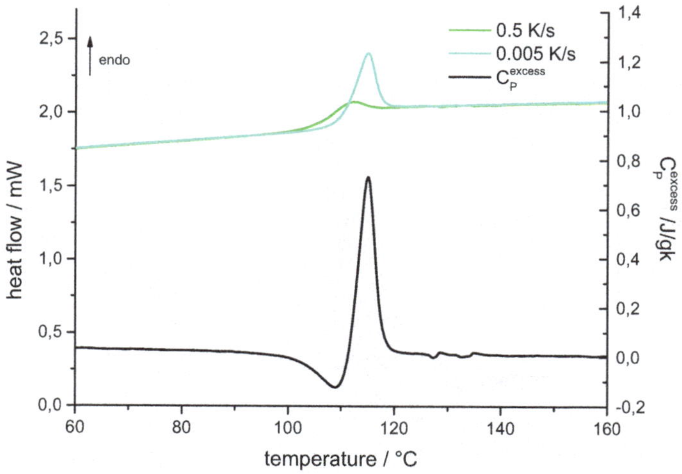

Fig. 9 Example treatment of a single experimental data set using the described procedure and Eq. (4) [8]. The upper two curves correspond to the heat flow from the sample during heating at 0.5 K/s after cooling at 0.5 and 0.005 K/s. The lower curve represents the resulting data via Eq. (4) excess heat capacity for heating after cooling at 0.005 K/s

capacity removes any influences of the different slopes of glassy and liquid heat capacities outside the glass transition range. This way, operator-dependent subjective influences that are common for the C_p^{red} determination, as outlined below, are excluded when C_p^{norm} is employed for comparison of the model and experimental data. One example, for cooling rate C1 equal to $q = 5 \times 10^{-3}$ K/s, is shown in Fig. 9. The excess heat capacity curves were normalized by the step height of PS glass transition. The procedure can be expressed by the relation:

$$C_p^{\text{excess}}(T) = C_p^{H1}(T) - C_p^{H2}(T). \tag{4}$$

For subsequent comparisons of measurement and model computations, the glass transition step height, ΔC_p, was used as a normalization factor as described by the relation:

$$C_p^{\text{norm}}(T) = \frac{C_p^{H1}(T) - C_p^{H2}(T)}{\Delta C_p}, \tag{5}$$

where $\Delta C_{\text{p}} = C_{\text{p}}^{\text{liq}} - C_{\text{p}}^{\text{gl}}$ is the difference between liquid and glass heat capacity. As a result, the resulting heat capacity values are substantially distinct from zero only in the glass transition region (Fig. 9).

The ranges in temperature change rates employed in previous works (e.g. 0.021–0.66 K/s in the work by Hodge [105], or 6.2×10^{-5} to 1.6×10^{-2} K/s in [106]) are extended in our measurements to 5×10^{-6} to 2 K/s. Thus, almost six

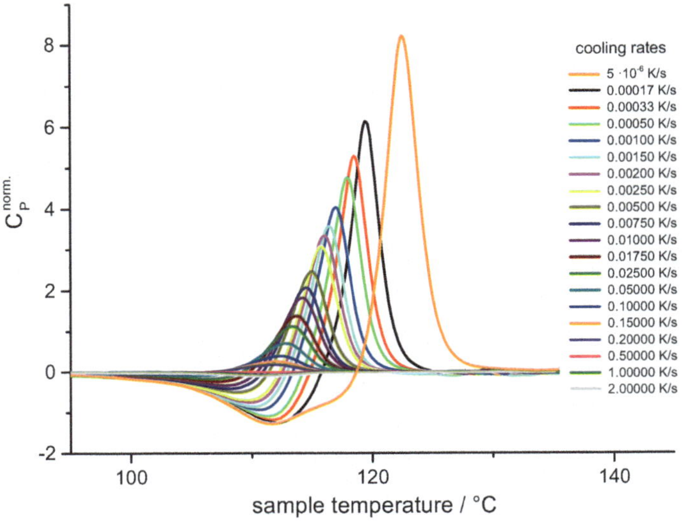

Fig. 10 Experimentally measured excess heat capacity curves for the polystyrene PS168N sample at cooling rates ranging from 2 to 5×10^{-6} K/s [8]

orders of magnitude of cooling rates were covered. These results are comparable with the measurements by Schawe [7], yet they are also qualitatively different. This difference is important for the modelling of kinetics and will be discussed below. The resulting experimental heat capacity curves are presented in Fig. 10. With increasing cooling rates q, the peaks shift to lower temperatures, the height of peak maxima decreases whereas their width increases. The overall shift in temperature of the peak maxima is on the order of 15 K.

Similar experiments were performed on a fast-scanning calorimeter (FSC) [22] for higher cooling rates covering the range from 10^{-1} K/s up to 10^4 K/s with a reference heating rate of 5000 K/s. The sample mass in this case is almost three orders of magnitude less. The obtained heat capacity curves were not fitted in this work; thus we have only used the measured $T_g(q_c)$ dependencies for obtaining the VFT parameters and testing modelling results.

One more additional experiment at $q = \pm 10$ K/min (0.1667 K/s) was performed to verify how well these common curves are fitted. This basic experiment corresponds to conventional rates of cooling/heating for calorimetry on polymer samples and will be used as an additional reference curve for clarifying and discussing the modelling results. The measured curves are presented in Fig. 11a, where the black-coloured lines represent respective C_p curves for polystyrene from ATHAS (The Advanced Thermal Analysis System) Data Bank [107, 108].

We will compare theoretical data with experiment in terms of reduced heat capacity curves:

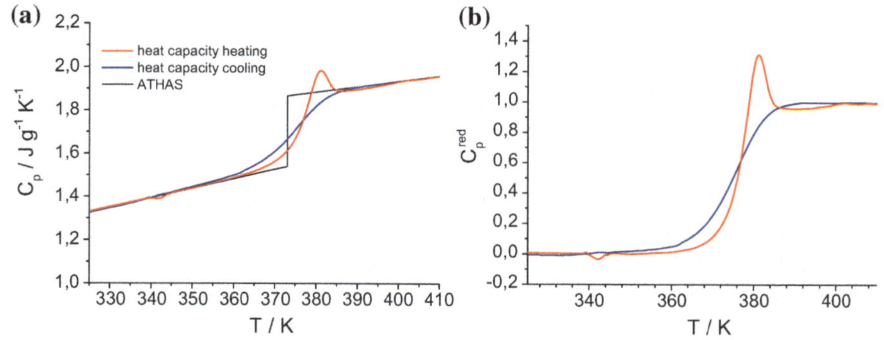

Fig. 11 C_p curves of polystyrene at linear cooling and heating at 10 K/min (0.1667 K/s). **a** Initial experimental curves compared with ATHAS (The Advanced Thermal Analysis System) Data Bank data for polystyrene. **b** Resulting reduced heat capacity for cooling and heating as obtained via Eq. (6)

$$C_p^{red}(T) = \frac{C_p(T) - C_p^{glass}(T)}{C_p^{liquid}(T) - C_p^{glass}(T)}. \tag{6}$$

For example, the reduced C_p curves are presented in Fig. 11b. In one of the theoretical approaches utilized in this work, the Gutzow–Schmelzer approach (Sect. 3), C_p^{liq} does depend on T, so this linear dependence will be substituted into Eq. (6) in order to compare theoretical results with experimental curves.

To compare modelled kinetics with experimental C_p^{norm} curves, two C_p^{red}-curves corresponding to C_p^{H1} and C_p^{H2} in Eq. (2) must be known. Each of these curves for the calculations is considered as a separate experiment (C1-H1 and C2-H2 on Fig. 8), because the measurements start and end in liquid state at sufficiently high temperatures to remove any sample "memory" of previous thermal history. Below, we show that model parameter sets may differ considerably for different cooling rates. Thus, for direct comparison, it is necessary to obtain experimental heating C_p^{red}-curves for each separate measurement. To do this, the $C_{\mathrm{p}}^{\mathrm{red}}$ curve, corresponding to 0.5 K/s heating (C_p^{H2}) was obtained from experiments and added to each $C_{\mathrm{p}}^{\mathrm{norm}}$ curve on Fig. 10. The obtained heat capacities will be used in some parts of this work when variation of model parameters on cooling rates is considered.

The measured heat capacity curves reflect the kinetics of the thermal glass transition of polystyrene for a wide range of cooling rates. In addition, one can evaluate the general and integral properties of glass transition as well as obtain their q dependencies. They are the limiting fictive temperature $T_{f'}$, referred to as the calorimetric T_g, and the parameters of the positive C_p peaks—maxima C_p^{max} and positions T_{max}.

The T_f value represents a structural state of the system and can be calculated from the experimental C_{p} curves with the expression given, for example, in Ref. [100]. Using the normalized excess heat capacity curves, additional computations are required to calculate these T_f values. The slopes of the liquid and the glass dependencies can be expressed from the respective triangles in Fig. 12:

$$C_p^{liq} = \frac{\Delta H + \Delta H^*}{T_{f1} - T_{f2}}, \quad C_p^{gl} = \frac{\Delta H*}{T_{f1} - T_{f2}}, \tag{7}$$

where T_{f1} and T_{f2} are the limiting values of the fictive temperature upon cooling at different rates, q_1 and q_2, and ΔH is the corresponding enthalpy change on reheating of the glass. The value of $\Delta C_p = C_p^{liq} - C_p^{gl}$ is measured experimentally, and thus from (7) follows the relation for T_{f2}:

$$T_{f2} = T_{f1} - \frac{\Delta H}{\Delta C_p}. \tag{8}$$

To estimate the limiting fictive temperature from each curve in Fig. 10 using Eq. (8), one must calculate the value of ΔH integrating the whole curve, and use the value of T_{f1} calculated from C_p for a run at constant heating and cooling, 0.5 K/s, used as the baseline during measurements. As an example, the obtained dependencies of T_f and T_{max} on q are show on Fig. 13.

3 Theoretical Modelling

In this part of the chapter, we report results of modelling of the kinetics of glass transition with several different theoretical approaches. These methods are first briefly characterized, followed by testing their applicability for modelling a single calorimetric experiment. Finally, we demonstrate their non-applicability to the measurements in a wide range of cooling rates. All of the methods below are based on defining

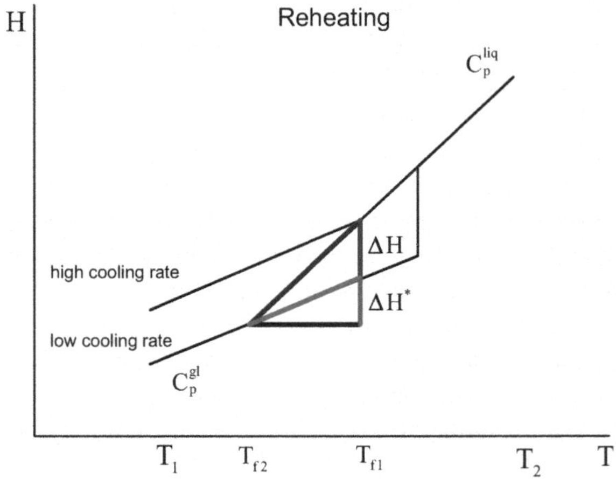

Fig. 12 Schematic representation of the enthalpy behaviour of two glasses, obtained at different cooling rates, on heating at a constant rate. C_p^{gl}, C_p^{liq} are the heat capacities in glassy and liquid state, respectively

a single structural state parameter that reflects the system's non-equilibrium state. Although the Kovacs–Aklonis–Hutchinson–Ramos (KAHR) method exists [109], which is initially a multi-parameter approach for the same problem, it has been studied thoroughly elsewhere [110].

Irreversible thermodynamics approach

This approach is based on irreversible thermodynamics, developed for describing macroscopically out-of-equilibrium systems, especially kinetic transitions such as chemical reactions or phase transformations [111]. Overall, it is suitable for describing the glass transition as dynamic freezing-in of general systems' properties. Using this approach, one works within a framework that combines the universality of thermodynamics with the possibility of detailed definitions of specific systems properties via statistical physics models [3].

Here, we shall restrict ourselves to the general approach and introduce ξ—a structural parameter, reflecting the system's departure from its equilibrium state, and ξ_{eq}—its value at equilibrium. The temporal evolution of ξ is expressed by the equation [112]:

$$\frac{d\xi}{dt} = -\frac{1}{\tau(T,\xi)}\left(\xi - \xi_{eq}\right), \tag{9}$$

where $\tau(T,\xi)$ is the relaxation time of the system, which, due to the nonlinearity of the glass transition, depends also on ξ, in addition to temperature T and pressure p. The pressure dependence is not considered in this work and will be omitted. By defining the expressions for $\tau(T,\xi)$, one can investigate the kinetics of change of systems'

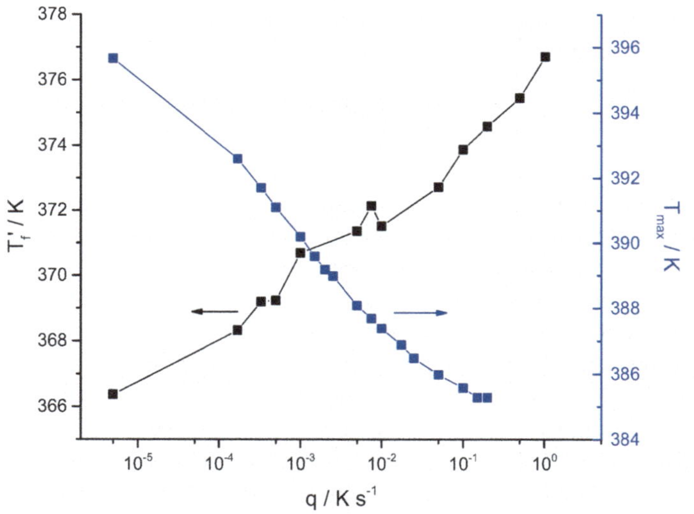

Fig. 13 Dependence of curve properties—limiting fictive temperature T_f' (an integral curve parameter) and position of the peak T_{max} on cooling rate q

properties during the glass transition via Eq. (9). These different expressions will be discussed below for both definitions of the structural parameter.

We also need to define the dependence of ξ_{eq} on p and T. Here, this will be done within the lattice-hole model of a liquid, where $\xi = N_0/(N_A + N_0)$, N_0 being the number of holes per mole. For the configurational part of entropy, we can write the mixing expression as

$$S_{conf} = -R\left[\ln(1-\xi) + \frac{\xi}{1-\xi}\ln\xi\right] \tag{10}$$

and using further some phenomenological expression for enthalpy dependence on ξ [112–114], for Gibbs free energy, we obtain:

$$G_{conf} = \xi\chi RT_m + RT\left[\ln(1-\xi) + \frac{\xi}{1-\xi}\ln\xi\right], \tag{11}$$

where $\chi \approx 3.3$ for many liquids, T_m is the melting temperature and R is the universal gas constant. The expression for ξ_{eq} is obtained from the condition of $\left(\frac{\partial G_{conf}}{\partial \xi}\right) = 0$.

One should also account for the non-exponentiality of the relaxation for the glass-forming process. To obtain the non-exponential relaxation character similar to the Kohlrausch law, Eq. (9) must be modified accordingly [115]:

$$\frac{d\xi}{dt} = -\frac{1}{\tau(T,\xi)}\frac{A_0}{t^{1-\beta}}(\xi - \xi_{eq}), \tag{12}$$

where A_0 is a model parameter and β defines the stretching of the exponential, $0 < \beta < 1$, $\beta = 1$ corresponding to exponential relaxation.

The configuration part of heat capacity is expressed as:

$$C_{p,conf}(T) = \left(\frac{dH_{conf}(T,\xi)}{dT}\right)\frac{d\xi}{dT}. \tag{13}$$

The first multiplier in Eq. (13) is constant in our case, but could have another dependence in other models. This method will be referred to via abbreviation GS (Gutzow–Schmelzer method) in the following sections.

Fictive temperature-based approaches

The most popular structural parameter nowadays is the so-called fictive temperature T_{f}, introduced originally by Tool [99]. It corresponds by definition to the temperature at which the equilibrium state of the system is the same as the current non-equilibrium state (characterized by T and T_{f}, respectively). In his general work published in 1971, Narayanaswamy derived expressions for the relaxation of system properties based on three suppositions [9]. The first one stated structural relaxation to proceed via a single non-exponential mechanism with constant activation energy. Second, he supposed that the shape of the equilibrium response function does not depend on current values of T and T_f (extended thermorheological simplicity).

Finally, it was supposed that the relation between force and flow is linear; in other words, the only cause of nonlinearity is the change of a property due to relaxation of the fictive temperature.

These suppositions allowed for introducing the reduced time scale and writing the integral to calculate the overall relaxation of a property. In further works, Moynihan and co-workers [100, 116] have proposed numerical schemes for calculating the evolution of system properties on different thermal histories, including temperature steps and linear changes in T (cooling/heating). Finally, one can write a general expression for fictive temperature for any given thermal experiment. To do this, the timescale of the experiment is divided into sufficiently small intervals Δt_j, and the value of fictive temperature at the end of each interval is equal to [116, 117]:

$$T_{f,n} = T_0 + \sum_{i=1}^{n} \Delta T_i \left(1 - \exp\left(-\left(\sum_{j=i}^{n} \frac{\Delta t_j}{\tau(T_j, T_{f,j})}\right)^{\beta}\right)\right), \tag{14}$$

where T_0 is the initial temperature, and $\tau(T_j, T_{f,j})$ are the values of relaxation time at given T_j and $T_{f,j}$. Thus, one can extend Narayanaswamy's approach considering different values for the relaxation time, possibly even beyond his supposition.

Defined this way, the fictive temperature allows for the direct calculation of reduced heat capacity via the relation:

$$C_p^{red} = \frac{dT_f}{dT}. \tag{15}$$

From (15), one can see that $C_p^{red} = 0$ occurs in the glassy state (where T_f is constant, frozen-in) and $C_p^{red} = 1$ occurs in the liquid state (where $T_f = T$, system is at equilibrium). Thus, to model the experimental data, one must either use Eq. (6) or reverse it by using predefined $C_{p,\text{liquid}}(T)$ and $C_{p,\text{glass}}(T)$ dependencies.

3.1 Expressions for the Relaxation Time

To proceed further with modelling, one must provide an expression for the relaxation time dependence on temperature and structural parameters. Several of these expressions were proposed during the last 50 years, in addition to classical laws originally defined even before them.

The basic expressions for τ are the Arrhenius and the Vogel–Fulcher–Tamman (VFT) laws. In both cases, τ does not depend on structural parameters (nonlinearity is not introduced). The first law holds for the so-called "strong" glass-forming materials, whereas the second is true for "fragile" ones (polymers included). The Arrhenius expression is

$$\tau^{Ars}(T) = \tau_0 e^{\frac{A}{RT}}, \tag{16}$$

where A is a certain activation energy corresponding to structural relaxation and τ_0 is the time constant at infinite temperature. The VFT law bears the following expression:

$$\tau^{VFT}(T) = \tau_0 e^{\frac{A}{R(T-T_0)}}, \tag{17}$$

where T_0 is a certain reference temperature, at which systems relaxation time asymptotically grows to infinity. Using a straightforward criterion for the glass transition as

$$(q\tau(T))|_{T=T_g} \cong C, \tag{18}$$

one can obtain the Bartenev–Ritland equation for systems obeying the Arrhenius relaxation law [118, 119], whereas for VFT-type systems, a modified expression should be used:

$$\lg q = \lg \tau_0' - \frac{A'}{T_g - T_0}, \quad \tau_0' = \frac{\tau_0}{C}, \quad A' = A \lg e. \tag{19}$$

It can be seen from (19) that by fitting the cooling rate dependence of glass transition temperature, one can directly obtain values of the parameters A and T_0, and have a quite accurate estimate of τ_0. This estimate is not exact, because the value of C is unknown. Yet, as an approximation, $C \sim 1$ can be considered (in comparison with the big value of τ_0).

Another expression for the relaxation time that may be used is a consequence of the well-known Adam–Gibbs (AG) theory of glass transition [13]. The authors developed a molecular-kinetic theory of glass-forming liquids, supposing that relaxation involves a cooperative rearrangement of subsystems (cooperatively rearranging regions, CRRs). As a result, an exponential dependence of relaxation time on macroscopic configurational entropy, $S_c(T)$, is obtained:

$$\tau^{AG}(T) = \tau_0 \exp\left(\frac{A}{TS_c(T)}\right). \tag{20}$$

The configurational entropy can be estimated via

$$S_c(T) = \int_{T_2}^{T} \frac{\Delta C_p(T)}{T} dT, \tag{21}$$

where $\Delta C_p(T) = C_{p,liq}(T) - C_{p,glass}(T)$, and T_2 is a model parameter reflecting the temperature at which $S_c(T_2) = 0$ and a second-order phase transition occurs. Generally, the definition of ΔC_p may differ (with respect to the second term). The linear dependence of ΔC_p on T is often considered as $\Delta C_p = C_0 + C_1 T$ and one obtains [120]:

$$\tau^{AG}(T) = \tau_0 \exp\left(\frac{Q/RT}{\ln(T/T_2) + C(T - T_2)}\right), \tag{22}$$

where $Q{=}AR/C_0$ and $C{=}C_1/C_0$.

When modelling the kinetics of glass transition, one must also account for the nonlinearity of the process using the modified versions of expressions (16), (17) or (22) by introducing an additional parameter x, where $0{<}x < 1$. Thus, it is supposed that a certain part of the material relaxes not in accordance with the external temperature, but with the structural state parameter (i.e. eventually frozen-in). We present the VFT modified expression:

$$\tau^{VFT}(T, T_f) = \tau_0 \exp\left\{\frac{A}{R(T - T_0)}x + \frac{A}{R(T_f - T_0)}(1 - x)\right\} \tag{23}$$

with the nonlinear Arrhenius expression being similar. When ξ is the structural order parameter, we may use an expression [112]:

$$\tau^{VFT}(T, \xi) = \tau_0 \exp\left(\frac{A}{R(T - T_0)} + \frac{b_2}{RT}(\xi - \xi_{eq})\right), \tag{24}$$

where b_2 is the nonlinearity parameter. In [8], it was shown that for modelling the polystyrene glass transition at different cooling rates, the modified nonlinear expression for relaxation time is quite effective, where nonlinearity is introduced via the term proportional to the configurational entropy (in mixing form):

$$\tau^{I}(T, \xi) = \tau_0 \exp\left(\frac{A}{T - T_0} - \frac{B}{T\left(\ln(1 - \xi) + \frac{\xi}{1-\xi}\ln\xi\right)}\right). \tag{25}$$

Concerning the AG theory, Scherer in [120] noticed that configurational entropy should correspond to the current structure of a system, but not its equilibrium state. The argument of the S_c function in (20) was substituted by T_f, leading to the modified expression for relaxation time:

$$\tau^{AG}(T) = \tau_0 \exp\left(\frac{Q/RT}{\ln(T_f/T_2) + C(T_f - T_2)}\right). \tag{26}$$

On the other hand, in [121] Hutchinson proposed that nonlinearity should be introduced in a way similar to (23) by partitioning the main configurational function. Thus, the expression for S_c was introduced in the form:

$$S_c(T, T_f) = x_s S_c(T) + (1 - x_s)S_c(T_f), \tag{27}$$

where x_s is the alternative nonlinear parameter.

A number of advanced methods have been proposed basing on the aforementioned described approaches. For example, in [122, 123], an alternative configurational entropy-based method for describing structural relaxation and ageing of different polymers was developed. The authors took advantage of a recently discussed two-step relaxation effect observed for polymers [124], and supposed that there exists an intermediate state of entropy relaxation (above equilibrium line) at which the system stays during ageing. The structural parameter within this method is the configurational entropy, and the model introduces one additional parameter δ, reflecting departure of the "intermediate plateau" from the equilibrium line in terms of heat capacity C_p. In calculations, one obtains the evolution of configurational entropy and enthalpy of the system, which are directly connected with C_p.

3.2 Fitting the T_g Dependence

Prior to modelling the kinetics of glass transition, by which we mean the whole $C_p(T)$ curves, we shall start with fitting the cooling rate dependencies of T_g. This is considered a relatively easy task since one works with a single value for each q_c instead of the specific shape of the heat capacity curve. A number of fits of $T_g(q)$ have been made for polystyrene samples. Hodge accumulated some of PS fit results in [98, 105], yet the Arrhenius-type relaxation was considered. Later, the activation diagram reflecting a dynamic T_g versus frequency dependence was fitted in [117] via the VFT law. A wide cooling rate range was covered recently in our work in [8], and also by Schawe in [7]. These data are summarized in Table 2 and visualized in Fig. 14, with our updated measurements via fast-scanning DSC. In [7], Schawe fitted the T_g dependence with the fixed value of $T_0 = 333$ K. We have performed fits of these data with all parameters varied (thus T_0 is slightly different).

The agreement between different measurement results, in sense of the comparison of respective VFT parameters, is acceptable, yet one should notice from Fig. 14 that the values of T_g vary by several degrees: the disagreement is growing with increasing lg|q|. For example, while for $q = 1$ K/s one can calculate $\Delta T_g = 0.4$ K, whereas for $q = 10^{-5}$ K/s and $q = 10^5$ K/s, the differences are already $\Delta T_g = 2.5$ K and $\Delta T_g = 5.7$ K, respectively.

Table 2 VFT parameters of the polystyrene glass transition from three different sources

Source	$-\lg(\tau_0/s)$	A′ in K	T_0 in K
$T_g(q)$, this work and [8]	13.5 ± 1.4	460 ± 100	342 ± 4
$T_g(q)$, [7]	12.6 ± 1.4	500 ± 130	336 ± 5
$T_g(w)$, [117]	12 ± 1	570 ± 90	333 ± 4

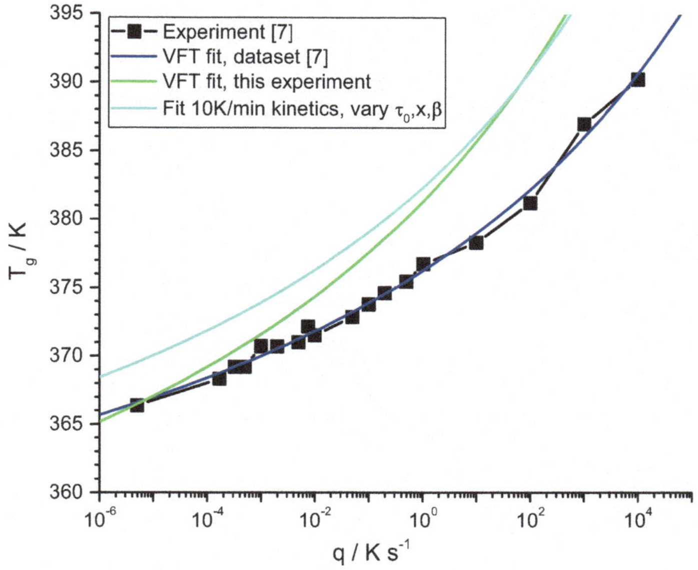

Fig. 14 Comparison of dependencies of glass transition temperature on cooling rate, q, obtained in different works [7, 8]. Each data was fitted via the VFT law; parameters are given in Table 2

For future work, one should mention the differences between the resulting VFT parameters and the parameters governing the relaxation time. Let us use, instead of Eq. (18), the general glass transition criterion [118, 119, 125]:

$$\left(\frac{1}{T}q\tau(T)\right)\bigg|_{T=T_g} = C, \quad C \cong 1. \tag{28}$$

In this way, a modified expression for the T_g cooling rate dependence is obtained:

$$\lg q = \lg T_g - \lg \tau_0' - \frac{A'}{T_g - T_0}, \tag{29}$$

where τ_0' and A' are defined as in Eq. (19), but now τ_0' is very close to τ_0 because C is almost unity. Moreover, because lgT_g varies only slightly, it can be considered as a constant ~2.55 correction to lgτ_0. Equations (19) and (29) can be the basis for comparing the results of kinetic modelling and fitting the T_g dependencies.

One should also keep in mind that the VFT parameters obtained from the activation diagram correspond to measurements of the dynamic glass transition with respect to frequency. While these parameters could be used for modelling the glass transition kinetics [117], direct comparison of $T_g(\omega)$ and $T_g(q)$ cannot be made on a single scale without taking into account their difference. This topic has been carefully discussed, for example, in [103] with respect to polystyrene.

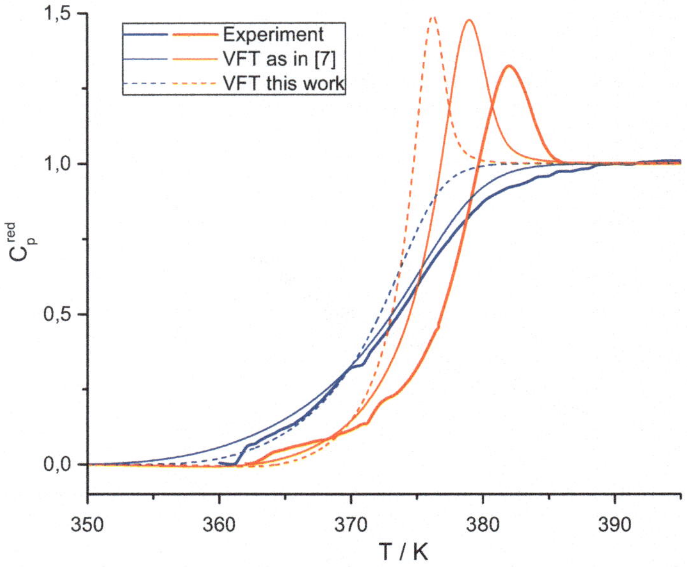

Fig. 15 Modelling and fitting results of a single cooling/heating experiment via the TNM method at scanning rate of 10 K/min. Two parameter VFT sets were used (from [7], and from this work), both bearing considerable differences from measured kinetics, especially for the heating part. Results of parameter minimization with respect to τ_0, x and β improve overall agreement

3.3 Fitting a Single Cooling/Heating Experiment

A first approach to modelling the glass transition course would be to model a single cooling/heating experiment on polystyrene (Fig. 11). This has been done in [8, 126] for the conventional 10 K/min scanning rate. In Fig. 15, we show the results of modelling with VFT parameter sets from Table 2. The disagreement is quite significant. The conclusion is that even a single experiment is not modelled well when one uses the VFT parameters obtained from the $T_g(q)$ dependence. The variation of model parameters is required. A standard Nelder–Mead minimization procedure was applied [127], taking account of possible parameter correlations. The criterion of fit quality, σ, was the root-mean-square deviation in significant part of the curves with step of 0.01 K, divided by the number of points.

The general applicability of different models was verified as a next step of modelling. The resulting parameter values and the fit quality, σ, are presented in Table 3, and quality of the fits is visualized for some of the models in Fig. 16.

The quality of the obtained fits is comparable in most cases, with the GS method showing the worst agreement. Interestingly, this disagreement comes mostly from the undershoot appearing before the peak C_p dependence, whereas the agreement with other parts of the curve is better.

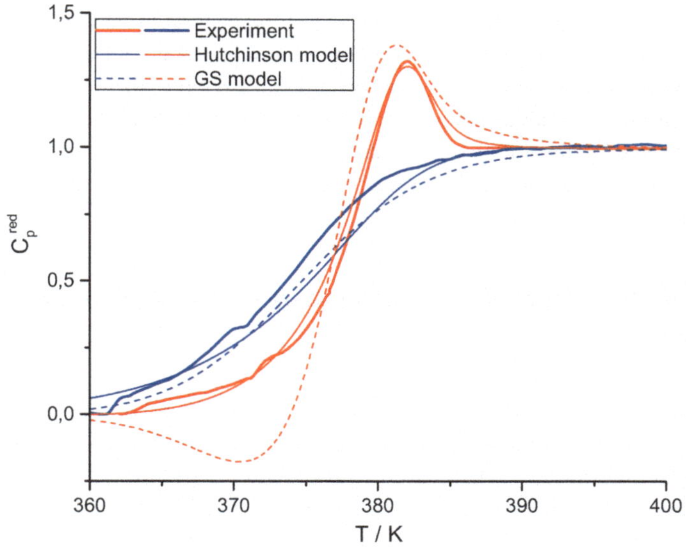

Fig. 16 Fitting the 10 K/min cooling/heating calorimetric experiment with different methods. Examples of applying Hutchinson model and the GS-I method are presented

A correlation between model parameters is evident. Variation of any values of the parameters changes the C_p-dependencies. For example, increasing τ_0 results in all peaks (heating run) shifting to higher temperatures, in parallel to an increase in height (C_p-value in maximum), whereas their width and temperature shift between peaks corresponding to different cooling rates do not change. Increasing T_∞ results in all peaks also shifting to higher temperatures; however, their height decreases, their width increases, and the peak shift (ΔT between different peaks) decreases.

Table 3 Parameter values and fit quality for the modelling of the 10 K/min (0.1667 K/s) cooling–heating curves of PS168N [126]

Method	Parameter values					σ*10^3
	ln(τ_0/s)	β	T_0, T_2 in K	A in K	Additional	
TNM-VFT	−15.8	0.514	343.6	632	$x = 0.19$	2.0
TNM-AG	−93.3	0.55	290	1610	$C = -0.0025$	2.1
GS-VFT	−8.18	–	328.2	500.2		44
Hutchinson	−65.4	0.559	226.5	13,200	$x_s = 0.217$	2.1
Conf. entropy	−37.4	0.556	264.9	1000	$\delta = 0.247$	1.8
GS I	−21.4	–	282.1	2210	B/R≈0	14

The definitions of the models and their parameters were given in a previous section of the chapter

Changing x and β weakly affects the position of the peak, mostly varying its height. A variation of at least two model parameters is required for adjusting the kinetic curve to experimental data. The general trend for VFT parameters, as reflected in Table 3, consists in sizeable changes of the τ_0 and A parameters. The basic relaxation time is increased by several orders of magnitude, whereas A decreases by a factor of almost two. This trend, as we shall see, continues for multiple curve fitting.

3.4 Modelling Multiple Measurements in a Wide Cooling Rates Range

The thermal scheme of measurements in a wide range of cooling rates that was explained above and shown in Fig. 8 has a specific character in the sense that cooling rate q is different from q_{heating} in most cases. The modelling of the C_p^{norm} curves via the TNM method using VFT parameters described in Table 2 bears results that disagree with experiment (see Fig. 17). The disagreement increases with decreasing cooling rate, especially on the value of the peak maximum, which grows exponentially (see Fig. 18).

Several different minimization procedures were performed and compared in [8]. Basically, minimization runs were performed to simultaneously fit all of the C_p curves in the cooling rate range 1.7×10^{-4} to 2 K/s: an equal number of ~500 points per curve were selected in an approximate temperature range of 360–410 K with a step

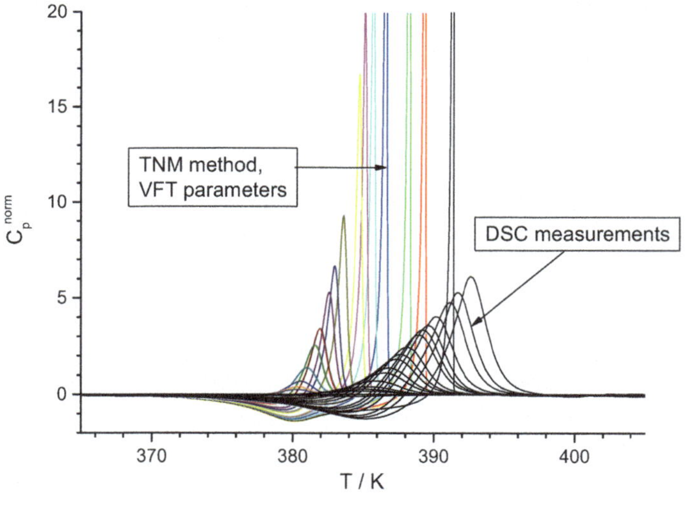

Fig. 17 Modelling the kinetics of glass transition in a wide range of cooling rates (1.7×10^{-4}–2 K/s, successive heating with 0.5 K/s) using the TNM method and VFT parameters as in [117]. Disagreement between model and experiment increases with decreasing cooling rate [8]

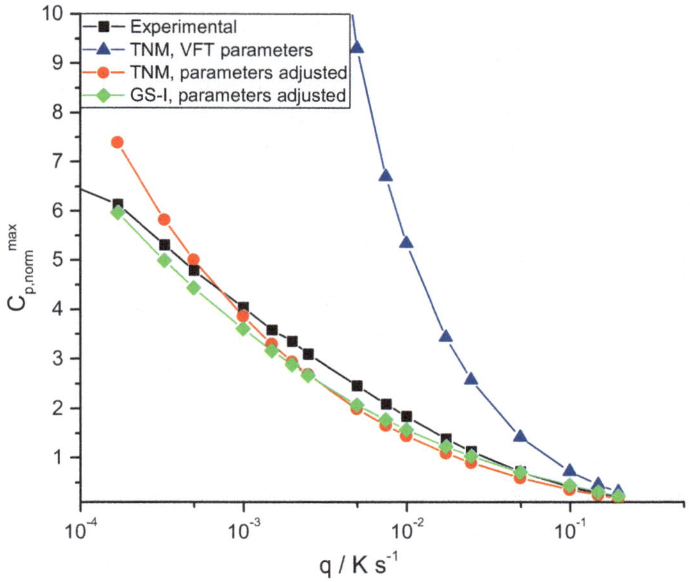

Fig. 18 Comparison of the maximum value of C_p^{norm} curves measured in experiment (black) to modelled results using the TNM methods with a common set of model parameters [117] (blue triangles) and with their values obtained after parameter readjustment for agreement with experiment (red squares). Green diamonds—irreversible thermodynamics approach with Eq. (26)

Table 4 Values of the model parameters for simultaneously fitting the C_p^{norm} curves of glass transition in a wide range of cooling rates via different methods [8]

Method	Parameter values				
	$\ln(\tau_0/s)$	β	T_0, T_2 in K	A in K	Additional
TNM-VFT	−4.66	0.845	354.1	174.2	$x = 0.446$
TNM-AG	−66.1	0.585	291.4	1070	$C = -2.53$ kK^{-1}
GS, Eq. (25)	−21.23	–	235	2540	$B/R = 160$

of 0.1 K. Again, equal weights for each curve were used at first. In the next attempt, the fit of all curves was performed with different weights for each cooling rate. These values were inversely proportional to C_p^{max} for the given q, with a weight value of unity for the largest peak in the group ($q = 1.7 \times 10^{-4}$ K/s). The resulting curves are presented in Fig. 19 employing the parameters given in Table 4. In this case, each experimental measurement has the same "value" during the fit. The improved quality of fit, for example, is reflected in Fig. 18 where it can be seen that heat capacity maxima are fitted reasonably well, though certain disagreements at all cooling rates still exists.

Comparing the resulting fit parameters, it can be seen that all of them are very different from the experimental parameters obtained from the activation diagram fits, or

$T_g(q)$ fits. The value of T_0 changes by approximately 20 K. The nonlinearity parameter x for the modelled curves increases twice, the values of β are also increased. The values of τ_0 and A vary strongly. The difference between model and experimental values for τ_0 and A is of concern for the simultaneous fit of all curves (the value of τ_0 is larger than 10 ms). Detailed modelling shows that this decrease of the values of τ_0 and A is caused by fitting of the curves corresponding to the slowest cooling rates. Apparently, exclusion of these curves results in an increase of the obtained values of τ_0 and A parameters, moving towards agreement with their experimentally obtained values (Table 2). A series of different fits, including fitting with constant value of A and fit with only τ_0 and β as free parameters, were performed [8]. In any of these cases, the fit of experiment is worse than the one presented on Fig. 19, especially with respect to the height of positive peaks and the shape of negative peaks.

Compared to the values reported in the literature for different measurements on polystyrene samples, it should be noted again that an Arrhenius law for relaxation time has frequently been used, resulting in incomparable values of A and $\ln\tau_0$. In a late work by Simon and Koh, a comparison of the TNM parameters for modelling different measurements on the enthalpy relaxation of polystyrene are given [128]. The value of the nonlinearity parameter x ~ 0.4–0.5 is in agreement with the range of 0.44–0.52 mentioned in [128], whereas the values of β are slightly above the reported value of 0.55–0.8. Again, we attribute this difference to the fit of exceptionally slow cooling rates. Different values of A and $\ln\tau_0$, but comparable values of x and β, are presented in [129]. An increased values of x and β were used for the fits in [130]. In another recent work [131], where a heterogeneous modification of the

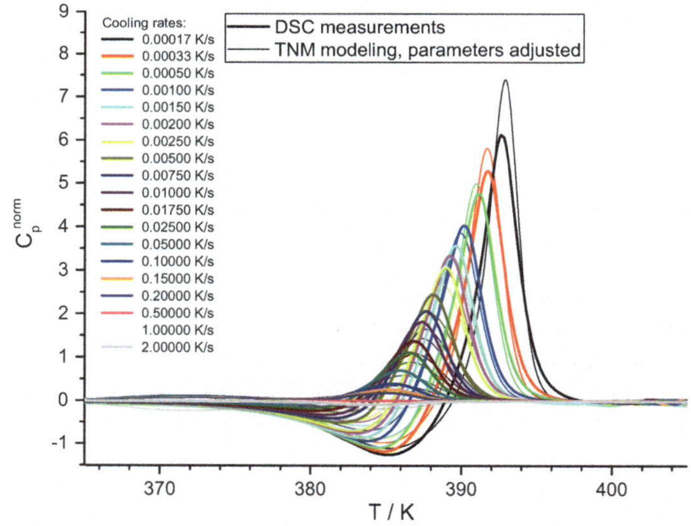

Fig. 19 Simultaneously fitting the measured C_p^{norm} curves by the TNM method with all of the parameters readjusted to minimize the root-mean-square deviation [8]. Thin coloured lines—model results; thick coloured lines—experimental data

TNM model is proposed and applied to the experimental data from [117] and others, model parameters are closer to those reported here (B = 570 K, $\ln\tau_0 = -11.8$ s, T_0 = 333 K). In [132], the value of x is similar (0.410), whereas the value of β lies in the range defined in [128].

The application of irreversible thermodynamics for simultaneous modelling of heat capacity curves meets similar difficulties. As mentioned in the previous section, a search for a suitable expression for the relaxation time was performed [8]. This was due to poor results of applying common expressions to modelling. Equation (26) introduces nonlinearity connected in a way with configurational entropy approaches (the second term is of the Adam–Gibbs type). During the search for an adequate expression, it was found that a strong dependence on model parameters, especially for AG expressions and similar forms, affects the performance of the Nelder–Mead procedure. Further work with these expressions for $\tau(T)$ is required to improve accuracy. Furthermore, for this method and expressions therein, no procedure exists for obtaining of a whole set of model parameters from the T_g versus q dependence (different expressions and the structural parameters). No direct convenient relations between model parameters have been obtained so far. As a result, all of the parameter values were obtained from the minimization procedure to acquire the best agreement of modelled heat capacity curves with experiment. The initial values for $\ln\tau_0$, B/R and T_0 were taken from VFT with the initial value of A being 0.5. Modelling results are presented in Fig. 20. The fitting was performed for $q = 1.7 \times 10^{-4}$ K/s. After minimization, the resulting parameters are as follows: $\ln(\tau_0/\mathrm{s}) = -21.23$, $B/\mathrm{R} = 2.54$ kK, $A/\mathrm{R} = 160$ K, $T_0 = 235$ K. These values are different as compared to other model results, yet the values of B/R and T_0 are comparable to some of the values of T_2 and Q parameters in AG-based approaches, including the value of $\ln\tau_0$. The agreement shown in Fig. 20 is relatively good (in comparison with TNM). Peak heights are reproduced slightly better, and similar widths are obtained. The worst agreement is for the position of the peaks.

A remark concerning the application of the lattice-hole model and the phenomenological expressions should be made. The T_m temperature used here arises in the expression for the configurational enthalpy of the material considered. Within the lattice-hole model approach for the liquid, H_{conf} is proportional to the heat of evaporation at T_m, while this heat itself is proportional to the value of T_m (Trouton's rule). For atactic polystyrene, as for all non-crystalline materials, there is no melting in the sense of a crystal-to-liquid phase transition (indeed, there is also no such thing as the heat of evaporation for a polymer), yet in the present work, we stay with this approach in order to only clarify its general applicability for this type of problem in comparison with other methods. The value of T_m was selected as ~516 K for melting isotactic polystyrene. The results of this section show that the considered method can also be applied for modelling the glass transition in polymers with varying parameters. Modifications of the theory would require considering a statistical model of the polymer melt, which can also be formulated within the periodic lattice formalism.

Application of the AG expression from Eq. (22) for modelling the experimental curves in a wide range of cooling rates results in the same fit quality as for the

VFT expression. The respective figures are presented elsewhere [8], whereas the parameters and the root-mean-square deviation are reported in Table 4.

Furthermore, we should discuss the parameter variations with cooling rate. Indeed, a better agreement of the modelled and measured C_p curves will be better in this case. Similar variations have been done, for example, in [8, 98, 105, 130, 133]. Each single kinetic curve can be fitted using different values of all model parameters. Yet, it was already mentioned that a variation of two parameters (τ_0 and β) should be enough. The increase in β results in the increase of peak heights and does not change their positions, whereas increasing τ_0 shifts the peak to the right and slightly decreases the peak height. For a single heat capacity curve, and almost for any q, there exist certain values of τ_0 and β that would produce a good fit with reasonable values of the other three model parameters. Thus, additional minimization for each respective C_p^{red} curve could be performed in order to obtain a good reproduction of experimental data. Let us also mention that the discussion in the previous section shows that the value of τ_0 is the most undefined (compared to A and T_0) from the $T_g(q)$ dependence, or from the activation diagram. The non-exponentiality parameter β is not obtained from this type of measurement at all.

We performed this procedure for all different sets of values of the model parameters A, T_0 and x, as obtained previously for the VFT, AG and GS approaches. The resulting curves are similar for different methods; we thus present a single example for AG approach in Fig. 21. Each modelled C_p^{norm} curve is a subtraction of two C_p^{red} curves, modelled with different values of τ_0 and β. Fitting was performed with various conditions—by points or by curve properties (peak height and position), all

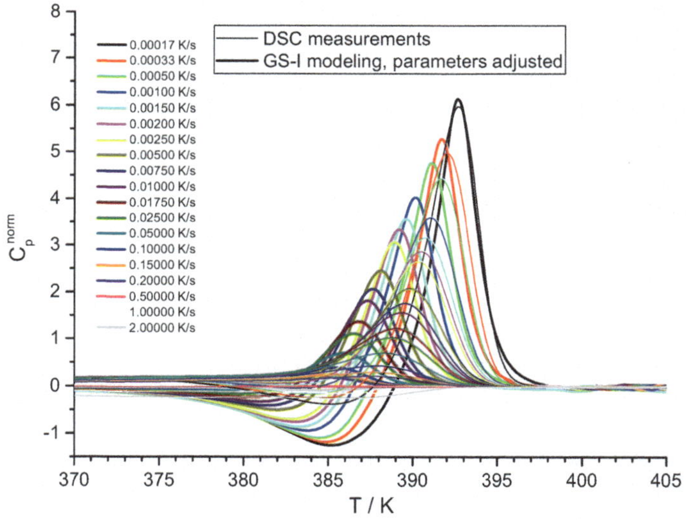

Fig. 20 Results of modelling the DSC measurements of glass transition of polystyrene via the irreversible thermodynamics approach with expression (26). The agreement on peak height is slightly better than for TNM method, peak width is comparable, while peaks position are fitted badly

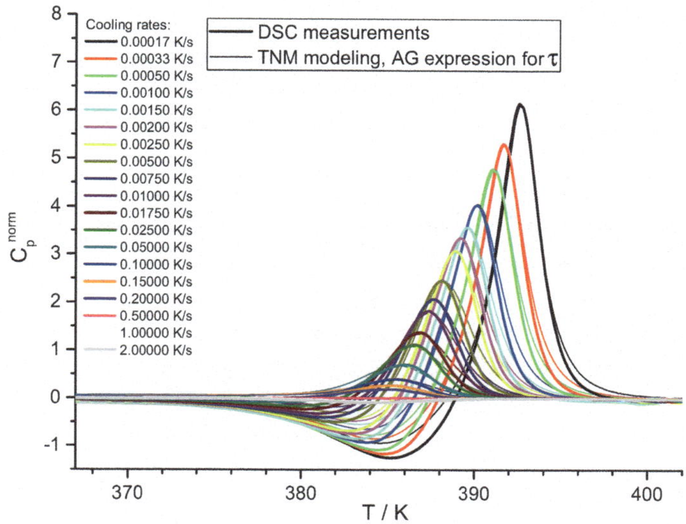

Fig. 21 Experimental (thick) and modelled (thin) DSC curves of polystyrene glass transition kinetics for a wide range of cooling rates. The corresponding C_p reduced curves were fitted with variation of τ_0 and β parameters. Other parameter values were taken from best simultaneous fit results

bearing similar results. A visible discrepancy in the fit of the undershoots is seen, which may be connected with the model of the melt.

A representation of the dependencies of the varied parameters on cooling rate q is given in Figs. 22 and 23. The values of unvaried parameters are taken from previous fits, excluding the irreversible thermodynamics approach, where all of the parameters had to be changed. The dependence of τ_0 and β on cooling rate q is similar for all cases, being monotonous for slow cooling ($q < 10^{-2}$ K/s) and more complicated for the higher rates in the region where q approaches and exceeds the constant value of the heating rate (0.5 K/s). A remark should be made concerning the GS method where four parameters were varied, namely τ_0, A, B and T_0. The dependencies of last two are not shown, because they have no clear character. Moreover, they are almost constant in the given q-range (changes by up to ~20% when $q > 0.1$ K/s occur).

The results demonstrate the dramatic inconsistencies that arise in modelling the glass transition kinetics for a wide range of cooling rates. This conclusion will be further supported in the following discussion section.

4 Discussion

The quality of the model fits to the experimental heat capacity curves is usually verified by comparing the root-mean-square deviations σ. The big difference between calculated curves and experiment for simultaneous fits with actual VFT parameters

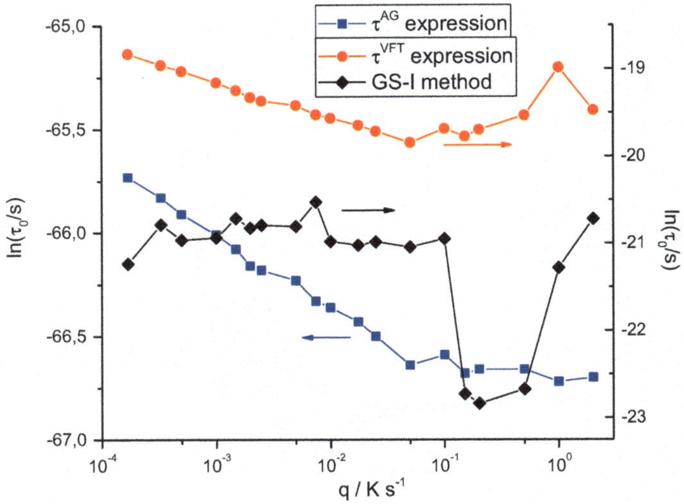

Fig. 22 Dependence of the varied τ_0 parameter for separately fitting the C_p curves via TNM method using VFT and AG expressions (red circles and blue squares, respectively) and the irreversible thermodynamics approach, using Eq. (26) (black diamonds). The coloured arrows indicate the axis corresponding to each line

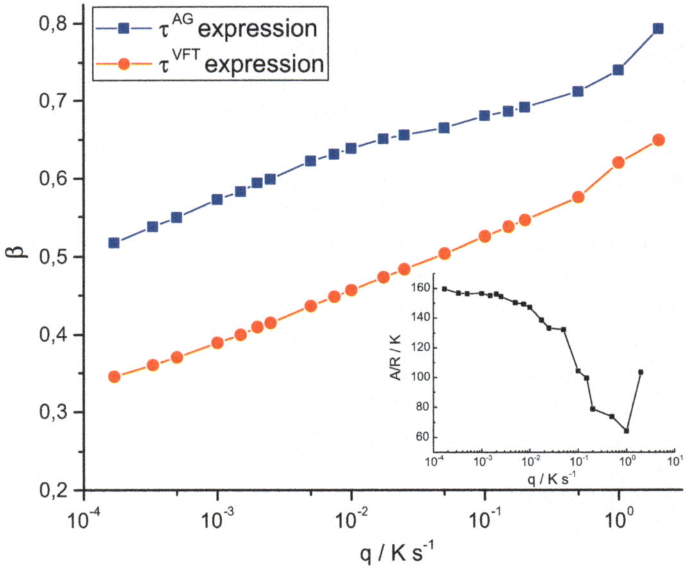

Fig. 23 Dependence of the varied β parameter for separately fitting the C_p curves via TNM method using VFT and AG expressions (blue squares and red circles, respectively). The main parameter of the irreversible thermodynamics approach, A/R, is presented in the inset

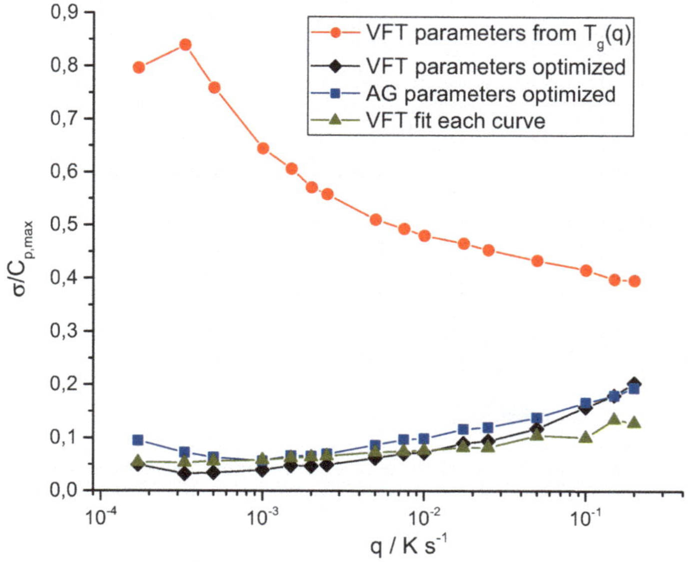

Fig. 24 Cooling rate dependence of the root-mean-square deviation for different methods of modelling the C_{p} curves of the polystyrene glass transition. Red circles—VFT parameters taken from $T_{\text{g}}(q)$ dependence fit, C_{p} modelled in Fig. 17; black diamonds—VFT model parameters optimized as to fit the C_{p} curves (see Fig. 19); blue squares—AG model parameters optimized the same way; dark yellow triangles—VFT parameters from $T_{\text{g}}(q)$ dependence, except τ_0 and β, which were optimized for each curve (see Fig. 21)

is well visualized in both Figs. 17 and 18. For comparison of σ values for each curve, the $\sigma(q)$ dependence is presented in Fig. 24. To account for the differences in C_p values corresponding to different rates (which grow substantially as q decreases), σ was divided by the corresponding value of the C_{p} maximum at a given cooling rate. It is seen that the fit quality is sufficiently improved by either varying all five model parameters, or by varying some of them for each curve. Another conclusion from Fig. 24 is that the experimental VFT parameters result in worse fits for slower cooling rates. Indeed, apart from the positions of C_p curves, a big deviation of their width and height with experiment is evident—the modelled peaks are very narrow and high (see Fig. 17).

To further clarify the reasons for this inconsistency, we want to refer to the experiments with equal cooling/heating rates. To do this, the $q_{\text{c}} = q_{\text{h}}$ reduced heat capacity curves were calculated for the common set of polystyrene VFT parameters (Table 2). The resulting curves are presented in Fig. 25. The heating curves in this figure are visually better than the modelled curves in Fig. 17, having adequate width and height at any rate q. Whereas these should be further compared to experiment, we can compare T_{g} defined from the cooling curves, for example, with the mid-point temperature of C_p from experimental data [7]. This comparison is done in Fig. 26, and the agreement is quite good. One can conclude that most of the deviations from modelling

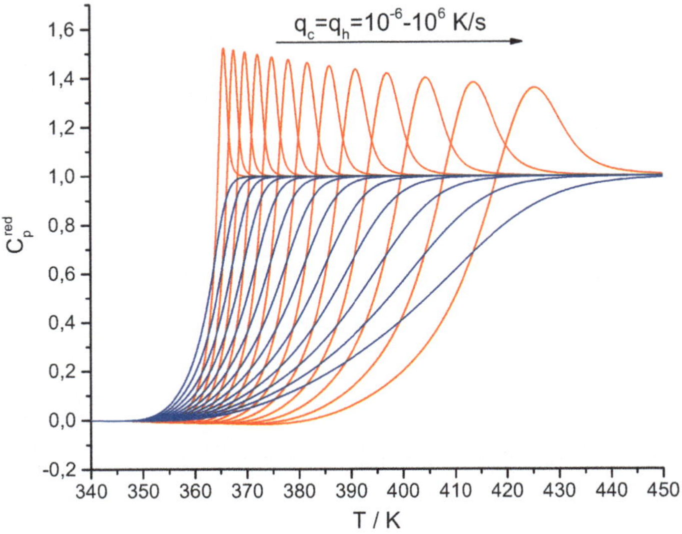

Fig. 25 Modelled heat capacity curves of the polystyrene glass transition at equal cooling/heating rates in the range from 10^{-6} to 10^{6} K/s. Blue lines—cooling curves, red lines—heating curves

occur when the cooling rate q is considerably different from the heating rate in the experimental cycle. Moreover, since for $q_c \gg q_h$, the height of the overshoot peaks is very small, the effect is attributed only to thermal histories when $q_c \ll q_h$.

To demonstrate the general character of this inconsistency, we further model two different experiments. The first one is made for the same polystyrene material (with VFT parameters from the previous calculations), with the scanning rates shifted to higher values. In Fig. 27, the modelled normalized heat capacity curves (similar to experiments with low q) are presented. The constant heating rate was set to 5000 K/s, whereas the cooling rate varied in the range q = 0.1–10^4 K/s. Indeed, the same disagreement of modelled C_p with experiment is seen in Fig. 27. We can generally conclude that, for polystyrene, the inconsistencies on the modelled overshoot width and height become large at approximately $q/q_h < 10^{-2}$.

To demonstrate that this type of behaviour does not depend on the specific system, we further consider similar heat capacity curves for glycerol. The VFT parameters were taken from Refs. [98, 134] and a similar experiment with constant heating rate of 0.5 K/s and a wide range of cooling rates was modelled. The results are presented in Fig. 28 in terms of normalized heat capacity curves C_p^{norm}. The very similar type of behaviour of the peaks is observed. Thus, this general inconsistency should be accounted for the TNM method.

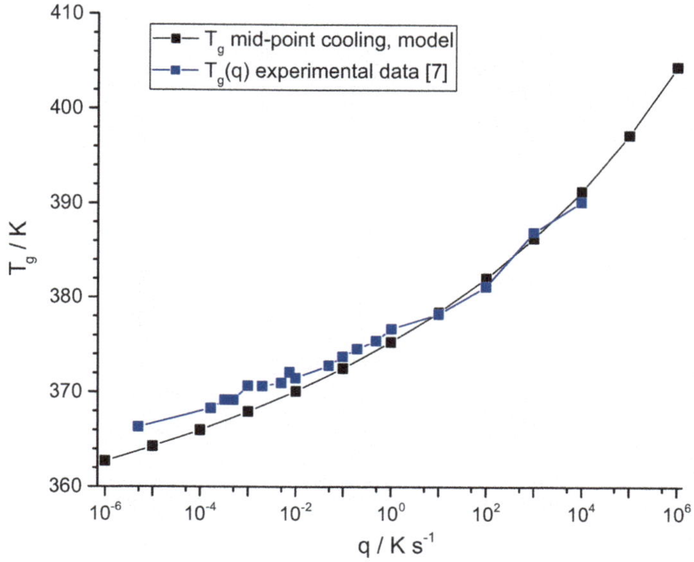

Fig. 26 Glass transition temperature obtained from modelled cooling curves as the mid-point of C_p. Comparison with VFT dependence obtained in [7]

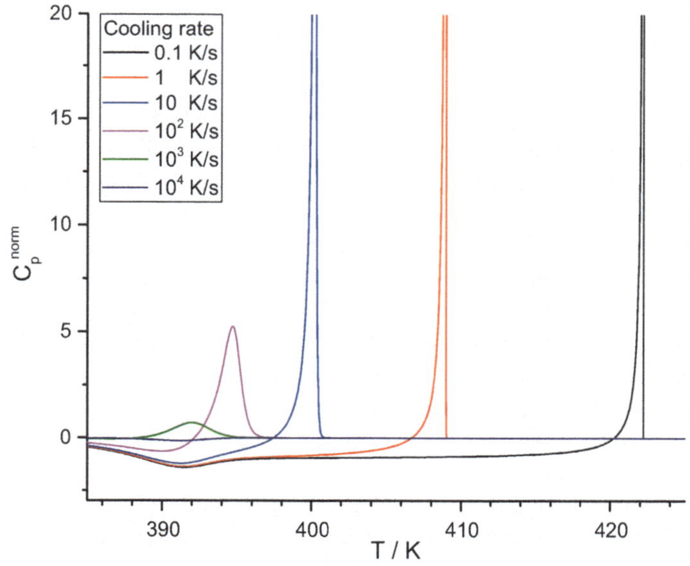

Fig. 27 Modelled C_p^{norm} curves for polystyrene glass transition at high scanning rates: cooling rates in the range 0.1–10^4 K/s with further heating at a constant rate of 5000 K/s

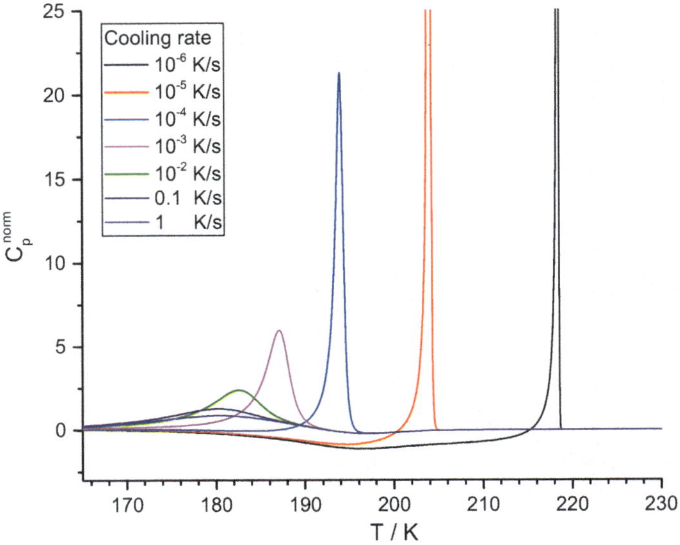

Fig. 28 Modelled C_p^{norm} curves for the glass transition kinetics of glycerol in a wide range of cooling rates (10^{-6}–1 K/s). VFT parameters taken from the literature [98, 134]

5 Conclusions

In the present chapter, experimental DSC data on the kinetics of the glass transition of atactic polystyrene were presented for a wide range of cooling rates ($q = 5 \times 10^{-6}$–2 K/s) and for constant heating rate 0.5 K/s. In addition, the preliminary T_g values, which were measured for higher cooling rates on a fast-scanning chip calorimeter, were used ($q = 1 \times 10^4$ K/s, heating at 5000 K/s). The respective dependence of the glass transition temperature $T_g(q)$ was obtained and fitted with the Vogel–Fulcher–Tammann law. The theoretical aim of the work was to explore the capabilities of common theoretical methods to model heat capacity curves for the experiments performed. To our knowledge, these are the first attempts to do this for such a wide dynamic rate of control parameters. Several common methods of modelling were used. The structural order parameters were thus the fictive temperature T_f and the number of lattice-holes ξ.

The results of simultaneous modelling of all of the heat capacity curves with common VFT parameters (obtained from the $T_g(q)$ or $T_g(\omega)$ activation diagram) were used. The disagreement with experiment is clearly seen in Fig. 17. This disagreement becomes dramatic when the cooling rate q is much less than corresponding heating rate q_h. For approximately equal values, the dependencies are acceptable.

Similar experiments were modelled for fast-scanning rates ($q_h = 5000$ K/s, see Fig. 27) resulting in the same behaviour. In addition, we demonstrate the general character of the problem by calculating the C_p curves for the glass transition of glycerol (see Fig. 28). Modelling of the complex heat capacity with similar models is possible but was not performed because of the insufficient agreement between model and experiment for experiments at constant scan rates. The proper update of the methods of modelling is the subject of ongoing research.

Acknowledgements CS acknowledges the financial support from the Ministry of Education and Science of the Russian Federation, grant 14.Y26.31.0019. T. V. and J. W. P. acknowledge the financial support by the Heisenberg-Landau program of the German Federal Ministry of Education and Research (BMBF, Germany).

References

1. Stillinger FH, Debenedetti PG (2013) Glass Transition Thermodynamics and Kinetics. Annu Rev Condens Matter Phys 4:263–285. https://doi.org/10.1146/annurev-conmatphys-030212-184329
2. Gutzow IS, Schmelzer JWP (2013) The vitreous state: thermodynamics, structure, rheology, and crystallization. Springer, Berlin. https://doi.org/10.1007/978-3-642-34633-0
3. Schmelzer JWP, Gutzow IS, Mazurin OV, Priven AI, Todorova SV, Petroff BP (2011) Glasses and the glass transition. Wiley-VCH Verlag GmbH & Co. KGaA, Weinheim. https://doi.org/10.1002/9783527636532
4. Cavagna A (2009) Supercooled liquids for pedestrians. Phys Rep 476:51–124. https://doi.org/10.1016/j.physrep.2009.03.003
5. Lubchenko V (2015) Theory of the structural glass transition: a pedagogical review. Adv Phys 64:283–443. https://doi.org/10.1080/00018732.2015.1057979
6. Minakov AA, Schick C (2007) Ultrafast thermal processing and nanocalorimetry at heating and cooling rates up to 1 MKs. Rev Sci Instrum 78 https://doi.org/10.1063/1.2751411
7. Schawe JEK (2015) Measurement of the thermal glass transition of polystyrene in a cooling rate range of more than six decades. Thermochim Acta 603:128–134. https://doi.org/10.1016/j.tca.2014.05.025
8. Tropin TV, Schulz G, Schmelzer JWP, Schick C (2015) Heat capacity measurements and modeling of polystyrene glass transition in a wide range of cooling rates. J Non Cryst Solids 409:63–75. https://doi.org/10.1016/j.jnoncrysol.2014.11.001
9. Narayanaswamy OS (1971) A model of structural relaxation in glass. J Am Ceram Soc 54:491–498. https://doi.org/10.1111/j.1151-2916.1971.tb12186.x
10. Vogel H (1921) Das Temperaturabhängigkeit Gesetz der Viskosität von Flüssigkeiten. Phys Zeitschrift 22:645
11. Fulcher GS (1925) Analysis of recent measurements of the viscosity of glasses. J Am Ceram Soc 8:339–355. https://doi.org/10.1111/j.1151-2916.1925.tb16731.x
12. Tammann G, Hesse W (1926) Die Abhängigkeit der Viscosität von der Temperatur bie unterkühlten Flüssigkeiten. Zeitschrift Für Anorg Und Allg Chemie 156:245–257. https://doi.org/10.1002/zaac.19261560121
13. Adam G, Gibbs JH (1965) on the temperature dependence of cooperative relaxation properties in glass-forming liquids. J Chem Phys 43:139. https://doi.org/10.1063/1.1696442
14. Pijpers TFJ, Mathot VBF, Goderis B, Scherrenberg RL, van der Vegte EW (2002) High-speed calorimetry for the study of the kinetics of (de)vitrification, crystallization, and melting of macromolecules. Macromolecules 35:3601–3613. https://doi.org/10.1021/ma011122u
15. Adamovsky S, Minakov A, Schick C (2003) Scanning microcalorimetry at high cooling rate. Thermochim Acta 403:55–63. https://doi.org/10.1016/S0040-6031(03)00182-5
16. Adamovsky S, Schick C (2004) Ultra-fast isothermal calorimetry using thin film sensors. Thermochim Acta 415:1–7. https://doi.org/10.1016/j.tca.2003.07.015

17. Minakov AA, Adamovsky SA, Schick C (2005) Non-adiabatic thin-film (chip) nanocalorimetry. Thermochim Acta 432:177–185. https://doi.org/10.1016/j.tca.2005.01.073
18. Kolesov IS, Androsch R, Radusch H-J (2004) Non-isothermal crystallization of polyethylenes as function of cooling rate and concentration of short chain branches. J Therm Anal Calorim 78:885–895. https://doi.org/10.1007/s10973-004-0455-y
19. Mettler T, Flash DSC, (n.d.). http://au.mt.com/au/en/home/products/Laboratory_Analytics_Browse/TA_Family_Browse/Flash_DSC.html
20. Mathot V, Pyda M, Pijpers T, Vanden Poel G, van de Kerkhof E, van Herwaarden S, van Herwaarden F, Leenaers A (2011) The flash DSC 1, a power compensation twin-type, chip-based fast scanning calorimeter (FSC): First findings on polymers. Thermochim Acta 522:36–45. https://doi.org/10.1016/j.tca.2011.02.031
21. van Herwaarden AW (2005) Overview of calorimeter chips for various applications. Thermochim Acta 432:192–201. https://doi.org/10.1016/j.tca.2005.04.027
22. Zhuravlev E, Schick C (2010) Fast scanning power compensated differential scanning nanocalorimeter: 1. The device. Thermochim Acta 505:1–13. https://doi.org/10.1016/j.tca.2010.03.019
23. Zhuravlev E, Schick C (2016) Non-adiabatic scanning calorimeter for controlled fast cooling and heating. In: Fast Scanning Calorimetry. Springer International Publishing, Cham, pp 81–104. https://doi.org/10.1007/978-3-319-31329-0_2
24. Minakov AA, van Herwaarden AW, Wien W, Wurm A, Schick C (2007) Advanced nonadiabatic ultrafast nanocalorimetry and superheating phenomenon in linear polymers. Thermochim Acta 461:96–106. https://doi.org/10.1016/j.tca.2007.03.020
25. Minakov AA, Schick C (2015) Dynamics of the temperature distribution in ultra-fast thin-film calorimeter sensors. Thermochim Acta 603:205–217. https://doi.org/10.1016/j.tca.2014.05.030
26. Minakov AA, Roy SB, Bugoslavsky YV, Cohen LF (2005) Thin-film alternating current nanocalorimeter for low temperatures and high magnetic fields. Rev Sci Instrum 76:43906. https://doi.org/10.1063/1.1889432
27. Zhuravlev E, Schick C (2010) Fast scanning power compensated differential scanning nano-calorimeter: 2. Heat capacity analysis. Thermochim Acta 505:14–21. https://doi.org/10.1016/j.tca.2010.03.020
28. Minakov AA, Mordvintsev DA, Schick C (2004) Melting and reorganization of poly(ethylene terephthalate) on fast heating (1000 K/s). Polymer 45:3755–3763. https://doi.org/10.1016/j.polymer.2004.03.072
29. Gradys A, Sajkiewicz P, Minakov AA, Adamovsky S, Schick C, Hashimoto T, Saijo K (2005) Crystallization of polypropylene at various cooling rates. Mater Sci Eng, A 413–414:442–446. https://doi.org/10.1016/j.msea.2005.08.167
30. Minakov AA, Mordvintsev DA, Schick C (1000) Isothermal reorganization of poly(ethylene terephthalate) revealed by fast calorimetry (1000 K s^{-1} ; 5 ms). Faraday Discuss 128(2005):261–270. https://doi.org/10.1039/b403441d
31. De Santis F, Adamovsky S, Titomanlio G, Schick C (2006) Scanning nanocalorimetry at high cooling rate of isotactic polypropylene. Macromolecules 39:2562–2567. https://doi.org/10.1021/ma052525n
32. Minakov AA, Mordvintsev DA, Tol R, Schick C (2006) Melting and reorganization of the crystalline fraction and relaxation of the rigid amorphous fraction of isotactic polystyrene on fast heating (30,000 K/min). Thermochim Acta 442:25–30. https://doi.org/10.1016/j.tca.2005.11.032
33. Pyda M, Nowak-Pyda E, Heeg J, Huth H, Minakov AA, Di Lorenzo ML, Schick C, Wunderlich B (2006) Melting and crystallization of poly(butylene terephthalate) by temperature-modulated and superfast calorimetry. J Polym Sci, Part B: Polym Phys 44:1364–1377. https://doi.org/10.1002/polb.20789
34. Tol RT, Minakov AA, Adamovsky SA, Mathot VBF, Schick C (2006) Metastability of polymer crystallites formed at low temperature studied by ultra fast calorimetry: polyamide 6 confined in sub-micrometer droplets versus bulk PA6. Polymer 47:2172–2178. https://doi.org/10.1016/j.polymer.2006.01.052

35. De Santis F, Adamovsky S, Titomanlio G, Schick C (2007) Isothermal nanocalorimetry of isotactic polypropylene. Macromolecules 40:9026–9031. https://doi.org/10.1021/ma071491b
36. Gradys A, Sajkiewicz P, Adamovsky S, Minakov A, Schick C (2007) Crystallization of poly(vinylidene fluoride) during ultra-fast cooling. Thermochim Acta 461:153–157. https://doi.org/10.1016/j.tca.2007.05.023
37. Minakov AA, Wurm A, Schick C (2007) Superheating in linear polymers studied by ultrafast nanocalorimetry. Eur Phys J E 23:43. https://doi.org/10.1140/epje/i2007-10173-8
38. Ray VV, Banthia AK, Schick C (2007) Fast isothermal calorimetry of modified polypropylene clay nanocomposites. Polymer 48:2404–2414. https://doi.org/10.1016/j.polymer.2007.02.055
39. Silvestre C, Cimmino S, Duraccio D, Schick C (2007) Isothermal crystallization of isotactic poly(propylene) studied by superfast calorimetry. Macromol Rapid Commun 28:875–881. https://doi.org/10.1002/marc.200600844
40. Krumme A, Lehtinen A, Adamovsky S, Schick C, Roots J, Viikna A (2008) Crystallization behavior of some unimodal and bimodal linear low-density polyethylenes at moderate and high supercooling. J Polym Sci, Part B: Polym Phys 46:1577–1588. https://doi.org/10.1002/polb.21494
41. Schick C (2009) Differential scanning calorimetry (DSC) of semicrystalline polymers. Anal Bioanal Chem 395:1589–1611. https://doi.org/10.1007/s00216-009-3169-y
42. Brucato V, Piccarolo S, La Carrubba V (2002) An experimental methodology to study polymer crystallization under processing conditions. The influence of high cooling rates. Chem Eng Sci 57:4129–4143. https://doi.org/10.1016/S0009-2509(02)00360-3
43. Janeschitz-Kriegl H (2010) Crystallization modalities in polymer melt processing. Springer Vienna, Vienna. https://doi.org/10.1007/978-3-211-87627-5
44. Cebe P, Partlow BP, Kaplan DL, Wurm A, Zhuravlev E, Schick C (2015) Using flash DSC for determining the liquid state heat capacity of silk fibroin. Thermochim Acta 615:8–14. https://doi.org/10.1016/j.tca.2015.07.009
45. Lopeandía AF, Cerdó LI, Clavaguera-Mora MT, Arana LR, Jensen KF, Muñoz FJ, Rodríguez-Viejo J (2005) Sensitive power compensated scanning calorimeter for analysis of phase transformations in small samples. Rev Sci Instrum 76: 65104. https://doi.org/10.1063/1.1921567
46. Lopeandia AF, Valenzuela J, Rodríguez-Viejo J (2008) Power compensated thin film calorimetry at fast heating rates. Sensors Actuators A Phys 143:256–264. https://doi.org/10.1016/j.sna.2007.11.006
47. Merzlyakov M (2006) Method of rapid (100,000 Ks^{-1}) controlled cooling and heating of thin samples. Thermochim Acta 442:52–60. https://doi.org/10.1016/j.tca.2005.11.018
48. van Herwaarden S, Iervolino E, van Herwaarden F, Wijffels T, Leenaers A, Mathot V (2011) Design, performance and analysis of thermal lag of the UFS1 twin-calorimeter chip for fast scanning calorimetry using the Mettler-Toledo Flash DSC 1. Thermochim Acta 522:46–52. https://doi.org/10.1016/j.tca.2011.05.025
49. Iervolino E, van Herwaarden AW, van Herwaarden FG, van de Kerkhof E, van Grinsven PPW, Leenaers ACHI, Mathot VBF, Sarro PM (2011) Temperature calibration and electrical characterization of the differential scanning calorimeter chip UFS1 for the Mettler-Toledo Flash DSC 1. Thermochim Acta 522:53–59. https://doi.org/10.1016/j.tca.2011.01.023
50. Poel GV, Sargsyan A, Mathot V, Assche GV, Wurm A, Schick C, Krumme A, Zhou D (2011) Recommendation for temperature calibration of fast scanning calorimeters (FsCs) for sample mass and scan rate. Beuth Verlag GmbH, Berlin
51. Vanden Poel G, Istrate D, Magon A, Mathot V (2012) Performance and calibration of the Flash DSC 1, a new, MEMS-based fast scanning calorimeter. J Therm Anal Calorim 110:1533–1546. https://doi.org/10.1007/s10973-012-2722-7
52. Baur H, Wunderlich B (1998) About complex heat capacities and temperature-modulated calorimetry. J Therm Anal Calorim 54:437–465. https://doi.org/10.1023/A:1010126005720
53. Schäfer K (1940) Die Stoßanregung intramolekularer Schwingungen in Gasen und Gasmischungen. VII. Theorie der Schalldispersion bei Vorhandensein mehrerer Normalschwingungen, Zeitschrift Für Phys Chemie 46B. https://doi.org/10.1515/zpch-1940-4613

54. Jeong Y-H (2001) Modern calorimetry: going beyond tradition. Thermochim Acta 377:1–7. https://doi.org/10.1016/S0040-6031(01)00538-X
55. Alig I (1997) Ultrasonic relaxation and complex heat capacity. Thermochim Acta 304–305:35–49. https://doi.org/10.1016/S0040-6031(97)00174-3
56. Khalife A, Pathak U, Richert R (2011) Heating liquid dielectrics by time dependent fields. Eur Phys J B 83:429–435. https://doi.org/10.1140/epjb/e2011-20599-5
57. Richert R (2011) Calorimetry based on energy absorbed from time-dependent fields. J Non Cryst Solids 357:726–730. https://doi.org/10.1016/j.jnoncrysol.2010.05.088
58. Richert R (2011) Reverse calorimetry of a supercooled liquid: propylene carbonate. Thermochim Acta 522:28–35. https://doi.org/10.1016/j.tca.2010.09.016
59. Birge N (1986) Specific-heat spectroscopy of glycerol and propylene glycol near the glass transition. Phys Rev B 34:1631–1642. https://doi.org/10.1103/PhysRevB.34.1631
60. Suga H (2001) Adiabatic calorimeter as an ultra-low frequency spectrometer: interplay between phase and glass transitions in solids. Thermochim Acta 377:35–49. https://doi.org/10.1016/S0040-6031(01)00540-8
61. Suga H (2005) Ultra-slow relaxation in ice and related substances. Proc Japan Acad Ser B 81:349–362. https://doi.org/10.2183/pjab.81.349
62. Schick C, Tanneberger H, Donth E (1980) Ergebnisse direkter messungen der enthalpieretardation in PVC und ihr zusammenhang mit dem glasübergang und der struktur, Die Makromol. Chemie, Rapid Commun 1:407–409. https://doi.org/10.1002/marc.1980.030010611
63. Schick C, Tanneberger H, Donth E (1982) Zeitabhängigkeit der Enthalpie im Glasübergangsbereich von Polyvinylchlorid. Acta Polym 33:163–168. https://doi.org/10.1002/actp.1982.010330301
64. Hemminger W, Höhne GWH (1984) Calorimetry - fundamentals and practice. Vch., Weinheim. https://doi.org/10.1002/bbpc.19850891130
65. Höhne GWH, Hemminger WF, Flammersheim H-J (2003) Differential scanning calorimetry. Springer, Berlin. https://doi.org/10.1007/978-3-662-06710-9
66. Boller A, Schick C, Wunderlich B (1995) Modulated differential scanning calorimetry in the glass transition region. Thermochim Acta 266:97–111. https://doi.org/10.1016/0040-6031(95)02552-9
67. Schawe JEK (1995) Principles for the interpretation of modulated temperature DSC measurements. Part 1. Glass transition. Thermochim Acta 261:183–194. https://doi.org/10.1016/0040-6031(95)02315-S
68. Merzlyakov M, Schick C (1999) Complex heat capacity measurements by TMDSC Part 1. Influence of non-linear thermal response. Thermochim Acta 330:55–64. https://doi.org/10.1016/S0040-6031(99)00040-4
69. Höhne GWH, Merzlyakov M, Schick C (2002) Calibration of magnitude and phase angle of TMDSC: Part1: basic considerations. Thermochim Acta 391:51–67. https://doi.org/10.1016/S0040-6031(02)00163-6
70. Merzlyakov M, Höhne GWH, Schick C (2002) Calibration of magnitude and phase angle of TMDSC: Part 2. Calibration practice. Thermochim Acta 391:69–80. https://doi.org/10.1016/S0040-6031(02)00164-8
71. Kamasa P, Merzlyakov M, Pyda M, Pak J, Schick C, Wunderlich B (2002) Multi-frequency heat capacity measured with different types of TMDSC. Thermochim Acta 392–393:195–207. https://doi.org/10.1016/S0040-6031(02)00102-8
72. Schawe JEK, Hütter T, Heitz C, Alig I, Lellinger D (2006) Stochastic temperature modulation: A new technique in temperature-modulated DSC. Thermochim Acta 446:147–155. https://doi.org/10.1016/j.tca.2006.01.031
73. Merzlyakov M, Schick C (2001) Simultaneous multi-frequency TMDSC measurements. Thermochim Acta 377:193–204. https://doi.org/10.1016/S0040-6031(01)00554-8
74. Merzlyakov M, Schick C (2001) Step response analysis in DSC — a fast way to generate heat capacity spectra. Thermochim Acta 380:5–12. https://doi.org/10.1016/S0040-6031(01)00631-1

75. Sullivan P, Seidel G (1966) An ac temperature technique for measuring heat capacities. Ann Acad Sci Fenn. A VI 58–62
76. Kraftmakher Y (2004) Modulation calorimetry. Springer, Berlin. http://www.springer.com/gp/book/9783540210825
77. Christensen T, Olsen NB, Dyre JC, Tokuyama M, Oppenheim I, Nishiyama H (2008) Can the frequency dependent isobaric specific heat be measured by thermal effusion methods? In: AIP Conference Proceedings, AIP, pp 139–141. https://doi.org/10.1063/1.2897769
78. Jakobsen B, Olsen NB, Christensen T (2010) Frequency-dependent specific heat from thermal effusion in spherical geometry. Phys Rev E 81:61505. https://doi.org/10.1103/PhysRevE.81.061505
79. Christensen T, Olsen NB, Dyre JC (2007) Conventional methods fail to measure cp(w) of glass-forming liquids. Phys Rev E 75:41502. https://doi.org/10.1103/PhysRevE.75.041502
80. Birge NO, Nagel SSR (1985) Specific-heat spectroscopy of the glass transition. Phys Rev Lett 54:2674–2677. https://doi.org/10.1103/PhysRevLett.54.2674
81. Christensen T (1985) The frequency dependence of the specific heat at the glass transition. Le J Phys Colloq 46: C8-635-C8-637. https://doi.org/10.1051/jphyscol:19858102
82. Glorieux C, Nelson KA, Hinze G, Fayer MD (2002) Thermal, structural, and orientational relaxation of supercooled salol studied by polarization-dependent impulsive stimulated scattering. J Chem Phys 116:3384–3395. https://doi.org/10.1063/1.1445749
83. Bentefour EH, Glorieux C, Chirtoc M, Thoen J (2003) Broadband photopyroelectric thermal spectroscopy of a supercooled liquid near the glass transition. J Appl Phys 93:9610–9614. https://doi.org/10.1063/1.1576300
84. Merzlyakov M (2003) Integrated circuit thermopile as a new type of temperature modulated calorimeter. Thermochim Acta 403:65–81. https://doi.org/10.1016/S0040-6031(03)00083-2
85. Huth H, Minakov AA, Schick C (2006) Differential AC-chip calorimeter for glass transition measurements in ultrathin films. J Polym Sci, Part B: Polym Phys 44:2996–3005. https://doi.org/10.1002/polb.20921
86. Huth H, Minakov AA, Serghei A, Kremer F, Schick C (2007) Differential AC-chip calorimeter for glass transition measurements in ultra thin polymeric films. Eur. Phys. J. Spec. Top. 141:153–160. https://doi.org/10.1140/epjst/e2007-00033-y
87. Ahrenberg M, Shoifet E, Whitaker KR, Huth H, Ediger MD, Schick C (2012) Differential alternating current chip calorimeter for in situ investigation of vapor-deposited thin films. Rev Sci Instrum 83:33902. https://doi.org/10.1063/1.3692742
88. Shoifet E, Chua YZ, Huth H, Schick C (2013) High frequency alternating current chip nano calorimeter with laser heating. Rev Sci Instrum 84:73903. https://doi.org/10.1063/1.4812349
89. Svanidze AV, Huth H, Lushnikov SG, Kojima S, Schick C (2009) Phase transition in tetragonal hen egg-white lysozyme crystals. Appl Phys Lett 95:263702. https://doi.org/10.1063/1.3275858
90. Svanidze AV, Huth H, Lushnikov SG, Schick C (2012) Study of phase transition in tetragonal lysozyme crystals by AC-nanocalorimetry. Thermochim Acta 544:33–37. https://doi.org/10.1016/j.tca.2012.06.013
91. Ahrenberg M, Chua YZ, Whitaker KR, Huth H, Ediger MD, Schick C (2013) In situ investigation of vapor-deposited glasses of toluene and ethylbenzene via alternating current chip-nanocalorimetry. J Chem Phys 138:24501. https://doi.org/10.1063/1.4773354
92. Weyer S, Hensel A, Schick C (1997) Phase angle correction for TMDSC in the glass-transition region. Thermochim Acta 304–305:267–275. https://doi.org/10.1016/S0040-6031(97)00180-9
93. Boucher VM, Cangialosi D, Yin H, Schönhals A, Alegría A, Colmenero J (2012) Tg depression and invariant segmental dynamics in polystyrene thin films. Soft Matter 8:5119. https://doi.org/10.1039/c2sm25419k
94. Priestley RD, Cangialosi D, Napolitano S (2015) On the equivalence between the thermodynamic and dynamic measurements of the glass transition in confined polymers. J Non Cryst Solids 407:288–295. https://doi.org/10.1016/j.jnoncrysol.2014.09.048

95. Cangialosi D, Alegría A, Colmenero J (2016) Effect of nanostructure on the thermal glass transition and physical aging in polymer materials. Prog Polym Sci 54–55:128–147. https://doi.org/10.1016/j.progpolymsci.2015.10.005
96. Perez-de-Eulate NG, Di Lisio V, Cangialosi D (2017) Glass transition and molecular dynamics in polystyrene nanospheres by fast scanning calorimetry. ACS Macro Lett. 6:859–863. https://doi.org/10.1021/acsmacrolett.7b00484
97. Schick C, Lexa D, Leibowitz L (2012) Differential scanning calorimetry and differential thermal analysis. In: Kaufmann EN (ed) Characterization Materials, 1st ed. Wiley Inc, New York, pp 483–495
98. Hodge IM (1994) Enthalpy relaxation and recovery in amorphous materials. J Non Cryst Solids 169:211–266. https://doi.org/10.1016/0022-3093(94)90321-2
99. Tool AQ (1946) Relation between inelastic deformability and thermal expansion of glass in its annealing range*. J Am Ceram Soc 29:240–253. https://doi.org/10.1111/j.1151-2916.1946.tb11592.x
100. Moynihan CT, Easteal AJ, DeBolt MA, Tucker J (1976) Dependence of the fictive temperature of glass on cooling rate. J Am Ceram Soc 59:12–16. https://doi.org/10.1111/j.1151-2916.1976.tb09376.x
101. Sarge SM, Hemminger W, Gmelin E, Höhne GWH, Cammenga HK, Eysel W (1997) Metrologically based procedures for the temperature, heat and heat flow rate calibration of DSC. J Therm Anal 49:1125–1134. https://doi.org/10.1007/BF01996802
102. Hao N, Böhning M, Schönhals A (2007) Dielectric properties of nanocomposites based on polystyrene and polyhedral oligomeric phenethyl-silsesquioxanes. Macromolecules 40:9672–9679. https://doi.org/10.1021/ma071777g
103. Chua YZ, Schulz G, Shoifet E, Huth H, Zorn R, Scmelzer JWP, Schick C (2014) Glass transition cooperativity from broad band heat capacity spectroscopy. Colloid Polym Sci 292:1893–1904. https://doi.org/10.1007/s00396-014-3280-2
104. Beiner M, Garwe F, Schröter K, Donth E (1994) Dynamic shear modulus in the splitting region of poly(alkyl methacrylates). Colloid Polym Sci 272:1439–1446. https://doi.org/10.1007/BF00654174
105. O'Reilly JM, Hodge IM (1991) Effects of heating rate on enthalpy recovery in polystyrene. J Non Cryst Solids 131–133:451–456. https://doi.org/10.1016/0022-3093(91)90338-7
106. Hadač J, Slobodian P, Říha P, Sáha P, Rychwalski RW, Emri I, Kubát J (2007) Effect of cooling rate on enthalpy and volume relaxation of polystyrene. J Non Cryst Solids 353:2681–2691. https://doi.org/10.1016/j.jnoncrysol.2007.05.017
107. Wunderlich B (1995) The ATHAS database on heat capacities of polymers. Pure Appl Chem 67:1019–1026. https://doi.org/10.1351/pac199567061019
108. Pyda M, Wunderlich B (2013) ATHAS Data Bank. http://www.springermaterials.com
109. Kovacs AJ, Aklonis JJ, Hutchinson JM, Ramos AR (1979) Isobaric volume and enthalpy recovery of glasses. II. A transparent multiparameter theory. J Polym Sci Polym Phys Ed 17:1097–1162. https://doi.org/10.1002/pol.1979.180170701
110. Mazinani SKS, Richert R (2012) Enthalpy recovery in glassy materials: heterogeneous versus homogenous models. J. Chem. Phys. 136:174515. https://doi.org/10.1063/1.4712032
111. Prigogine I, Defay R (1954) Chemical thermodynamics. Longmans, London
112. Möller J, Gutzow I, Schmelzer JWP (2006) Freezing-in and production of entropy in vitrification. J Chem Phys 125:94505. https://doi.org/10.1063/1.2346673
113. Tropin TV, Schmelzer JWP, Schick C (2011) On the dependence of the properties of glasses on cooling and heating rates I. Entropy, entropy production and glass transition temperature. J Non Cryst Solids 357: 1291–1302. https://doi.org/10.1016/j.jnoncrysol.2010.11.111
114. Tropin TV, Schmelzer JWP, Schick C (2011) On the dependence of the properties of glasses on cooling and heating rates II. Prigogine-Defay ratio, fictive temperature and fictive pressure. J Non Cryst Solids 357: 1303–1309. https://doi.org/10.1016/j.jnoncrysol.2010.12.005
115. Gutzow I, Grigorova T, Avramov I, Schmelzer JWP (2002) Generic phenomenology of vitrification and relaxation and the Kohlrausch and Maxwell equations. Phys Chem Glas 43C: 477–486. http://www.ipc.bas.bg/PPages/Avramov/Edinb_477-486.pdf

116. Moynihan CT, Macedo PB, Montrose CJ, Gupta PK, DeBolt MA, Dill JF, Dom BE, Drake PW, Easteal AJ, Elterman PB, Moeller RP, Sasabe H, Wilder JA (1976) Structural relaxation in vitreous materials. Ann NY Acad Sci 279: 15–35. https://doi.org/10.1111/j.1749-6632.1976.tb39688.x
117. Weyer S, Merzlyakov M, Schick C (2001) Application of an extended Tool—Narayanaswamy–Moynihan model. Part 1. Description of vitrification and complex heat capacity measured by temperature-modulated DSC. Thermochim Acta 377:85–96. https://doi.org/10.1016/S0040-6031(01)00543-3
118. Schmelzer JWP (2012) Kinetic criteria of glass formation and the pressure dependence of the glass transition temperature. J Chem Phys 136:74512. https://doi.org/10.1063/1.3685510
119. Schmelzer JWP, Tropin TV (2015) Kinetic criteria of glass-formation, pressure dependence of the glass-transition temperature, and the Prigogine-Defay ratio. J Non Cryst Solids 407. https://doi.org/10.1016/j.jnoncrysol.2014.07.049
120. Scherer GW (1984) Use of the Adam-Gibbs equation in the analysis of structural relaxation. J Am Ceram Soc 67:504–511. https://doi.org/10.1111/j.1151-2916.1984.tb19643.x
121. Hutchinson JM, Montserrat S, Calventus Y, Cortés P (2000) Application of the Adam–Gibbs equation to the non-equilibrium glassy state. Macromolecules 33:5252–5262. https://doi.org/10.1021/ma992015r
122. Brunacci A, Cowie JMG, Ferguson R, Gómez Ribelles JL, Vidaurre Garayo A (1996) Structural relaxation in polystyrene and some polystyrene derivatives. Macromolecules 29:7976–7988. https://doi.org/10.1021/ma960336m
123. Gómez JL, Ribelles M, Monleón Pradas M, Vidaurre Garayo A, Romero Colomer F, Mas Estelles J, Meseguer Duenas JM (1995) Structural relaxation of glass-forming polymers based on an equation for configurational entropy. 2. Structural relaxation in polymethacrylates. Macromolecules 28:5878–5885. https://doi.org/10.1021/ma00121a026
124. Cangialosi D, Boucher VM, Alegría A, Colmenero J (2013) Direct evidence of two equilibration mechanisms in glassy polymers. Phys Rev Lett 111:95701. https://doi.org/10.1103/PhysRevLett.111.095701
125. Schmelzer JWP, Tropin TV (2013) Dependence of the width of the glass transition interval on cooling and heating rates. J Chem Phys 138:34507. https://doi.org/10.1063/1.4775802
126. Tropin TV, Schmelzer JWP, Aksenov VL (2017) On the possibility of modeling of polymers glass transition in a wide range of cooling and heating rates. J Mol Liq 235:172–177. https://doi.org/10.1016/j.molliq.2016.12.009
127. Nelder JA, Mead R (1965) A simplex method for function minimization. Comput J 7:308–313. https://doi.org/10.1093/comjnl/7.4.308
128. Koh YP, Simon SL (2013) Enthalpy recovery of polystyrene: does a long-term aging plateau exist? Macromolecules 46:5815–5821. https://doi.org/10.1021/ma4011236
129. Liu G, Zuo Y, Zhao D, Zhang M (2014) Study on enthalpy relaxation of polystyrene by assuming the existence of an intermediate aging plateau. J Non Cryst Solids 402:160–165. https://doi.org/10.1016/j.jnoncrysol.2014.06.002
130. Liu G, Li L, Zheng Y, Zuo Y (2013) Temperature gradient in sample and its effect on enthalpy relaxation model fitting of polystyrene. J Non Cryst Solids 365:13–22. https://doi.org/10.1016/j.jnoncrysol.2013.01.017
131. Richert R (2011) Heat capacity in the glass transition range modeled on the basis of heterogeneous dynamics. J Chem Phys 134:144501. https://doi.org/10.1063/1.3577580
132. Liu G, Zuo Y, Lin J, Zhao D (2014) Study on enthalpy relaxation of glassy polystyrene using a structure-dependent Kohlrausch stretch exponent combined with coupling model. Eur Phys J E Soft Matter 37:63. https://doi.org/10.1140/epje/i2014-14063-8
133. Simon SL, Sobieski JW, Plazek DJ (2001) Volume and enthalpy recovery of polystyrene. Polymer 42:2555–2567. https://doi.org/10.1016/S0032-3861(00)00623-6
134. Wang LM, Velikov V, Angell CA (2002) Direct determination of kinetic fragility indices of glassforming liquids by differential scanning calorimetry: kinetic versus thermodynamic fragilities. J Chem Phys 117:10184–10192. https://doi.org/10.1063/1.1517607

Dipolar Correlations in 1,4-Polybutadiene Across the Timescales: A Numerical Molecular Dynamics Simulation Investigation

Mathieu Solar and Wolfgang Paul

Abstract Broadband dielectric spectroscopy is able to follow the time correlation of its observable, the polarization of a sample, over an unprecedented range of frequency or timescales, respectively. Features in the dielectric susceptibility as a function of frequency are assigned to different molecular motions, which, for polymers in the melt, range from high-frequency vibrations over a possible Johari-Goldstein β-relaxation, the segmental or α-relaxation, a possible normal mode for chains with a net dipole moment along the end-to-end vector of the chain to conductivity contributions on the low-frequency side. This assignment is a statement about correlations between the local dipole moments making up the sample polarization and their behavior across length and timescales. Such correlations can depend on temperature and on density as well as on confinement effects. A molecular dynamics simulation of a chemically realistic model of 1,4-polybutadiene confined between graphite walls allows for a study of such correlations over a broad temperature range in the bulk-like center of the film, as well as in the wall regions where confinement-induced correlations are present. We have already shown that confinement induces a new type of normal mode into this polymer which does not possess such a mode in the bulk. Here, we discuss the scaling of the dielectric relaxation process in this system and additionally analyze the temperature and confinement dependence of high-frequency vibrations.

Keywords Molecular dynamics simulations · Vibration-relaxation crossover Polymer films

1 Introduction

Broadband dielectric spectroscopy (BDS) has been developed into a method covering dynamic processes in condensed matter over an unparalleled range of frequency (or

M. Solar
INSA, Institut Charles Sadron, 23 rue du Loess, 67034, Strasbourg Cedex 2 , France
e-mail: Mathieu.Solar@ics-cnrs.unistra.fr

W. Paul (✉)
Institut für Physik, Martin-Luther-Universität, von Seckendorff Platz 1, 06120 Halle, Germany
e-mail: Wolfgang.Paul@physik.uni-halle.de

F. Kremer and A. Loidl (eds.), *The Scaling of Relaxation Processes*, Advances in Dielectrics, https://doi.org/10.1007/978-3-319-72706-6_11

time) scales of up to 21 orders of magnitude. This makes this technique ideally suited to study the glass transition of materials, a phenomenon accompanied by an increase of typical relaxation times by 14 orders of magnitude without any obvious structural changes. Of course, this huge range of timescales can only be covered combining a range of different techniques and measurement approaches [1–3]. It is a non-trivial question in how far all these approaches measure the same object, which, in the simplest version of a dielectric experiment, is the dielectric susceptibility scanned in the Fourier frequency space. In contrast, during MD simulations of the dielectric relaxation the time autocorrelation function of the polarization of the sample needs to be computed as a measure of the dielectric response of the polymer. As the total polarization of a sample is made up from molecular, local dipole moments, this response depends on the absence or presence of correlations between these local moments on the different time scales. For polymers, these local dipole moments are carried by chain segments (repeat units) [4].

We will address the dielectric response of one polymer, 1,4-polybutadiene, in this contribution. While such questions about correlations are already relevant in the bulk, they become even more interesting for confined sytems, such as thin polymer films [5], due to the influence of the confining interfaces. Confining walls restructure a polymer film over a distance of 2–3 nm, but today experiments can be performed on films only ten or several tens of nanometers thick [5], so one could expect a strong confinement effect on the dielectric signal.

In Sect. 2, we will discuss the theoretical background of the analysis of our simulation trajectory with respect to the dielectric properties of the simulated sample. Section 3 will then present technical information on our model and simulation technique. Sections 4 and 5 will discuss our analysis of correlation effects on the relaxational and vibrational dynamics, respectively, and present the temperature scaling of relaxation processes in these time windows. Finally, Sect. 6 will present some conclusions.

2 Theoretical Background

Dielectric spectroscopy [6] is typically performed using a sample confined by two parallel electrodes as depicted in Fig. 1. A sinusoidal electric field perpendicular to the electrodes is generated and polarizes the medium. For small applied fields, this only leads to an orientation of the permanent dipole moments present in the molecules. For polymeric systems, local dipole moments $\boldsymbol{\mu}_i$ are carried by some or all of the segments making up the chains, so the total dipole moment of the system, denoted $\boldsymbol{\mu}_{\text{tot}}$, is a sum of the segmental ones $\boldsymbol{\mu}_{\text{tot}} = \sum_i \boldsymbol{\mu}_i$. The macroscopic polarization in a volume V is then given as $\mathbf{P} = (1/V) \cdot \sum_i \boldsymbol{\mu}_i$. In the frame of linear response theory of any causal system, the dielectric permittivity tensor $\boldsymbol{\varepsilon} = \varepsilon_{\alpha\beta}$ links the polarization to the external electric field $\mathbf{E}$, by the following relation $\mathbf{P} = \varepsilon_0 \cdot (\boldsymbol{\chi} \bar{\otimes} \mathbf{E})$, where $\boldsymbol{\chi} = (\boldsymbol{\varepsilon} - \mathbf{1})$ is the dielectric susceptibility tensor, ε_0 is the dielectric permittivity of vacuum, and the symbol $\bar{\otimes}$ is the once-contracted product between a second order tensor and a vector.

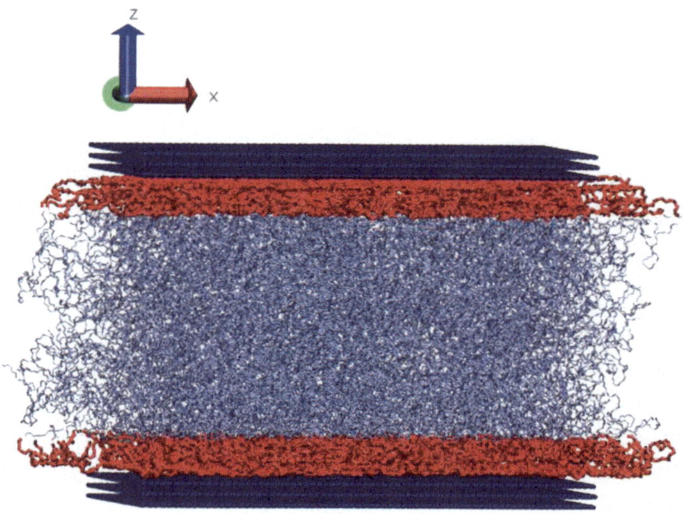

Fig. 1 Snapshot of a MD configuration of 1,4-polybutadiene confined between graphite walls. In a dielectric experiment, the confining interfaces would act as electrodes generating an electric field perpendicular to the walls making the interfaces. These walls perturb the structure and the dynamics in the melt over a region of a few nanometer in thickness, where any physical property gradually evolves to a bulk-like behavior far away from the confining surface. This effect depends on temperature as well

When $\boldsymbol{E} = E\cdot\hat{\mathbf{e}}_z$, and $\boldsymbol{\varepsilon}$ (or $\boldsymbol{\chi}$) can be assumed to be a diagonal tensor, one obtains $\boldsymbol{P} = P\cdot\hat{e}_z$, and only the component $\varepsilon_{zz} = \varepsilon^\star$ is relevant.[1]

According to the fluctuation–dissipation theorem [7], the complex dielectric function $\varepsilon^\star$ can be calculated from the Fourier transform of the normalized dipole correlation function, which in our geometry is given by the z-component relaxation and the dielectric signal in the time domain is proportional to $\langle \mu_{\mathrm{tot,z}}(t)\cdot\mu_{\mathrm{tot,z}}(0)\rangle$, according to,

$$\frac{\varepsilon^\star(\omega)-\varepsilon_\infty}{\Delta\varepsilon} = 1 - \mathrm{i}\omega\cdot\int_0^{+\infty}\phi_{\mathrm{z}}(\mathrm{t})\cdot\mathrm{e}^{-\mathrm{i}\omega\mathrm{t}}\,\mathrm{dt} \quad \text{with,} \quad \phi_{\mathrm{z}}(\mathrm{t}) = \frac{\langle \mu_{\mathrm{tot,z}}(\mathrm{t})\cdot\mu_{\mathrm{tot,z}}(0)\rangle}{\langle[\mu_{\mathrm{tot,z}}(0)]^2\rangle}, \tag{1}$$

where i is the imaginary unit, and where the angular brackets indicate a thermodynamic average, the index z denoting the projection onto the direction of the applied electric field.

The complex dielectric function $\varepsilon^\star$ consists of a real ε' and an imaginary ε'' part, where one has by definition that $\varepsilon^\star(\omega) = \varepsilon'(\omega)-\mathrm{i}\cdot\varepsilon''(\omega)$, $\Delta\varepsilon$ being the dielectric strength of the medium. Consequently, the dielectric loss is given by

[1] In a cartesian basis like $(\hat{\mathbf{e}}_x; \hat{\mathbf{e}}_y; \hat{\mathbf{e}}_z)$, a vector $\mathbf{P}$ is written as $\mathbf{P} = P_i\cdot\hat{\mathbf{e}}_i$, the sum over the repeated dummy indice i being implicit with $i = x; y; z$. A tensor $\boldsymbol{\varepsilon}$ is written as a sum of dyads $\boldsymbol{\varepsilon} = \varepsilon_{\alpha\beta}\cdot(\hat{\mathbf{e}}_\alpha\otimes\hat{\mathbf{e}}_\beta)$. The identity tensor is denoted by $\mathbf{1} = \delta_{\alpha\beta}\cdot(\hat{\mathbf{e}}_\alpha\otimes\hat{\mathbf{e}}_\beta)$.

$$\frac{\varepsilon''(\omega)}{\Delta\varepsilon} = \omega\cdot \int_0^{+\infty} \phi_z(t)\cdot e^{-i\omega t}\, \mathrm{dt}\ . \quad (2)$$

In many polymers (so-called type B polymers), one can assume that the individual segmental dipoles relax essentially independently, so that the relaxation function $\langle \mu_{\mathrm{tot,z}}(t)\cdot\mu_{\mathrm{tot,z}}(0)\rangle$ relaxes like $\langle \mu_{\mathrm{seg,z}}(t)\cdot\mu_{\mathrm{seg,z}}(0)\rangle$, averaged over all segments: The experiment probes then local segmental relaxation. In the so-called type A polymers (e.g., poly-isoprene), the charge repartition inside the group of atoms making a segment makes the segmental dipole moment having components tangent and perpendicular to the chain backbone. The perpendicular segmental dipole moments are locally normal to the chain backbone and are able to relax independently. However, the tangent segmental contributions add up along the chain backbone, which represents a dipole moment parallel to the chain end-to-end vector. As a consequence, in these polymers dielectric spectroscopy experiments can address local (segmental) as well as global/normal (end-to-end vector) conformational relaxation of polymer chains. Bulk systems—or the center of a confined film, when the film is thick enough—are homogeneous and isotropic, and here, the susceptibility tensor is proportional to the unit tensor, $\boldsymbol{\chi} = \chi\cdot\mathbf{1}$. The dipolar correlations along each Cartesian direction are equal on average, and thus, the dielectric observable is proportional to the autocorrelation function of the complete dipole vector, $\langle \mu_{\mathrm{tot,z}}(t)\cdot\mu_{\mathrm{tot,z}}(0)\rangle \propto \langle \boldsymbol{\mu}_{\mathrm{tot}}(t)\cdot\boldsymbol{\mu}_{\mathrm{tot}}(0)\rangle$, not only its z-component. However in a confined system, close enough to the interfaces, this is not the case, and one has to be aware that the measured quantity is only the relaxation of the normal component of the dipole moment. An electric field applied perpendicular to the surface probes other molecular motions than a field applied parallel to the surface. In confinement, one can write the susceptibility tensor as $\boldsymbol{\chi} = \chi_\alpha\cdot\mathbf{e}_\alpha \otimes \mathbf{e}_\alpha$ (sum over α), with a Cartesian basis $\hat{\mathbf{e}}_\alpha$ $\alpha = x, y, z$. Due to the rotational symmetry, one has $\chi_x = \chi_y \neq \chi_z$, and correspondingly for the dielectric permittivity. The orientation-dependent dielectric relaxation of a polymer film was first analyzed carefully for a computer simulation in a study of Peter et al. [7], who considered, however, a coarse-grained bead-spring polymer model without any torsion stiffness in the chain backbone representation.

Careful experimental [8] as well computer simulation [9] studies reveal that even for type B polymers in the bulk, the assumption of uncorrelated relaxation of different segments discussed above seems to be only an approximation for the dielectric behavior. Furthermore, in confined systems the walls introduce an additional structuring in the polymer layers next to them. For 1,4-polybutadiene, for example, it was found that the walls lead to an orientation of the cis-segments which carry the dipolar response in this polymer, and, consequently, to a dipolar layer next to the walls. Also, the gyration ellipsoids of the instantaneous chain configurations get oriented in such a way that a chain with its center of mass in proximity to the walls has its long axis parallel to the wall. This generates a chain conformation with a residual chain dipole moment oriented perpendicular to the end-to-end vector of the chains, which is preferentially oriented along the long half-axis on average, and thus a new type of normal mode. This effect has been termed type A' polymers [10], as here the chain dipole is not along (type A) but perpendicular to the end-to-end vector.

It is therefore not at all clear in all cases, which dipolar correlations are actually making up the response in a dielectric experiment. Furthermore, different correlations can be important on different timescales—or frequency scales—of the dielectric response. This in turn might give rise to a very different scaling with temperature of the features in these timescale or frequency scale. Let us write the total dipole moment as $\boldsymbol{\mu}_{\text{tot}}(t) = \sum_{k=1}^{K} \sum_{n=1}^{N} \boldsymbol{\mu}_{kn}(t)$, where $\boldsymbol{\mu}_{kn}(t)$ is the dipole moment of segment n in chain k at time t. We then have

$$\langle \boldsymbol{\mu}_{\text{tot}}(t) \cdot \boldsymbol{\mu}_{\text{tot}}(0) \rangle = \sum_{k=1}^{K} \sum_{n,m=1}^{N} \langle \boldsymbol{\mu}_{kn}(t) \cdot \boldsymbol{\mu}_{km}(0) \rangle + \sum_{k=1}^{K} \sum_{l \neq k} \sum_{n,m=1}^{N} \langle \boldsymbol{\mu}_{kn}(t) \cdot \boldsymbol{\mu}_{lm}(0) \rangle \,. \tag{3}$$

K being the number of chains in the considered volume (In our simulation, the complete film or a layer of 1.2 nm thickness contains $K = 720$ chains or $K = 90$ chains, respectively). In a homogeneous system, all K chains are equivalent (in a confined system, we assume that all K chains in a given layer located at a certain distance from the wall are equivalent). We obtain,

$$\langle \boldsymbol{\mu}_{\text{tot}}(t) \cdot \boldsymbol{\mu}_{\text{tot}}(0) \rangle = K \cdot \Bigg[\sum_{n,m=1}^{N} \langle \boldsymbol{\mu}_{\alpha n}(t) \cdot \boldsymbol{\mu}_{\alpha m}(0) \rangle + \frac{1}{2}(K-1) \cdot \sum_{n,m=1}^{N} \langle \boldsymbol{\mu}_{\alpha n}(t) \cdot \boldsymbol{\mu}_{\beta m}(0) \rangle \Bigg] \,, \tag{4}$$

where we consider two arbitrarily selected chains α and β in the given volume. Analysis of dielectric experiments on polymers generally considers correlations between the dipole moments of different chains not to be contributing to the measured signal. However, this has to be considered as an open question. The same approach had been followed for decades for the analysis of NMR experiments, where it recently became clear [11] that these are susceptible to correlations between magnetic dipoles on different chains. Following this assumption, one has

$$\langle \boldsymbol{\mu}_{\text{tot}}(t) \cdot \boldsymbol{\mu}_{\text{tot}}(0) \rangle = K \cdot \sum_{n,m=1}^{N} \langle \boldsymbol{\mu}_{n}(t) \cdot \boldsymbol{\mu}_{m}(0) \rangle = K \cdot \langle \boldsymbol{\mu}_{\text{chain}}(t) \cdot \boldsymbol{\mu}_{\text{chain}}(0) \rangle \,, \tag{5}$$

with the dipole moment of one chain, $\boldsymbol{\mu}_{\text{chain}}$. For type A polymers, the time correlation of the chain dipoles decays due to vibrations, fast segmental motions, and relaxation of the overall chain orientation. For type B polymers, one furthermore assumes that all segmental dipoles in a chain are uncorrelated and has

$$\langle \boldsymbol{\mu}_{\text{tot}}(t) \cdot \boldsymbol{\mu}_{\text{tot}}(0) \rangle = K N_{\text{d}} \cdot \langle \boldsymbol{\mu}_{seg}(t) \cdot \boldsymbol{\mu}_{seg}(0) \rangle \,, \tag{6}$$

where N_{d} is the number of segments per chain carrying a dipole moment (assuming identical dipoles as for PBD). After presenting the model in the next section, we will contrast the predictions of (5) and (6) for the temperature scaling of the dielectric spectrum in different frequency ranges, and for bulk chains and chains next to a wall.

3 Model and Simulation Technique

We simulate a chemically realistic model for a 1,4-polybutadiene (PBD) melt, with 55% trans and 45% cis content, with $N_c = 720$ chains containing $N_p = 116$ united atoms. These united atoms represent either CH or CH_2 groups along the chain backbone (in the chain interior) or CH_3 groups (at the chain ends). In the model, no distinction was made between CH_3 and CH_2 groups to save computational efforts, which are supposed to differ only by their masses as a first approximation. In addition, CH and CH_2 groups are distinguished by differences in the non-bonded (Lennard-Jones) interaction parameters in the force field that was used [12]. For bulk properties, this force field for PBD had been carefully validated by detailed comparisons of simulation results with NMR spin lattice relaxation times [13], neutron spin echo data probing the chain dynamics in the time domain [14], and dielectric relaxation data [9]. The glass transition temperature T_g for this polymer has been estimated as $T_g = 178$ K by differential scanning calorimetry (DSC) [12], and as $T_g = 175$ K from rheological measurements [15]. For the choice of $N_p = 116$, the chains have 29 repeat units (a repeat unit denoted as trans or cis consists of 2 CH and 2 CH_2 groups, respectively). This chain length still belongs to the regime of Rouse chain lengths for this polymer. For the calculation of the dielectric response, it is important to note that the force field [12] does not include partial charges since these are very small, and hence are not expected to lead to significant correlations between the positions of the carbon atoms: This assumption in the polymer modeling allows us to save computational efforts. This expectation is corroborated by the excellent agreement with the relevant experimental data [13, 14]. Dielectric properties are estimated on a stored trajectory of configurations by reinserting the appropriate partial charges into the atomic configurations [9, 16, 17]. Each chain in the melt is a random copolymer of 13 cis and 16 trans conformers, of which the cis units carry the dipole moments [18], i.e., $N_\mathrm{d} = 13$.

The force field for graphite was taken from the literature [19], and for the interaction potential between the carbon atoms of graphite and the united atoms of PBD, we used Lorentz-Berthelot combining rules. At a high temperature $T = 353$ K, which is about two times the glass transition temperature of this polymer and which we chose as a reference temperature for the further cooling of the system, the lateral linear dimensions commensurate with the graphite crystal structure are $L_x = 14.91$ nm and $L_y = 14.76$ nm, and periodic boundary conditions in both x and y directions are imposed. We place three lattice planes of the graphite crystal at the coordinate positions $z = -0.34$ nm, $z = 0$ nm, and $z = D = 9.97$ nm (see Fig. 1), using a periodic boundary condition in z-direction at 10.31 nm. This choice of boundary condition, together with the fact that a distance cutoff is applied to the Lennard-Jones interactions, effectively creates a system of a PBD melt confined by two half spaces of graphite crystals. The wall distance at each temperature is adapted until the density in the center of the film reaches the bulk density at this temperature and ambient pressure [20]. In total, our simulated systems contains 83,520 united atoms, and we simulate trajectories over times up to one microsecond (i.e., 10^9 time steps of $\delta t = 1$ fs, using the GROMACS package [21]). Most of the simulations were carried out using 2048 cores on a IBM Blue Gene architecture.

From our MD simulation results, one obtains dielectric signals in the time domain $\phi_z(t)$. According to the fluctuation–dissipation theorem, the complex zz-dielectric function $\varepsilon^\star$ is calculated from the Fourier transform of this normalized relaxation function. From the mathematical point of view, the Fourier transform of an analytical square-integrable function is a well-defined operation, unlike the Fourier transform of a signal sampled with the sampling frequency $\Delta_t = q \cdot \delta t$, q being a positive-definite integer. Without entering into the details, most of the Fourier transforms presented in this contribution were computed using the "Fast Fourier Transform (FFT)" algorithm applied to signals sampled with Δ_t over a time window t_∞, but regularly and linearly re-sampled to a signal made of 2^{26} or 2^{27} points. The limitations, in the frequency spectrum range, match also the restrictions imposed by the Nyquist–Shannon sampling theorem, to avoid misunderstanding of the results due to aliasing. Then, one can analyze the relaxational parts of the dipolar correlations just by Fourier transforming the data from the MD simulations, or by fitting a phenomenological function capturing the stretched exponential α-decay as well as the desorption process and the short-time dynamics, when they are present. This function is the following,

$$\phi_z(t) = w_0 \cdot e^{-(t/\tau_\beta)^{\beta_0}} + w_1 \cdot e^{-(t/\tau_\alpha)^{\beta_1}} + w_2 \cdot e^{-(t/\tau_{\text{des}})^{\beta_2}} , \qquad (7)$$

where τ_β, τ_α, τ_{des}, β_i, and w_i are the fitting parameters. The weights w_i of the various relaxation processes, ranging between 0 and 1, satisfy the normalization condition $w_0 + w_1 + w_2 = 1$.

In Fig. 2 (top), we illustrate a dielectric relaxation at $T = 353$ K, just in a layer at the interface of 1.2 nm thickness. The data from the MD simulations are compared to a regression of the data to two models, one being the sum of 2 stretched exponentials and the second the sum of 3. Clearly, the first model can capture the α- and the desorption (at long time) processes, but fails to describe the short-time dynamics. Adding a third relaxation decay counteracts this problem and, clearly, captures the main decay induced by the vibrational motions, but also fails to describes the small sinusoidal oscillations taking place around the main vibrational decay. In Fig. 2 (bottom), we show the dielectric loss calculated from these dielectric relaxation functions. The α- and the desorption processes are well captured by the models. Surprisingly, the model made of 3 stretched exponential functions, even if it fails to describe the Fourier spectrum at high magnitudes of the frequency, helps to understand what is observed: The first peak at $\omega \simeq 40$ THz corresponds to the principal decay of the vibrational modes (although slightly shifted in frequency), whereas the second peak at $\omega \simeq 100$ THz, not present in the spectrum of a 3-KWW like function, indicates clearly the frequency of vibration for the damped oscillations around the principal decay in the short-time regime.

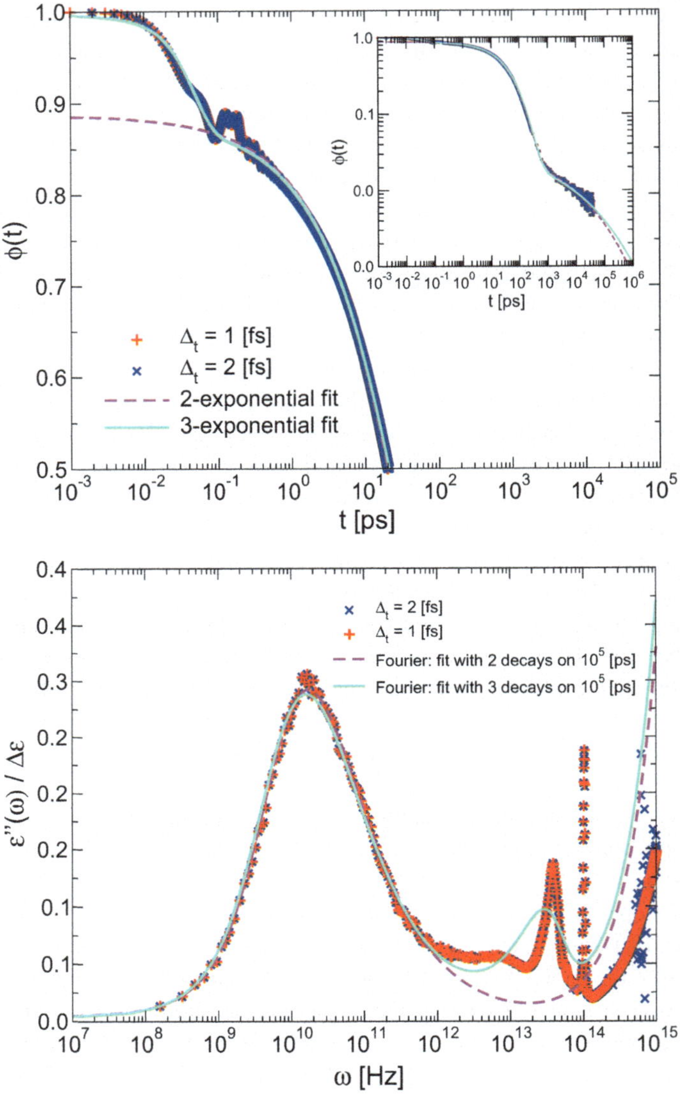

Fig. 2 Dielectric relaxation at $T = 353$ K (top), in a layer at the interface and being 1.2 nm thick. A comparison is made between the data from the MD simulations and the regression of the data to two models, one being the sum of 2 stretched exponential and the second the sum of 3; see Eq. (7). We resolved the relaxation function every integration timestep (being $\delta t = 1$ fs), respectively every 2 integration steps used in the MD simulations. Dielectric loss at $T = 353$ K (bottom) for the same layer as a function of frequency

4 Scaling of Relaxational Dynamics

In carbon-based polymer chains, there exists a rather universal timescale for the crossover between vibrational dynamics and relaxational dynamics, which is around 1 ps [22]. For shorter timescales, bond vibrations and vibrations in the minima of the torsional potentials dominate the response of the system, but these are damped vibrations decorrelating at the picosecond scale (see also Fig. 2). For larger times, the dynamics is dominated by transitions between different minima of the torsional potential [23], inducing conformational relaxation and diffusion of the monomers or the molecules. In this section, we will now first focus on the dielectric response in the relaxation domain, that is for time scales larger than about 10 ps, and for frequencies up to a few hundred Gigahertz. As one can see in Fig. 3, there is basically no difference in the dielectric response in the center of the film whether one assumes uncorrelated segments following (6) or uncorrelated chain dipoles following (5), as one would generaly assume for a type B polymer.

Both types of dipolar correlation function agree down to below the 1% level, as one can see on the double logarithmic plot following the decay of dipolar correlations over about six orders of magnitude in time. In this time regime, and for all temperatures shown, the structural relaxation (α-relaxation) is the main contributor to the decay of the dipolar correlations in the center of the film. However, on the percent level, there are indications of a change of slope in the correlation functions at the lowest temperatures. At these temperatures (240 K and perhaps 253 K), it seems that the

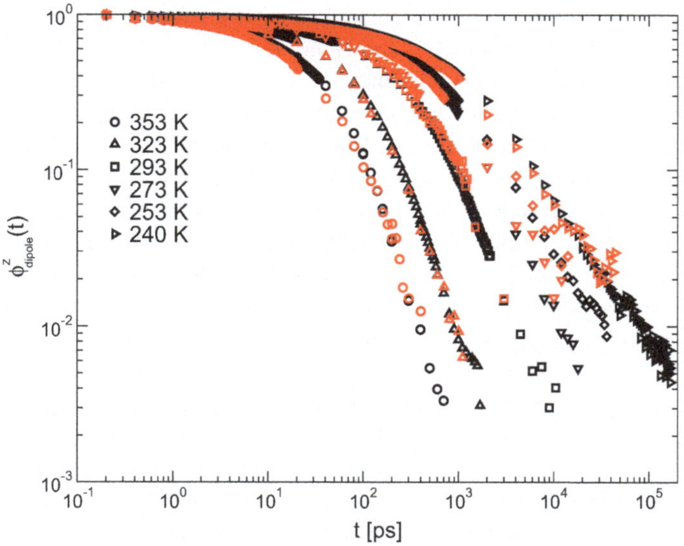

Fig. 3 Dipolar autocorrelation function of the normal component of the dipole moment in the center of the film. For the black symbols, this is calculated assuming uncorrelated segmental dipoles (6), whereas for the red symbols, the chain dipole moments are used (5)

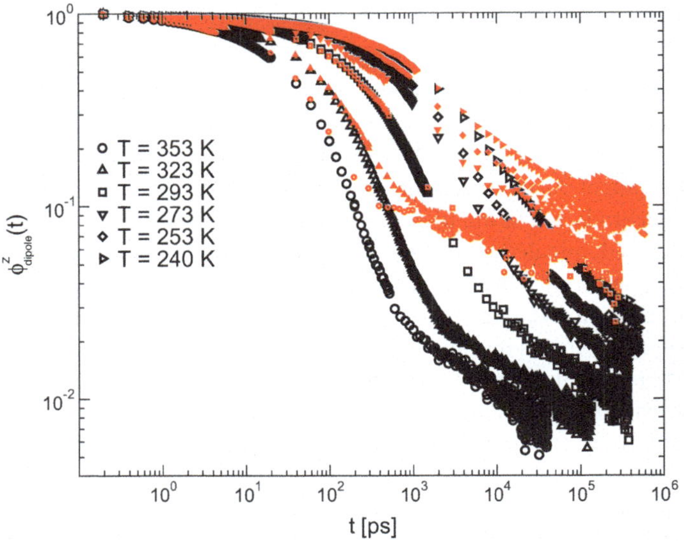

Fig. 4 Dipolar autocorrelation function of the normal component of the dipole moment in a layer of 1.2 nm thickness next to the walls. For the black symbols, this is calculated assuming uncorrelated segmental dipoles (6), whereas for the red symbols, the chain dipole moments are used (5)

orientational correlations induced by the walls start to be felt even in the center of the 10 nm thick films studied here. These correlations are dramatic, when one compares the dipolar correlation functions calculated using (6) and (5) for a layer of 1.2 nm thickness next to the walls, which is shown in Fig. 4.

Here even the dipolar correlation functions calculated assuming uncorrelated segmental dipoles, show a clear additional relaxation process at long times present for all temperatures, i.e., even at $T \simeq 2T_g$. This process contributes at a 1% level at high temperatures, but by up to 10% at lower temperatures. We already identified this process as the desorption process of chains at the walls [10, 24–26]. Its temperature scaling is determined by the adsorption strength to the walls, which in our case is given by relatively weak Lennard-Jones interactions. However, for the binding of a chain to the walls, these have to be scaled up by the number of contacts that a chain in a random coil configuration has with a wall, when it is adsorbed. This number scales as the square root of the chain length. In our case, this creates a long-time scale for desorption even at high temperatures. At high temperatures, this timescale still fits into the simulation time window, being up to 1 μs. An Arrhenius temperature dependence for this process was found [20], so at lower temperatures, the Vogel–Fulcher temperature dependence of the α-relaxation dominates, and the clear distinction between α-relaxation and desorption process vanishes (as one can observe in Fig. 4). In other cases, like long entangled polystyrene chains adsorbed onto silica [27], the individual adsorption strengths are already much larger, being of Coulombic nature. Also, the chains are much longer, leading to a desorption timescale which is

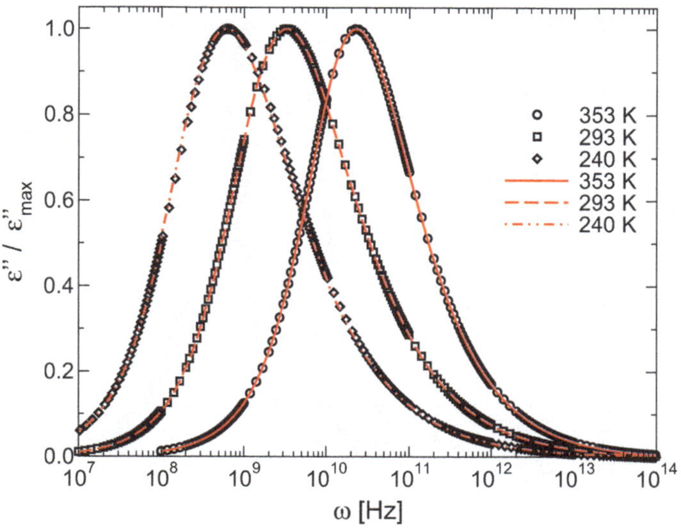

Fig. 5 Dielectric susceptibility calculated at three different temperatures for the center of the film, based on the dipolar correlation functions shown in Fig. 3. Black symbols are obtained using (6), red lines using (5)

so much larger than all experimentally accessible timescales, that the chains can be considered irreversibly adsorbed. As we discussed before, due to the orientational effect of the walls, the end-to-end vector of the chains in the first layer at the walls is oriented preferentially parallel to the walls, while the dipole moment of these chains is perpendicular to the walls (i.e., these chains possess a new type of normal mode [10]). These additional correlations between the dipole moments of one chain, which are induced by the walls, lead to a huge difference in the dielectric response calcuted using (5), shown by the red symbols in Fig. 4, compared to the prediction using (6). The contribution of the desorption process to the total decay now is above the 10% level at all temperatures, and the decay of correlations in the long-time regime becomes much slower. Furthermore, there is also a visible change of the correlation decay in the α-relaxation regime at smaller times; i.e., the structural relaxation as observed by dielectrics is different here, whether one assumes uncorrelated segments or not.

Since only the frequency dependence of the dielectric permittivity can be addressed experimentally, the Fourier transform[2] of the data in Figs. 3 and 4 was performed. From Fig. 5 we can conclude, that the dipolar relaxation functions in the center of the film agree, whether calculated for independent segmental or chain dipole moments. Furthermore, the frequency spectrum covers the range from megahertz to terahertz, with one dominating peak denoting the α-relaxation of PBD.

[2]The spectra shown in Figs. 5 and 6 were actually calculated by Fourier transforming the fit function of Eq. (7) adjusted to the data. The location of the α-peaks and of the desorption peaks for the wall layers are presented in the activation diagram of Fig. 7.

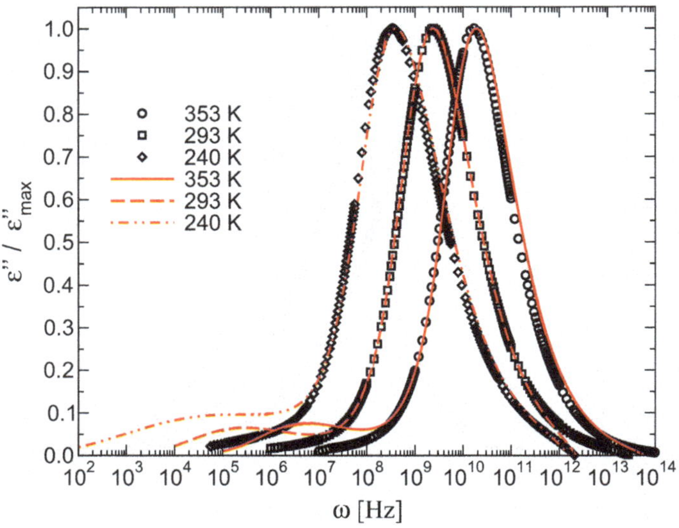

Fig. 6 Dielectric susceptibility at three selected temperatures calculated for the layers at the walls, based on the dipolar correlation functions shown in Fig. 4. Black symbols are obtained using (6), red lines using (5)

For the layers next to the walls, however, the dielectric spectrum calculated using the assumption of uncorrelated chain dipoles (5), shown in Fig. 6 by the red symbols, displays a more pronounced structure than the one based on independent segments. The strong contribution of the desorption process visible at large times in Fig. 4 now shows up as a separate peak at low frequencies, when the susceptibility is calculated based on chain dipoles. In the spectra calculated assuming independent segmental dipoles, the desorption process only shows up as an excess in the low-frequency wings of the α-peak, visible in a time–temperature superposition plot [26].

Instead of calculating the dielectric response layer resolved (like we show for the wall and center layers in Figs. 5 and 6), we can also apply this method for layers of growing thicknesses at the walls. The resulting timescales are shown by the blue symbols in Fig. 7. Clearly, a Vogel–Fulcher–Tamman fit to these data is compatible to the bulk behavior shown with black symbols and extrapolates to the bulk-glass transition temperature for all thicknesses. From these data, we thus conclude that we cannot see a glass transition shift with decreasing film thickness from the analysis of dielectric correlations in PBD. We also include the temperature dependence of the position of the desorption peak in the chain dielectric spectrum, which shows an Arrhenius dependence. The α timescale therefore increases faster than the desorption timescale. If we would extrapolate the Arrhenius fit in Fig. 7, we would predict that they match about 15 K above T_g; however, this involves quite some extrapolation from our data. We next turn to an analysis of the short-time behavior of the dielectric response.

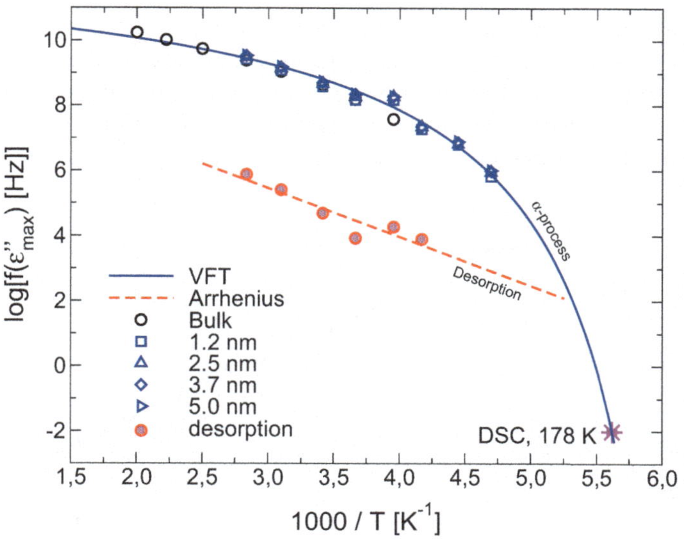

Fig. 7 Activation plot for the dielectric α-peak positions calculated for films of different thicknesses at the walls. The VFT fit agrees with the bulk behavior shown as black circles. Additionally, the desorption timescale was included, which follows an Arrhenius temperature dependence

5 Scaling of Vibrational Dynamics

In the time window below 1 ps, the dynamics in polymer melts of carbon-based chains is governed by vibrational degrees of freedom. These degrees of freedom consist of bond length and bond angle vibrations and librations (anharmonic vibrations) in the minima of the dihedral potentials. In our polymer model, however, the short-time dynamics only consists of bond angle vibrations and dihedral librations, since the covalent bonds were kept fixed at their equilibrium length. All these degrees of freedom undergo damped vibrations, dying out around the picosecond timescale. The dynamics of a segmental dipole moment in this time window is thus governed by the vibrational degrees of freedom in this segment, and perhaps by the vibrational degrees of freedom of the neighboring segments as well. Let us assume for simplicity, that these movements add up to some effective damped oscillation with a frequency ω_{seg}. Adding up such damped oscillations in their effect on the dipole moment of the complete sample, one has to take into account that all these individual oscillations have a random phase. Already in the nineteenth century, Lord Rayleigh showed that the resulting dynamics of a large number of vibrations of the same frequency, but of arbitrary phase, is diffusive with a Gaussian probability distribution [28] for the fluctuations, at a given time scale. We would therefore assume that, while the segmental dipolar correlation function shows damped oscillations, these should be at least reduced in amplitude for the whole box, if not completely vanish. This was observed experimentally for the incoherent scattering function in Ref. [22], where one measured a Debye function (i.e., a decay like $e^{-t/\tau}$) for short times, where actually all individual scatterers display a damped sinusoidal oscillatory motion.

Inside a chain, the neighboring segmental dipoles are coupled through the dihedral degrees of freedom in the segments, so it is a priori not clear whether their oscillations show constructive or destructive interference, when added. As one can see in Fig. 8,

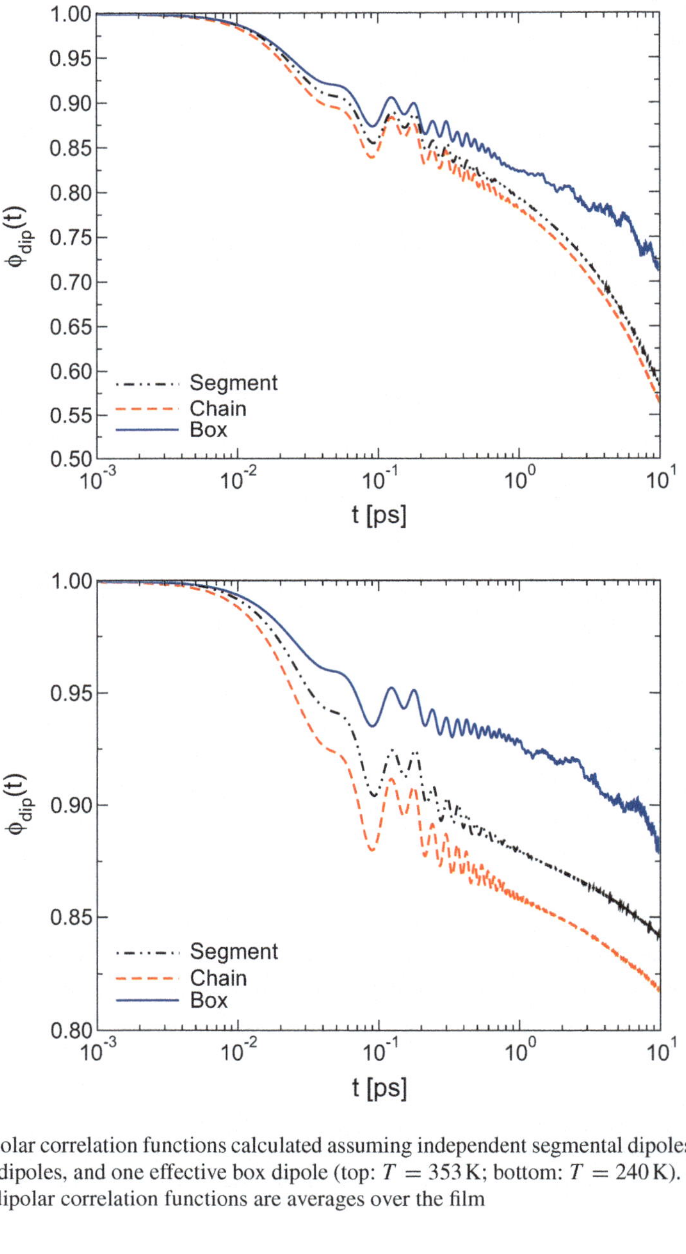

Fig. 8 Dipolar correlation functions calculated assuming independent segmental dipoles, independent chain dipoles, and one effective box dipole (top: $T = 353\,\mathrm{K}$; bottom: $T = 240\,\mathrm{K}$). Segmental and chain dipolar correlation functions are averages over the film

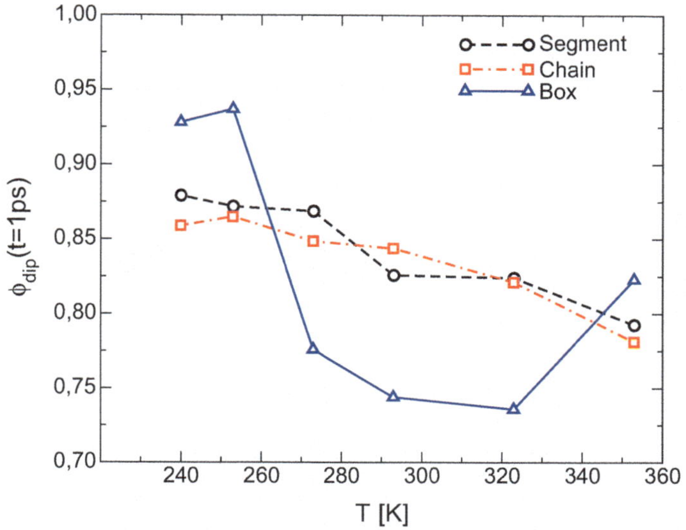

Fig. 9 Value of the dipolar correlation function at the time $t = 1$ ps (i.e., at the vibration–relaxation crossover), as a function of the temperature. The segmental correlations and chain correlations show a monotonous decrease in the dielectric relaxation strength of the vibration processes with decreasing temperature, while the box value, which contains correlations between different chains, shows non-monotonic behavior. The origin of this effect is unclear to us, but we can not exclude that the poor statistics, in time and in space, of the box-averaged relaxation function may be a possible explanation

all three short-time correlation functions, which are shown here for times up to 10 ps, behave rather similarly, although the reduction in amplitude for the oscillations in the box dipole correlation function is clearly visible. The argument of Lord Rayleigh, which we were using above, relies on the law of large numbers: seemingly, our box size of about 2000 nm^3 is not large enough to reach the limit of purely diffusive behavior. One furthermore notices that the amount of decay at 1 ps is reduced at the lower temperature, shown in the bottom plot compared to the high temperature in the top plot. These two temperatures bracket the temperature interval where we have reliable long-time data for these correlation functions. For the dipolar correlation function calculated based on the chain dipoles, we do not see a reduction in the oscillation amplitude, and there may be even a slight increase. This may be due to the strong coupling between dihedral fluctuations of next-nearest neighbor angles along a polymer chain. This coupling allows correlated dihedral oscillations, which are not strongly susceptible to packing effects in a dense polymer melt, as the chain segment between the two angles basically twists within its own space.

Significantly, the positions of the maxima and minima in the oscillatory part are equal among the three correlation functions, and also the period of the oscillations. Also, both do not depend on temperature in the temperature interval studied, which reaches down to about $1.35 \cdot T_g$. Finally, we would like to notice that the oscillatory behavior in all the correlation functions is damped out on the 1 ps timescale, as expected.

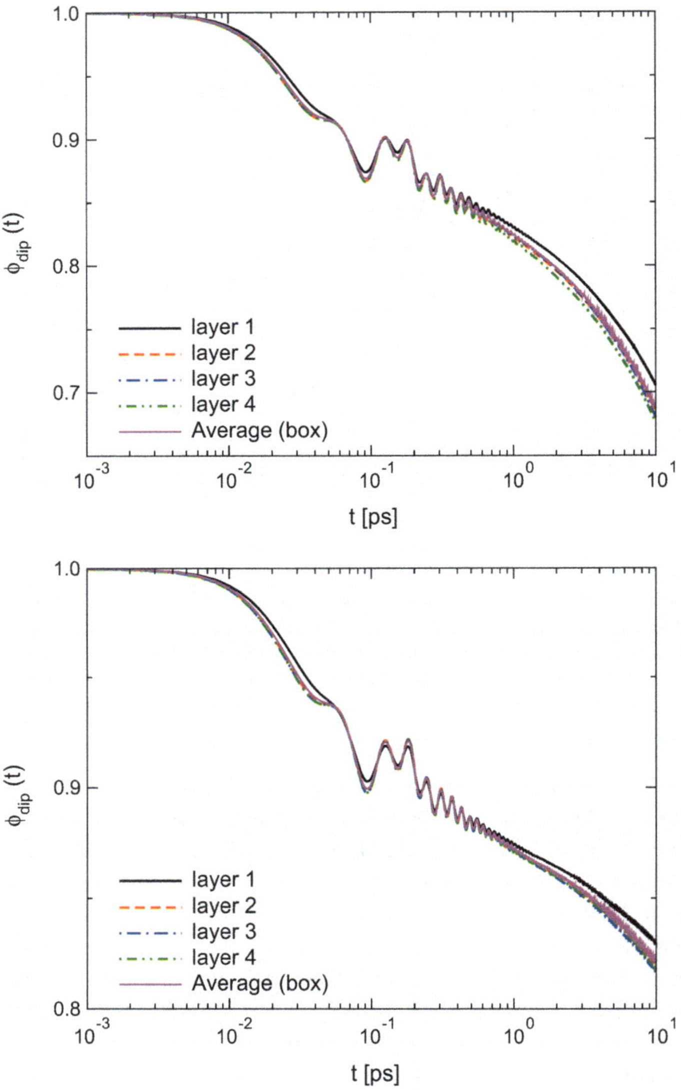

Fig. 10 Layer-resolved dipolar segmental correlation functions calculated assuming independent segmental dipoles (top: $T = 323\,\text{K}$; bottom: $T = 253\,\text{K}$). The layer 1 is at the walls, and the layer 4 is in the middle. Both halves of the film are averaged over. The layer-resolved functions are also compared to the average over the complete film

When we look at the amount of decay induced by the vibrational degrees of freedom (i.e., the corresponding dielectric strength), we observe similar behavior, when we study segmental or chain correlation functions in Fig. 9, whereas the box-dipole correlation function behaves differently. In the area of polymeric confined systems a question still remains under active debate: What is the effect of correlations induced by the presence of the interfaces? With this in mind, we assessed the segmental short-time dynamics, resolved with respect to the layer within which the segment is at time zero. The results are plotted in Fig. 10, which compares these relaxation functions at the temperatures of $T = 323$ K and $T = 253$ K, as typical examples of the short-time relaxation behavior. By comparing the two temperatures, basically no qualitative effect of the temperature is visible. Also, for all layers, except for the first layer at the walls, an identical behavior is observed, which agrees with what we found for the relaxational dynamics at later times. For the first layer, the dynamics at very short times is slightly slowed down compared to the other layers, but leads to the same amount of decay at around 60 fs. The ensuing oscillations are slightly reduced in amplitude, and the decay after 1 ps is slightly less than for the other layers. All this can be assigned to the increased density in this first layer. We note, however, that only the amplitude of the decay is affected, but the oscillatory pattern is identical between the layers: The same high-frequency peak(s) in the dielectric loss has to be expected.

This detailed analysis of the short-time vibrational dynamics needs now to be linked to the analysis of the relaxational behavior discussed in the previous chapter. For this purpose, the segmental autocorrelation functions were determined at all temperatures, with a high resolution able to resolve the short-time oscillatory behavior and extending this deep into the relaxational regime. The resulting correlation functions are then Fourier transformed. From the Fourier transform, the dielectric loss is determined [26]. We present this loss function in Fig. 11, over a frequency range unprecedented for MD simulations of polymer systems. In this figure (top), the low-frequency peaks again denote the structural relaxation in the film. Here, they are not scaled to the same height, so one can see that the peak height decreases, while the width of the peak increases revealing a broader spectrum of contributing relaxation times with decreasing temperature. As already anticipated, the high-frequency peak at around $\omega = 100$ THz, due to the vibrational processes, does not depend on temperature. From the bottom plot in the same figure, this conclusion remains valid at fixed temperature, when one resolves the different layers. The vibrational processes seem to be an intrinsic characteristic of the molecular force field, but are neither shifted by a temperature variation nor influenced by the confinement. Furthermore, also this peak increases in width as the temperature decreases. We also included a data set for a simulation at 353 K using repulsive interaction with the walls only, denoted by "(RW)." Here, we observe a speedup of the α-relaxation compared to the real graphite surface at the same temperature, but, again, the positions of the high-frequency peaks are unaffected. In between this vibrational peak and the α-relaxation peak, a minimum develops with decreasing temperature, indicative of a loss of relaxation modes in this part of the frequency spectrum. This is a widening timescale separation between vibrational modes and relaxational modes, generally connected with the glass transition in a system.

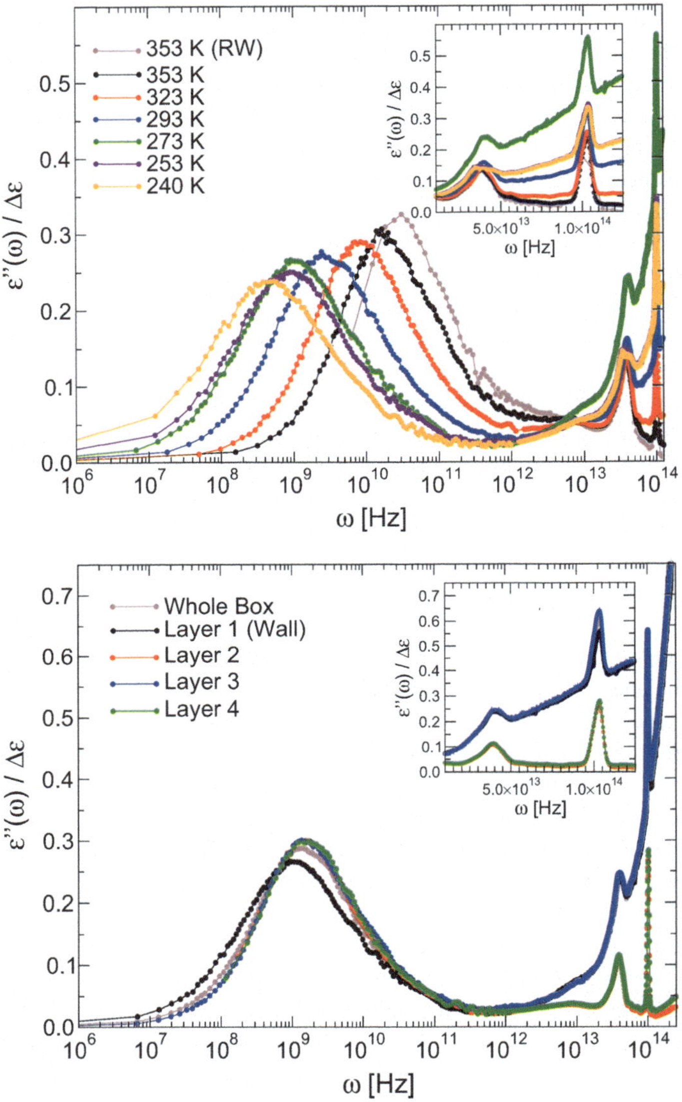

Fig. 11 Dielectric loss as a function of the frequency for all the temperatures investigated (top) and calculated from the dielectric relaxation in a layer being 1.2 nm thick, just located at the interface (averaging being performed over both sides of the polymer film). Dielectric loss as a function of the frequency for all the layers (from the interface to the bulk-like region) at a temperature of 273 K (bottom). The insets in both figures show the high-frequency regime only for clarity

For polymers however, there is another source for this timescale separation, which are the barriers in the torsional potential. For PBD in this temperature window, which is still far above the glass transition temperature and also above the mode-coupling temperature of this polymer, the timescale separation is due to these intramolecular effects, not to packing effects as was demonstrated some time ago [29, 30]. Crossing over the barrier into a different dihedral state is an activated process, whose timescale increases exponentially in the barrier height over $k_B T$. These barrier crossings are the elementary motions leading to the structural relaxation peak. On the other hand, as one can see in Fig. 11, the timescale for the vibrational degrees of freedom is temperature independent, and thus, a timescale separation ensues. Another feature existing in the terrahertz frequency regime is a kink around $\omega \simeq 6$ Thz ($f \simeq 1$ THz), which becomes more pronounced with decreasing temperature. This is the timescale of 1 ps, over which the vibrational motions are damped out. Thus, the most interesting behavior related to the short-time dynamics is found in the crossover frequency regime to the structural relaxation where temperature affects the largest changes.

6 Conclusions

In this contribution, we have discussed results from chemically realistic MD simulations of 1,4-PBD films confined by graphite walls. These simulations followed the dynamics in the polymer film over a wide temperature range approaching the glass transition of the polymer. We studied the dielectric response in different frequency windows (i.e., time windows), ranging from the THz vibrational processes, to the α-relaxation, and the desorption process of chains from the walls, which contributes to the dielectric relaxation in the first layer of adsorbed chains (a layer of thickness about the radius of gyration R_g). The high-frequency vibrational peak shows no temperature dependence in its position but a broadening with lowering temperature. In contrast, the α-relaxation peak has strongly temperature dependent location and moves by a bit more than 2 decades in frequency in the temperature window of our simulations. It also decreases in height and slightly broadens upon lowering the temperature. The temperature dependence of the α-relaxation of the confined film (between 10 and 20 nm in thickness) is compatible with the Vogel–Fulcher–Tamman behavior found for bulk polymers. We find no shift of the glass transition temperature for our film, even when we analyse the structural relaxation timescale for layers of different thicknesses close to the walls. In contrast to the α timescale, the desorption process shows an Arrhenius temperature dependence. Its energy scale is set by the strength of the Lennard-Jones interactions with the carbon atoms in the walls, and the average number of monomers per chain which are attached to the wall for a chain in the first layer. Since both do not change with temperature, there is no collective effect on the desorption process, and it shows an Arrhenius temperature dependence. While the corresponding desorption timescale is much longer than the Rouse time of the chains at high temperatures, this difference in activation behavior means that the desorption timescale becomes smaller than the structural relaxation timescale at some temperature above the glass transition of the film. Toward this temperature, the desorption process can no longer be resolved as a separate regime in the temporal correlation functions but merges with the α-process. In principle, MD simulations at low temperatures extending up to 10 μs should be performed to further confirm this latter observation.

We also discussed the question which correlations are contributing to the dielectric signal, and how this might depend on temperature and confinement effects. Experimentally, this is a relevant question as different experimental techniques are used to measure different frequency windows, and these techniques have different observables. For the terahertz regime, a measurement addressing the dipole moment of the complete box cannot be used, as this does no longer show the oscillatory behavior of the individual segments. Here, infrared spectroscopy is used. In the α-relaxation regime, parallel plate capacitors are used, which address the relaxation of the complete dipole moment (more precisely, the component of the box dipole perpendicular to the walls) and both measurements have to be normalized to each other to yield the complete experimental spectrum. We have provided here the complete spectrum covering both frequency regimes, assuming that the box dipole consists of independently relaxing segmental dipoles. Our comparison to an assumption of independently relaxing chain dipoles in the time domain shows that these would yield the same behavior in the high-frequency regime. From this combined description of the vibrational and relaxational regime, one can see how the timescale separation of these types of processes upon cooling a polymer melt leads to a deepening minimum in the dielectric loss at the terahertz scale, where there are no molecular motions providing modes in this frequency window at low temperatures. We also explained that in polymers this timescale separation is strongly influenced by the presence of dihedral rotation barriers, as the timescale to overcome them increases exponentially in $1/k_BT$ upon reducing the temperature. We argued that at the walls segmental dipole correlations within a chain are important, because the chains are oriented and develop a chain dipole moment perpendicular to the walls. The wall desorption process visible for the chain dipole correlation function should be visible in experiments on ultra-thin films; however, our example of PBD is not the best one in this respect, as the dipole moments in PBD are weak, and the strength of this process is small, which makes it hard to measure for PBD. Type B polymers of not too high molecular weight with larger dipole moment adsorbed at an apolar surface would be better candidates to observe this desorption process, as long as it still falls into experimentally accessible time or frequency windows.

Acknowledgements The authors are grateful for funding received within the Focused Research Program SPP 1369 of the German Science Foundation. A generous grant of CPU time at the Jülich Supercomputer Center (JUQUEEN and JUROPA) is gratefully acknowledged. We thank K. Binder (University of Mainz, Germany), F. Kremer (University of Leipzig, Germany) for stimulating discussions and H. Meyer (Institut Charles Sadron, France) for providing tools for the data cosmetics.

References

1. Kremer F (2002) Dielectric spectroscopy yesterday, today and tomorrow. J Non-Cryst Solids 305:1
2. Schneider U, Lunkenheimer P, Pimenov A, Brand R, Loidl A (2001) Wide range dielectric spectroscopy on glass-forming materials: an experimental overview. Ferroelectrics 249:89
3. Kaatze U (2013) Measuring the dielectric properties of materials. Ninety-year development from low-frequency techniques to broadband spectroscopy and high-frequency imaging. Meas Sci Technol 24:012005
4. Boyd RH, Smith GD (2007) Polymer dynamics and relaxation. Cambridge University Press, Cambridge
5. Kremer F (ed) (2014) Dynamics in geometrical confinement. Springer, Cham
6. Kremer F, Schönhals A (2003) Broadband dielectric spectroscopy. Springer, Berlin
7. Peter S, Napolitano S, Meyer H, Wübbenhorst M, Baschnagel J (2008) Modeling dielectric relaxation in polymer glass simulations - dynamics in the bulk and in supported polymer films. Macromolecules 41:7729
8. Bahar I, Erman B, Kremer F, Fischer EF (1992) Segmental motions of cis-polyisoprene in the bulk state: interpretation of dielectric relaxation data. Macromolecules 25:816
9. Smith GD, Borodin O, Paul W (2002) A molecular-dynamics simulation study of dielectric relaxation in a 1,4-polybutadiene melt. J Chem Phys 117:10350
10. Solar M, Paul W (2017) Chain relaxation in thin polymer films: turning a dielectric type-B polymer into a type-A one. Soft Matter 13:1646
11. Lozovoi A, Petrova L, Mattea C, Stapf S, Rößler EA, Fatkullin N (2017) On the theory of the proton dipolar-correlation effect as a method for investigation of segmental displacement in polymer melts. J Chem Phys 147:074904
12. Smith GD, Paul W (1998) A united atom force field for molecular dynamics simulations of 1,4-polybutadiene based upon quantum chemistry calculations on model molecules. J Phys Chem A 102:1200
13. Smith GD, Borodin O, Bedrow D, Paul W, Qiu X, Ediger MD (2001) ^{13}C NMR spin-lattice relaxation and conformational dynamics in a 1,4-polybutadiene melt. Macromolecules 34:5192
14. Smith GD, Paul W, Monkenbusch M, Richter D (2001) On the non-gaussianity of chain motion in unentangled polymer melts. J Chem Phys 114:4285
15. Zorn R, Mopsik FI, McKenna GB, Willner L, Richter D (1997) Dynamics of polybutadienes with different microstructures. 2. Dielectric response and comparisons with rheological behavior. J Chem Phys 107:3645
16. Priestley RD, Cangialosi D, Napolitano S (2015) On the equivalence between the thermodynamic and dynamic measurements of the glass transition in confined polymers. J Non-Cryst Solids 407:288
17. Solar M, Mapesa EU, Kremer F, Binder K, Paul W (2013) The dielectric -relaxation in polymer films: a comparison between experiments and atomistic simulations. Europhys Lett 704:66004
18. Solar M, Binder K, Paul W (2014) Dielectric relaxation of a polybutadiene melt at a crystalline graphite surface: atomistic molecular dynamics simulations. In: Kremer F (ed) Advances in dielectrics: dynamics in geometrical confinement. Springer, Cham
19. Steele WA (1973) The physical interaction of gases with crystalline solids - I, Gas-solid energies and properties of isolated adsorbed atoms. Surf Sci 36:317
20. Solar M, Binder K, Paul W (2017) Relaxation processes and glass transition of confined polymer melts: a molecular dynamics simulation of 1,4-polybutadiene between graphite walls. J Chem Phys 146:203308
21. Hess B, van der Sproel D, Lindahl E (2008) GROMACS 4: algorithms for highly efficient, load-balanced, and scalable molecular simulation. J Chem Theory Comput 4:435. see also http://www.gromacs.org
22. Colmenero J, Arbe A, Alegria A (1993) Crossover from Debye to non-Debye dynamical behavior of the α-relaxation observed by quasi-elastic neutro-scattering in a glass-forming polymer. PRL 71(16):2603–2606

23. Paul W, Smith GD (2004) Structure and dynamics of amorphous polymers: computer simulations compared to experiment and theory. Rep Prog Phys 67:1117
24. Yelash L, Virnau P, Binder K, Paul W (2010) Slow process in confined polymer melts: layer exchange dynamics at a polymer solid interface. Phys Rev E 82:050801**R**
25. Yelash L, Virnau P, Binder K, Paul W (2012) Three-step decay of time correlation at polymer-solid interfaces. Europhys Lett 98:28006
26. Solar M, Paul W (2015) Dielectric α-relaxation of 1,4-polybutadiene confined between graphite walls: numerical molecular dynamics simulations. Eur Phys J E 38:37
27. Napolitano S, Glynos E, Tito NB (2017) Glass transition of polymers in bulk, confined geometries, and near interfaces. Rep Prog Phys 80:036602
28. Lord Rayleigh FRS (1880) On the resultant of a large number of vibrations of the same pitch and of arbitrary phase. Phil Mag 10:73; On James Bernoulli's Theorem in Probatilities (1899) Phil Mag 47:246
29. Krushev S, Paul W (2003) Intramolecular caging in polybutadiene due to rotational barriers. Phys Rev E 67:021806
30. Bedrov D, Smith GD (2006) A molecular dynamics simulation study of segmental relaxation processes in miscible polymer blends. Macromolecules 39:8526

Linear Viscoelasticity of Polymers and Polymer Nanocomposites: Molecular-Dynamics Large Amplitude Oscillatory Shear and Probe Rheology Simulations

Theodoros Davris, Alexey V. Lyulin, Arlette R. C. Baljon, Victor M. Nazarychev, Igor V. Volgin, Sergey V. Larin and Sergey V. Lyulin

Abstract In this chapter, we discuss coarse-grained and atomistic molecular-dynamics simulation studies of the rheological properties of bulk polymer systems and polymer nanocomposites. Both systems contain monodispersed and non-crosslinked chain molecules. A multiscale strategy is applied to characterize the rheological behavior on different length scales of the systems structural organization. Fully atomistic simulations provide insights in rheological properties on smaller length scales than those accessible through coarse-grained simulations. Different approaches are utilized to obtain rheological moduli at these different length scales. At both levels of description, cyclic shear deformation is performed to characterize macroscopic properties of the systems before and after filler insertion. In the fully atomistic simulations of polyimide R-BAPB, passive microrheology approach is employed in addition to active rheology. To this end, a probe particle is immersed into the atomistic polymer matrix. Then, local rheological properties on the length scales at and beyond the Kuhn length are estimated. Results are compared with macroscopic rheological properties obtained by shear deformation. Additionally, the influence of the strain amplitude on the resulting rheological properties is examined. The reported coarse-grained simulations show a strong decrease of the nanocomposites storage modulus with increasing strain amplitude,

T. Davris · A. V. Lyulin (✉)
Theory of Polymers and Soft Matter, Department of Applied Physics,
Technische Universiteit Eindhoven, 5600 MB Eindhoven, The Netherlands
e-mail: a.v.lyulin@tue.nl

A. R. C. Baljon
Department of Physics, San Diego State University, San Diego, CA 92128, USA

V. M. Nazarychev · I. V. Volgin · S. V. Larin · S. V. Lyulin
Institute of Macromolecular Compounds,
Russian Academy of Sciences, Bolshoy pr. 31, 199004 St. Petersburg, Russia

F. Kremer and A. Loidl (eds.), *The Scaling of Relaxation Processes*,
Advances in Dielectrics, https://doi.org/10.1007/978-3-319-72706-6_12

which is accompanied by a maximum in the loss modulus (the so-called Payne effect); the onset of the softening is observed in the linear regime of deformation at strain amplitude of about 0.01. Moreover, the dependence of the storage modulus on the instantaneous strain exhibits both softening and hardening regimes, in agreement with recently reported [22] Large Amplitude Oscillatory Shear (LAOS) experiments. The simulations suggest that the observed hardening is caused by the shear-induced decrease of the non-affine diffusion of the polymer segments due to filler particles acting as effective crosslinks between polymeric chains and, hence, hindering diffusion. Moreover, the formation of "glassy" immobile layers at the nanoparticle interface strongly increases the storage modulus at low strain amplitudes. The strain softening with increasing strain amplitude is connected to the mobilization of these glassy layers and an increase in the dynamic heterogeneity of the polymer matrix. A breakup of the network structure plays a role as well.

1 Introduction

The application of polymer-based composite materials has grown significantly worldwide in areas as diverse as automotive industry, rail and naval transport, construction, infrastructure, defense, aerospace, medicine, health care, electronics, and telecommunication [1, 2]. This growth has been driven by innovative new techniques to synthesize polymers. Most importantly, by adding inorganic fillers light-weight and high-strength materials are obtained [3]. The market of filled plastics could easily grow further, provided one can remediate the lack of understanding of the correlation between the microscopic structure and macroscopic properties of particle-filled polymer matrices, in particular under dynamic conditions. Such understanding can result in unique, new composite materials, as well as improved performance of existing materials.

Polymer nanocomposites—that is composites in which the filler particles are of nanosize—are just at their infancy of development. Given that, everything else being equal, the effectiveness of a filler increases with its surface-to-volume ratio [4–6], replacing micro- with nanosized fillers in polymer-based composites is an important branch of the emerging field of nanotechnology [7]. Rheological experiments [8] show that the mechanical reinforcement upon the addition of filler particles is primarily controlled by the inter-particle interactions, mediated to some extent by the polymer matrix. Atomistic-level interactions involving nanoparticles and polymer segments dictate macroscopic properties. Hence, it is necessary to understand how *molecular-level* interactions correlate with spatial structuring of filler nanoparticles and the polymer matrix at the *mesoscale* and, from there, how structure and dynamics at the mesoscale give rise to macroscopic properties. Such understanding is needed for fine-tuning of the molecular-level interactions and the overall composition of the material in order to obtain the desired macroscopic response. A wide range of length and time scales must be addressed in such investigations.

A key to resolving this puzzle is the molecular origin of the so-called Payne effect [9]: a strong drop in the storage modulus upon increasing the strain amplitude of the applied oscillatory shear deformation. What is the molecular origin of the change in the polymeric nanocomposite that causes such a drastic drop and, hence, the loss of the desired mechanical reinforcement? Answering this question is crucial to control the performance of polymer materials by adding rationally designed nanostructured fillers [10]. The way to do this is through computer simulations.

Several groups have already investigated the molecular mechanisms underlying the mechanical reinforcement of polymer nanocomposites. Hagita et al. [11], Chen et al. [12], and the Liu and Lyulin groups [13–18] investigated the role of direct particle–particle interactions and that of the segmental orientation at the particle—matrix interface. Still, the dominant molecular mechanism that causes the reinforcement or loss of strength remains elusive. Kutvonen et al. [19, 20] have argued that nanoparticles serve as crosslinks in the formation of temporary polymeric networks by attachment to multiple polymeric segments. By varying the polymer–nanocomposite interaction strength, they found that the reinforcement is inversely correlated with the mobility of the filler particles with respect to the polymer segments. In all of the above studies more or less spherical nanoparticles were employed; hence, additional studies with more realistic filler shapes are needed to verify the general validity of the conclusions [21].

Experiments that impose Large Amplitude Oscillatory Shear (LAOS) on polymer nanocomposites have been performed recently by Mermet-Guyennet et al. [22]. They show an intriguing apparent contradiction: Although the moduli exhibited strain softening (Payne effect [9]), within a single oscillatory cycle strain hardening is observed. Mermet-Guyennet et al. explain this apparent contradiction by noting that the apparent strain hardening is due to an increase in the slope of the elastic component of the stress as a function of strain curve. Hence, it is a local effect, and the overall rheology remains strain softening. Softening is a consequence of the molecular rearrangements due to the applied stress amplitude. The deformation rate is the *dominant time scale* responsible for the systems relaxation, not an intrinsic time scale. This observation is in agreement with the observed strain-rate frequency superposition rheology results of Wyss et al. [23] for micellar systems. This superposition—that is, the fact that relaxation processes are dictated by strain rate rather than strain amplitude or frequency—was observed in simulations of temporary crosslinked polymeric networks by Wilson et al. [24, 25]. A drop in the elastic modulus upon increasing the shear amplitude is observed as well and attributed to rearrangement in the polymeric network structure, as a sparser network forms. These networks resemble those of nanocomposites in which nanoparticles act as temporary crosslinks. However, since they do not contain the actual nanoparticles that (as we will show below) affect the dynamics of the polymeric matrix, they lack the molecular detail needed to explain everything contributing to the Payne effect.

In this chapter, we examine polymeric nanocomposites by means of molecular-dynamics computer simulations at a coarse-grained, as well as at an atomistic level. The coarse-grained model probes only length scales larger than the typical Kuhn-length values of polymer chains (in general approximately 1 nm or larger) [26].

Despite the lack of the molecular detail, coarse-grained molecular-dynamics simulations offer good opportunities to explore macroscopic, large-scale properties of the polymer nanocomposites. In addition, given that the chemical specificity of the polymer matrix is responsible for the resulting macroscopic effects, atomistic-level simulations are performed to investigate this smaller length scale.

At the coarse-grained and fully atomistic level, moduli are obtained from strain sweep experiments. They show strain softening—that is, a decrease in the elastic modulus upon increasing the amplitude (Payne effect). On the other hand, a Lissajous analysis—i.e., the stress–strain dependence—of a single oscillation cycle indicates that the material strain hardens. To explain this seeming contradiction between a rheological and a Lissajous analysis in the nonlinear regime, we compare different methods for calculating rheological properties. Both filled and unfilled polymer melts are studied and compared. Moreover, we examine how interactions (attractive or repulsive) among the nanoparticles affect the observed Payne effect. Since we are interested in possible polymer segmental relaxation contribution to strain softening in addition to network remodeling, we study how molecular rearrangements lead to the Payne effect by inspecting segmental motion in the polymer matrix during applied deformation.

For the fully atomistic nanocomposite model, we use a passive microrheological simulational approach to extract local rheological polymer melt properties. These results are compared with those obtained from simulations of the same atomistic system subjected to the active shear deformation (see Sect. 5 for more details).

The remaining part of this chapter is organized as follows. In Sect. 2 the models and the simulation methods are described. In Sect. 3 the results of the shear deformation modeling and the simulated mechanical moduli are summarized. The observed strain hardening is discussed in Sect. 4, followed by the results of the fully atomistic simulations presented in Sect. 5. The Payne effect is discussed in Sect. 6, and our conclusions follow in Sect. 7.

2 Models and Methods

A multiscale strategy is implemented in this study. The polymer matrix of the simulated systems consists of linear, coarse-grained or atomistic, non-crosslinked and non-entangled polymer chains. In all cases, the filler particles are inserted explicitly.

At the coarse-grained level, the polymer matrix contains 100 bead-spring chains with 50 monomers each. The nanoparticles (NPs) consist of 256 Lennard-Jones (LJ) beads placed randomly in a sphere of radius 4σ, which is comparable to the gyration radius of the polymer chains (here σ is the characteristic scale of the Lennard-Jones interactions, see below, Eq. 1). A filler volume fraction was fixed at $\varphi_f = 28.5\%$. To create the nanocomposite system, the initial simulation box containing the equilibrated unfilled polymer melt is expanded until the density is low enough for the NPs not to overlap with the polymer chains after their placement in the simulation box, as shown in Fig. 1. For the fully atomistic level, bulk samples of thermoplastic

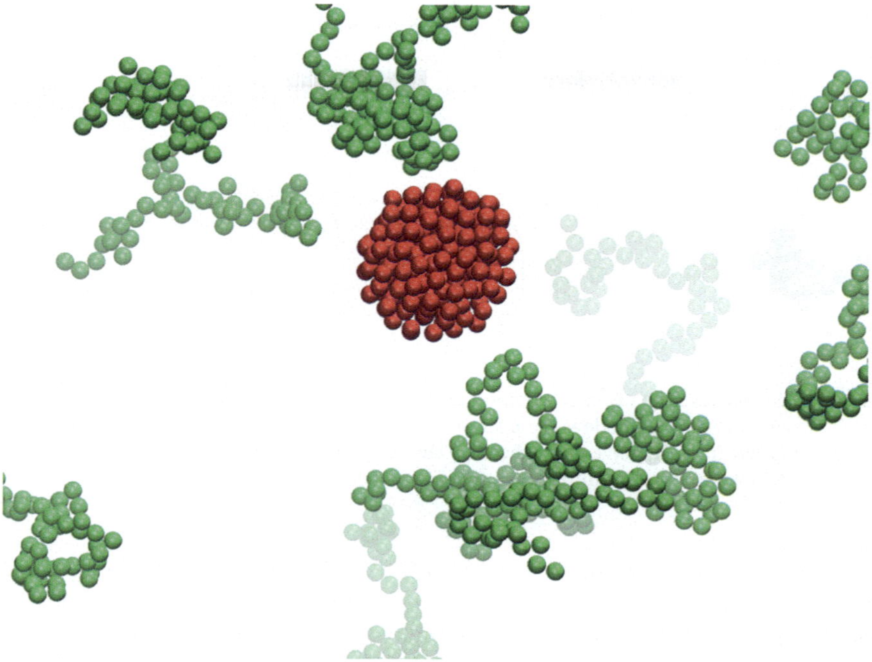

Fig. 1 A snapshot from a simulation of the coarse-grained nanocomposite model. The filler beads are shown in red, and the polymer chains in green. For clarity, only some of the polymer chains and the nanoparticles are shown. The depicted nanoparticle has an average radius of about four times the diameter of the monomer beads

polyimide R-BAPB filled with the fullerene nanoparticles are modeled, as shown in Fig. 2.

All the MD simulations are performed with periodic boundary conditions implemented in all three dimensions of the simulation box. The LAMMPS software package [27] is used for coarse-grained simulations and the Gromacs [28, 29] package for the fully atomistic simulations. All equilibration and production runs are performed above the glass-transition temperature of the system at hand. Below the simulation methods are discussed in more detail.

Fully atomistic model A model of the thermoplastic polyimide R-BAPB based on 1,3-bis-(3′,4-dicarboxyphenoxy)benzene (dianhydride R) and 4,4′-bis-(4″-aminophenoxy)diphenyl (BAPB diamine) is employed, as shown in Fig. 2. All interactions are parameterized within the Gromos53a5 force-field [30, 31], which has proved its effectiveness for investigating the structural, thermal, and mechanical properties of the aromatic polyimides. In the simulations at hand, partial charges have not been taken into account. The chain length varies from one to eight monomers.

The initial configuration of the bulk polymer system is obtained in several steps [32–52]. First, the initial box containing randomly placed partly coiled 27 polyimide chains is compressed for 26 ns. To eliminate the influence of this fast compression,

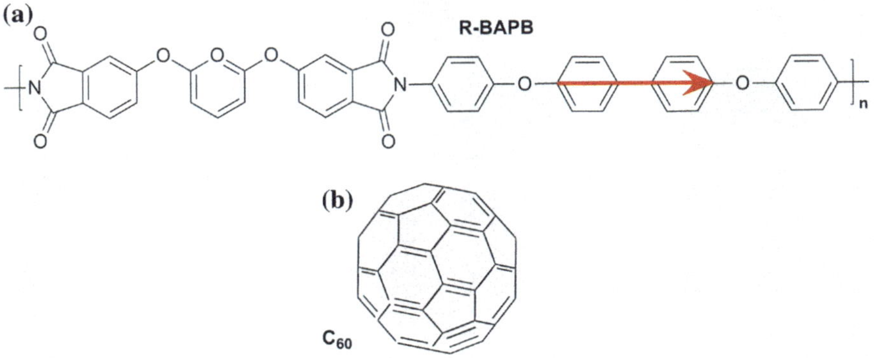

Fig. 2 Chemical structure of the **a** repeating unit of the R-BAPB polyimide and **b** C_{60} fullerene which was used as probe particle in the microrheology simulations. For more details, we refer the reader to the paragraph "**Passive microrheology simulations**" in this section. Red arrow marks the polymer chain fragment which diffusive motion was investigated in Sect. 4

cyclic thermal annealing from $T = 600$ K to $T = 290$ K is performed, followed by the 2-μs equilibration run. During these long simulations, the average size of polymer chains (gyration radius) reaches its equilibrium value, which is found to be in a good agreement with theoretical estimations.

After the equilibration run, an additional 1-μs long production run at $T = 600$ K is performed. To obtain independent starting configurations for the oscillatory shear simulations, we choose five configurations separated by 100 ns from this trajectory. During all the atomistic simulations, a Berendsen thermostat and barostat are used with time constants $\tau_T = 0.1$ ps and $\tau_P = 0.5$ ps, respectively.

Coarse-grained model In the coarse-grained simulations, Newton's equations of motion are integrated using the velocity-Verlet algorithm with a time step of $\delta t = 0.005\tau$. The interactions between the non-bonded beads are modeled with a modified LJ 12-6 potential,

$$U_{nb}(r) = 4\varepsilon\left[\left(\frac{\sigma}{r}\right)^{12} - \left(\frac{\sigma}{r}\right)^{6}\right] + S_f(r); \quad r < r_{cut}^{out} \tag{1}$$

Here $S_f(r)$ is a switching function that smoothly ramps the energy to zero from $r_{cut}^{in} = 2.5\sigma$ to $r_{cut}^{out} = 3.5\sigma$ [53–55]. The values of the LJ parameters σ and ε are chosen in accordance with previous studies of similar model systems [53–55]. The Lennard-Jones units of measurement are used throughout the text; i.e., m is the unit of mass, ε is the unit of energy, σ is the unit of length, and $\tau = \sigma\sqrt{m/\varepsilon}$ is the unit of time. The LJ parameters for the monomer beads were $\sigma_{\mathrm{m}} = \sigma$, $\varepsilon_{mm} = \varepsilon$, $m_{\mathrm{m}} = m$. The diameter of the filler beads comprising the NPs is equal to the diameter of the monomer beads, $\sigma_{\mathrm{f}} = \sigma_{\mathrm{NP}} = \sigma_{\mathrm{m}}$. For filler beads belonging to the same NP, $\varepsilon_{ff} = 50\varepsilon$. This setting will keep the beads of a nanoparticle together throughout the simulation. For filler beads that belong to different NPs, $\varepsilon_{\text{NP-NP}} = \varepsilon$. In most simulations, both the repulsive and attractive parts of the LJ potential are included in the interactions as indicated by Eq. 1. However, to study the effect of NP-NP interactions on the Payne effect, a purely repulsive NP-NP interaction has, in Sect. 6, been studied too. Finally, the interaction strength among filler beads and monomer beads was set as $\varepsilon_{fm} = 2\varepsilon$. This interaction is needed to avoid the aggregation of NPs, and all the parameter settings are summarized in Table 1.

Covalently bonded beads interact through a combination of an attractive Finite-Extensible-Nonlinear-Elastic (FENE) potential, and a repulsive and truncated LJ 12-6 potential, [53–55]

$$U_{bn}(r) = -0.5k_{FENE}R_{max}^2 \ln\left[1 - \left(\frac{r}{R_{max}}\right)^2\right] + 4\varepsilon\left[\left(\frac{\sigma}{r}\right)^{12} - \left(\frac{\sigma}{r}\right)^6\right] + \varepsilon. \quad (2)$$

The parameters k_{FENE} and R_{max} denote the stiffness and the maximum elongation of the spring, respectively. The stiffness and maximum elongation of the bonded potential are set equal to $k_{\text{FENE}} = 30\varepsilon/\sigma^2$ and $R_{\text{max}} = 1.5\sigma$, respectively. The LJ parameters ε and σ have the same values with their non-bonded counterparts (this particular choice of bonded and non-bonded interactions has been shown to prevent unphysical chain crossings [53–55]).

We allow the systems to equilibrate under NPT conditions at $T = 0.6\varepsilon/k_B$ and $P = 0\varepsilon/\sigma^3$. At these settings, the system is above the glass-transition temperature [56]. During both the equilibration and production runs, we use values recommended in the LAMMPS manual for the thermostat and barostat parameters; i.e., the temperature and the pressure are allowed to relax to the specified average value over a time interval of 0.5τ and 5.0τ, respectively [27].

Cyclic shear deformation The SLLOD equations of motion [18] are used for the coarse-grained and fully atomistic simulations. For each value of strain amplitude γ_{max} data are obtained from averages over 10 oscillatory cycles. For the fully atomistic simulations, the Gromacs routine *deform* was used. The value of the strain amplitude γ_{max} is varied between 0.01 and 0.45 in the coarse-grained and between 0.09 and 0.31 in the atomistic simulations. For the highest strain amplitudes, all the simulated systems are in the nonlinear regime. The constant frequency $\omega = 0.0063\tau^{-1}$ was used in the coarse-grained simulations and that of the fully atomistic simulations

Table 1 Lennard-Jones energy parameters among the various types of beads comprising the simulated coarse-grained models

Energy	Description	Value
ε_{mm}	Monomer–monomer	1
ε_{ff}	Filler–filler (same NP)	50
$\varepsilon_{NP\text{-}NP}$	Filler–filler (distinct NP)	1
ε_{fm}	Filler–monomer	2

All potentials are cutoff at 3.5σ, except the one between NPs, which is sometimes purely repulsive was varied in the range from ~10^8 to ~10^{11} Hz. At these fast deformation rates, the non-entangled and non-crosslinked systems display rubber-like properties. Therefore, our conclusions concerning the observed Payne effect and strain hardening should also apply to the rubbers and elastomer-based nanocomposites. Note that the deformation is sinusoidal in the coarse-grained simulations, whereas in the atomistic ones it follows a symmetric sawtooth wave pattern (for more details see Sect. 3).

Passive microrheology simulations The methodology described above enables us to measure macroscopic rheological characteristics and mechanical moduli. However, the rheological behavior on smaller length scales—of the order of a Kuhn segment—has to be addressed using a different approach, namely passive microrheology. The main idea of this method is that rheological properties can be obtained by tracking the thermal diffusion of a probe particle. The complex shear modulus G^* can be estimated from the generalized Stokes–Einstein relation (GSER) [57],

$$G^*(\omega) = \frac{k_B T}{\pi \cdot R_{particle} \cdot i\omega \cdot F\{\Delta r^2(\Delta t)\}}. \quad (3)$$

Here $\Delta r^2(\Delta t)$ is the mean-squared displacement (MSD) of the center-of-mass of the probe particle, $R_{particle}$ is the radius of the probe particle, T is the temperature, and $F\{\}$ stands for the Fourier transform.

Compared to cyclic deformation, the microrheology method has as a unique advantage due to its mobility to determine rheological properties in a wide frequency range without having to apply an oscillatory shear, which can be problematic at high temperatures. Hence, microrheology has been used extensively to study complex polymer systems [58–64], such as biological polymers [58, 59], polymer networks [60, 61], and hydrogels [62]. In addition, microrheology has been applied in computer simulations to characterize rheological properties of the polymer melts [65–69] and polymer–liquid interfaces [70, 71]. Most of these studies employed coarse-grained models, given that they aimed at answering fundamental questions and were used to validate the microrheological approach.

Below we report on the use of the passive microrheology approach to obtain moduli from simulations of an R-BAPB polymer melt. A C_{60} fullerene is used as the probe particle. The fullerene is modeled on a fully atomic level and parameterized with the same force-field as used for the polymer. In order to insert a fullerene particle after equilibration of the polymer, a spherical void is carved out in the center of the simulation box and increased until it reaches the size of the C_{60} nanoparticle, which fills 0.5% of the systems' volume. After that, the system is equilibrated during a 100-ns run at a range of temperatures between 500 and 700 K. This is followed by a production run of 500 ns, during which the mean-squared center-of-mass displacement of the fullerene is monitored.

In a recent study [48], we have shown that C_{60} displays the sub-diffusive behavior at intermediate time scales, due to the coupling between the nanoparticle and polymer segments, which also move sub-diffusively. Moreover, the comprehensive analysis of the fullerene diffusive behavior presented in [48] has proven that its sub-diffusive motion is an indication of the polymer local viscoelasticity.

This passive microrheology approach allows us to study the effects underlying reinforcement in the nanocomposite systems at small length scales. They complement coarse-grained and fully atomistic simulation studies in which moduli are obtained from cyclic shear deformation that probe larger length scales. A quantitative link between the coarse-grained and atomistic results will be established.

3 Computing Moduli Using LAOS Simulations

Polymer elastomer-based nanocomposites exhibit *strain softening* behavior among subsequent oscillation cycles of increasing strain amplitude; i.e., the modulus decreases with increasing strain amplitude [9, 72]. This is the so-called Payne effect, first discovered in the 1960s. However, during a single oscillation cycle of constant strain amplitude, the same materials exhibit *strain hardening* at high values of strain. This seeming discrepancy is exasperated by the fact that in the nonlinear regime of stress response—where the Payne effect is observed—the definition of the moduli is not unique. The conclusion of recent experiments [22] aimed at elucidating this seemingly paradoxical strain softening/hardening behavior, is that the strain hardening observed in polymer nanocomposites is a local effect; that it depends on the employed method of calculating the moduli.

Here we compare two methods of extracting the moduli of our simulated systems. The first method is commonly used in experiments of viscoelastic materials. Moduli are obtained from the instantaneous stress response of the material using Eqs. 4 or 5 below. The second method is based on the analysis of stress–strain plots using Eq. 9. We now discuss both methods in more detail.

The components of the complex modulus as calculated by a rheometer [22]—as well as by typical MD simulations [13–17]—can be extracted by fitting the instantaneous stress response using Chebyshev polynomials of the first kind, T_{n}, as

$$\sigma(t) = \gamma_{max} \sum_{n,odd} \left[G'_n \cdot T_n(\sin(n\omega t)) + G''_n \cdot T_n(\cos(n\omega t))\right]. \tag{4}$$

From this equation, the storage modulus $G' = G'_1$ and loss modulus $G'' = G''_1$ can be derived.

To evaluate the elastic and storage moduli in the atomistic simulations (note that in atomistic simulations the applied deformation changes linearly with time due to its sawtooth nature), the strain–time dependence is fitted using the following relation:

$$\sigma(t) = \sum_{n,odd}^{11} \gamma_{\max}^{n} \left[G'_n \sin(n\omega t) + G''_n \cos(n\omega t)\right], \tag{5}$$

where corresponding γ^n_{max} for each harmonic is estimated from a fit to:

$$\gamma(t) = \sum_{n,odd}^{11} \gamma^n_{\max} \sin(n\omega t). \quad (6)$$

This fitting procedure is illustrated in Fig. 3, which shows the instantaneous stress response at different values of γ_{max}. Panels (a) and (b) show results obtained from coarse-grained simulations and (c) and (d) do the same for fully atomistic ones. The observed difference between the stress–time dependencies in the coarse-grained and atomistic systems is attributed to the specifics of the shear deformation procedure in the Gromacs and LAMMPS package. As was mentioned earlier, in the atomistic simulations the strain was varied as a sawtooth wave (linearly with time), whereas in the coarse-grained simulations the strain was varied sinusoidal.

Data in Fig. 3 are averaged over 10 oscillatory cycles. We note that the stress response is almost identical during all cycles. Hence, the simulations do not show an irreversibly higher elastic modulus in the first cycle than in the subsequent ones (the so-called *Mullins effect)*. Nevertheless, the *Payne effect* is observed in the coarse-

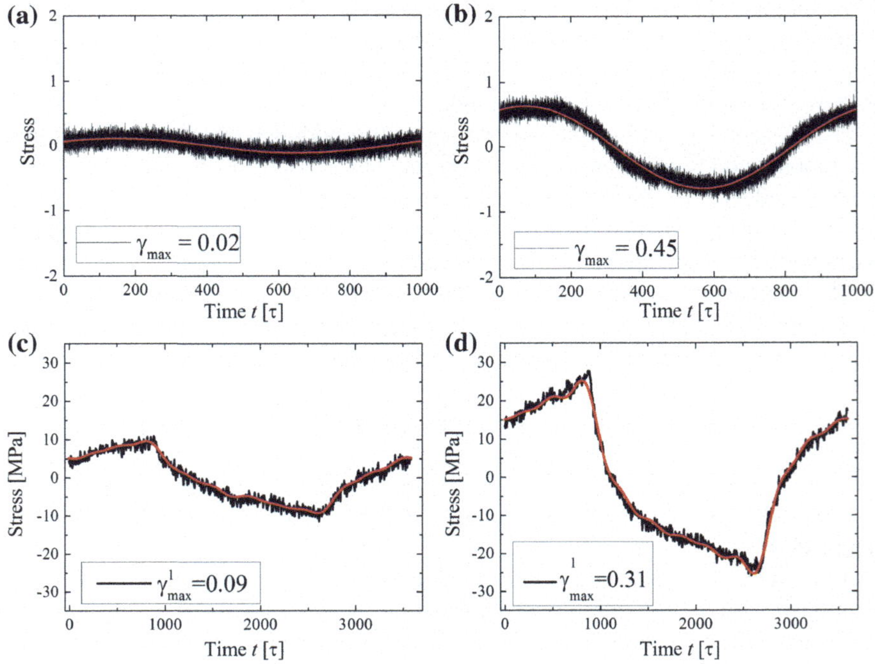

Fig. 3 Time dependence of the instantaneous stress $\sigma(t)$ at $\gamma_{\max} = 0.02$ (**a**) and 0.45 (**b**). The red lines are the fitting curves based on Eq. 4 using $n = 1, 3$. The time dependence of the instantaneous stress for the R-BAPB atomistic system calculated with different strain amplitudes at $T = 530$ K, for $\gamma^1_{\max} = 0.09$ (**c**) and $\gamma^1_{\max} = 0.31$ (**d**). The red lines are the fitting curves based on Eq. 5

grained system, as will be discussed in detail in Sect. 6, regardless of the fact that irreversible changes—such as covalent-bond rupture—do not take place.

When fitting the coarse-grained data shown in Fig. 3, only the first two odd harmonics are employed. For the coarse-grained model, the fit is clearly better for $\gamma_{\max} = 0.02$ (a) than for $\gamma_{\max} = 0.45$ (b). A better fit would have been obtained by including harmonic contributions of higher order, $n > 3$. Nevertheless, the inclusion of just the $n = 3$ harmonics yields paradoxical results, as can be seen in Fig. 4. One would expect the third harmonic to be close to zero at small strain amplitudes and to increase with higher amplitude, but observed is the opposite. For this reason, we use a second alternate method to calculate the moduli, as explained below.

The first method described above follows the approach used by a typical rheometer's software. A second method is based on the analysis of stress–strain (Lissajous) curves. In a nonlinear system, like the one we investigate, moduli do not have a unique definition [73]. In fact, two distinct elastic moduli can be defined as G' and G'_M, given in Eqs. 7 and 8 below,

$$G' = \frac{\omega}{\pi \gamma_{\max}^2} \oint \sigma(t)\gamma(t)dt. \tag{7}$$

$$G'_M = \left(\frac{d\sigma}{d\gamma}\right)_{\gamma=0} = G' - 3G'_3 \tag{8}$$

Equation 7 is used in the first approach, whereas Eq. 8 gives the elastic modulus as the slope of the stress–strain curve at zero strain [22], which can be obtained from a Lissajous plot. From Eqs. 4 and 8 a relation between the elastic stress σ_e and strain γ can be derived that is valid for infinitesimal strain:

$$\sigma_e(\gamma) = G'_M\gamma + 4G'_3\frac{\gamma^3}{\gamma_{max}^2} \tag{9}$$

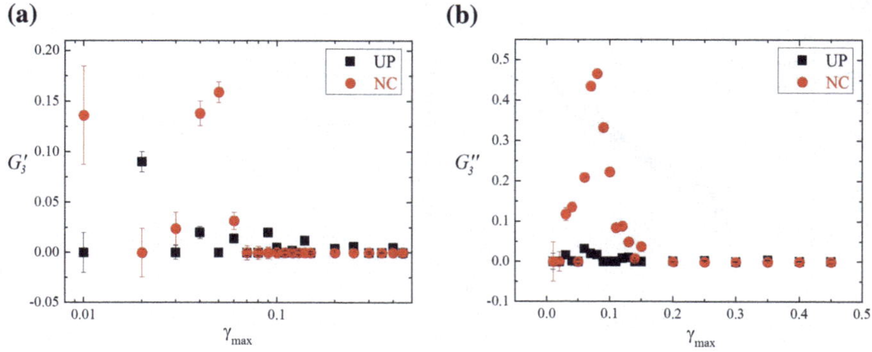

Fig. 4 The (higher) harmonic elastic G'_3 (**a**) and viscous G''_3 (**b**) contributions to the stress versus the strain amplitude. The black and red data points correspond to the unfilled coarse-grained polymer (UP) and the coarse-grained nanocomposite (NC), respectively

In Eq. 9 both γ and σ_e depend on time. The elastic component of the stress $\sigma_e(t)$ and viscous component $\sigma_\nu(t)$ are defined by the following equation

$$\sigma(t) = \frac{\sigma(t) - \sigma(-t)}{2} + \frac{\sigma(t) + \sigma(-t)}{2} = \sigma_e(t) + \sigma_v(t) \tag{10}$$

The separation of the total stress in elastic and viscous components defined by Eq. 10 is a generalization [55] of the well-known relation $\sigma(t) = G'\gamma(t) + G''\dot{\gamma}(t)/\omega = \sigma_e(t) + \sigma_v(t)$, which is only valid in the linear regime.

These (elastic) stress–strain curves are shown in Fig. 5 for both coarse-grained nanocomposites and unfilled fully atomistic systems. The stress response at each strain amplitude is again averaged over the ten oscillation cycles. G'_M and G'_3 can now be obtained from fits of Eq. 9 to these stress–strain curves at zero strain [22].

In Fig. 5 strain softening, that is, a decrease in the slope G'_M with increasing strain amplitude, is clearly observed in all cases. Moreover, strain hardening, or a slight upturn of the curves is observed in all cases at the higher strain amplitudes. We did observe strain hardening in both unfilled atomistic and unfilled coarse-grained

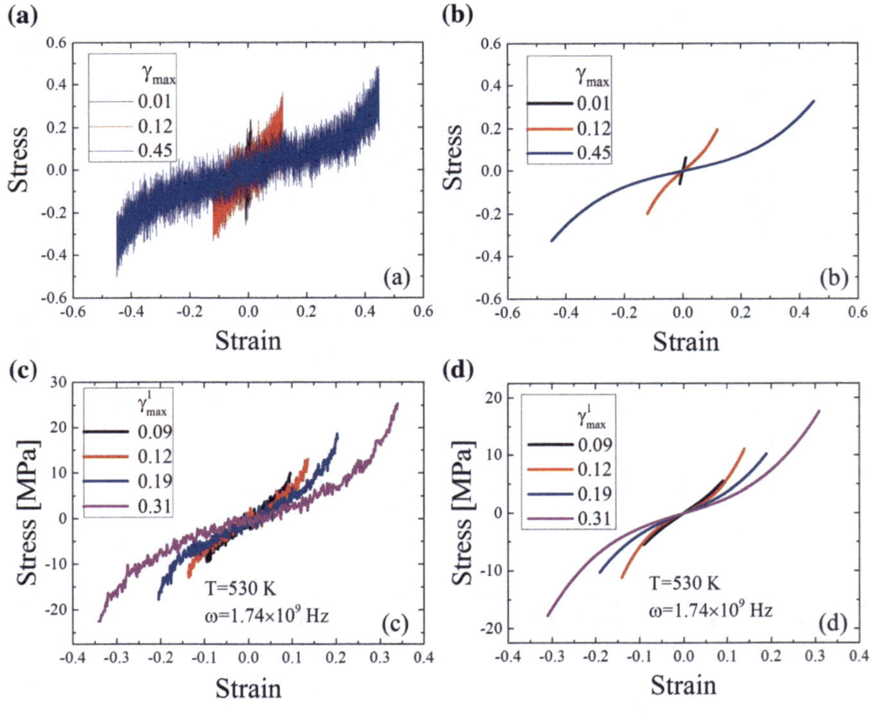

Fig. 5 Elastic stress σ_e—instant strain dependencies for different strain amplitudes $\gamma_{\max}$, for the coarse-grained nanocomposite (**a**, **b**). **a** Raw simulation data; **b** simulation data fitted to Eq. 9; Elastic stress σ_e—instant strain dependences for different strain amplitudes $\gamma^1_{\max}$ for the unfilled fully atomistic polymer melt (**c**, **d**). **c** Raw simulation data; **d** simulation data from (**c**) fitted to Eq. 9

systems (not shown). However, the hardening is weaker in the coarse-grained than in the atomistic system, most likely since the latter is closer to the corresponding glass transition. We conclude that strain hardening is an intrinsic property of the polymer matrix, enhanced by the presence of nanoparticles.

Figure 6 shows the elastic modulus G'_M and the third-order elastic harmonic G'_3 obtained from fits to Eq. 9. The moduli G' obtained using the first method are added for comparison.

In Fig. 6, $G' \approx G'_M$ for the coarse-grained simulations (a), but their values differ significantly in the atomistic ones (c). As expected, the elastic moduli decrease with higher strain amplitude in all cases. Panels (b) and (d) compare the value of the third-order elastic harmonics obtained using both approaches (Eqs. 4 and 5 are used for the first, Eq. 9 for the second). Equation 9 gives the expected physically consistent result; i.e., G'_3 is equal to zero in the linear regime of stress response and increases under strain-hardening conditions.

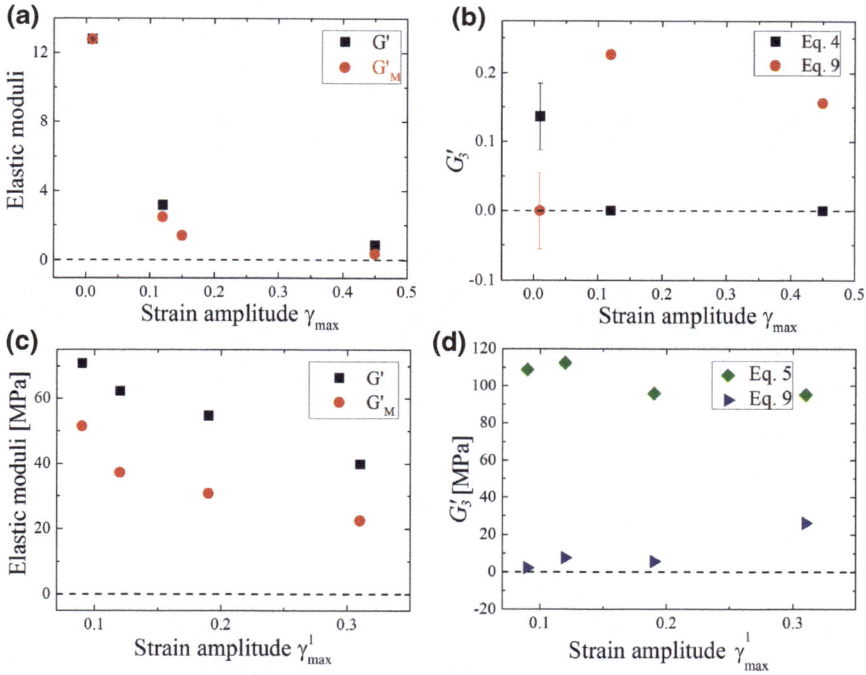

Fig. 6 **a** Elastic moduli G' and G'_M and **b** third-order elastic harmonic G'_3 versus the strain amplitude, for the coarse-grained nanocomposite model with $\varepsilon_{NP} = 1$. The elastic modulus G' was calculated by fitting the simulation data with Eq. 4. The G'_M was obtained by fitting the data with Eq. 8. Equation 4 gives seemingly unphysical results for the third-order elastic harmonics. The problem was solved when the moduli were extracted by fitting data in Fig. 5b, d with Eq. 9. **c** The dependence of the elastic moduli on the strain amplitude for the unfilled atomistic system at $T = 530$ K calculated using Eqs. 4 and 9. **d** Third-order elastic harmonics G'_3 dependence on the strain amplitude calculated using Eqs. 5 and 9

4 Microscopic Mechanisms of the Strain Hardening

In the previous section, it was concluded that strain hardening is an intrinsic property of the polymer matrix, further enhanced by the presence of nanoparticles. In this section, it will be further discussed why this is the case. Figure 7 shows normalized stress–strain curves for both the coarse-grained nanocomposite and the coarse-grained unfilled polymer. The elastic stress is normalized by the stress amplitude, $\sigma_{\max}$, the strain by the strain amplitude, $\gamma_{\max}$.

At $0.01 < \gamma_{\max} < 0.20$, both the composite and the unfilled matrix exhibit strain hardening, but this is more pronounced in the nanocomposite than in the unfilled polymer. Hence, we conclude that the strain hardening produced in the simulations is an intrinsic property of the matrix, greatly enhanced by the presence of the NPs. However, the unfilled polymer surpassed its yield point for $\gamma_{\max} > 0.20$, whereupon it exhibits liquid behavior (i.e., the stress is unaffected by the strain, the horizontal line in Fig. 7b).

In order to elucidate the microscopic causes of the strain hardening, we calculate the non-affine displacement (NAD) of the monomer beads. The NAD is obtained by subtracting the displacement of each bead, due to the convective velocities imposed on them by the shear deformation of the simulation box, from the observed total mean-squared displacement. This can be expressed as follows

$$g_0^a(\gamma) = \left\langle \left| \tilde{\boldsymbol{r}}_j(\gamma(t)) - \tilde{\boldsymbol{r}}_j(\gamma(t=0)=0) \right|^2 \right\rangle \tag{11}$$

where $\tilde{\boldsymbol{r}}_j(t)$ is the coordinate of bead j at time t obtained after subtraction. $\tilde{\boldsymbol{r}}_j(0)$ is the coordinate of the same bead at the starting time of the deformation; a denotes the constituent; that is to say, $a = m$ for the monomer segments of the matrix, $a = P$ for the center-of-mass (COM) of the polymer chains, and $a = f$ for the nanoparticles. The averaging in Eq. 11 is performed over different monomers, not over different time intervals.

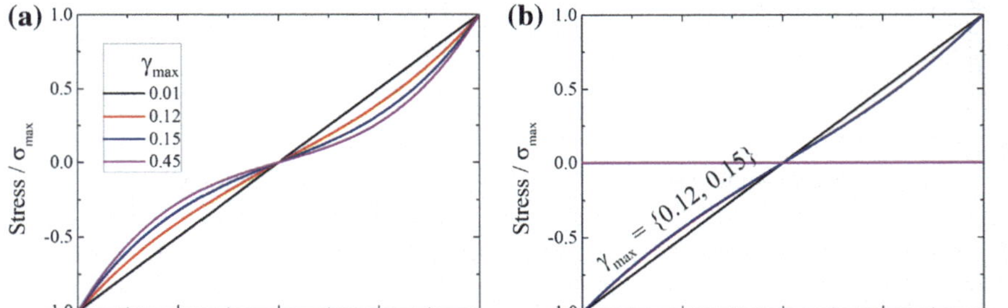

Fig. 7 Elastic stress σ_e of the coarse-grained simulations, normalized by the stress amplitude $\sigma_{\max}$ versus the normalized strain for **a** the nanocomposite and **b** the unfilled polymer matrix. Each curve refers to a different value of the strain amplitude $\gamma_{\max}$

Since the imposed shear deformation at time t is the same as that at $t + T/2$ (T is the period of the cyclic shear), thus the corresponding averaging was performed, $\langle g_0^m(\gamma)\rangle = \frac{1}{2}\left(g_0^m(\gamma(t)) + g_0^m\left(\gamma\left(t + \frac{T}{2}\right)\right)\right)$. In Fig. 8 the average NAD $\langle g_0^m\rangle$ of the monomer beads is depicted at $0.01 \leq \gamma_{max} \leq 0.45$ for the nanocomposite (a) and for the unfilled polymer (b).

In the unfilled polymer (b), the NAD of the monomer beads is hardly affected by either the imposed strain or the strain amplitude. This fact seems to correspond well with the constant elastic modulus as a function of strain and strain amplitude observed experimentally. On the other hand, nanocomposites show a strong dependence of NAD on strain and strain amplitude, as can be seen in Fig. 8a. For a given strain amplitude, the NAD of the monomer beads is higher during the eccentric (from $\gamma = 0$ to $\gamma = \gamma_{\max}$) than during the concentric (from $\gamma = \gamma_{\max}$ to $\gamma = 0$) contraction of the nanocomposite. This is most clearly seen for $\gamma_{\max} = 0.45$: As the strain increases, the NAD of the monomer beads increases as well until the strain attains a value close to 0.15. Increasing the strain to higher values leads to the decrease of the NAD of the monomer beads. This is most prominent at the highest strain amplitude, $\gamma_{\max} = 0.45$. This behavior was also observed for the non-affine displacements g_0^f of the NPs and for the NAD displacements g_0^P of the center-of-mass (COM) of the polymer chains, the only difference being a smaller displacement (not shown).

The NAD results indicate that in a nanocomposite an increase in strain amplitude results in a higher non-affine mobility of the monomer beads near zero strain, which corresponds to the observed strain softening (i.e., the Payne effect). On the other hand, at $\gamma = \gamma_{max}$, the NAD of the monomer beads is independent of the strain-amplitude value. As a consequence, the monomer-bead mobility at $\gamma = \gamma_{max}$ is similar for all values of the strain amplitude. However, the higher the value of γ_{max}, the more abrupt is the drop in the NAD of the monomer beads near γ_{max}. The decrease of the NAD and, hence, the mobility of the beads with increasing strain—most likely—causes the observed strain hardening.

Next we investigate the radius of gyration of the coarse-grained polymer chains, which is shown in Fig. 9. During the equilibrium simulations, before the application

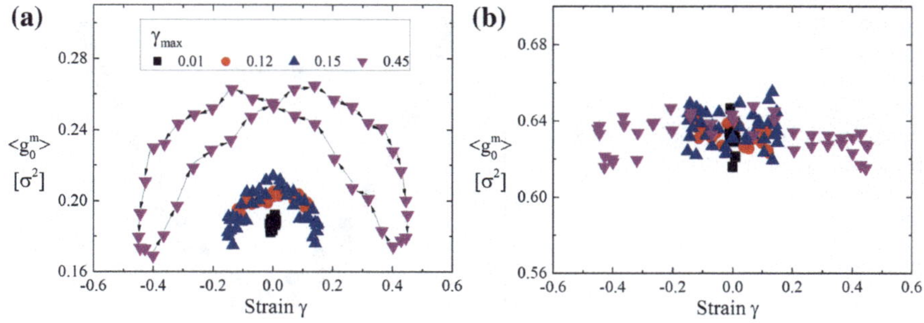

Fig. 8 Averaged non-affine displacements (NAD) $\langle g_0^m\rangle$ of the monomer beads for the nanocomposite (**a**) and for the unfilled polymer (**b**). The dependence of the NAD on strain is depicted for four different values $\gamma_{\max}$ of the strain amplitude

of shear, the average radius of gyration of the polymer chains (not shown) was approximately 3.5–4σ. For the chains in the nanocomposites the radius of gyration is slightly higher, though. This difference, however small, might imply that the NPs act as (weak) crosslink junctions that inhibit to a small degree the stretching of polymer chains. During oscillatory shear in the X-direction, the lateral components of R_g^2 (i.e., R_{gy}^2 and R_{gz}^2) are not affected, but the polymer chains are stretched in the X-direction. The radius of gyration attains a maximum value of $R_{gx}^2 \approx 16\sigma^2$ at $\gamma_{max} = 0.45$. A comparison of Figs. 8 and 9 shows that strain hardening occurs in a strain region where R_{gx}^2 is larger than the value $8.3\sigma^2$ of the corresponding Gaussian coil [74]. We conclude that strain hardening in both unfilled and filled polymers is partially due to the stretching of the molecules at the highest strains. The enhanced strain hardening of filled polymers is related to a strong NAD near zero strain, which softens the system.

Since the monomer–nanoparticle interactions are strongly attractive in the reported simulations, we expect that the relaxation times of the monomer beads are higher in the composite than in the unfilled polymer matrix. To verify this, we calculated the time dependence of the instantaneous NAD g_0^a for both the monomer beads and the NPs in coarse-grained simulations (Fig. 10a). Also calculated has been the time dependence of the NAD of the R-BAPB fragment (see Fig. 1b) obtained from fully atomistic simulations (Fig. 10b). In general, at very short times (0.01τ–0.1τ for CG systems and below 0.001 ns for atomistic systems), a ballistic regime is observed with a slope close to two. This is followed by a plateau where the tagged monomer is trapped in a cage of neighboring monomers. After some characteristic time τ_α (the so-called α-relaxation time, [75, 76]), the monomer escapes from its cage and a diffusive Rouse with a (theoretical) exponent of 0.5 follows; see Fig. 10a.

In the present simulations, the slope of the NAD for the monomer beads is approximately equal to 0.68 in the coarse-grained simulations, Fig. 10a, for both the unfilled polymer and composite. On the other hand, the slope of the NAD of the polyimide fragment in the atomistic simulations is lower and has a value close to 0.6 (Fig. 10b).

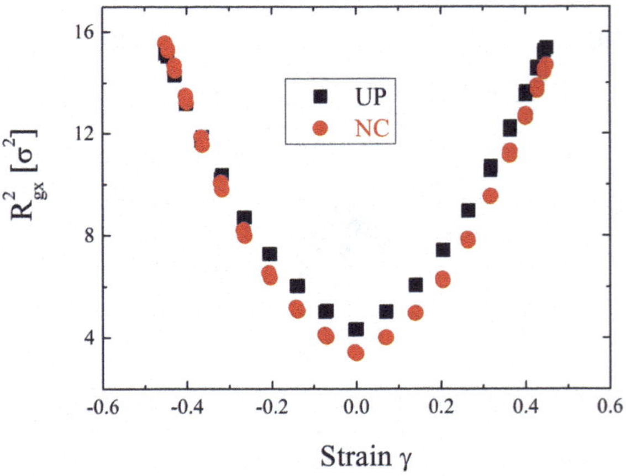

Fig. 9 Squared radius of gyration of the coarse-grained polymer chains in the deformation direction in the nanocomposite (NC) and in the unfilled polymer (UP) as a function of strain at $\gamma_{max} = 0.45$

Thus, diffusion in the coarse-grained system is slightly faster than in the atomistic one. The exponent 0.68 of the coarse-grained simulations is larger than the expected Rouse exponent likely due to the small length of the simulated chains. Furthermore, the time scale of the cage-escape is apparently less than 10τ; the Rouse time, i.e., the time within which linear (not sub-linear) time dependence should be expected, is longer than $10{,}000\tau$. The characteristic deformation time at $\omega = 0.0063\tau^{-1}$ is close to 300τ. The same is observed in the atomistic simulations. In this case, the overall length of each deformation trajectory is equal to 36 ns, which is smaller than the time scale where a linear time dependence of the polymer chain fragments is observed. At the same time, the frequency of the deformations (which is varied from 1.74×10^8 to 1.74×10^{11}) corresponds to deformation times laying in the range from 5 ps to about 6 ns.

Hence, for the both coarse-grained and atomistic system the relaxation time of the monomer beads is longer than the characteristic deformation time, and therefore we conclude that the relaxation time of the polymer chains' COM is much longer than the characteristic time of deformation. This means that the system is always out of equilibrium during the deformation (rather than being close to equilibrium), and the controlling time of the simulations is the characteristic deformation time, not an intrinsic time of the deformed polymer system (in agreement with Ref. [21]). Last, we note that the statistical quality of the curves in Fig. 10 is worse than what it would have been had we taken into account multiple initial times in the calculation of the NAD.

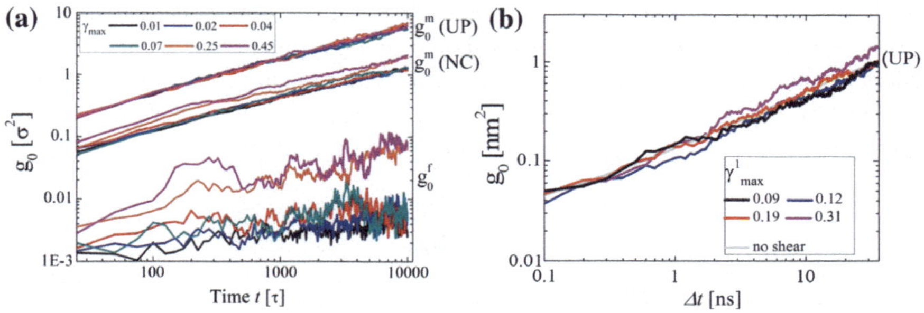

Fig. 10 **a** Instantaneous NAD of the monomer beads g_0^m, and of the filler beads g_0^f, in the unfilled polymer (UP) and in the nanocomposite (NC) for different strain-amplitude values. **b** Instantaneous NAD of the atomistic R-BAPB polyimide fragment calculated at different strain amplitudes at $T = 530$ K

5 Macroscopic and Local Rheological Properties of the Polymer Matrix

The macroscopic reinforcement of polymer nanocomposites stems from the effects occurring on the small length scales. Thus, it is important to understand the rheological behavior of a polymer matrix on the length scales of structural organization corresponding to the nanoparticle–polymer interface.

First of all, the rheological properties of the atomistic bulk polymer melt above its glass-transition temperature ($T_g \sim 470$ K, [41]) are simulated within the cyclic shear deformation. However, performing the computational shear experiment at elevated temperatures is a challenging task, due to the high fluidity of the polymer matrix. Hence, only data between 500 and 530 K are shown in Fig. 11. An alternative route to deal with this problem is to characterize the rheological properties by the passive microrheology approach; this enables to enlarge significantly the temperature range of simulations. Figure 11 presents the results for storage G' and loss G'' moduli obtained within the generalized Stokes–Einstein relation (Eq. 3 in Sect. 2). This required embedding the fullerene C_{60} particle in the R-BAPB polymer melt and measuring its mean-squared displacement.

The most evident and intriguing effect, which is immediately seen from Fig. 11, is the difference between the results of two methods. To characterize the ratio between the storage and loss moduli in both methods, we plotted the ratio between the values of G' and G'' at $T = 530$ K calculated by cycling shear deformation (CSD) and passive microrheology (PM), as a function of frequency (Fig. 12).

Though the overall qualitative agreement between the functional trends of the curves is observed, Fig. 11, the macroscopic moduli from shear simulations are orders of magnitude (depending on frequency) larger than the corresponding values obtained from passive microrheology (Fig. 12). However, the difference remains nearly constant up to the frequencies approaching that of the sound wave in the

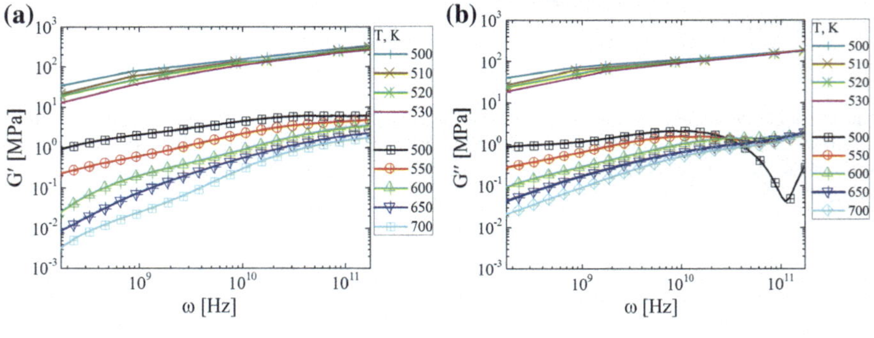

Fig. 11 Frequency dependence of the storage G' (**a**) and loss G'' (**b**) moduli for R-BAPB atomistic polymer melt at temperatures above the glass transition. Shear rheology has been performed at $\gamma^1_{max} = 0.12$. The results are calculated using cycling shear simulations (asterisks) and passive rheology (open symbols). Lines serve as guide to the eye

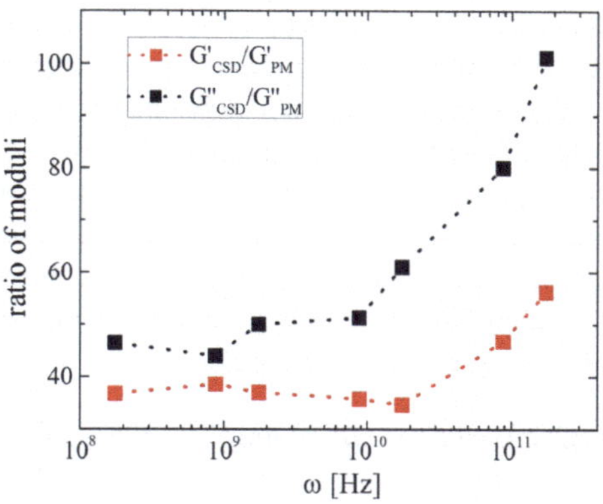

Fig. 12 Ratio between the storage G' (red symbols) moduli and the loss G'' (black symbols) at $T =$ 530 K obtained using cyclic shear deformation (CSD) and passive microrheology (PM). Dotted lines serve as guide to the eye

medium. Surprisingly, the observed difference in the rheological moduli could be easily eliminated if one plots the temperature dependence of the loss tangent, $tg(\delta) = G''/G'$ (Fig. 13).

Good quantitative agreement is observed inherent to the results of the shear deformation at elevated temperatures. Moreover, the loss tangent increases when the temperature is increasing. At the same time, the fluidity of the sample at the fixed temperature is lowered while increasing the frequency of deformation.

In what follows we will discuss several factors which may contribute to the observed differences in the rheological moduli produced by two different methods.

Influence of the probe particle properties Passive rheology measurements are very sensitive to the nature of interactions between the filler nanoparticles and the polymer chains. In the present simulations, the spherical shape of the C_{60} nanoparticle implies partial charges to be negligibly small, and thus, specific Coulombic interactions between the polymer matrix and the nanoparticle are avoided. Another important issue which should also be mentioned is the influence of nanoparticle size on the resulting properties. For this reason, we have also performed simulations with the

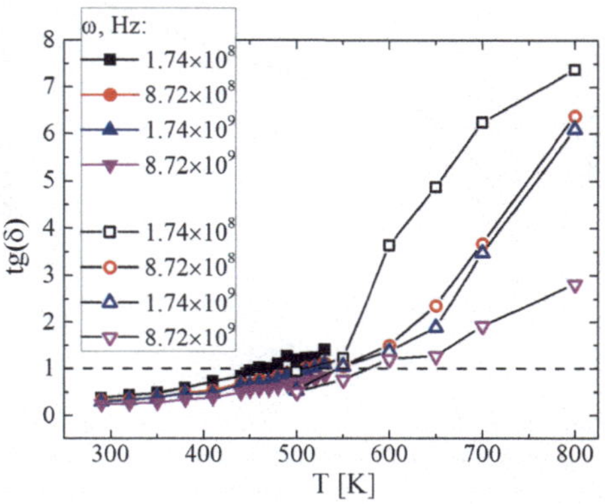

Fig. 13 Temperature dependence of the loss tangent tg(δ) calculated using cycling shear deformation (closed symbols) and passive microrheology (open symbols) at different frequencies. Lines serve as guide to the eye

spherical C_{240} nanoparticle, which diameter is twice larger than that for the C_{60}. The corresponding mean-squared displacements are presented in Fig. 14.

As it can be seen, by increasing the size of the nanoparticle the mean-squared displacements become smaller. However, if one scales the MSD functions with the corresponding size of the nanoparticle, both curves collapse onto a single master curve, at least in the simulated frequency window of 10^9–10^{11} Hz. This implies the applicability of the generalized Stokes–Einstein relations (given by Eq. 3 in Sect. 2) for calculating rheological moduli in this frequency range, and their independence of the nanoparticle size [61].

Influence of the strain amplitude on the rheological properties Another possible explanation for the observed difference between LAOS and microrheology results may be the influence of the shear deformation amplitude γ_{max}. As it has been discussed above, shear deformation has been performed in both linear and nonlinear stress–strain regimes, in the range from 0.09 to 0.31 in the atomistic simulations, with angular frequency $\omega = 1.74 \times 10^9$ Hz. The results are depicted in Fig. 15.

As it can be seen, the results are quite independent on the strain amplitude. Both moduli slightly decrease with increase of γ^1_{max}, but the resulting decrease of the storage and loss moduli is not enough to gain consistency between two simulational approaches.

Influence of the polymer chain length We have carried out atomistic simulations on the systems containing R-BAPB chains of different length, varying the number of monomers from 1 to 8. To preserve the consistency between all the systems studied, the overall number of monomers was chosen to be 216 in each case. The resulting storage and loss moduli measured at $\omega = 1.74 \times 10^9$ Hz and $\gamma_{max} = 0.12$ are presented in Fig. 16.

Better consistency for the storage moduli is gained for the oligomeric fluid (N_p = 1). However, for the loss moduli the agreement between the results of the shear

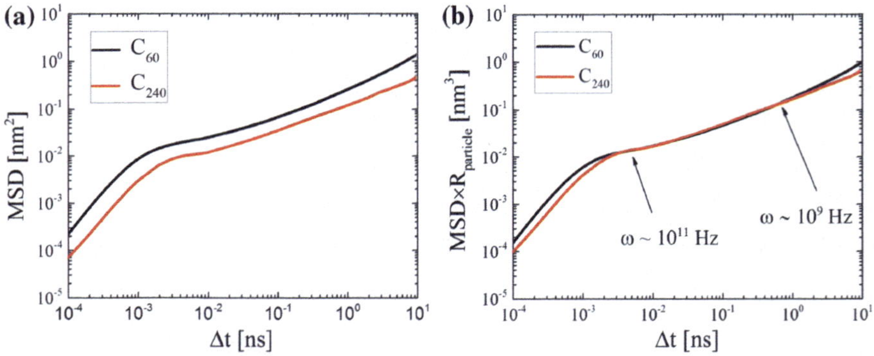

Fig. 14 **a** Mean-squared displacements of the center-of-mass of the fullerene C_{60} (black line) and C_{240} (red line) at T = 600 K. **b** The mean-squared displacements of the fullerenes scaled with the nanoparticle size. The MSD of C_{60} is taken from [48]

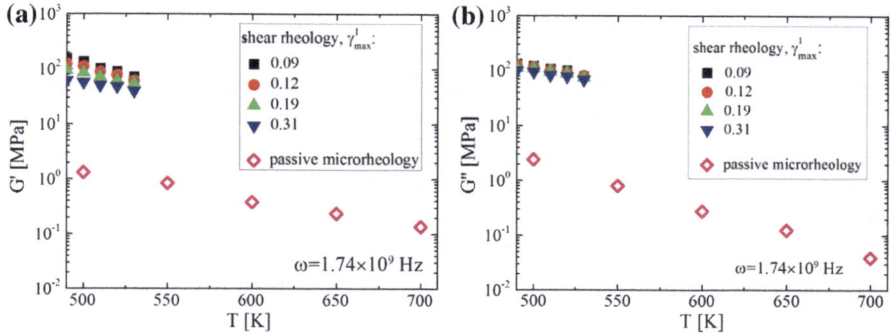

Fig. 15 Temperature dependence of the storage G' (**a**) and loss G'' (**b**) moduli for the atomistic R-BAPB melt obtained from the cyclic shear deformation at different strain amplitude $\gamma^1_{\max}$ (closed symbols) and from the passive microrheology measurements (open symbols)

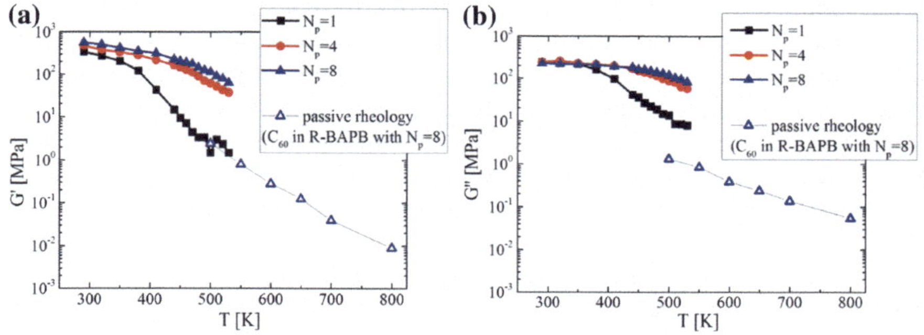

Fig. 16 Temperature dependence of the storage G' (**a**) and the loss G'' (**b**) moduli for the atomistic model of unfilled R-BAPB calculated using cycling shear deformation (closed symbols) at different polymerization degree N_p. The G' and G'' dependences calculated using passive microrheology (open symbols) for $N_p = 8$ are added for comparison. Lines serve as guide to the eye

deformation simulations and passive rheology is much worse. All in all, neither the shear amplitude variation, nor the possible chain length influence has provided us with the desired consistency between two approaches.

Finally, another possible explanation for the observed differences comes from the recent theoretical studies by Yamamoto and Schweizer on nanoparticle diffusion in polymer melts [77, 78]. In these papers, by applying statistical dynamical theories Yamamoto and Schweizer investigate a violation of the Stokes–Einstein ratio for predicting diffusion coefficients of the nanoparticles in polymer melts. The most intriguing outcome is that the degree of violation of the Stokes–Einstein predictions decreases significantly for the nanoparticles, which size approaches the entanglement length of polymer chains in the system. In this case Stokes–Einstein gives the correct diffusion coefficient of the nanoparticle if the viscosity and the size of the nanoparticle are known. From the point of view of passive microrheology, this means that such large nanoparticles may be used to probe macroscopic rheological properties of the polymer melts, which are typically measured in shear experiments. This assumption is clearly validated by the recent coarse-grained simulations by Karim et al. [68, 69]. In these studies, good correspondence between the results of two methods (shear rheology and passive microrheology) was achieved for nanoparticles which size was commensurate with polymer R_g or d_T (reputation tube diameter). However, in our atomistic simulations the size of the fullerene used as a probe particle was sufficiently smaller; i.e., the ratio between characteristic sizes of nanoparticle and the polymer was about 0.1. Thus, the observed discrepancy between the results on rheological properties obtained by shear rheology and passive microrheology may be due to this reason.

Yet another important finding of these studies was that the size of the system should also be properly chosen to compare the active and passive rheology. Considering the passive rheology approach, one should take into account that the distance between the periodic images of the probe particle should be larger than the penetration depth Δ (according to the notation of Karim et al. [68, 69]) of the shear wave induced by the nanoparticle in the polymer medium. Thus, in this case, the elastic wave will decay before reaching the particle's periodic image in the neighboring box. We have measured the mean-squared displacements of the C_{60} fullerene in a box which size was increased by ~25% (all lengths were increased from 6 to 7.5 nm). No differences in the MSD functions were found. However, if one aims to obtain the rheological moduli at the frequencies approaching the terminal zone ($G' \sim \omega^2$, $G'' \sim \omega$), as it is the case in the present fully atomistic simulations, the necessary box size should be chosen appropriately. Following the methodology proposed in [68, 69], we estimated that in this case the size of the fully atomistic system should be 10^6 times larger than that used in the current simulations! This fact could also be used to explain the observed differences between the results of the two rheological approaches discussed above.

6 Microscopic Mechanisms of the Payne Effect

In this section, we finally discuss the effect of NP interactions on the reinforcement loss, that is the Payne effect [9, 72]. Figure 17 shows the storage modulus G' and loss modulus G'' for the *coarse-grained model* of the nanocomposite using both attractive and repulsive NP-NP interactions. These data are compared with those for the unfilled polymer. Clearly the value of the storage modulus G' at low strain amplitude is much higher when attractive rather than repulsive interactions are used. It is more or less close to zero for unfilled polymers. Given that the elastic modulus reaches zero for high strain amplitude, the drop is most pronounced in nanocomposites with attractive interactions. Hence, these have the highest reinforcement before shear and exhibit the strongest Payne effect. The maximum value of the viscous modulus occurs at higher strain amplitude and is larger when attractive NP-NP interactions rather than repulsive ones are used.

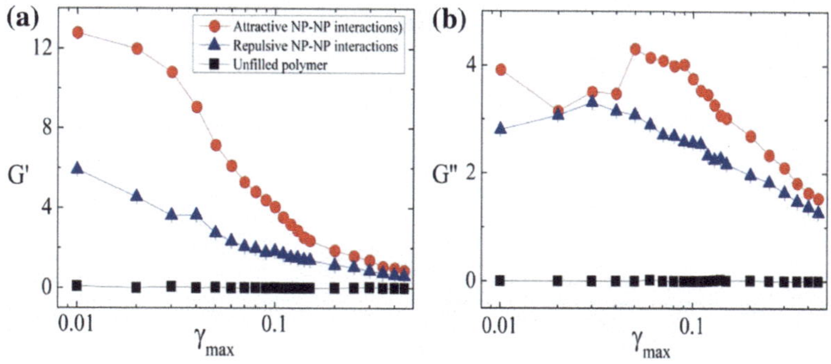

Fig. 17 Payne effect [the strain magnitude dependence of the viscoelastic moduli G' (**a**) and G'' (**b**)] for the unfilled coarse-grained polymer and for the coarse-grained nanocomposite with attractive and repulsive interactions between nanoparticles

To investigate the microscopic origin of these data, we obtain the mean-squared displacement of the polymer beads. Figures 18 and 19 show spatial profiles of the instantaneous mean-squared displacements g_m containing both random and convective motions,

$$g_m = \left\langle \left| \boldsymbol{r}_j(t) - \boldsymbol{r}_j(0) \right|^2 \right\rangle \tag{12}$$

Depending on the value of the strain amplitude, the monomers between NPs were arranged in few (typically 4–5) different dynamical layers. Mean-squared displacement for the monomers is layer-averaged and presented separately.

In accordance with experimental observations in unfilled, glassy polymers, increasing the strain amplitude leads to a narrower spatial distribution of relaxation times of the polymer segments [79], as indicated by the narrower distribution of g_m between mobility layers.

Evidently, with attractive NP-NP interactions several layers near the interface are immobile at very low strain, while such immobile layers are not present in the system with only repulsive interactions. At $\gamma_{\max} = 0.01$, this glassy immobile layer has an average width of about 2σ. At larger strain amplitudes, it disappears.

Given that the mobility of monomers near the NPs is larger in the repulsive than in the attractive system, whereas further away it differs less, the mobility among the layers is less in the system with repulsive NP-NP interactions. Consequently, higher dynamic heterogeneity is observed in the nanocomposite with attractive interactions. This effect is more pronounced at larger strain amplitude, and we believe that this heterogeneity causes the increased drop of the modulus of "attractive" systems compared to "repulsive" ones, that is observed in Fig. 17. It should be noted though that both systems, that with repulsive and that with attractive interactions, show a drop in the modulus with increasing strain rate, which is absent in unfilled polymers. We believe that this drop is caused by network formation that reinforces the system at small strain amplitude, but not at the large shear amplitudes that break down such a

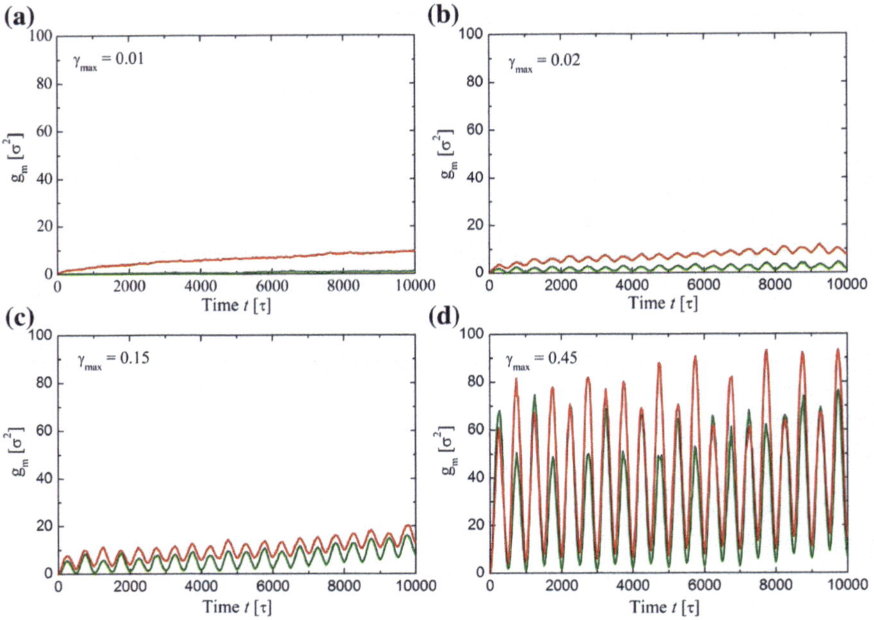

Fig. 18 Spatial profiles of the instantaneous MSD (consisting of the sums of random and convective motion) of the monomer beads in the composite with attractive NP-NP interactions for different strain magnitudes. Only the first (green line, closest to the NP) and the last (red line) dynamic polymer layers between nanoparticles are shown, so as to better illustrate the effect of the strain amplitude on the dynamic heterogeneity of the polymer matrix

network, as is found by others [19, 20, 24]. It also shows that the existence of glassy layers, which contribute to dynamical heterogeneity, plays an important role in the Payne effect, as pointed out above.

7 Conclusions

The main aim of the simulations that we discussed in this chapter has been to elucidate the contradictory behavior of polymer nanocomposites to exhibit both strain-hardening and strain-softening behavior under LAOS experiments of coarse-grained and fully atomistic systems. In addition, we investigate the difference between results obtained from active cyclic shear simulations and those obtained using passive microrheology within the fully atomistic model. We found that strain hardening arises from a decrease in the non-affine displacements of the polymer segments of the matrix, which—in turn—were caused by the extreme elongation of the polymer chains in the deformation direction. Hardening is observed in both unfilled and filled polymeric systems; however, the effect is stronger in the latter. The Payne effect was

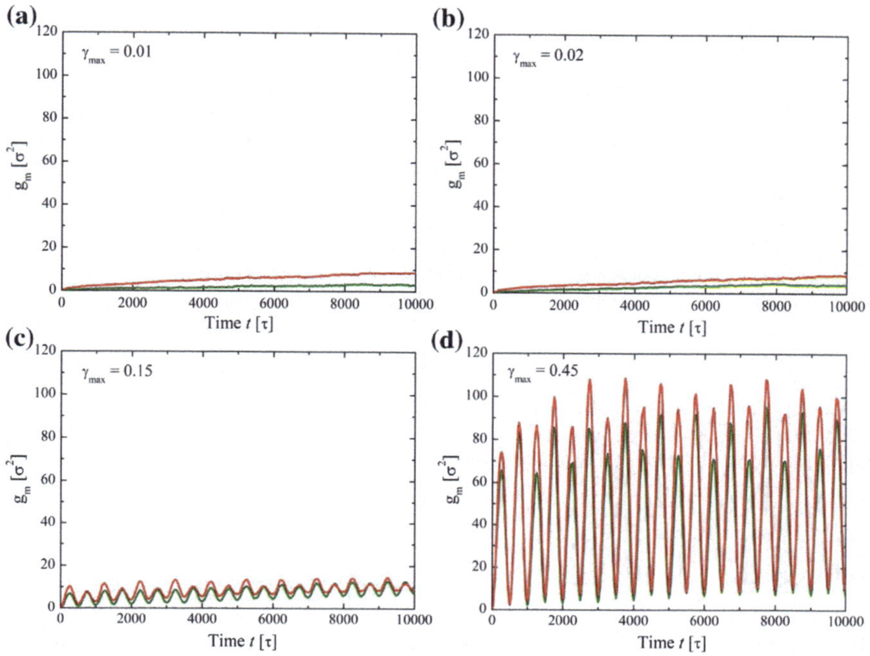

Fig. 19 Spatial profiles of the instantaneous MSD (consisting of the sums of random and convective motions) of the monomer beads in the composite with repulsive NP-NP interactions for different strain magnitudes. Only the first (green line) and last (red line) layers are shown, so as to better illustrate the effect of the strain amplitude on the dynamic heterogeneity of the polymer matrix

studied by comparing the storage modulus as a function of frequency for an unfilled to that with attractive and repulsive interactions among the nanoparticles. These show that the mechanical reinforcement at low strain amplitude—and, hence, its drop with increasing strain amplitude—is higher in a nanocomposite with attractive NP-NP interactions than in the same with repulsive ones. For the unfilled polymers, there is no reinforcement or drop. We believe that the reinforcement is partly due to the formation of a network structure in which the nanoparticles act as crosslinks. At high strain amplitudes, some of the crosslinks are broken. However, an additional mechanism explains the dependence on the NP-NP interaction strength. Probing the mean-squared displacement of the monomer beads as a function of the distance from the NPs' surface, we find that a glassy polymer layer is present near the NPs when attractive NP-NP interactions are used, but not when these interactions are repulsive. Hence, attractive NP-NP interactions slightly enhance the dynamic heterogeneity of the polymer matrix. This apparently leads to higher reinforcement at low shear amplitude and a stronger Payne effect. Further studies should involve the variations of the filler volume fraction and the radius of the NPs, to further affirm the general applicability of the aforementioned results. The difference between two methods,

LAOS and microrheology, in atomistic simulations can be due to the ratio between the entanglement distance and the size of a probe particle.

Acknowledgements T. Davris and A. Lyulin acknowledge the FOM Foundation for the support of the presented research. It was also sponsored by the Stichting Nationale Computerfaciliteiten (National Computer Facilities Foundation, NCF) through the usage of its supercomputer facilities, with financial support from the Nederlandse Organisatie voor Wetenschappelijk Onderzoek (Netherlands Organization from Scientific Research, NWO). V. M. Nazarychev, I. V. Volgin, S. V. Larin, and S. V. Lyulin acknowledge the financial support from the Ministry of Education and Science of the Russian Federation under the Contract no. 14.Z50.31.0002 (megagrant of the Government of the Russian Federation according to the Resolution no. 220 of April 9, 2010). The atomistic simulations have been performed using the computational resources of the Institute of Macromolecular Compounds, Russian Academy of Sciences, the equipment of the shared research facilities of HPC computing resources at Lomonosov Moscow State University, and resources of the federal collective usage center Complex for Simulation and Data Processing for Mega-science Facilities at NRC "Kurchatov Institute." We thank Daniel Bonn, Doros Theodorou, Thijs Michels, Rajesh Khare, as well as the industrial partners at SKF and Michelin for very fruitful discussions.

References

1. Giannelis EP (1996) Polymer layered silicate nanocomposites. Adv Mater 8:29–35. https://doi.org/10.1002/adma.19960080104
2. Sinha Ray S, Okamoto M (2003) Polymer/layered silicate nanocomposites: a review from preparation to processing. Prog Polym Sci 28:1539–1641. https://doi.org/10.1016/j.progpolymsci.2003.08.002
3. Tjong SC (2006) Structural and mechanical properties of polymer nanocomposites. Mater Sci Eng R Rep 53:73–197. https://doi.org/10.1016/j.mser.2006.06.001
4. Zhu A-J, Sternstein S (2003) Nonlinear viscoelasticity of nanofilled polymers: interfaces, chain statistics and properties recovery kinetics. Compos Sci Technol 63:1113–1126. https://doi.org/10.1016/S0266-3538(03)00032-0
5. Varol HS, Sánchez MA, Lu H, Baio JE, Malm C, Encinas N, Mermet-Guyennet MRB, Martzel N, Bonn D, Bonn M, Weidner T, Backus EHG, Parekh SH (2015) Multiscale effects of interfacial polymer confinement in silica nanocomposites. Macromolecules 48:7929–7937. https://doi.org/10.1021/acs.macromol.5b01111
6. Allegra G, Raos G, Vacatello M (2008) Theories and simulations of polymer-based nanocomposites: from chain statistics to reinforcement. Prog Polym Sci 33:683–731. https://doi.org/10.1016/j.progpolymsci.2008.02.003
7. Moll JF, Akcora P, Rungta A, Gong S, Colby RH, Benicewicz BC, Kumar SK (2011) Mechanical reinforcement in polymer melts filled with polymer grafted nanoparticles. Macromolecules 44:7473–7477. https://doi.org/10.1021/ma201200m
8. Long D, Sotta P (2007) Stress relaxation of large amplitudes and long timescales in soft thermoplastic and filled elastomers. Rheol Acta 46:1029–1044. https://doi.org/10.1007/s00397-007-0187-6
9. Payne AR (1965) Effect of dispersion on the dynamic properties of filler-loaded rubbers. J Appl Polym Sci 9:2273–2284. https://doi.org/10.1002/app.1965.070090619
10. Kumar SK, Jouault N, Benicewicz B, Neely T (2013) Nanocomposites with polymer grafted nanoparticles. Macromolecules 46:3199–3214. https://doi.org/10.1021/ma4001385
11. Hagita K, Morita H, Doi M, Takano H (2016) Coarse-grained molecular dynamics simulation of filled polymer nanocomposites under uniaxial elongation. Macromolecules 49:1972–1983. https://doi.org/10.1021/acs.macromol.5b02799

12. Chen Y, Li Z, Wen S, Yang Q, Zhang L, Zhong C, Liu L (2014) Molecular simulation study of role of polymer–particle interactions in the strain-dependent viscoelasticity of elastomers (Payne effect). J Chem Phys 141:104901. https://doi.org/10.1063/1.4894502
13. Wang L, Zheng Z, Davris T, Li F, Liu J, Wu Y, Zhang L, Lyulin AV (2016) Influence of morphology on the mechanical properties of polymer nanocomposites filled with uniform or patchy nanoparticles. Langmuir 32:8473–8483. https://doi.org/10.1021/acs.langmuir.6b01049
14. Liu J, Zhang L, Cao D, Wang W (2009) Static, rheological and mechanical properties of polymer nanocomposites studied by computer modeling and simulation. Phys Chem Chem Phys 11:11365–11384. https://doi.org/10.1039/b913511a
15. Liu J, Gao Y, Cao D, Zhang L, Guo Z (2011) Nanoparticle dispersion and aggregation in polymer nanocomposites: insights from molecular dynamics simulation. Langmuir 27:7926–7933. https://doi.org/10.1021/la201073m
16. Wang W, Hou G, Zheng Z, Wang L, Liu J, Wu Y, Zhang L, Lyulin AV (2017) Designing polymer nanocomposites with a semi-interpenetrating or interpenetrating network structure: toward enhanced mechanical properties. Phys Chem Chem Phys 19:15808–15820. https://doi.org/10.1039/C7CP01453H
17. Wang W, Zhang Z, Davris T, Liu J, Gao Y, Zhang L, Lyulin AV (2017) Simulational insights into the mechanical response of prestretched double network filled elastomers. Soft Matter 13:8597–8608. https://doi.org/10.1039/C7SM01794D
18. Liu J, Wang Z, Zhang Z, Shen J, Chen Y, Zheng Z, Zhang L, Lyulin AV (2017) Self-assembly of block copolymer chains to promote the dispersion of nanoparticles in polymer nanocomposites. J Phys Chem B 121:9311–9318. https://doi.org/10.1021/acs.jpcb.7b08670
19. Kutvonen A, Rossi G, Ala-Nissila T (2012) Correlations between mechanical, structural, and dynamical properties of polymer nanocomposites. Phys Rev E 85:41803. https://doi.org/10.1103/PhysRevE.85.041803
20. Kutvonen A, Rossi G, Puisto SR, Rostedt NKJ, Ala-Nissila T (2012) Influence of nanoparticle size, loading, and shape on the mechanical properties of polymer nanocomposites. J Chem Phys 137:214901. https://doi.org/10.1063/1.4767517
21. Litvinov VM, Orza RA, Klüppel M, van Duin M, Magusin PCMM (2011) Rubber–filler interactions and network structure in relation to stress–strain behavior of vulcanized, carbon black filled EPDM. Macromolecules 44:4887–4900. https://doi.org/10.1021/ma2007255
22. Mermet-Guyennet MRB, Gianfelice de Castro J, Habibi M, Martzel N, Denn MM, Bonn D (2015) LAOS: the strain softening/strain hardening paradox. J Rheol 59:21–32. https://doi.org/10.1122/1.4902000
23. Wyss HM, Miyazaki K, Mattsson J, Hu Z, Reichman DR, Weitz DA (2007) Strain-rate frequency superposition: a rheological probe of structural relaxation in soft materials. Phys Rev Lett 98:238303. https://doi.org/10.1103/PhysRevLett.98.238303
24. Wilson M, Baljon A (2017) Microstructural origins of nonlinear response in associating polymers under oscillatory shear. Polymers 9:556. https://doi.org/10.3390/polym9110556
25. Wilson M, Rabinovitch A, Baljon ARC (2015) Computational study of the structure and rheological properties of self-associating polymer networks. Macromolecules 48:6313–6320. https://doi.org/10.1021/acs.macromol.5b00885
26. Mark JE (2007) Physical properties of polymers handbook. Springer Science & Business Media
27. Plimpton S (1995) Fast parallel algorithms for short-range molecular dynamics. J Comput Phys 117:1–19. https://doi.org/10.1006/JCPH.1995.1039
28. Hess B, Kutzner C, van der Spoel D, Lindahl E (2008) GROMACS 4: algorithms for highly efficient, load-balanced, and scalable molecular simulation. J Chem Theory Comput 4:435–447. https://doi.org/10.1021/ct700301q
29. Van Der Spoel D, Lindahl E, Hess B, Groenhof G, Mark AE, Berendsen HJC (2005) GROMACS: fast, flexible, and free. J Comput Chem 26:1701–1718. https://doi.org/10.1002/jcc.20291
30. Oostenbrink C, Villa A, Mark AE, Van Gunsteren WF (2004) A biomolecular force field based on the free enthalpy of hydration and solvation: the GROMOS force-field parameter sets 53A5 and 53A6. J Comput Chem 25:1656–1676. https://doi.org/10.1002/jcc.20090

31. Oostenbrink C, Soares TA, Van Der Vegt NFA, Van Gunsteren WF (2005) Validation of the 53A6 GROMOS force field. Eur Biophys J 34:273–284. https://doi.org/10.1007/s00249-004-0448-6
32. Lyulin SV, Gurtovenko AA, Larin SV, Nazarychev VM, Lyulin AV (2013) Microsecond atomic-scale molecular dynamics simulations of polyimides. Macromolecules 46:6357–6363. https://doi.org/10.1021/ma4011632
33. Lyulin SV, Larin SV, Gurtovenko AA, Nazarychev VM, Falkovich SG, Yudin VE, Svetlichnyi VM, Gofman IV, Lyulin AV (2014) Thermal properties of bulk polyimides: insights from computer modeling versus experiment. Soft Matter 10:1224–1232. https://doi.org/10.1039/c3sm52521j
34. Nazarychev VM, Larin SV, Lukasheva NV, Glova AD, Lyulin SV (2013) Evaluation of the characteristic equilibration times of bulk polyimides via full-atomic computer simulation. Polym Sci Ser A 55:570–576. https://doi.org/10.1134/S0965545X1308004X
35. Lyulin SV, Larin SV, Gurtovenko AA, Lukasheva NV, Yudin VE, Svetlichnyi VM, Lyulin AV (2012) Effect of the SO2 group in the diamine fragment of polyimides on their structural, thermophysical, and mechanical properties. Polym Sci Ser A 54:631–643. https://doi.org/10.1134/S0965545X12070048
36. Falkovich SG, Lyulin SV, Nazarychev VM, Larin SV, Gurtovenko AA, Lukasheva NV, Lyulin AV (2014) Influence of the electrostatic interactions on thermophysical properties of polyimides: molecular-dynamics simulations. J Polym Sci Part B: Polym Phys 52:640–646. https://doi.org/10.1002/polb.23460
37. Falkovich SG, Larin SV, Lyulin AV, Yudin VE, Kenny JM, Lyulin SV (2014) Influence of the carbon nanofiller surface curvature on the initiation of crystallization in thermoplastic polymers. RSC Adv 4:48606–48612. https://doi.org/10.1039/C4RA07438F
38. Larin SV, Falkovich SG, Nazarychev VM, Gurtovenko AA, Lyulin AV, Lyulin SV (2014) Molecular-dynamics simulation of polyimide matrix pre-crystallization near the surface of a single-walled carbon nanotube. RSC Adv 4:830–844. https://doi.org/10.1039/C3RA45010D
39. Falkovich SG, Larin SV, Nazarychev VM, Volgin IV, Gurtovenko AA, Lyulin AV, Lyulin SV (2014) Computer simulation of the heat-resistant polyimides ULTEM^{TM} and EXTEM^{TM} with the use of GROMOS53a6 and AMBER99 force fields. Polym Sci Ser A 56:558–567. https://doi.org/10.1134/S0965545X14040063
40. Larin SV, Glova AD, Serebryakov EB, Nazarychev VM, Kenny JM, Lyulin SV (2015) Influence of the carbon nanotube surface modification on the microstructure of thermoplastic binders. RSC Adv 5:51621–51630. https://doi.org/10.1039/C5RA07851B
41. Nazarychev VM, Larin SV, Yakimansky AV, Lukasheva NV, Gurtovenko AA, Gofman IV, Yudin VE, Svetlichnyi VM, Kenny JM, Lyulin SV (2015) Parameterization of electrostatic interactions for molecular dynamics simulations of heterocyclic polymers. J Polym Sci Part B: Polym Phys 53:912–923. https://doi.org/10.1002/polb.23715
42. Nazarychev VM, Lyulin AV, Larin SV, Gofman IV, Kenny JM, Lyulin SV (2016) Correlation between the high-temperature local mobility of heterocyclic polyimides and their mechanical properties. Macromolecules 49:6700–6710. https://doi.org/10.1021/acs.macromol.6b00602
43. Falkovich SG, Nazarychev VM, Larin SV, Kenny JM, Lyulin SV (2016) Mechanical properties of a polymer at the interface structurally ordered by graphene. J Phys Chem C 120:6771–6777. https://doi.org/10.1021/acs.jpcc.5b11028
44. Borzdun NI, Larin SV, Falkovich SG, Nazarychev VM, Volgin IV, Yakimansky AV, Lyulin AV, Negi V, Bobbert PA, Lyulin SV (2016) Molecular dynamics simulation of poly (3-hexylthiophene) helical structure in vacuo and in amorphous polymer surrounding. J Polym Sci Part B: Polym Phys 54:2448–2456. https://doi.org/10.1002/polb.24236
45. Nazarychev VM, Lyulin AV, Larin SV, Gurtovenko AA, Kenny JM, Lyulin SV (2016) Molecular dynamics simulations of uniaxial deformation of thermoplastic polyimides. Soft Matter 12:3972–3981. https://doi.org/10.1039/C6SM00230G
46. Glova AD, Falkovich SG, Larin SV, Mezhenskaia DA, Lukasheva NV, Nazarychev VM, Tolmachev DA, Mercurieva AA, Kenny JM, Lyulin SV (2016) Poly(lactic acid)-based nanocomposites filled with cellulose nanocrystals with modified surface: all-atom molecular dynamics simulations. Polym Int 65:892–898. https://doi.org/10.1002/pi.5102

47. Lyulin SV, Larin SV, Nazarychev VM, Fal'kovich SG, Kenny JM (2016) Multiscale computer simulation of polymer nanocomposites based on thermoplastics. Polym Sci Ser C 58:2–15. https://doi.org/10.1134/s1811238216010082
48. Volgin IV, Larin SV, Abad E, Lyulin SV (2017) Molecular dynamics simulations of fullerene diffusion in polymer melts. Macromolecules 50:2207–2218. https://doi.org/10.1021/acs.macromol.6b02050
49. Glova AD, Larin SV, Falkovich SG, Nazarychev VM, Tolmachev DA, Lukasheva NV, Lyulin SV (2017) Molecular dynamics simulations of oligoester brushes: the origin of unusual conformations. Soft Matter 13:6627–6638. https://doi.org/10.1039/C7SM01419H
50. Lukasheva NV, Tolmachev DA, Nazarychev VM, Kenny JM, Lyulin SV (2017) Influence of specific intermolecular interactions on the thermal and dielectric properties of bulk polymers: atomistic molecular dynamics simulations of Nylon 6. Soft Matter 13:474–485. https://doi.org/10.1039/c6sm02169g
51. Nazarychev V, Larin S, Lyulin A, Dingemans T, Kenny J, Lyulin S (2017) Atomistic molecular dynamics simulations of the initial crystallization stage in an SWCNT-polyetherimide nanocomposite. Polymers 9:548. https://doi.org/10.3390/polym9100548
52. Nazarychev VM, Dobrovskiy AY, Larin SV, Lyulin AV, Lyulin SV (2017) Simulating local mobility and mechanical properties of thermostable polyimides with different dianhydride fragments. J Polym Sci Part B: Polym Phys. https://doi.org/10.1002/polb.24550
53. Davris T, Lyulin AV (2015) Coarse-grained molecular-dynamics simulations of capped crosslinked polymer films: equilibrium structure and glass-transition temperature. Polym Compos 36:1012–1019. https://doi.org/10.1002/pc.23413
54. Davris T, Lyulin AV (2015) A coarse-grained molecular dynamics study of segmental structure and mobility in capped crosslinked copolymer films. J Chem Phys 143:74906. https://doi.org/10.1063/1.4928961
55. Davris T, Mermet-Guyennet MRB, Bonn D, Lyulin AV (2016) Filler size effects on reinforcement in elastomer-based nanocomposites: experimental and simulational insights into physical mechanisms. Macromolecules 49:7077–7087. https://doi.org/10.1021/acs.macromol.6b00844
56. Batistakis C, Lyulin AV, Michels MAJ (2012) Slowing down versus acceleration in the dynamics of confined polymer films. Macromolecules 45:7282–7292. https://doi.org/10.1021/ma300753e
57. Mason TG (2000) Estimating the viscoelastic moduli of complex fluids using the generalized Stokes-Einstein equation. Rheol Acta 39:371–378. https://doi.org/10.1007/s003970000094
58. Weihs D, Mason TG, Teitell MA (2006) Bio-microrheology: a frontier in microrheology. Biophys J 91:4296–4305. https://doi.org/10.1529/biophysj.106.081109
59. Wirtz D (2009) Particle-tracking microrheology of living cells: principles and applications. Annu Rev Biophys 38:301–326. https://doi.org/10.1146/annurev.biophys.050708.133724
60. Sarmiento-Gomez E, Santamaría-Holek I, Castillo R (2014) Mean-square displacement of particles in slightly interconnected polymer networks. J Phys Chem B 118:1146–1158. https://doi.org/10.1021/jp4105344
61. Narita T, Indei T (2016) Microrheological study of physical gelation in living polymeric networks. Macromolecules 49:4634–4646. https://doi.org/10.1021/acs.macromol.6b00745
62. Narita T, Mayumi K, Ducouret G, Hébraud P (2013) Viscoelastic properties of poly(vinyl alcohol) hydrogels having permanent and transient cross-links studied by microrheology, classical rheometry, and dynamic light scattering. Macromolecules 46:4174–4183. https://doi.org/10.1021/ma400600f
63. Pommella A, Preziosi V, Caserta S, Cooper JM, Guido S, Tassieri M (2013) Using optical tweezers for the characterization of polyelectrolyte solutions with very low viscoelasticity. Langmuir 29:9224–9230. https://doi.org/10.1021/la4015948
64. Abdala AA, Amin S, van Zanten JH, Khan SA (2015) Tracer microrheology study of a hydrophobically modified comblike associative polymer. Langmuir 31:3944–3951. https://doi.org/10.1021/la504904n
65. Kuhnhold A, Paul W (2015) Active one-particle microrheology of an unentangled polymer melt studied by molecular dynamics simulation. Phys Rev E 91:42601. https://doi.org/10.1103/PhysRevE.91.042601

66. Kuhnhold A, Paul W (2014) Passive one-particle microrheology of an unentangled polymer melt studied by molecular dynamics simulation. Phys Rev E 90:22602. https://doi.org/10.1103/PhysRevE.90.022602
67. Kuhnhold A, Paul W (2014) Temperature dependent micro-rheology of a glass-forming polymer melt studied by molecular dynamics simulation. J Chem Phys 141:124907. https://doi.org/10.1063/1.4896151
68. Karim M, Indei T, Schieber JD, Khare R (2016) Determination of linear viscoelastic properties of an entangled polymer melt by probe rheology simulations. Phys Rev E 93:1–12. https://doi.org/10.1103/PhysRevE.93.012501
69. Karim M, Kohale SC, Indei T, Schieber JD, Khare R (2012) Determination of viscoelastic properties by analysis of probe-particle motion in molecular simulations. Phys Rev E 86:51501. https://doi.org/10.1103/PhysRevE.86.051501
70. Song Y, Dai LL (2010) Two-particle interfacial microrheology at polymer−polymer interfaces. Langmuir 26:13044–13047. https://doi.org/10.1021/la102171k
71. Song Y, Luo M, Dai LL (2010) Understanding nanoparticle diffusion and exploring interfacial nanorheology using molecular dynamics simulations. Langmuir 26:5–9. https://doi.org/10.1021/la901902t
72. Raos G, Moreno M, Elli S (2006) Computational experiments on filled rubber viscoelasticity: what is the role of particle−particle interactions? Macromolecules 39:6744–6751. https://doi.org/10.1021/ma061008h
73. Hyun K, Wilhelm M, Klein CO, Cho KS, Nam JG, Ahn KH, Lee SJ, Ewoldt RH, McKinley GH (2011) A review of nonlinear oscillatory shear tests: analysis and application of large amplitude oscillatory shear (LAOS). Prog Polym Sci 36:1697–1753. https://doi.org/10.1016/j.progpolymsci.2011.02.002
74. Rubinstein M, Colby RH (2003) Polymer physics. OUP Oxford, Oxford
75. Leutheusser E (1984) Dynamical model of the liquid-glass transition. Phys Rev A 29:2765–2773. https://doi.org/10.1103/PhysRevA.29.2765
76. Hansen JP, Levesque D, Zinn-Justin J (1991) Liquids, freezing and the glass transition. North Holland, Amsterdam
77. Yamamoto U, Schweizer KS (2011) Theory of nanoparticle diffusion in unentangled and entangled polymer melts. J Chem Phys 135:224902. https://doi.org/10.1063/1.3664863
78. Yamamoto U, Schweizer KS (2015) Microscopic Theory of the long-time diffusivity and intermediate-time anomalous transport of a nanoparticle in polymer melts. Macromolecules 48:152–163. https://doi.org/10.1021/ma501150q
79. Lee H-N, Paeng K, Swallen SF, Ediger MD (2009) Direct measurement of molecular mobility in actively deformed polymer glasses. Science 323:231–234

Index

Symbols
1-phenyl-1-propanol, 218, 221, 225
1-propanol, 217, 221
5-methyl-2-hexanol, 225
α-relaxation, 62, 176, 247, 281, 283
β-relaxation, 62, 220

A
AC-calorimetry, 309
AC chip calorimetry, 288
AC conductivity, 29
Activation energy, 82, 83, 88, 283
Activation plot, 365
Activation volume, 83
Active shear deformation, 378
Adam-Gibbs theory, 307
Adam-Gibbs model, 4, 5, 90, 94
Adsorption strength, 362
Angell plot, 37
Anisotropic scattering, 208
Anomalous diffusion, 256
Arrhenius, 362, 364, 365, 371
Arrhenius expression, 327
Arrhenius law, 83, 87, 88
Arrhenius-like, 62
Arrhenius-type temperature dependence, 2, 3
Atomistic simulations, 64
Average relaxation time, 25, 28, 30, 35–37, 39, 40, 53, 259

B
Bartenev-Ritland equation, 328
Bertrand, 6
Binary glass-forming, 175
Blue shift, 64
Boson Peak (BP), 281
Broad band dynamic calorimetry, 309
Broadband Dielectric Spectroscopy (BDS), 61, 68, 279
Broadband loss-spectra, 31, 49
Bulk density, 66

C
Calorimetric glass transition, 307
Calorimetry, 309
Carbon-based polymer chains, 361
Cation, 230
Chemically realistic model, 353, 358
Coarse-grained, 378, 380
Coaxial reflectometer, 285
Coherence area, 209
Cole-Cole (CC), 10, 28, 29
Cole-Davidson, 261
Collective dipolar relaxation, 254
Complex heat capacity, 315
Component-selective NMR, 191
Conductive and convection, 310
Configurational entropy, 329
Confinement, 353, 354, 356, 369
Confinement effects, 372
Cooling rate, 307, 309
Cooperatively Rearranging Regions (CRRs), 328
Correlation function, 282
Coupling Model (CM), 82
Cross correlations, 206, 215, 217
Crystallization, 69–71

D
DC- and AC- Calorimetry, 68

F. Kremer and A. Loidl (eds.), *The Scaling of Relaxation Processes*, Advances in Dielectrics, https://doi.org/10.1007/978-3-319-72706-6

DC-conductivity, 229
Deborah number, 24
Debye function, 366
Debye liquids, 159
Debye relaxation, 203, 219
Debye-process, 217
Debye-Stokes-Einstein, 235
Debye-Waller factor, 291
Decoupling of time scales, 154
Degree of heterodyning, 210
Density fluctuations, 83–85
Density scaling, 77–88
Depolarized dynamic light scattering, 205
Derivative method, 294
Desorption process, 359, 362–364, 371, 372
Dielectric constant, 28, 29
Dielectric loss, 355, 359, 360, 369, 370, 372
Dielectric permittivity, 247
Dielectric response layer resolved, 364
Dielectric spectroscopy, 175
Dielectric susceptibility, 353, 354, 363, 364
Differential AC chip calorimeter, 316
Differential Scanning Calorimeters (DSC), 310
Diffusion, 271
Diffusion constant, 235
Dihedral fluctuations, 367
Dipolar correlations, 356, 357, 359, 361
Discotic liquid crystals, 281
Distribution of correlation times, 173
Distribution of relaxation times, 28, 51, 52, 253, 269
Distributions of mobilities, 267
Donth, 5, 16
Double differential cross section, 289
Dynamic coupling parameter, 254
Dynamic heterogeneities, 174
Dynamic light scattering, 207
Dynamic modulus, 86, 87
Dynamically "asymmetric" binary glass formers, 175
Dyre, 6, 39, 78, 107, 229, 233, 235

E
Elastic fixed window scans, 290
Elastic stress, 385
Electrode polarization, 229
Electro-optical measurements, 158
End-to-end vector, 218
Energy landscape
Enthalpy, 324
Especially synthesized high-T_g components, 175
Excess heat capacity, 320
Excess wing, 3, 10, 23, 33, 34, 37–40, 53–55, 176, 178, 204, 261
Exchange NMR, 173
External control parameters, 309

F
Fast degrees of freedom, 312
Fast Fourier Transform (FFT), 359
Fast Scanning Chip Calorimetry (FSC), 309
Fast-β process, 41
Field autocorrelation, 209
Finesse, 211
Finite-Extensible-Nonlinear-Elastic (FENE) potential, 381
Fluctuation dissipation theorem, 207, 355, 359
Fourier correlation analysis, 285
Fourier transformation, 315
Fourier-Transform Infrared (FTIR) - Spectroscopy, 61
Fragility, 78, 82, 84–86
Free Spectral Range (FSR), 211
Frequency-dependent heat capacity, 312
Frequency-temperature superposition (FTS), 174
Fullerene, 382
Fully atomistic, 378

G
Gardner transition, 23, 48–51, 54
Gaussian approximation, 248
Generalized diffusion coefficient, 272
Generalized Kirkwood parameter, 247
Generalized Stokes–Einstein relation (GSER), 382
Generalized susceptibility, 207, 210, 211
Gibbs-Di Marzio model, 2
Glass transition, 77, 78, 80, 83–87, 280, 354, 358, 364, 371, 372
Glass transition temperature coefficient, 86
Glass-forming systems, 247
Glassy and liquid state, 324
Glassy dynamics, 174, 279
Glassy solidification, 62, 70, 71
Glassy state, 279, 308
Glycerol, 61
Graphite walls, 355, 371
Gromacs, 379
Gutzow-Schmelzer approach, 323
Gyration ellipsoids, 356

H
H NMR^2, 178
H_2O—LiCl solutions, 36

Havriliak-Negami (HN) function, 65, 261, 286
Heat capacity, 308
Heat capacity curves, 344
Heat flow rate, 312
Heat-balance equation, 310
HEMIM-BF4, 233
Hexagonal columnar liquid crystalline phase, 279
High frequency wing, 67
Homodyne scattering, 209
Hydrogen bonding, 113, 125, 131, 133, 160, 204, 206, 207, 215, 217
Hydrogen-bonds, 64, 217

I
Ideal mixing, 121, 123, 139, 144, 148, 149, 151, 152, 156, 161
Incoherent neutron scattering, 262
Inelastic neutron scattering, 290
Inflection point, 87, 88
Infrared and Raman-spectroscopy, 68
Infrared spectroscopy, 130, 372
Intensity correlation, 209
Intermediate scattering function, 206
Inter-molecular, 61, 254
Intra-molecular, 61, 64
Ionic liquids, 203, 229
Irreversible thermodynamics, 307
Isochronal states, 81
Isochronal temperature scans, 316
Isomorph invariant, 81
Isothermal bulk modulus, 86
Isothermal compressibility, 86

J
Johari-Goldstein (JG) relaxations, 31, 38, 82
Johari-Goldstein β-process, 203

K
Kauzmann paradox, 2
Kinetics of the glass transition, 307
Kirkwood-Fröhlich correlation factor, 127, 128, 143, 152-154, 162, 215-217, 229, 233
Kohlrausch-Williams-Watts, 255
Kovacs-Aklonis-Hutchinson-Ramos (KAHR), 308

L
Lamb-Mössbauer factor (LMF), 255
LAMMPS, 379
Large Amplitude Oscillatory Shear (LAOS), 377
Laser heating, 316
Legendre Polynomial, 206
Lennard-Jones interactions, 362, 371
Lennard-Jones potential, 77, 79, 81
Librational fluctuations, 71
Librations, 67
Limiting fictive temperature, 323
Liquid crystals, 279
Liquid-glass transition, 78
Liquid-like (isotropic) reorientation, 173
Lissajous analysis, 378
Local oscillator, 210
Loss modulus, 383
Loss tangent, 393
Low-Molecular Weight (LMW) glass-forming systems, 258

M
Main relaxation, 176
Mauro, 6, 8, 9, 11, 12, 13, 36
Maxwell field, 214
Maxwell-Wagner-Sillars polarization, 229
MD simulations, 79–82
Mean squared displacement, 255, 292, 394
Mechanical reinforcement, 376
Mechanical Spectroscopy, 68
Microscopic correlation function, 214
Mode-Coupling Theory (MCT), 1, 7, 9, 12, 13, 41
Modelling of kinetics, 322
Molecular dipole moment, 206
Molecular Dynamics simulation, 353
Molecular mobility, 82, 84
Molecular polarizability, 206
Molecular reorientation, 205
Molecular-dynamics computer simulations, 377
Momentum transfer, 248
Monohydroxy Alcohols (MAs), 217, 203
Morse – potential, 65
Multi-component liquids, 175
Multi-frequency measurements, 314
Multispeckle light scattering, 211

N
Nanostructured filler, 377
Nematic liquid crystal, 279
Neutron scattering, 279
Non-Affine Displacement (NAD), 388
Non-Arrhenius behavior, 82
Non-exponential, 261
Non-exponentiality, 308

Nonlinear shear mechanical measurements, 127
Non-linearity, 308
Normal mode, 353, 356, 363
Normalization factor, 321
Nuclear Magnetic Resonance (NMR), 68
Nuclear magnetic resonance spectroscopy, 175
Nyquist–Shannon sampling theorem, 359

O
OMIM-NTf_2, 238
Orientational effect of the walls, 363
Oscillator strength, 69
Oscillatory shear, 377

P
Partial heterodyning, 209
Passive microrheology, 382
Payne effect, 377
Photon Correlation Spectroscopy (PCS), 68, 203, 208
Physical aging, 46, 48, 50, 121, 133
Pitch-drop experiment, 25, 26
Plastic crystals, 281
Polarization, 353, 354
Poly(ethylene-glycol)-dibenzoate, 61
Poly(ethylene-glycol)methyl-ether-acrylate, 61
Poly(ethylene-glycol)phenyl-ether-acrylate, 61
Poly(propylene glycol), 11, 15, 17, 141
Polybutadiene, 353–356, 358
Polymer-based composite, 376
Polymer-plasticizer systems, 175
Polymers, 247
Polystyrene, 307
Prepeak, 134–137, 139, 140, 144, 147–149
Propylene carbonate, 35, 36, 40, 42, 111, 261
Propylene glycol, 11, 14, 15, 17, 36, 67
Propylene glycol oligomers, 67, 113

Q
Quasielastic Neutron Scattering (QENS), 247, 290

R
Random large-angle jumps, 237
Red shift, 65
Reduced units, 81, 82
Relaxation map, 309
Relaxation time, 308, 327
Reorientational correlation function, 178
Reorientational fluctuations, 174
Ring-chain equilibria, 152, 157
Root-mean-square deviations, 339
Rotational diffusion, 237, 268

S
Salol, 61
Scaling exponent, 77–84, 86–88
Scattering vector, 249
Secondary-relaxation, 217, 218
Secondary β relaxation, 62
Segmental relaxation, 255, 378
Self-correlation function, 247
Shoving model, 6, 7
Siegert relation, 209
Simultaneous modelling, 344
Single-particle rotational relaxation time, 254
Size disparity of their components, 175
Slower degrees of freedom, 312
Small angle libration, 70
Sorbitol, 61
Souletie, 6, 8, 9, 11–13
Specific heat spectroscopy, 279
Static Kirkwood parameter, 254
Static structure factor, 257
Storage modulus, 383
Strain hardening, 377, 383
Strain softening, 377, 383
Strain-rate frequency superposition, 377
Structural relaxation, 77–84, 86–88
Structural relaxation time, 174
Structure of glass, 308
Supramolecular modes, 125
Susceptibility minimum, 42

T
Tandem Fabry Perot Interferometry (TFPI), 203, 210, 211
Temperature oscillations, 316
Temperature-Modulated Differential Scanning Calorimetry (TMDSC), 288, 309
Theoretical methods, 307
Theoretical modelling, 324
Theory of isomorphs, 80
Thermal history, 309
Thermodynamic fluctuation model,
Thermodynamic modulus, 86
Thermodynamic scaling, 77, 78, 87
Thermoplastic polyimide R-BAPB, 379
Thermorheological simplicity, 326
Thermotropic liquid crystals, 282
Threitol, 61
Time correlation, 367
Time correlation functions, 206
Time dependency, 312
Time domain, 313

Time domain relaxation function, 253
Time-temperature superposition (TTS), 174
Tool-Narayanswamy-Moynihan approach, 307
Trans and gauche conformers, 75
Trans and gauche conformations, 74
Translational diffusion, 268
Transmission function, 211
Tripropylene glycol,
Tumbling mode, 286
Two glass transition temperatures, 175
Type A polymers, 356
Type-A glass formers, 33, 34, 38, 40, 53, 176
Type-B glass formers, 32, 38, 40, 53

U
Unfilled polymer, 389

V
Van der Waals liquid, 68, 78, 80, 82
Van Hove formalism, 248
Van Hove-singularity, 281
Vibration-relaxation crossover, 367
Vibrational degrees of freedom, 365, 369, 371
Vibrational Density of States (VDOS), 281
Vibrational frequencies, 62
Vibrational spectroscopy, 133
Viscosity, 1, 2, 23–27, 30–32, 34–36, 44, 48, 53, 55, 77, 87, 108, 125, 126, 140, 150–152, 155, 204, 219, 229, 268, 270, 280, 395
Viscosity enhancement, 151, 152, 154, 155
Vogel-Fulcher-Tammann (VFT) law, 63, 260, 280, 309, 364, 371

W
Water solutions, 247
Williams-Watts approach, 221
Wobbling-on-a-cone model, 177

X
X-ray diffraction, 121, 134, 135, 137–139, 144, 148, 150, 161
X-ray scattering, 68
Xylitol, 61

Z
Zero-shear viscosity, 125, 126, 140, 150–152, 154

The manufacturer's authorised representative in the EU is Springer Nature Customer Service Centre GmbH, Europaplatz 3, 69115 Heidelberg, Germany. If you have any concerns regarding our products, please contact ProductSafety@springernature.com

Printed and bound by CPI Group (UK) Ltd, Croydon, CR0 4YY
30/09/2026
02225654-0013